Université Joseph Fourier

Les Houches

Session LXXXVIII

2007

Dynamos

Lecturers who contributed to this volume

Thierry Alboussière
Bérengère Dubrulle
Mathieu Dumberry
Christopher C. Finlay
Chris A. Jones
Peter Olson
Yannick Ponty
Anvar Shukurov
Dmitry Sokoloff
Sabine Stanley

ÉCOLE D'ÉTÉ DE PHYSIQUE DES HOUCHES

SESSION LXXXVIII, 30 JULY–24 AUGUST 2007

ÉCOLE THÉMATIQUE DU CNRS

DYNAMOS

Edited by

Ph. Cardin and L.F. Cugliandolo

ELSEVIER

Amsterdam – Boston – Heidelberg – London – New York – Oxford
Paris – San Diego – San Francisco – Singapore – Sydney – Tokyo

Elsevier
Radarweg 29, PO Box 211, 1000 AE Amsterdam, The Netherlands
Linacre House, Jordan Hill, Oxford OX2 8DP, UK

First edition 2008

Library of Congress Cataloging-in-Publication Data
A catalog record for this book is available from the Library of Congress

British Library Cataloguing in Publication Data
A catalogue record for this book is available from the British Library

ISBN: 978-0-0805-4812-8
ISSN: 0924-8099

For information on all Elsevier publications
visit our website at elsevierdirect.com

Transferred to Digital Printing in 2010

ÉCOLE DE PHYSIQUE DES HOUCHES

Service inter-universitaire commun
à l'Université Joseph Fourier de Grenoble
et à l'Institut National Polytechnique de Grenoble

Subventionné par le Ministère de l'Éducation Nationale,
de l'Enseignement Supérieur et de la Recherche,
le Centre National de la Recherche Scientifique,
le Commissariat à l'Énergie Atomique

Previous sessions

Publishers:

- Session VIII: Dunod, Wiley, Methuen
- Sessions IX and X: Herman, Wiley
- Session XI: Gordon and Breach, Presses Universitaires

- Sessions XII–XXV: Gordon and Breach
- Sessions XXVI–LXVIII: North Holland
- Session LXIX–LXXVIII: EDP Sciences, Springer
- Session LXXIX–LXXXVII: Elsevier

Organizers

CARDIN Philippe, Observatoire de Grenoble, Université Joseph-Fourier, France

CUGLIANDOLO Leticia, Université Pierre et Marie Curie, Paris VI, France

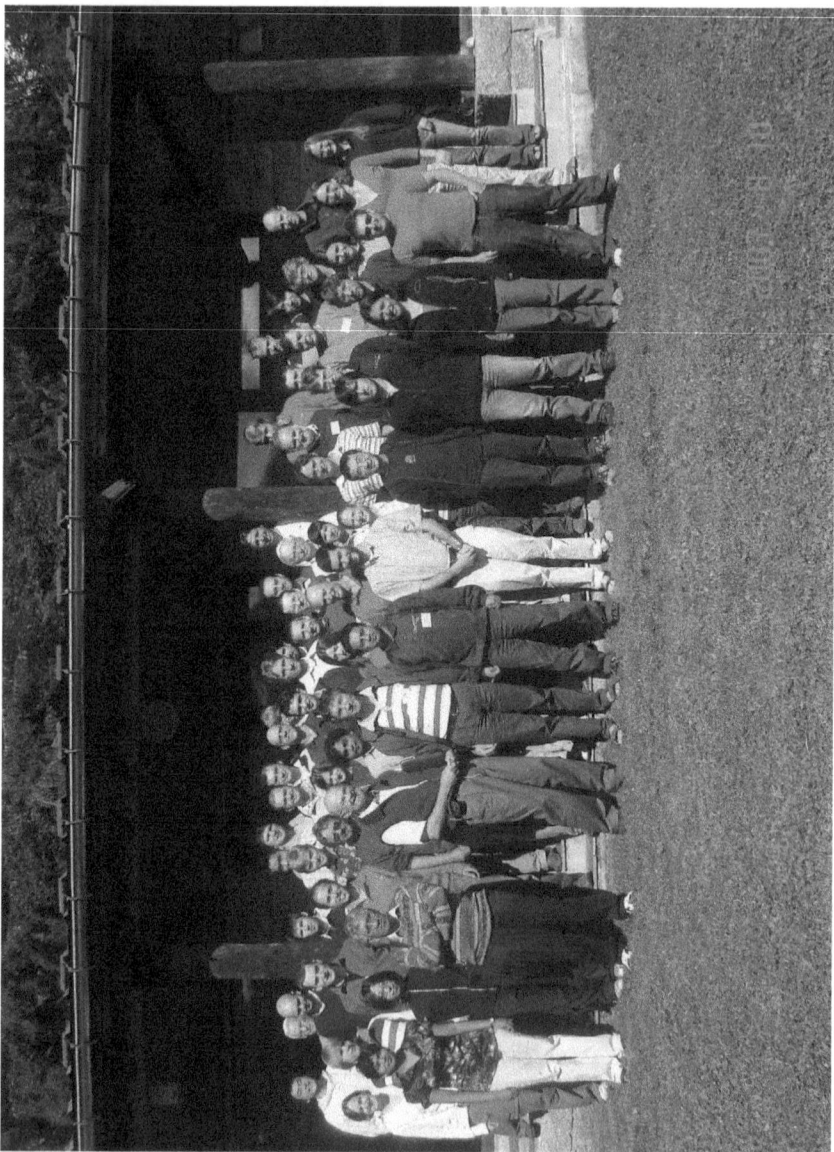

General Lecturers

ALBOUSSIERE Thierry, Observatoire de Grenoble, Université Joseph-Fourier, France

JONES Christopher, Department of Mathematics, University of Leeds, UK

OLSON Peter, Earth and Planetary Sciences, Johns Hopkins University, USA

SHUKUROV Anvar, School of Mathematics and Statistics, Newcastle University, UK

Research Lecturers

AUBERT Julien, Institut de Physique du Globe de Paris, France

AURNOU Jonathan, Earth and Space Sciences, UCLA, USA

DUBRULLE Bérengère, Service de Physique de l'État Condensé, CEA, France

DUMBERRY Mathieu, Department of Physics, University of Alberta, Canada

FINLAY Christopher, ETH, Zurich, Switzerland

FRICK Peter, Institute of Continuous Media Mechanics, Perm, Russia

GLATZMAIER Gary, University of California, Santa Cruz, USA

GOLUB Leon, Harvard-Smithsonian Center for Astrophysics, Cambridge, USA

JACKSON Andy, ETH, Zurich, Switzerland

KAGEYAMA Akira, Earth Simulator Center, JAMSTEC, Japan

HULOT Gauthier, Institut de physique du globe de Paris, France

PINTON Jean-François, Ecole Normale Supérieure de Lyon, France

PONTY Yannick, Observatoire de la Cote d'Azur, Nice, France

SAUR Joachim, University of Cologne, Germany

SCHEKOCHIHIN Alexander, DAMTP/Imperial College, UK

STANLEY Sabine, Department of Physics, University of Toronto, Canada

SUBRAMANIAN Kandaswamy, Inter-University for Astronomy and Astrophysics, Pune, India

THESS Andre, Ilmenau University of Technology, Germany

TILGNER Andreas, Institute of Geophysics, University of Göttingen, Germany

TITLE Alan, Stanford-Lockheed Institute for Space Research, Palo Alto, USA

TOBIAS Steve, Applied Maths, Leeds, UK

TOOMRE Juri, University of Colorado, Boulder, USA

Participants

AMIT Hagay, Institut de Physique du Globe de Paris, France

AUGUSTON Kyle, JILA, Boulder, USA

BAGGALEY Andrew, School of Mathematics and Statistics, Newcastle University, UK

BERHANU Michael, Laboratoire de Physique Statistique École Normale Supérieure de Paris, France

CALKINS Michael, UCLA, USA

CANET Elisabeth, Observatoire de Grenoble, Université Joseph-Fourier, France

CARAMETE Laurentiu-Ioan, Max-Planck-Institut Radioastronomie, Bonn, Germany

CARDIN Philippe, Observatoire de Grenoble, Université Joseph-Fourier, France

CHUPIN Anton, Institute of Continuous Media, Perm, Russia

DAVIAUD François, Service de Physique de l'État Condensé, CEA Saclay, France

DAVIES Chris, University of Leeds, UK

DE LA TORRE Alberto, Departamento de Física y Matemática, Universidad de Navarra, Pamplona, Spain

DEGUEN Renaud, Observatoire de Grenoble, Université Joseph-Fourier, France

DENISOVA Natalia, Mechanics and Mathematics, Moscow University, Russia

DHARMARAJ Girija, Department of Physics, University of Toronto, Canada

DRISCOLL Peter, Department of Earth & Plantary Sciences, Johns Hopkins University, Baltimore, USA

EVONUK Martha, ETH Zürich, Switzerland

FANJAT Grégory, Ecole Normale Supérieure de Lyon, France

FERREIRA Barbara, University of Cambridge, UK

FISCHER Michael, Forschungszentrum Dresden, Germany

GAGNIERE Nadège, Laboratoire de Géophysique Interne et Tectonophysique, Université Joseph-Fourier, France

GISSINGER Christophe, École Normale Supérieure de Paris, France

GOMEZ PEREZ Natalia, Department of Physics, Alberta, Canada

GUERVILLY Céline, Laboratoire de Géophysique Interne et Tectonophysique, Université Joseph-Fourier, France

GUSAKOV Mikhail, Institute Ioffe, St Petersburg, Russia

HERREMAN Wietze, Institut de Recherche sur les Phénomènes Hors Equilibre, Marseille, France

HORI Kumiko, Department of Earth & Planetary Sciences, Nagoya, Japan

KANTOR Elena, Astrophysics Department, Institute Ioffe, St Petersburg, Russia

KELLEY Douglas, University of Maryland, USA

KING Eric, UCLA, USA

LI Kuan, ETH Zürich, Switzerland

LIPSBERGS Guntis, University of Latvia, Latvia

MARSENIC Alexandra, Geophysical Institute Bratislava, Slovakia

MARTI Philippe, ETH Zürich, Switzerland

MIYAGOSHI Takehiro, Earth Simulator Center, Yokohama, Japan

MOUHALI Waleed, Laboratoire Univers et Théorie, Meudon, France

RESHETNYAK Maxim, Russia

REUTER Klaus, Max-Planck-Institut Plasmaphysik, Garching, Germany

RIBEIRO Adolfo, Laboratoire d'Informatique pour la Mécanique et les Sciences de l'Ingénieur Orsay, France

ROSIN Mark, DAMPT, Cambridge, UK

SAHOO Ganapati, Department of Physics, Bangalore, India

SHAH Nikhil, DAMTP Cambridge, UK

SILVA Luis, Institut de Physique du Globe de Paris, France

SODERLUND Krista, UCLA, USA

SOLTIS Tomas, Department of Astrophysics, Bratislava, Slovakia

STERENBORG Glenn, Harvard University, USA

SUR Sharanya, Department of Astronomy & Astrophysics, Pune, India

TRINH, Antony, Royal Observatory of Belgium, Belgium

TRUMPER Tobias, Institut Geophysik Dresden, Germany

VALETTE Bernard, Université de Savoie, France

WOLTANSKI Dominik, Torun Center for Astronomy, Poland

Preface

Our knowledge of the Earth's magnetic field has developed over many centuries. Since the first compass used in China in the begining of the Christian era [1], the tendency of the needle to point North has been an enduring mystery. Not only scientists, but also poets have been intrigued by the "virtue" of the polar star to attract magnets. Here is an exerpt from one of the first poems written in French, in 1202 by Guyot de Provins, troubadour and monk:

"... / Un art font qui mentir ne peut / Par la vertu de la marinette / Une pierre laide et brunette / Où li fers volontiers se joint / Ainsi regardent le droict point / Puis, qu'une aiguile l'ait touchie / Et en festu l'ont fichie / En l'eau la mettent sans plus / Et le festus la tient dessus / Puis se tourne la pointe toute / Contre l'estoile; si sans doute /... ".

Explanations for why magnetic needles point North began to appear as early as the thirteenth century. Around 1269 Pierre de Maricourt wrote one of the first scientific treatises in the Western World, *Epistola de magnete*, on this problem and in 1600, William Gilbert proposed his famous explanation in *De Magnete, Magneticisque Corporibus, et de Magno Magnete Tellure*, suggesting the Earth behaves like a permanent magnet. Neither of these proved to be consistent with magnetic observations accumulated by observatories over the next three centuries, however.

A new idea, the so-called *dynamo theory*, was launched by Joseph Larmor [2] in 1919 to explain the magnetic fields of the sun and the Earth; According to Larmor, the motion of an electrically conducting fluid would be at the origin of the observed magnetic fields. The development of this proposal was rather slow. More than 70 years were necessary to produce a complete numerical simulation of a self-sustaining fluid dynamo [3] and an experimental demonstration of the same effect in a vigorous flow of sodium [4,5]. These two successes opened the way to a new research era on geophysics, astrophysics and cosmological realisations of the dynamo. Indeed, after the numerical and experimental proofs of the existence of the dynamo process, the scientific community has set out to explain all the natural magnetic fields produced on the Earth and other planets, on the sun and other stars, on galaxies and galaxy clusters by this mechanism.

More than 50 PhD students and young researchers attended the Les Houches summer school on Dynamos. During a month they learnt the recent advances in the field and they defined new challenges in dynamo studies. The participants belong to different scientific communities: geophysics, astrophysics, physics, applied mathematics. Members of these communities seldom interacted in the past despite their common interest on dynamos. Traditionally, astrophysicists and geophysicists focused on the detailed observation of their favorite objects, whereas physicists concentrated in understanding the self-induced process itself, and applied mathematicians tended to be attracted by the beauty of the mathematical structure of the (complicated!) set of differential magneto-hydrodynamic equations. The aim of putting together young representatives of each group was to encourage a new generation of scientists to overcome these traditional disciplinary barriers.

The high scientific quality of the school was the result of the effort put by the lecturers in the preparation of their courses. To select them was not an easy task and we apologise to those who would have liked to come but could not. The four main courses consisted of 8 hours lectures on basic knowledge of dynamos. In addition, there were a number of 3 hours lectures on currently active specialized topics. These proceedings collect the lecture notes of the four main courses and a selection of the shorter lectures. Thierry Alboussière taught the fundamentals on Magneto-hydrodynamic (MHD), Peter Olson and Anvar Shukurov lectured on planetary and astrophysical dynamos, respectively, and Chris Jones was given the hard task of summarizing more than 50 years research on dynamo theory. We wish to express our gratitude to Elisabeth Canet, Renaud Deguen, Peter Driscoll, Barbara Ferreira, Nadège Gagnière, Céline Guervilly, Wietze Herreman, Douglas Kelley and Marc Rosin who helped in preparing their written contributions.

Many participants expressed to us how much they enjoyed the summer school. The ambiance was extremely pleasant even though the weather was quite rigorous (all week-ends were just perfect!). We shall certainly remember how Anvar Shukurov lost a bottle of "Absinthe" in a bet on the mathematical development of perturbations for $\alpha - \omega$ dynamos. Chris Jones was very happy to return to Les Houches, where he first came with his father, Harry Jones, who taught quantum mechanics at the VI summer school (1956). Chris even remembers ringing the famous bell! Possibly, François Cardin (6 months old in the photograph) will have similar memories 50 years from now.

The organisation of the Session 88 of the Les Houches Summer School and the publication of this book have been made possible thanks to the financial aid that we received from the Les Houches School of Physics, the CNRS, the GdR

"Dynamos" and the Laboratoire de Géophysique Interne et Tectonophysique de l'Université Joseph-Fourier, Grenoble, France.

Philippe Cardin
Leticia F. Cugliandolo

References

[1] J. Needham, in: *Science and civilisation in China Vol. 4*, Cambdridge University Press, Cambridge, 1962.

[2] J. Larmor, Reports of British Association for the advancement of Science (1919) 159.

[3] G.A. Glatzmaier and P.H. Roberts, Nature **377**, 203 (1995).

[4] A. Gailitis, O. Lielausis, S. Dement'ev, E. Platacis, A. Cifersons, G. Gerbeth, T. Gundrum, F. Stefani, M. Christen, H. Hänel and G. Will, Physical Review Letters **84**, 4365 (2000).

[5] R. Stieglitz and U. Müller, Physics of Fluids **13**, 561 (2001).

CONTENTS

Contents

Contents

Contents

Course 1

FUNDAMENTALS OF MHD

Thierry Aboussière

Ph. Cardin and L.F. Cugliandolo, eds.
Les Houches, Session LXXXVIII, 2007
Dynamos
© *2008 Published by Elsevier B.V.*

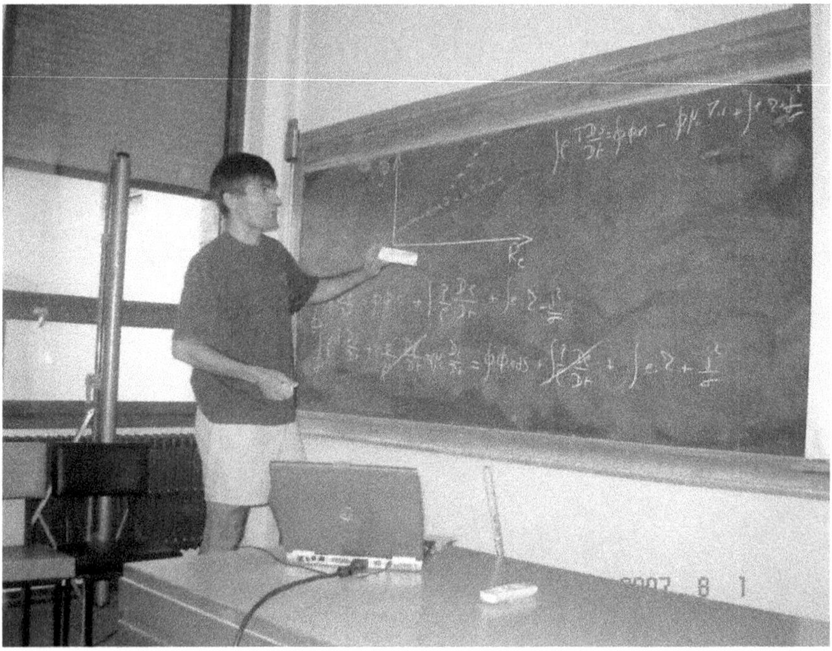

Contents

3

Introduction

Dynamo action takes its origin in the coupling between dynamics and electromagnetism. In most case it concerns the dynamics of fluids with a good electrical conductivity and the magneto-quasi-static limit of electromagnetism is used. Magnetohydrodynamics (subsequently MHD), i.e. the coupling between dynamics and electromagnetism in an electrically conducting fluid, can then be modelled by the coupling between the Navier–Stokes equation and the induction equation.

Fluid dynamics as a whole and a big chunk of electromagnetism are then contained in MHD, but in addition MHD contains novel features, not restricted to the dynamo effect. Basic aspects of this rich coupling will be presented here and can be referred to in the course of this *École d'été* on dynamos.

The lectures will be organized as follows:

1. Historical development of MHD and various fields of application.

2. Set of governing equations.

3. Energetics, dissipation. Entropy balance.

4. Useful dimensionless parameters and equations.

5. Hierarchy of MHD approximations.

6. Identify main physical mechanisms (2D tendency, Hartmann layers, parallel layers, pseudo-diffusion, Alfvén waves, skin effect and field expulsion).

7. Linear and non-linear stability of MHD flows.

8. Introduce experimental and theoretical results on MHD turbulence.

1. A short history of MHD

Magnetohydrodynamics results from the coupling between electromagnetic effects and the dynamics of electrically conducting fluids. This culminates with the Navier–Stokes and induction equations, of relatively similar structure, but the history leading to them is quite different. Let us examine separately the fluid dynamics historical development first in an arbitrary selection.

1.1. The hydrodynamics chain

The theory of dynamics remains rather descriptive until Newton.

Newton (1642–1727)
He spent his life as an academic in Cambridge (UK). He published his *Principia mathematica philosophiænaturalis* in 1687 and we are interested here in his second law of dynamics: $\mathbf{f} = m\mathbf{a}$, relating the acceleration $\mathbf{a}$ of a particle of mass m to the force $\mathbf{f}$ exerted on it.

Euler (1701–1783)
He was a great mathematician from Switzerland and left contributions in many area of science, working mainly in St Petersburg and Berlin. Euler equations (published in *Mémoires de l'académie des sciences de Berlin*, 1757) are a general version of Newton's second law to a continuous medium with no internal shear stress $\rho\partial\mathbf{u}/\partial t + \rho(\mathbf{u}\cdot\nabla)\mathbf{u} = -\nabla p$. They result from taking into account the continuity equation, another discovery by Euler: $\partial\rho/\partial t + \mathrm{div}(\rho\mathbf{u}) = 0$.

Navier (1785–1836)
He was a French engineer (teaching at *École des Ponts et Chaussées* and then *École Polytechnique*), with practical interests such as suspension bridges. He derived the Navier–Stokes equations in 1821 while Stokes was two years old! However, he did not understand the concept of internal shear stress and his idea was only to modify Euler's equations to account for dissipation: $\rho\partial\mathbf{u}/\partial t + \rho(\mathbf{u}\cdot\nabla)\mathbf{u} = -\nabla p + \mu\nabla^2\mathbf{u}$ for a fluid of uniform (dynamical) viscosity μ ($\mathrm{kg\,m^{-1}\,s^{-1}}$).

Stokes (1819–1903)
Like Newton, he was an academic in Cambridge (UK). He introduced the concept of internal shear stress and created the mathematical tools necessary to describe them, thus bringing firm foundations to the Navier–Stokes equations (*On the theories of the internal friction of fluids in motion*, published in 1845).

1.2. The electromagnetism chain

Before Volta, our knowledge is restricted to electrostatic phenomena (Amber), lightening (lightening rod, Franklin, 1752) and "animal electricity" (Galvani, 1790s). Magnetism (magnetite) is thought to be of a different nature.

Volta (1745–1827)

He was an Italian physicist (Roma, Pavia), and was the inventor of the "voltaic pile" or battery (1800, letter to the Royal Society), which made possible further research in electricity.

Ampère (1775–1836)

He was a French mathematician and made a first link between electricity and magnetism (1820), stating how electrical currents would produce a magnetic field.

Ohm (1789–1854)

He was a German scientist, whose main achievement was to discover the law of electrical conduction, in his book *Die galvanische Kette, mathematisch bearbeitet* (1827): $\mathbf{j} = \sigma \mathbf{E}$.

Faraday (1791–1867)

He was a modest English bookbinder who happened to read the books he was working on. He then went on to be involved in electricity and made great discoveries. The most important concerns the law of induction (1831), whereby the variation of the magnetic flux across a loop produces a difference in electric potential. This constitutes an important theoretical progress and also a powerful way of producing electricity.

Maxwell (1831–1879)

He was Scottish but made his career in Cambridge, where he extended the work of Faraday and produced a synthesis of electromagnetism under the form of four equations ("On Faraday's lines of force" was read to the Cambridge Philosophical Society in two parts, 1855 and 1856).

1.3. *A short history of magnetohydrodynamics*

Let us now identify a few milestones in the history of magnetohydrodynamics or hydromagnetics.

Faraday (1791–1867)

Faraday must be mentioned again for his dynamo disk. A rotating disk in a magnetic field produces an electric potential difference between its centre and outer radius. It is possible to take advantage of this tension to produce an electrical current when these two points are connected to the ends of a wire through sliding contacts (see Fig. 1). Faraday had also tried to investigate the electromagnetic signature of the flow of the Thames in the Earth magnetic field.

Fig. 1. Faraday dynamo disk. Faraday's attempt to measure the volume flow rate of the Thames.

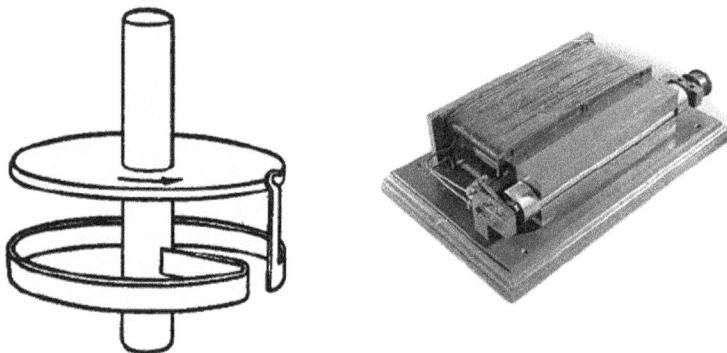

Fig. 2. Bullard model (left) and Siemens self-excited dynamo (right).

Siemens (1816–1892)
In 1866, Siemens (German inventor and industrialist) made a change to Faraday's disk. The electrical current is now forced to flow in a loop, thus generating the magnetic field and removing the need of permanent magnets. This is the first self-excited dynamo, operating when the disk rotation is strong enough (and the loop wound in the correct orientation, see Fig. 2). It is not homogeneous though, as the shape and arrangement of the electrical conductors has been carefully designed.

Ànyos Jedlik (1800–1895)
This Hungarian scientist has apparently built a self-excited dynamo in 1861, but failed to attract attention on his discovery.

Larmor (1857–1942)
In 1919, this British scientist put forward the idea of dynamo action for the Sun and the Earth.

Hartmann

In two papers published in 1937, this Danish scientist studied theoretically and experimentally the effect of a steady magnetic field on the motion of a liquid metal. He discovered thin boundary layers (subsequently called Hartmann layers) and the stabilizing effect of externally imposed magnetic fields on turbulence.

Alfvén (1908–1995)

In 1942, Alfvén (United States) published a paper in Nature describing the mechanism of Alfvén waves. This was a major discovery that would enable dynamo action deeper than the skin effect. According to the legend, his work remained unnoticed until a meeting in Chicago, 1948, when Fermi said "of course . . .".

Alfvén wave demonstration: Lundquist (1949), Lehnert (1953) and Jameson (1964) in liquid sodium. Bostik and Levine (1952), Allen et al. (1959), De Silva (1961) and Spillman (1963) in plasmas.

Elsasser (1904–1991)

He was a German scientist who worked in France (briefly) and in the United States. He is considered as the father of Earth's dynamo magnetism. Although the suggestion of dynamo action for the Earth had been made previously by Larmor, he developed a quantitative analysis, relying on the existence of a strong toroidal magnetic field inside the core.

Bullard (1907–1980)

Proponent of dynamo theory and inventor of a simple model for dynamo. He had many contributions to geomagnetism.

Shercliff (1927–1983)

He was a British scientist and developed the understanding of flows under an imposed magnetic field: parallel shear layers, stability under a magnetic field.

Kulikovskii (1933–)

The Russian scientist has been working on various subjects, and has written two papers on MHD flows. He discovered the equivalent of geostrophic contours for rotating flows, the so-called "characteristic surfaces" in the presence of a strong imposed magnetic field.

Demonstration of dynamo action: Lowes and Wilkinson (1963) in an homogeneous solid with rotating cylinders. Gailitis (1999) and Mühler (1999) with liquid sodium. The VKS (Von Karman Sodium) team (2007) with liquid sodium and rotating ferro-magnetic disks produces dynamo action in a chaotic regime including magnetic field reversals.

2. Governing equations

As we have seen above, the starting points for MHD are Navier–Stokes and Maxwell equations. If a mutual interaction is to take place between dynamics and electromagnetics, it must appear in the equations. This is the role of the additional Lorentz force, $q\mathbf{E} + \mathbf{j} \times \mathbf{B}$, in the Navier–Stokes equations while the corresponding action of dynamics on electromagnetics is entirely due to the change of electrical and magnetic fields in the reference system where matter is at rest and where Ohm's law is applicable.[1]

Let us start with Navier–Stokes. Let $\mathbf{u}$ be the velocity field of an electrically conducting fluid, which will be supposed to be non-magnetic and non-dielectric. The magnetic field, electrical field, electric charge density, electrical current density, thermodynamic pressure, density, kinematic viscosity and electrical conductivity are denoted $\mathbf{B}$, $\mathbf{E}$, q, $\mathbf{j}$, p, ρ, ν, σ respectively. In the local reference system moving with the fluid, the electric and magnetic fields are changed into:

$$\mathbf{E}' = \mathbf{E} + \mathbf{u} \times \mathbf{B}, \qquad \mathbf{B}' = \mathbf{B} \tag{2.1}$$

in the classical limit, $u/c \ll 1$. Ohm's law, $\mathbf{j} = \sigma \mathbf{E}'$, is expressed in the original reference system:

$$\mathbf{j} = q\mathbf{u} + \sigma(\mathbf{E} + \mathbf{u} \times \mathbf{B}), \tag{2.2}$$

where $q\mathbf{u}$ is simply the contribution to the electric charge flux of the mean electric charge density moving with a velocity $\mathbf{u}$.

The Navier–Stokes equations, with Lorentz force, in a gravity field $\mathbf{g}$, can be written:

$$\rho \frac{\partial \mathbf{u}}{\partial t} + \rho(\mathbf{u}.\nabla)\mathbf{u} = -\nabla p + \rho\mathbf{g} + q\mathbf{E} + \mathbf{j} \times \mathbf{B} + \nabla \cdot [\rho\nu\nabla\mathbf{u}] \tag{2.3}$$

supplemented with the continuity condition $\partial\rho/\partial t + \nabla \cdot \rho\mathbf{u} = 0$. Maxwell equations are:

$$\nabla \cdot \mathbf{B} = 0, \tag{2.4}$$

$$\nabla \cdot \mathbf{E} = q/\epsilon, \tag{2.5}$$

$$\nabla \times \mathbf{B} = \mu\mathbf{j} + \mu\epsilon\frac{\partial \mathbf{E}}{\partial t}, \tag{2.6}$$

$$\nabla \times \mathbf{E} = -\frac{\partial \mathbf{B}}{\partial t}, \tag{2.7}$$

where μ and ϵ are the magnetic permeability and electrical susceptibility in vacuum. Electric charge conservation must be added to the equations above:

---

[1] Hall effect will be discussed elsewhere during this Summer school.

$$\frac{\partial q}{\partial t} + \nabla \cdot \mathbf{j} = 0. \tag{2.8}$$

MHD is entirely expressed with equations (2.2) to (2.8). Provided we are concerned with (relatively) good electrical conductors, these equations can be simplified to a large extent within the so-called magnetostatic approximation. This amounts to stating that the electric charge density is so small that it does not play any role. This is justified physically by the fact that electrostatic repulsion would quickly spread out electric charges in a conductor and reach a neutral state. This can be derived as follows. The divergence of equation (2.6), together with (2.2) and (2.5) leads to an equation for the electric charge density:

$$\frac{\epsilon}{\sigma}\left[\frac{\partial q}{\partial t} + (\mathbf{u} \cdot \nabla)q\right] + q = -\epsilon\nabla \cdot (\mathbf{u} \times \mathbf{B}). \tag{2.9}$$

Hence, one should compare the shortest typical timescale of the MHD problem under consideration with ϵ/σ. For a typical liquid metal, $\epsilon/\sigma \sim 10^{-16}$ to 10^{-18}s which implies that, provided all time-scales of the flow are much longer, the bracket on the left-hand side of equation (2.9) can be neglected and that the displacement current ($\epsilon \partial \mathbf{E}/\partial t$) can be neglected in equation (2.6). Hence $q = -\epsilon\nabla \cdot (\mathbf{u} \times \mathbf{B})$ implying that the component $q\mathbf{u}$ is negligible in Ohm's law and $\partial q/\partial t$ is negligible in the electric charge conservation equation (2.8). This condition of very short transient time for electric charge repulsion is the so-called magnetostatic approximation. When it is made, we are left with the following equations:

$$\frac{\partial \rho}{\partial t} + \nabla \cdot (\rho\mathbf{u}) = 0, \tag{2.10}$$

$$\nabla \cdot \mathbf{j} = 0, \tag{2.11}$$

$$\nabla \cdot \mathbf{B} = 0, \tag{2.12}$$

$$\nabla \cdot \mathbf{E} = q/\epsilon, \tag{2.13}$$

$$\nabla \times \mathbf{B} = \mu\mathbf{j}, \tag{2.14}$$

$$\nabla \times \mathbf{E} = -\frac{\partial \mathbf{B}}{\partial t}, \tag{2.15}$$

$$q = -\epsilon\nabla \cdot (\mathbf{u} \times \mathbf{B}), \tag{2.16}$$

$$\mathbf{j} = \sigma(\mathbf{E} + \mathbf{u} \times \mathbf{B}) \tag{2.17}$$

$$\rho\frac{\partial \mathbf{u}}{\partial t} + \rho(\mathbf{u} \cdot \nabla)\mathbf{u} = -\nabla p + \rho\mathbf{g} + \mathbf{j} \times \mathbf{B} + \nabla \cdot [\rho\nu\nabla\mathbf{u}]. \tag{2.18}$$

It will be seen that $\mathbf{B}$ plays a particular role, while $\mathbf{E}$ and q are auxiliary variables. Taking the curl of equation (2.14) and using the other equations for substitution leads to a governing equation for $\mathbf{B}$, the induction equation:

$$\frac{\partial \mathbf{B}}{\partial t} = \nabla \times (\mathbf{u} \times \mathbf{B}) + \eta\nabla^2\mathbf{B}, \tag{2.19}$$

which, given initial and boundary conditions, and for a given velocity field, enables one to determine **B** at all times. The other quantities can be derived if they are needed: **j** from (2.14), q from (2.16), **E** from (2.15).

In summary, many MHD problems can be expressed with two governing equations, Navier–Stokes (2.18) and the induction (2.19) equations.

References

[1] Mathematicians biography. http://www-gap.dcs.st-and.ac.uk/ history/BiogIndex.html.

[2] The Euler archive. http://math.dartmouth.edu/ euler/.

[3] History of electricity. http://www.english.upenn.edu/Projects/knarf.

[4] G. Rousseaux, Lorentz or Coulomb in Galilean electromagnetism? Europhysics Letters **71**(1), 15–20 (2005).

[5] P.H. Roberts, *An Introduction to Magnetohydrodynamics*, Longmans, 1964.

3. Energetics

Energetic aspects of MHD problems will often consist of a strong interplay between kinetic and magnetic energy. At the same time, the question of the relative importance of viscous and Joule dissipation will always be of importance. Those two aspects, energy and dissipation, are moreover quite independent from each other. Entropy sources and balance are a related but independent issue, which will prove useful in the analysis of planetary dynamos.

3.1. Mechanical energy

Taking the dot product of the Navier–Stokes equation with the velocity field **u** and integrating over a volume $\mathcal{V}$ provides the mechanical energy balance for that control volume. Using continuity (2.10) for inertial terms and integration by parts for viscous terms results in the following form:

$$\frac{d}{dt} \int_{\mathcal{D}(t)} \frac{\rho u^2}{2} = \int_{\mathcal{D}} \left[-\nabla p + \rho \mathbf{g} + \mathbf{j} \times \mathbf{B} \right] \cdot \mathbf{u} + \oint_{\mathcal{S}} \mathbf{u} \cdot \tau \cdot \mathbf{n} - \int_{\mathcal{D}} \mathbf{e} : \tau, \quad (3.1)$$

where **e** is the tensor of rate of deformation (in Cartesian coordinates, $e_{ij} = 1/2[\partial u_i/\partial x_j + \partial u_j/\partial x_i]$) and τ is the shear stress tensor ($\tau_{ij} = 2\rho \nu e_{ij}$). The left-hand side term is the material derivative $\partial/\partial t + \mathbf{u} \cdot \nabla$ of kinetic energy (time derivative of kinetic energy following the evolution of matter). On the right, we have the power of internal pressure gradients, power of gravitational forces, power of Lorentz forces, power of surface shear stress forces and viscous power dissipation, respectively.

3.2. Magnetic energy

An equation governing the evolution of magnetic energy is obtained from the dot product of **B** and the induction equation. Dividing by μ and integrating by parts both the electromotive and diffusion terms leads to:

$$\int_D \frac{\partial}{\partial t} \frac{B^2}{2\mu} = -\oint_S \frac{\mathbf{E} \times \mathbf{B}}{\mu} \cdot \mathbf{n} + \int_D (\mathbf{u} \times \mathbf{B}) \cdot \mathbf{j} - \int_D \frac{j^2}{\sigma}. \tag{3.2}$$

The rate of change of magnetic energy in the control volume is balanced by the sum of the flux of the Poynting vector $-\mathbf{E} \times \mathbf{B}/\mu$, the power of electromotive forces and Joule dissipation. It should be noted that the power of electromotive forces in (3.2) and the power of Lorentz forces in (3.1) are just equal and opposite. Physically the mechanical work done by Lorentz forces is directly supplied to magnetic energy. This term of energy exchange will make it difficult to evaluate magnetic and kinetic energy dissipations independently.

3.3. Balance of total energy

From first principle, the rate of change of the total energy in the control volume must be equal to the work done by external action on it.

$$\frac{d}{dt} \int_{D(t)} \left[\frac{\rho u^2}{2} + \rho e \right] + \int_D \frac{\partial}{\partial t} \frac{B^2}{2\mu}$$

$$= \int_D \rho \mathbf{g} \cdot \mathbf{u} + \oint_S -p \mathbf{u} \cdot \mathbf{n} + \mathbf{u} \cdot \tau \cdot \mathbf{n} - \frac{\mathbf{E} \times \mathbf{B}}{\mu} \cdot \mathbf{n} - \phi \cdot \mathbf{n}. \tag{3.3}$$

In the equation e denotes the specific internal energy of the fluid, and $\phi = -k\nabla T$ the conduction heat flux (T is the temperature field). Gravitational power is considered as external, as gravitational energy is not included on the left-hand side.

Comment on the Poynting vector: is it possible for the Earth magnetic field to lose energy into the empty space (i.e. through radiation)? Not for the main geomagnetic contributions, which satisfy curl(**B**) = **0**, because it is decreasing as a power law faster than r^{-3}, hence the flux of the associated Poynting vector would necessarily decrease towards zero with distance. Hence there is no net Poynting flux. The only part that can radiate Poynting flux is that part of the magnetic field obeying the electromagnetic equations, which is a minute fraction of the geomagnetic field owing to its large timescales. Let us derive a gross upper bound for this loss, i.e. by considering that the main dipolar geomagnetic field has a one-year timescale τ. From the point of view of electromagnetic radiation, the distance corresponding to this timescale is 1 light-year ($r \simeq 10^{16}$ m). We can estimate the Poynting flux at that distance and consider that it is lost by radiation.

From equation 2.15, $\mathbf{E} \sim r\mathbf{B}/\tau$. Hence, the Poynting vector is estimated as $r\mathbf{B}^2/(\mu\tau)$, where $\mathbf{B}$ decreases as r^{-3} and is of order 10^{-4} T on the surface of the Earth. When integrated over the sphere of radius 1 light-year, the loss of energy is found to be as small as 10^{-16} W. This reasoning would be valid if the Earth would be isolated in the universe. In reality the solar wind is generating the magnetosphere and its associated electric currents.

3.4. Balance of internal energy

It is obtained when the sum of kinetic and magnetic energy equations is subtracted to the total energy equation above. Using continuity and integration by parts for the power of pressure forces allows one to obtain:

$$\frac{d}{dt}\int_{\mathcal{D}(t)} [\rho e] = -\oint_{\mathcal{S}} \phi \cdot \mathbf{n} + \int_{\mathcal{D}} \frac{p}{\rho}\frac{D\rho}{Dt} + \int_{\mathcal{D}} \mathbf{e} : \tau + \frac{\mathbf{j}^2}{\sigma}. \tag{3.4}$$

Internal energy is changed by a supply of heat flux, internal compression power and viscous and Joule dissipations. These last two terms act now as positive sources, while they were sinks for mechanical and magnetic energies.

3.5. Entropy balance

Gibbs relationship of (local) equilibrium thermodynamics can be written $de = Tds - pd(1/\rho) + \mu_C dC$, where s is the specific entropy, C is the concentration of a solute element and μ_C its associated thermodynamic potential. It should be noted that:

$$\frac{d}{dt}\int_{\mathcal{D}(t)} [\rho e] = \int_{\mathcal{D}(t)} \left[\rho \frac{De}{Dt} \right]. \tag{3.5}$$

This is true not only for e but also for any other quantity. Using Gibbs relationship, it can be shown that:

$$\int_{\mathcal{D}(t)} \left[\rho T \frac{Ds}{Dt} \right] = -\oint_{\mathcal{S}} \phi \cdot \mathbf{n} + \int_{\mathcal{D}} \mu_C \nabla \cdot \mathbf{i} + \int_{\mathcal{D}} \mathbf{e} : \tau + \frac{\mathbf{j}^2}{\sigma}, \tag{3.6}$$

where the solute concentration governing equation has been used: $\rho DC/Dt = \nabla \cdot \mathbf{i}$, where the solute flux is $\mathbf{i} = -\alpha_D \nabla \mu_C$.

The local form of equation (3.6) can be written:

$$\rho T \frac{Ds}{Dt} = -\nabla \cdot \phi + \mu_C \nabla \cdot \mathbf{i} + \mathbf{e} : \tau + \frac{\mathbf{j}^2}{\sigma}. \tag{3.7}$$

Dividing by T, integrating again over the control volume provides the final entropy balance, after some integrations by parts:

$$\frac{d}{dt}\int_{\mathcal{D}(t)} \rho s = -\oint_{\mathcal{S}} \frac{\phi \cdot \mathbf{n}}{T} + \oint_{\mathcal{S}} \frac{\mu_C \mathbf{i} \cdot \mathbf{n}}{T}$$

$$+ \int_{\mathcal{D}} \left[\frac{k\nabla T \cdot \nabla T}{T^2} + \frac{\alpha_D \nabla \mu_C \cdot \nabla \mu_C}{T} + \frac{\mathbf{e} : \tau}{T} + \frac{\mathbf{j}^2}{\sigma T} \right]. \quad (3.8)$$

The surface integrals are the so-called reversible entropy exchange terms with the surrounding, while volume integrals on the right-hand side are entropy sources and must be positive according to the second principle. In fact, this condition is used to ascertain that k and α_D are positive scalars.

3.6. Gross Heat budget for the Earth's core

If one considers the Earth's core with a mean temperature T_c and a global heat flow Q_{cmb} lost to the mantle, it is possible to write the total energy and entropy balance for the core in the following approximate way:

$$mc_p \frac{dT_c}{dt} = -Q_{cmb},$$

$$\frac{mc_p}{T_c} \frac{dT_c}{dt} = -\frac{Q_{cmb}}{T_{cmb}} + \int_{\mathcal{D}} \frac{k\nabla T \cdot \nabla T}{T^2} + \frac{\mathbf{j}^2}{\sigma T_c},$$

where viscous dissipation has been neglected compared to ohmic dissipation. One can see here, at least in principle, how these two equations can provide an estimate for ohmic dissipation. From a value for Q_{cmb}, the energy equation can provide the secular cooling dT_c/dt. From the assumption of adiabatic thermal gradient in the liquid core, the entropy production term due to the heat flux can be determined. Finally, the entropy equation provides an estimate for the source of entropy due to Joule dissipation.

3.7. Joule versus viscous dissipation

Quite often, one would like to know how much energy dissipation can be attributed to electrical currents (Joule) and to velocity shear (viscous). This is clearly the case in numerical calculations, unless a turbulent model is used (then, one has to rely on the model to distinguish between viscous and Joule dissipations). In natural objects and in laboratory experiments, this is unfortunately not possible. As we have seen in the present section on energetics, global energy balance involves the sum of Joule and viscous dissipation and there is no way to determine them separately (except if one were to measure the whole velocity field down to the viscous dissipation scale, e.g. with PIV, in the whole fluid domain).

There have been attempts to estimate Joule dissipation in dynamo experiments, by taking the difference between the actual dissipation and the extrapolated dissipation from the hydrodynamical regime below dynamo threshold. This, in essence, assumes that turbulence will keep basically the same structure below and above dynamo threshold. There is no solid ground for such an assumption and it might well be that turbulence changes in such a way that the amount of viscous dissipation varies considerably across this threshold.

References

[1] P.H. Roberts, *An Introduction to Magnetohydrodynamics*, Longmans, 1964.
[2] B.A. Buffett, H.E. Huppert, J.R. Lister and A.W. Woods, On the thermal evolution of the Earth's core, JGR **101**(B4), 7989–8006 (1996).
[3] G.E. Backus, Gross thermodynamics of heat engines in deep interior of Earth, Proc. Nat. Acad. Sci. USA **72**(4), 1555–1558 (1975).

4. Dimensionless parameters

There are nearly a countless number of dimensionless parameters in MHD. However, they have been introduced for convenience when needed by various authors and they are not all independent. For each given problem under consideration it is important to figure out how many such parameters are required to specify the flow regime. Another task is to choose the most relevant set of dimensionless numbers: it is more arbitrary and subject to personal taste.

4.1. Fundamental units

In MHD there are generally four fundamental units, i.e. meter, kilogram, second, Ampere. The first three units pertain to pure dynamics, while the electromagnetic effects requires the addition of just one other unit, which has been chosen to be the Ampere.

The International Bureau of Weights and Measures (BIPM, for Bureau International des Poids et Mesures) in Paris maintains the International System. It is updated every few years by an international conference, the General Conference on Weights and Measures (CGPM, for Conférence Générale des Poids et Mesures).

It is interesting to see the evolution of the definition of each unit. Let us just mention the current definition for the four units mentioned above, plus temperature:

- **metre** (m): the metre is the length of the path travelled by light in vacuum during a time interval of 1/299 792 458 of a second (since 1983).
- **kilogram** (kg): the kilogram is the unit of mass; it is equal to the mass of the international prototype of the kilogram (since 1889).
- **second** (s): the second is the duration of 9 192 631 770 periods of the radiation corresponding to the transition between the two hyperfine levels of the ground state of the cesium 133 atom (since 1967).
- **ampere** (A): The ampere is that constant current which, if maintained in two straight parallel conductors of infinite length, of negligible circular cross-section, and placed 1 metre apart in vacuum, would produce between these conductors a force equal to 2×10^7 newton per metre of length (since 1948).
- **kelvin** (K): The kelvin, unit of thermodynamic temperature, is the fraction 1/273.16 of the thermodynamic temperature of the triple point of water (since 1967).

The definition of the metre needs that of the second. The definition of the ampere requires that of a newton (hence that of metre, kilogram and second).

All other units can be expressed in terms of the fundamental ones. Below is a list of popular units and a list of physical properties.

Coulomb, C	$\mathrm{A\,s}$
Tesla, T	$\mathrm{kg\,s^{-2}\,A^{-1}}$
Volt, V	$\mathrm{m^2\,kg\,s^{-3}\,A^{-1}}$
Farad, F	$\mathrm{m^{-2}\,kg^{-1}\,s^4\,A^2}$
Henry, H	$\mathrm{m^2\,kg\,s^{-2}\,A^{2}}$
Ohm, Ω	$\mathrm{m^2\,kg\,s^{-3}\,A^{-2}}$
Newton, N	$\mathrm{m\,kg\,s^{-2}}$

μ	$\mathrm{H\,m^{-1}}$	$\mathrm{m\,kg\,s^{-2}\,A^{-2}}$
σ	$\mathrm{\Omega^{-1}\,m^{-1}}$	$\mathrm{m^{-3}\,kg^{-1}\,s^3\,A^2}$
ρ		$\mathrm{m^{-3}\,kg}$
ν		$\mathrm{m^2\,s^{-1}}$

4.2. Dimensionless numbers

The Vashy-Buckingham theorem tells us how many parameters are needed to describe a problem.

Any dimensionless result of a problem, defined by n dimensional independent parameters in which m fundamental units appear, is a function of $n - m$ independent dimensionless parameters.

In MHD, considering only mechanical and electromagnetic effects (no temperature, chemistry, ...) the physical properties appearing in the Navier–Stokes and induction equations are ρ, ν, σ and μ: density, kinematic viscosity, electrical conductivity and magnetic permeability (in the absence of magnetic properties

$\mu = 4\pi\,10^{-7}\ \mathrm{H\,m^{-1}}$, as resulting from the definition of the ampere). Those dimensional parameters involve four fundamental units (m, kg, s, A). They are not dimensionally independent though, as $1/(\mu\sigma)$ has the same dimension of ν. It is called the magnetic diffusivity ($\mathrm{m^2\,s^{-1}}$). Therefore the ratio of kinematic and magnetic diffusivities is the first dimensionless number, the magnetic Prandtl number:

$$P_m = \mu\sigma\nu. \qquad (4.1)$$

Let us see how the Vashy-Buckingham theorem is applied in different cases with an increasing number of parameters. On top of the four physical properties (always present), we are going to investigate the possibility of having a length scale H, a velocity scale U or a magnetic field scale B or any combination of these three scales.

4.2.1. Zero dimensionless parameter
If we only have the four physical properties, ρ, ν, σ and μ, involving four fundamental units, m, kg, s, A, then any dimensionless output is a function of $4-4 = 0$ dimensionless parameter. In fact, nothing interesting can happen in this case. Without any length scale, one can envisage that the whole physical space is filled with the fluid, or a semi-infinite domain bounded by a plane. Nothing is going to happen so it is satisfactory that no dimensionless parameter is required to describe the outcome. It does not matter that the magnetic Prandtl number can be formed from the parameters and is dimensionless. The Vashy-Buckingham theorem tells us that we don't need it.

4.2.2. One dimensionless parameter
In addition to ρ, ν, σ and μ, we add just one of U, H or B in turn. In each case, there is one pertinent dimensionless parameter according to Vashy-Buckingham: not surprisingly, this is the magnetic Prandtl number identified previously.

The question whether these five parameters can describe an interesting problem can be asked. At least, when U is the additional parameter, the answer is definitely yes. Let us consider a semi-infinite domain with the fluid having a velocity U far from the boundary in a direction parallel to itself. In addition, the boundary is porous and there is a uniform suction U. Choosing the following scales:

$$\mathbf{x} \longleftrightarrow \frac{\nu}{U} \qquad \mathbf{u} \longleftrightarrow U$$

$$t \longleftrightarrow \frac{\nu}{U^2} \qquad p \longleftrightarrow \rho U^2$$

$$\mathbf{B} \longleftrightarrow U\sqrt{\rho\mu} \qquad \mathbf{j} \longleftrightarrow \frac{U^2}{\nu}\sqrt{\frac{\rho}{\mu}}$$

one can define dimensionless variables (with tildes)

$$\tilde{\mathbf{x}} \longleftrightarrow \frac{U\mathbf{x}}{\nu} \qquad \tilde{\mathbf{u}} \longleftrightarrow \frac{\mathbf{u}}{U}$$

$$\tilde{t} \longleftrightarrow \frac{U^2 t}{\nu} \qquad \tilde{p} \longleftrightarrow \frac{p}{\rho U^2}$$

$$\tilde{\mathbf{B}} \longleftrightarrow \frac{\mathbf{B}}{U\sqrt{\rho\mu}} \qquad \tilde{\mathbf{j}} \longleftrightarrow \frac{\nu\sqrt{\mu}\mathbf{j}}{U^2\sqrt{\rho}}$$

which are then substituted into Navier–Stokes and induction equation. Dividing Navier–Stokes by $\rho U^3/\nu$ and the induction equation by $U^3\sqrt{\rho\mu}/\nu$, one obtains:

$$\frac{\partial \tilde{\mathbf{u}}}{\partial \tilde{t}} + (\tilde{\mathbf{u}} \cdot \tilde{\nabla})\tilde{\mathbf{u}} = -\tilde{\nabla}\tilde{p} + \tilde{\mathbf{j}} \times \tilde{\mathbf{B}} + \tilde{\nabla}^2\tilde{\mathbf{u}},$$

$$\frac{\partial \tilde{\mathbf{B}}}{\partial \tilde{t}} = \tilde{\nabla} \times (\tilde{\mathbf{u}} \times \tilde{\mathbf{B}}) + P_m^{-1}\tilde{\nabla}^2\tilde{\mathbf{B}},$$

where $\tilde{\nabla}$ denotes the nabla derivation operator in the dimensionless spatial space spanned by $\tilde{\mathbf{x}}$. Finally, tildes are dropped and the dimensionless Navier–Stokes and induction equations take the usual form:

$$\frac{\partial \mathbf{u}}{\partial t} + (\mathbf{u} \cdot \nabla)\mathbf{u} = -\nabla p + \mathbf{j} \times \mathbf{B} + \nabla^2\mathbf{u}, \qquad (4.2)$$

$$\frac{\partial \mathbf{B}}{\partial t} = \nabla \times (\mathbf{u} \times \mathbf{B}) + P_m^{-1}\nabla^2\mathbf{B}. \qquad (4.3)$$

Although things happen simultaneously, one can imagine that the kinematic regimes develops with no free parameter, while the fate of the magnetic part depends on P_m. If P_m is large enough, one may enter a dynamo regime, which would in turn affect the flow regime. So, P_m is the single dimensionless parameter governing this problem. There is a little element of choice here, as one could have taken P_m^2 as the governing dimensionless parameter (or $2\,P_m$, or $\sqrt{P_m}$, ...).

4.2.3. *Two dimensionless parameters*
In addition to ρ, ν, σ and μ, we add two of U, H or B. With six parameters and four units, two independent dimensionless parameters must be identified.

parameters U and H: a common choice is P_m and the magnetic Reynolds number:

$$R_m = \mu\sigma U H. \qquad (4.4)$$

However, one can also take $(Re = UH/\nu, R_m)$ as governing parameters, or (Re, P_m), or $(Re^2, R_m + Re)$, ... With the following scales:

$$\mathbf{x} \longleftrightarrow H \qquad\qquad \mathbf{u} \longleftrightarrow U$$

$$t \longleftrightarrow \frac{H}{U} \qquad\qquad p \longleftrightarrow \rho U^2$$

$$\mathbf{B} \longleftrightarrow U\sqrt{\rho\mu} \qquad \mathbf{j} \longleftrightarrow \frac{U^2}{\nu}\sqrt{\frac{\rho}{\mu}}.$$

The Navier–Stokes and induction equations can be written:

$$\frac{\partial \mathbf{u}}{\partial t} + (\mathbf{u} \cdot \nabla)\mathbf{u} \;=\; -\nabla p + Re\,\mathbf{j} \times \mathbf{B} + Re^{-1}\nabla^2\mathbf{u}, \tag{4.5}$$

$$\frac{\partial \mathbf{B}}{\partial t} \;=\; \nabla \times (\mathbf{u} \times \mathbf{B}) + R_m^{-1}\nabla^2\mathbf{B}. \tag{4.6}$$

Many dynamo problems can be modelled with the two parameters R_m and Re. Starting from a non-MHD flow (the Lorentz force is zero), the Reynolds number is the only parameter governing the flow. Next, the question of dynamo instability is entirely related to R_m (see the induction equation). Finally both equations are coupled in the final saturated regime.

parameters U and B: it may be that there is no length scale but that a magnetic field is applied from the outside. This can apply to a flow in a semi-infinite domain, with no suction, but with a magnetic field applied perpendicular to the fluid plane boundary. In this case, in addition to the magnetic Prandtl number P_m, one can define the following parameter:

$$R = \frac{U}{B}\sqrt{\frac{\rho}{\nu\sigma}}, \tag{4.7}$$

who is a Reynolds number based on the thickness of the Hartmann layer (see later) $\delta = 1/B\,\sqrt{\rho\nu/\sigma}$. With the following scales:

$$\mathbf{x} \longleftrightarrow \sqrt{\frac{\rho\nu}{\sigma}}\frac{1}{B} \qquad\qquad \mathbf{u} \longleftrightarrow U$$

$$t \longleftrightarrow \sqrt{\frac{\rho\nu}{\sigma}}\frac{1}{BU} \qquad\qquad p \longleftrightarrow \rho U^2$$

$$\mathbf{B} \longleftrightarrow B \qquad\qquad \mathbf{j} \longleftrightarrow \sigma U B.$$

The Navier–Stokes and induction equations can be written:

$$\frac{\partial \mathbf{u}}{\partial t} + (\mathbf{u} \cdot \nabla)\mathbf{u} \;=\; -\nabla p + R^{-1}\mathbf{j} \times \mathbf{B} + R^{-1}\nabla^2\mathbf{u}, \tag{4.8}$$

$$\frac{\partial \mathbf{B}}{\partial t} \;=\; \nabla \times (\mathbf{u} \times \mathbf{B}) + R^{-1}P_m^{-1}\nabla^2\mathbf{B}. \tag{4.9}$$

It will be seen later that under the assumption of small magnetic Reynolds number (here the product $R\,P_m$), the single governing parameter is R.

parameters B and H: in this case finally, the two governing dimensionless parameters can be the magnetic Prandtl number (again) and the Lundquist number:

$$S = \sqrt{\frac{\mu}{\rho}} \sigma B H, \tag{4.10}$$

The following scales are chosen:

$$\mathbf{x} \longleftrightarrow H \qquad \mathbf{u} \longleftrightarrow \frac{1}{\mu \sigma H}$$

$$t \longleftrightarrow \mu \sigma H^2 \qquad p \longleftrightarrow \frac{\rho}{\mu^2 \sigma^2 H^2}$$

$$\mathbf{B} \longleftrightarrow B \qquad \mathbf{j} \longleftrightarrow \frac{B}{\mu H}.$$

Dimensionless Navier–Stokes and induction equations are:

$$\frac{\partial \mathbf{u}}{\partial t} + (\mathbf{u} \cdot \nabla)\mathbf{u} = -\nabla p + S^2 \mathbf{j} \times \mathbf{B} + P_m \nabla^2 \mathbf{u}, \tag{4.11}$$

$$\frac{\partial \mathbf{B}}{\partial t} = \nabla \times (\mathbf{u} \times \mathbf{B}) + \nabla^2 \mathbf{B}. \tag{4.12}$$

The role of the Lundquist number will become clear when we consider some MHD approximations in the next section.

4.2.4. Three dimensionless parameters

Let us consider the case when the size of the domain, the flow velocity and a magnetic field are imposed. There are seven parameters, ρ, ν, σ, μ, U, H and B. Thus 3 dimensionless parameters will be needed to specify the problem. For instance, P_m, R_m and S can be chosen. The following scales can be chosen:

$$\mathbf{x} \longleftrightarrow H \qquad \mathbf{u} \longleftrightarrow U$$

$$t \longleftrightarrow \frac{H}{U} \qquad p \longleftrightarrow \rho U^2$$

$$\mathbf{B} \longleftrightarrow B \qquad \mathbf{j} \longleftrightarrow \frac{B}{\mu H}.$$

The dimensionless governing equations can be written:

$$\frac{\partial \mathbf{u}}{\partial t} + (\mathbf{u} \cdot \nabla)\mathbf{u} = -\nabla p + \frac{S^2}{R_m^2} \mathbf{j} \times \mathbf{B} + \frac{P_m}{R_m} \nabla^2 \mathbf{u}, \tag{4.13}$$

$$\frac{\partial \mathbf{B}}{\partial t} = \nabla \times (\mathbf{u} \times \mathbf{B}) + R_m^{-1} \nabla^2 \mathbf{B}. \tag{4.14}$$

Many physical effects can manifest potentially out of these equations. It will be helpful to study some of them independently by considering approximations of the equations above in the vanishing limit of some dimensionless parameters.

References

[1] The international system SI, 8th edition 2006, http://www.bipm.org/fr/si/.
[2] R. Moreau, *Magnetohydrodynamics*, Kluwer Academic, 1990.

5. MHD approximations

There are a number of advantages in considering simplified versions of the full set of MHD equations. They are simpler and faster to solve numerically, and they help identify some MHD physical effects which also exist in the full version of the equations even when the conditions of validity of the simplified equations are not fulfilled.

5.1. Quasi-linear approximation

This approximation rests on the condition that the magnetic Reynolds number is small compared to unity: $R_m \ll 1$. Physically, this corresponds to a condition when the velocity field has no large impact on the main part of the magnetic field. This condition excludes dynamo action, so we assume that a constant magnetic field $\mathbf{B}_0$ is applied by external means. If the sources of current responsible for $\mathbf{B}_0$ lie outside the flow, this magnetic field is thus curl-free in the domain of the flow. From equation (4.14), the Laplacian term is dominant when $R_m \ll 1$ implying that there is little departure (of order R_m) from the magnetic field imposed in the absence of the flow. At this point, it is useful to decompose the magnetic field into the imposed field $\mathbf{B}_0$ and a small departure $\mathbf{b}$:

$$\mathbf{B} = \mathbf{B}_0 + \mathbf{b}. \tag{5.1}$$

This decomposition is substituted in equations (4.13) and (4.14) and $\mathbf{b}$ is neglected in each term unless the contribution of $\mathbf{B}_0$ vanishes exactly:

$$\frac{\partial \mathbf{u}}{\partial t} + (\mathbf{u} \cdot \nabla)\mathbf{u} = -\nabla p + \frac{S^2}{R_m^2} \mathbf{j} \times \mathbf{B}_0 + \frac{P_m}{R_m} \nabla^2 \mathbf{u}, \tag{5.2}$$

$$\frac{\partial \mathbf{b}}{\partial t} = \nabla \times (\mathbf{u} \times \mathbf{B}_0) + R_m^{-1} \nabla^2 \mathbf{b}. \tag{5.3}$$

This set of equations has been called the quasi-linear approximation (see Knaepen et al.) because the induction equation is linear. The single non-linear term is

$(\mathbf{u} \cdot \nabla)\mathbf{u}$. The electric current density in equation (5.2) is entirely due to $\mathbf{b}$ as $\mathbf{B}_0$ is curl-free.

In fact, as R_m is small, it does not play a key role anymore, the dimensionless equations (4.13) and (4.14) are not the best starting point and we shall express again the quasi-linear equations using dimensionless equations (4.11) and (4.12), based on a scale B related to the imposed magnetic field $\mathbf{B}_0$:

$$\frac{\partial \mathbf{u}}{\partial t} + (\mathbf{u} \cdot \nabla)\mathbf{u} = -\nabla p + S^2 \mathbf{j} \times \mathbf{B}_0 + P_m \nabla^2 \mathbf{u}, \tag{5.4}$$

$$\frac{\partial \mathbf{b}}{\partial t} = \nabla \times (\mathbf{u} \times \mathbf{B}_0) + \nabla^2 \mathbf{b}. \tag{5.5}$$

It is clear here that the Lundquist number S plays a key role. If it is larger than unity and larger than P_m, Alfvén waves will develop. If S is smaller than 1 or smaller than P_m, Alfvén waves will be suppressed by Ohmic diffusion (or viscous diffusion when $P_m > 1$). Alfvén waves can interact, and the non-linear term $(\mathbf{u} \cdot \nabla)\mathbf{u}$ can lead to a so-called weak turbulence, i.e. nearly linear Alfvén waves interacting weakly. The nature of turbulence is changed entirely.

5.2. *Quasi-static approximation*

When the Lundquist number is small compared to 1 or P_m, the equations can be further simplified. There can be no Alfvén wave propagation and the time dependent term in the induction equation can be removed. The remaining terms can be uncurled, leading to a version of Ohm's law where the electric field is a pure gradient:

$$\mathbf{j} = -\nabla\phi + \mathbf{u} \times \mathbf{B}_0, \tag{5.6}$$

where ϕ is the electric potential. Together with the unchanged equation (5.4), This is the so-called quasi-static MHD approximation. There can be no Alfvén waves in this approximation.

It is possible to write an relationship of order between the full Maxwell, Ohm and Navier–Stokes equations (FS), the set of MHD equations in the magnetostatic approximation (MS), the quasi-linear approximation (QL) and the quasi-static approximation (QS):

$$QS \subset QL \subset MS \subset FS, \tag{5.7}$$

where inclusion $\subset$ means that any physical effect existing on the left-hand side can also exist on the right-hand side.

5.3. A quick look at turbulence in the Earth core

For scales below 10 m, the magnetic Reynolds number R_m and the Lundquist number S are small compared to one. Hence, the quasi-static approximation constitute an appropriate model for turbulence.

However, for scales smaller than 10 km, and larger than 10 m, the magnetic Reynolds number is small and the Lundquist number is large. Hence the quasi-linear approximation is valid for these scales. In physical terms, this means that the flow will not generate a lot of magnetic field at these scales, but Alfvén waves can develop without much damping.

5.4. Remark

The quasi-static approximation is sometimes (in fact, very often) justified only by the fact that the magnetic Reynolds number is small (see for instance [1] and its title). This approximation requires also that the Lundquist number is small. This second condition is entirely independent from the first one and it is equally important to satisfy each of them.

When the magnetic Reynolds number is small and the Lundquist number is not small, it is appropriate to use the quasi-linear approximation.

References

[1] B. Knaepen, S. Kassinos and D. Carati, Magnetohydrodynamic turbulence at moderate magnetic Reynolds number, JFM **513**, 199–220 (2004).

[2] G. Backus, R.L. Parker and C. Constable, *Foundations of Geomagnetism*, Cambridge University Press, 1996.

6. MHD physical effects

The approximations will be considered starting from the most simplified quasi-static approximation, then the quasi-linear approximation and finally the full version of MHD. The quasi-static approximation (5.4) and (5.6) is written here again, with a different choice of dimensional scales:

$$\mathbf{x} \longleftrightarrow H \qquad \mathbf{u} \longleftrightarrow \frac{\nu}{H}$$

$$t \longleftrightarrow \frac{H^2}{\nu} \qquad p \longleftrightarrow \frac{\rho\nu^2}{H^2}$$

$$\mathbf{B} \longleftrightarrow B \qquad \mathbf{j} \longleftrightarrow \frac{\sigma B\nu}{H} \qquad \phi \longleftrightarrow \nu B.$$

The dimensionless governing equations for the divergence-less fields $\mathbf{u}$ and $\mathbf{j}$ can be written

$$\frac{\partial \mathbf{u}}{\partial t} + (\mathbf{u} \cdot \nabla)\mathbf{u} = -\nabla p + Ha^2 \mathbf{j} \times \mathbf{B}_0 + \nabla^2 \mathbf{u}, \tag{6.1}$$

$$\mathbf{j} = -\nabla \phi + \mathbf{u} \times \mathbf{B}_0, \tag{6.2}$$

where the dimensionless Hartmann number is defined as $Ha = \sqrt{\sigma \rho \nu} B H$.

In the analysis of physical effects arising from the presence of the imposed magnetic field $\mathbf{B}_0$, the assumption of a uniform magnetic field $\mathbf{B}_0$ in the z-direction will be made. Although this is not rigorous, it does not introduce any qualitative change.

6.1. Pseudo-diffusion

To obtain an equation on the velocity field $\mathbf{u}$ suitable for analysis, one can take the curl of the momentum equation twice and substitute $\nabla \times \mathbf{j}$ using the curl of Ohm's law:

$$\frac{\partial}{\partial t}\left(-\nabla^2 \mathbf{u}\right) + \nabla \times \nabla \times \left[(\mathbf{u} \cdot \nabla)\mathbf{u}\right] = Ha^2 \frac{\partial^2 \mathbf{u}}{\partial z^2} - \left(\nabla^2\right)^2 \mathbf{u}. \tag{6.3}$$

The asymptotic strong MHD regime is characterized by a large value of the Hartmann number ($Ha \longrightarrow \infty$). The dominant first term on the right-hand side of equation (6.3) can only be balanced by the time dependent first term on the left-hand side on a short timescale of order Ha^{-2}, i.e. the so-called Joule time $\rho/(\sigma B^2)$. These two terms define an equation of pseudo-diffusion for the velocity field. Its dispersion relationship takes the following form:

$$\omega = i Ha^2 \frac{(\mathbf{k}.\mathbf{B})^2}{k^2}. \tag{6.4}$$

This is called "pseudo-diffusion" as the speed of diffusion of a disturbance in the direction of the imposed magnetic field depends on the size of the disturbance.

6.2. Two dimensional core flow

When transient phenomena are absent, and when the non-linear term is weak, equation (6.3) reduces to:

$$Ha^2 \frac{\partial^2 \mathbf{u}}{\partial z^2} - \left(\nabla^2\right)^2 \mathbf{u} = 0, \tag{6.5}$$

from which it can be inferred that $\partial^2 \mathbf{u}/\partial z^2 = 0$ in the main part of the flow when $Ha \gg 1$. This does not exclude that the velocity field varies linearly in the z-direction. However this linear variation should be accompanied be a flow of electric current in the z-direction. When electrically insulating conditions are met at both ends of the magnetic line, we just get $\partial \mathbf{u}/\partial z = \mathbf{0}$. Hence MHD flows have a tendency to become two dimensional, like flows in a rotating system.

6.3. Hartmann layers

Along a wall, on a short length-scale, it is possible that the double Laplacian terms becomes comparable in magnitude to the other term. Where the magnetic field is not parallel to the wall, a so-called Hartmann boundary layer develops. Obviously, the direction of maximum variation is the normal direction to the wall $\mathbf{n}$ and equation (6.5) can be simplified into:

$$Ha^2 \, (\mathbf{B} \cdot \mathbf{n})^2 \, \frac{\partial^2 \mathbf{u}}{\partial n^2} - \frac{\partial^4 \mathbf{u}}{\partial n^4} = \mathbf{0}, \tag{6.6}$$

where n is the distance to the wall. Hartmann layers are solutions to this equation and the tangent velocity varies exponentially in them on a typical length-scale $Ha^{-1} \, (\mathbf{B} \cdot \mathbf{n})^{-1}$.

6.4. Parallel layers

Viscous terms represented by the bi-Laplacian term in (6.5) can also play a role in thin regions (boundary or free-shear layers) containing the magnetic field direction. In this case, equation (6.3) takes the following form with negligible inertia:

$$Ha^2 \, \frac{\partial^2 \mathbf{u}}{\partial z^2} - \frac{\partial^4 \mathbf{u}}{\partial n^4} = \mathbf{0}, \tag{6.7}$$

where n is again a coordinate perpendicular to the layer. The solutions to this equation are parallel layers. It is tedious to obtain a general solution for these layers, but a scaling analysis shows that if these layers stretch along a distance of order unity along the magnetic field lines, their thickness must be of order $Ha^{-1/2}$. Parallel layers were first analyzed by Shercliff (1953).

Interestingly, another result can be derived from equation (6.7) in the case when the cavity is of infinite length in the direction of the magnetic field. This gives a limit to the size of the core region where the flow is two-dimensional. There can be variations along the magnetic lines on a large length-scale: if the typical cavity length scale perpendicular to the magnetic lines is of order 1, both terms in (6.7) can balance provided the length-scale along the magnetic lines is of order Ha. This situation arises typically when an object of size of order unity is placed in an infinite region: its effect on a flow around it stretches along the magnetic lines as far as a distance of order Ha.

6.5. *Alfvén waves*

The description of Alfvén waves require that the quasi-linear approximation (5.4) and (5.5) is made instead of the quasi-static approximation. Once again, neglecting the non-linear term, and assuming that the imposed magnetic field is uniform in the z-direction, these equations can be curled to produce the following:

$$\frac{\partial \omega}{\partial t} = S^2 (\mathbf{B}_0 \cdot \nabla)\mathbf{j} + P_m \nabla^2 \omega, \tag{6.8}$$

$$\frac{\partial \mathbf{j}}{\partial t} = (\mathbf{B}_0 \cdot \nabla)\omega + \nabla^2 \mathbf{j}, \tag{6.9}$$

where ω is the curl of the velocity field **u**. When the Lundquist number is large, it is possible to neglect diffusion effects. Combining these equations then leads to an equation for the Alfvén waves:

$$\frac{\partial^2 \omega}{\partial t^2} = S^2 (\mathbf{B}_0 \cdot \nabla)^2 \omega. \tag{6.10}$$

The same differential operator applies to the induced magnetic field **b**. This wave operator has solutions in the form of travelling waves in the $\mathbf{B}_0$ and $-\mathbf{B}_0$ directions, with a dimensionless velocity S, i.e. a dimensional velocity $V_A = B_0/\sqrt{\rho\mu}$.

6.6. *Skin effect/field expulsion*

This effect can be understood in terms of the induction equation alone, hence it is not particular to MHD. Let us consider an electrically conducting solid submitted to an external alternating magnetic field of frequency f. The dimensional induction equation in a solid can be written:

$$\frac{\partial \mathbf{B}}{\partial t} = \frac{1}{\mu\sigma} \nabla^2 \mathbf{B}. \tag{6.11}$$

The scaling analysis of this "heat equation" tells us that the AC magnetic field enters a depth $\delta \sim 1/\sqrt{\mu\sigma f}$ into the solid conductor. This is the well-known skin effect.

Magnetic expulsion is of a similar nature and is pictured in a numerical simulation by Proctor and Weiss. The square cavity contains electrically conducting fluid with a steady rotating flow. The initial uniform magnetic field is expelled out of the fluid domain and remains confined to thin boundary layers. In a rotating frame attached to the fluid, this problem is very similar to the skin-effect problem. In the fixed system of reference, this can be seen as a large R_m effect, pertaining to the full MHD equations.

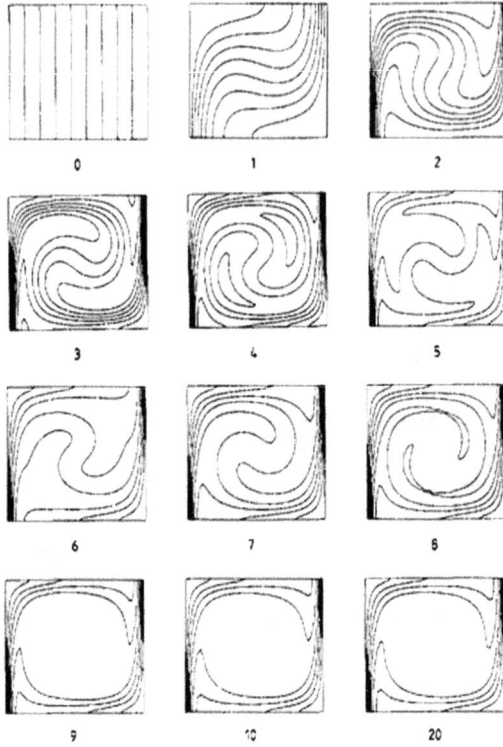

Fig. 3. Magnetic field expulsion (Proctor and Weiss) as a function of magnetic diffusion time.

6.7. Ideal MHD

Ideal MHD applies to perfectly conducting fluids $\sigma \longrightarrow \infty$. In this limit the dimensional induction equation can be written:

$$\frac{\partial \mathbf{B}}{\partial t} = \nabla \times (\mathbf{u} \times \mathbf{B}). \tag{6.12}$$

Expanding the right-hand side and taking into account mass conservation leads to the following equation for $\mathbf{B}/\rho$:

$$\frac{\partial}{\partial t}\left(\frac{\mathbf{B}}{\rho}\right) + (\mathbf{u} \cdot \nabla)\left(\frac{\mathbf{B}}{\rho}\right) = \left(\frac{\mathbf{B}}{\rho} \cdot \nabla\right)\mathbf{u}. \tag{6.13}$$

This equation for **B** is also the equation governing the evolution of little vectors δ having material end-points:

$$\delta(t+dt) = \delta(t) + dt\,(\mathbf{u}(\mathbf{x}+\delta, t) - \mathbf{u}(\mathbf{x}, t)) = \delta(t) + dt\,(\delta \cdot \nabla)\mathbf{u}. \qquad (6.14)$$

Hence the material derivative takes the following form

$$\frac{\partial \delta}{\partial t} + (\mathbf{u} \cdot \nabla)\delta = (\delta \cdot \nabla)\mathbf{u}. \qquad (6.15)$$

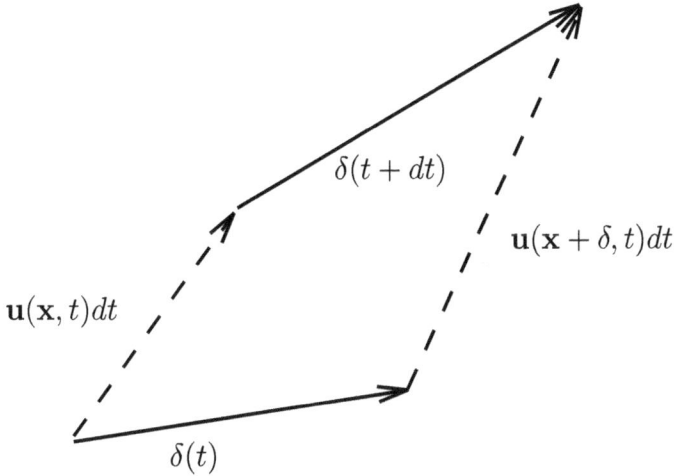

The consequences are the following:

HELMHOLTZ THEOREM: *a magnetic line is a material line.*

On a magnetic line, the magnetic field is parallel to the tangent vector (by definition). The tangent vector is parallel to the infinitesimal vector connecting to neighbouring points on the line. This small vector remains tangent as to the material line. The direction of the magnetic vector field obeying the same equation as this small vector remains also tangent.

ALFVÉN'S THEOREM: *a magnetic tube is material and its magnetic flux $\int_S \mathbf{B} \cdot d\mathbf{S}$ remains constant during its evolution.*

This can be shown first for magnetic tubes of infinitesimal cross-section S and length l. From the previous theorem, the magnetic tube is material, hence its mass ρSl is conserved. Then we can remark that the intensity of $\mathbf{B}/\rho$ is directly related to the stretching of the tube, like δ. Hence $B/(\rho l)$ remains constant. The product of these two conserved quantities BS is also conserved. The theorem applies to any magnetic tube by integration over infinitesimal tubes.

These theorems are useful to help understand some dynamo mechanisms like the stretch-twist-fold mechanism. They have a direct application on the surface on the Sun where magnetic lines have been visualized and where dissipative effect must be confined to very singular regions.

It should be noted also that these theorems have hydrodynamical counterparts in inviscid fluid dynamics, when magnetic field is replaced by vorticity field. Vortex lines are material lines and Kelvin's theorem (conservation of the flux of vorticity for a vorticity tube) is analogous to Alfvén's theorem.

References

[1] P.H. Roberts, *An Introduction to Magnetohydrodynamics*, Longmans, 1964.
[2] T. Alboussière, A geostrophic-like model for large Hartmann number flows, JFM **521**, 125–154 (2004).
[3] M.R.E. Proctor and N.O. Weiss, Magnetoconvection, Report on Progress in Physics **45**, 1317–1379 (1982).

7. Stability of MHD flows

All types of stability analyses used in classical fluid dynamics can be applied to MHD flows. At low magnetic Reynolds number, the effect of an applied magnetic field is to modify the stability threshold, and sometimes the nature of the most unstable mode. At large magnetic Reynolds number, a purely magnetic instability can arise without any initial magnetic field; this is the dynamo effect. Finally, an imposed magnetic field (or electrical current) can lead to an instability, with a base flow at rest: this is the case for the so-called pinch instability.

7.1. Linear stability

Infinitesimal disturbances are added to a basic solution of the governing equations and their immediate evolution is analyzed using linearized equations. This analysis is valid as long as the amplitude of the disturbances is small enough. Unfortunately, you cannot tell easily how small is "small enough"....

7.1.1. Small magnetic Reynolds number

Let us assume that the magnetic Reynolds and Lundquist numbers are small. In this case, the equations (6.1) and (6.2) for the quasi-static approximation apply. Most often, the electric potential ϕ is retained as the main electric quantity. An equation for ϕ is obtained by taking the divergence of (6.2) and taking benefit of the divergenceless condition for $\mathbf{j}$:

$$\nabla^2 \phi = \mathbf{B}_0 \cdot \nabla \times \mathbf{u}. \tag{7.1}$$

The Navier–Stokes equation can then be rewritten using ϕ

$$\frac{\partial \mathbf{u}}{\partial t} + (\mathbf{u} \cdot \nabla)\mathbf{u} = -\nabla p + Ha^2 \left(-\nabla \phi + \mathbf{u} \times \mathbf{B}_0 \right) \times \mathbf{B}_0 + \nabla^2 \mathbf{u}. \tag{7.2}$$

Let us consider a base flow $\mathbf{U}$ and Φ. The linearized evolution equations for infinitesimal disturbances $\mathbf{v}$ and φ added on top of the base flow is derived from the governing equations applied to the total flow minus those applied to the base flow:

$$\frac{\partial \mathbf{v}}{\partial t} = -(\mathbf{v} \cdot \nabla)\mathbf{U} - (\mathbf{U} \cdot \nabla)\mathbf{v}$$

$$- \nabla p + Ha^2 \left(-\nabla \varphi + \mathbf{v} \times \mathbf{B}_0 \right) \times \mathbf{B}_0 + \nabla^2 \mathbf{v}, \tag{7.3}$$

$$0 = -\nabla^2 \varphi + \mathbf{B}_0 \cdot \nabla \times \mathbf{v}. \tag{7.4}$$

If the base flow is stationary, these equations take the form of an eigenvalue problem, as disturbances can be written as the sum of eigenvalues of $\partial/\partial t$ as this linear operator commutes with the linear operator governing the disturbances. The disturbances are expanded as a function of space only (denoted with the same symbols $\mathbf{v}$ and φ) times $\exp(\omega t)$. This leads to the following eigenvalue problem:

$$\omega B \begin{bmatrix} \mathbf{v} \\ \varphi \end{bmatrix} = A \begin{bmatrix} \mathbf{v} \\ \varphi \end{bmatrix}, \tag{7.5}$$

where A denotes the operator governing the disturbances and B is the projection on the velocity field. The operator A depends on Ha and on Re, although this dependence is hidden in the magnitude of the base flow $\mathbf{U}$. For given values of Ha and Re, the eigenvalue with the largest real part is determined. If this real part is larger than zero, then the flow is unstable as this disturbance will grow without limit in the framework of the linear stability theory. In practice, a saturation regime will be reached in the form of a steady, periodic, pseudo-periodic or turbulent flow.

In the quasi-static approximation, there are not many specific problems attached to the MHD part of the study of the linear instability. A single additional function, φ has to be solved: it is governed by a Poisson equation with a source term proportional to the velocity disturbance. One has to worry about the nature of the electrical boundary conditions. This is straightforward: the Neuman condition $\partial \varphi / \partial n = 0$ applies for electrically insulating boundaries and the Dirichlet condition $\varphi = C^{ste}$ for perfectly conducting walls. Walls of finite thickness and finite electrical conductivity can be treated within the so-called "thin-wall" approximation:

$$\frac{\partial \varphi}{\partial n} = c \nabla_S^2 \varphi, \tag{7.6}$$

where ∇_S denotes the surface Laplacian and where $c = \sigma_w e_w / (\sigma H)$ is the relative conductance of the wall with respect to that of the fluid.

There is a number of experimental studies of flow instability in the presence of an externally imposed magnetic field. They have often been performed with a liquid metal (low Prandtl number 10^{-7}–10^{-5}) and the magnetic Reynolds number was small. In most cases, it has been shown that the application of a magnetic field has a stabilizing effect: this is true in particular when the base flow is insensitive to the applied magnetic field. Then things depend on the orientation of the vorticity with respect to the magnetic field. If they are perpendicular everywhere, the magnetic field has no effect on the instability (see 7.3 and 7.4). Otherwise its effect will be to increase the threshold value of the flow control parameter, Reynolds Re, Rayleigh Ra... It should be noted however that in a few cases, laminar stable flows without magnetic field applied can display an unstable behaviour when a magnetic field is applied (Lehnert 1953).

Let us examine again the specific case of an isolated Hartmann layer. Using again the dimensional scales leading to equations (4.8) and (4.9), the stability equations in the quasi-static approximation take the following form

$$\frac{\partial \mathbf{v}}{\partial t} = -(\mathbf{v} \cdot \nabla)\mathbf{U} - (\mathbf{U} \cdot \nabla)\mathbf{v}$$
$$\qquad -\nabla p + R^{-1}(-\nabla\varphi + \mathbf{v} \times \mathbf{B}_0) \times \mathbf{B}_0 + R^{-1}\nabla^2\mathbf{v}, \tag{7.7}$$
$$0 = -\nabla^2\varphi + \mathbf{B}_0 \cdot \nabla \times \mathbf{v}, \tag{7.8}$$

where the Reynolds number based on the thickness of the Hartmann layer $R = \sqrt{\rho/(\sigma v)}U/B$ is now the single dimensionless parameter (the magnetic Prandtl number is irrelevant in the quasi-static approximation). The linear stability analysis of the Hartmann layer goes back to Lock (1955) who found a critical threshold R_l for linear instability close to

$$R_l \simeq 5 \times 10^4. \tag{7.9}$$

However, many duct flow experiments with a transverse magnetic field have shown a drag increase when $Re/Ha \geq 300$, this ratio Re/Ha being precisely equal to R. This large discrepancy has led to the belief that this drag increase (and associated turbulence) was not related to the stability of the Hartmann layer. We shall see that energetic stability brings a different perspective.

7.1.2. Kinematic dynamo threshold

When the magnetic Reynolds number is large, dynamo instability can occur. This aspect will be covered in other lectures and will not be discussed here. Let us however mention some difficulties associated to the linear stability dynamo studies. The magnetic Prandtl number of a liquid metal is very small compared to one, so that if one is to observe dynamo instability ($R_m > 1$), the hydrodynamic Reynolds number is going to be very large $Re = R_m P_m^{-1}$. Presumably, the flow is already hydrodynamically unstable and is most likely in a state of turbulence. Theoretical studies have relied on the so-called kinematic dynamo analyses, whereby the velocity field is fixed, but experimental attempts to investigate the dynamo instability have had to address this question.

Let us mention a specific difficulty related to the boundary conditions of the magnetic field. At the boundaries of the electrically conducting domain, the magnetic field has to match a current-free magnetic field expanding infinitely. This magnetic field derives from a potential. There is no local explicit boundary condition to be applied to the magnetic field. There are two ways to overcome this difficulty. First, in the case of a simple geometry (planar, spherical, ...) the solutions can be expanded in orthogonal series of sine or spherical harmonic functions. Then, the matching condition becomes explicit for each wavenumber or degree. Alternatively, in the case of a complex geometry, one has to calculate the external magnetic field and ensure that it can be matched with its internal counterpart.

7.2. Energetic stability

The idea that energy is extracted from the base flow, then provided to disturbances and finally dissipated by viscous or Joule effects is implicit in the linear stability theory. This idea is at the heart of the dissipative energy stability theory. Things are a little different in the non-linear non-dissipative stability theory where other constraints than energy (momentum, angular momentum, helicity, ...) put a constraint on the flow such that a local minimum or maximum of energy cannot be escaped. Those flow configurations are then non-linearly stable with respect to small disturbances.

7.2.1. Dissipative energy stability

The starting point of the dissipative energy stability is the same as that of linear stability. The total flow is written as the sum of a base flow and disturbance. However, the non-linear term is retained entirely (no linear expansion). For instance, in the quasi-static approximation equation (7.7) the term $(\mathbf{v} \cdot \nabla)\mathbf{v}$ must be retained. Then the dot product of equation (7.7) minus R^{-1} times the dot product of φ with equation (7.8) is integrated over the volume $\mathcal{V}$ of the fluid:

$$\frac{d}{dt} \int_{\mathcal{V}} \frac{\mathbf{v}^2}{2} = -\int_{\mathcal{V}} \mathbf{v} \cdot \nabla \mathbf{U} \cdot \mathbf{v} - R^{-1} \int_{\mathcal{V}} \nabla \mathbf{v} : \nabla \mathbf{v} - R^{-1} \int_{\mathcal{V}} \mathbf{j}^2, \tag{7.10}$$

where $\mathbf{j}$ is derived from Ohm's law (5.6). The slowest possible decay rate of the energy of a disturbance is given by the following infimum over all divergence-free vectors fields that satisfy the boundary conditions

$$\inf \left(\frac{\int_{\mathcal{V}} \mathbf{v} \cdot \nabla \mathbf{U} \cdot \mathbf{v} + R^{-1} \int_{\mathcal{V}} \nabla \mathbf{v} : \nabla \mathbf{v} + R^{-1} \int_{\mathcal{V}} \mathbf{j}^2}{\int_{\mathcal{V}} \mathbf{v}^2} \right). \tag{7.11}$$

The Euler–Lagrange equation coming out of this extremal problem (with the divergence-free constraint) can be written

$$\omega \mathbf{v} = \mathbf{v} \cdot \mathbf{D} + \nabla p + R^{-1} \mathbf{j} \times \mathbf{B}_0 + R^{-1} \nabla^2 \mathbf{v}, \tag{7.12}$$

where $\mathbf{D}$ is the symmetrical deformation tensor $\mathbf{D} = 1/2(\nabla \mathbf{U} + \nabla \mathbf{U}^t)$. This equation is slightly different from the linear stability equation. The main difference is that the linear operator is now normal while the linear operator associated to the linear stability is not necessarily normal. Let us recall that an operator is normal when its eigenvectors are perpendicular to each other. The energy stability critical threshold Re_e is always lower than the linear stability threshold. Here is the qualitative reason why: let us consider a linear stable situation, say $Re < Re_l$. So all eigenvectors decay. However they do not decay with the same rate. Let us assume that there exist two eigenvalues, one having a much larger negative real part (both have a negative real part) and that their associated eigenvectors $e1$ and $e2$ are close. Then the fate of the small difference vector is clear. The fast decaying eigenvector will collapse immediately and the small difference vector will become similar to the slowly decaying eigenvector. Hence the small difference vector will temporarily grow in energy, i.e. $R > Re_e$.

A simple toy model corresponding to this situation of a non-normal operator is the following

$$\frac{\partial}{\partial t} \begin{bmatrix} p \\ q \end{bmatrix} = \begin{bmatrix} -\frac{1}{Re} & 0 \\ 1 & -\frac{2}{Re} \end{bmatrix} \begin{bmatrix} p \\ q \end{bmatrix}. \tag{7.13}$$

Its eigenvalues and associated eigenvectors are

$$-\frac{1}{Re} \qquad \begin{bmatrix} Re^{-1} \\ 1 \end{bmatrix},$$

$$-\frac{2}{Re} \qquad \begin{bmatrix} 0 \\ 1 \end{bmatrix}.$$

As the Reynolds number is made larger and larger, the eigenvectors become more and more similar. However the second one is always decaying faster than the first one by a factor two. Hence, the difference between these eigenvectors will experience a large transient growth when the Reynolds number is large.

For instance, the isolated Hartmann layer, and its single dimensionless parameter R (in the quasi-static approximation) can be analyzed in terms of energy stability. Now the energy threshold is $R_e \simeq 25.6$. This has to be compared to the linear threshold $R_l \simeq 5 \times 10^4$. There is a huge range of Reynolds numbers (based on the laminar thickness of the Hartmann layer) where some disturbances can grow temporarily, but where all infinitesimal disturbances must eventually decay. The fact that the flow appears to undergo transition to turbulence (resp. laminarization) when this Reynolds number exceeds $R \sim 300$ (resp. goes below 300) is consistent with the energy and linear stability results of Hartmann layers: $R_e < 300 < R_l$.

7.3. Stability of non-dissipative MHD

Some astrophysical flows have so large dimensionless Reynolds numbers that it may be appropriate to consider them as non-dissipative, at least from the point of view of their stability. A typical result is related to the stability of accretion disks. A rotating flow with angular momentum being a function increasing with the radius is shown to be hydrodynamically stable. If they were stable, accretion disks would not accrete, as fluid particles would never lose their angular momentum and would never converge towards the centre. A possible way out is to consider the stability of the MHD problem with an additional axial uniform magnetic field. The stability criterion is immediately changed as angular velocity must now increase with radius, a much more stringent condition that is not met by gravitational accretion disks.

The pinch effect was first analyzed by W.H. Bennett (1931). In its simplest form, an electrical current is passed through a cylindrical conductor. The magnetic field associated to this current is a toroidal magnetic field. The resulting Lorentz force $\mathbf{j} \times \mathbf{B}$ exerts a force contracting the conductor towards its axis. Alternatively, the magnetic field creates a magnetic pressure compressing the electrical conductor. However, this configuration of force equilibrium

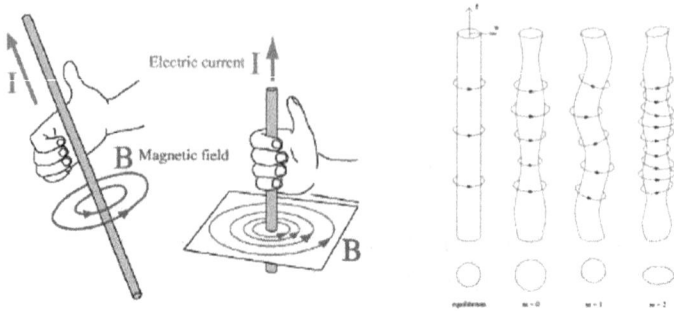

Fig. 4. Pinch and instabilities.

between magnetic forces and pressure can be very unstable: this is the pinch instability.

For instance, the axisymmetric instability can be understood qualitatively in the following terms. Suppose that the conductor has a slight local constriction. The current density and magnetic field will be stronger than at other places, leading to a larger constriction. This can eventually break the conductor into droplets (if fluid) or segments (solid wires).

Let us assume that an electrical current is flowing in a uniform direction and that its distribution is axisymmetrical and dependent on the radius. The question is whether this magnetostatic equilibrium is non-linearly stable or unstable. We are going to answer it only with respect to axisymmetrical disturbances. The underlying idea is to consider the variation of magnetic energy with all possible admissible displacements of the magnetic lines. Any displacement preserving the topology of the magnetic lines and preserving continuity are admissible. If it is found that the magnetic energy is minimum (i.e. increases for all admissible displacements), then non-linear stability of the magnetic configuration is ensured.

In a small displacement field η, as magnetic diffusion is neglected, the first variation of the magnetic field is obtained from the induction equation

$$\delta \mathbf{B} = \nabla \times (\eta \times \mathbf{B})$$

and the second order variation is

$$\delta^2 \mathbf{B} = \frac{1}{2} \nabla \times (\eta \times \delta \mathbf{B}).$$

The first and second order variations of the magnetic energy M can then be computed

$$\delta M = \int_{\mathcal{V}} \mathbf{B} \cdot \delta \mathbf{B} \, dV,$$

$$\delta^2 M = \int_{\mathcal{V}} (\delta \mathbf{B})^2 + 2\mathbf{B} \cdot \delta^2 \mathbf{B} \, dV.$$

When the configuration constitutes a magnetostatic equilibrium (the Lorentz force can be balanced by a pressure gradient), the first order variation δM is zero. Non-linear stability is then linked to the sign of $\delta^2 M$. If it is positive for all admissible displacements η the configuration is stable with respect to small finite amplitude disturbances as it cannot escape this equilibrium state from the energetically point of view.

The equilibrium configuration is defined in cylindrical coordinates by $\mathbf{B} = (0, b(r), 0)$. General axisymmetrical disturbances are considered $\eta = (\eta_r, 0, \eta_z)$, satisfying continuity for a constant uniform density. First and second order magnetic field disturbances can be readily computed

$$\delta \mathbf{B} = \left(0, -r\eta_r \frac{\partial}{\partial r} \left(\frac{b}{r} \right), 0 \right),$$

$$\delta^2 \mathbf{B} = \left(0, \frac{\partial}{\partial z} (\eta_r \eta_z) \, r \frac{\partial}{\partial r} \left(\frac{b}{r} \right) + \frac{\partial}{\partial r} \left[r\eta_r \eta_z \frac{\partial}{\partial r} \left(\frac{b}{r} \right) \right], 0 \right).$$

Then $\delta^2 M$ can be calculated as an integral in cylindrical coordinates

$$\delta^2 M = \int_{z=-\infty}^{\infty} \int_{r=0}^{\infty} \left[(\delta \mathbf{B})^2 + 2\mathbf{B} \cdot \delta^2 \mathbf{B} \right] 2\pi r \, dr \, dz.$$

This can be shown to be equal to

$$\delta^2 M = \int_{z=-\infty}^{\infty} \int_{r=0}^{\infty} -\pi r^2 \eta_r^2 \frac{\partial}{\partial r} \left(\frac{b}{r} \right)^2 dr \, dz.$$

Hence the equilibrium configuration defined by the magnetic distribution $b(r)$ is non-linearly stable provided $(b/r)^2$ is a decreasing function of the radius r.

Similar energy arguments can be invoked to analyze the stability of Euler flows and also the MRI (Magneto-Rotational Instability).

References

[1] D.D. Joseph, *Stability of Fluid Motions*, 1972.
[2] C.R. Doering and J.D. Gibbon, *Applied Analysis of the Navier–Stokes Equations*, Cambridge University Press, 1995.

[3] T. Alboussière and R.J. Lingwood, On the stability of the Hartmann layer, Phys. Fluids **118**, 2058–2068 (1999).

[4] W.H. Bennett, Magnetically self-focussing streams, Phys. Rev. **45**, 890 (1934).

[5] H.K. Moffatt, Magnetostatic equilibria and analogous Euler flows of arbitrarily complex topology. Part 2. Stability considerations, Journal of Fluid Mechanics **166**, 359–378 (1986).

8. MHD turbulence

Using the words of Shercliff, a magnetic field can destroy, redistribute or generate vorticity. In general, this is also true to MHD turbulence and it is difficult to assess whether a magnetic field has a stabilizing or destabilizing effect. There are cases (such as low magnetic Reynolds and Lundquist numbers in homogeneous turbulence) when it can be said that a magnetic field has a stabilizing effect: even then, it induces anisotropy and transport properties of MHD turbulence may be enhanced in the plane perpendicular to the applied magnetic field.

8.1. Elsasser equations

Elsasser has put the MHD equations into a symmetrical system using combinations of velocity and magnetic fields, $\mathbf{e}^{\pm} = \mathbf{u} \pm \mathbf{B}/\sqrt{\rho\mu}$, the so-called Elsasser variables. Combining Navier–Stokes and the induction equations with coefficients 1 and $\pm 1/\sqrt{\rho\mu}$ leads to:

$$\frac{\partial \mathbf{e}^{+}}{\partial t} + (\mathbf{e}^{-} \cdot \nabla)\mathbf{e}^{+} = -\nabla q + \nabla^2(\nu^{+}\mathbf{e}^{+} + \nu^{-}\mathbf{e}^{-}), \tag{8.1}$$

$$\frac{\partial \mathbf{e}^{-}}{\partial t} + (\mathbf{e}^{+} \cdot \nabla)\mathbf{e}^{-} = -\nabla q + \nabla^2(\nu^{-}\mathbf{e}^{+} + \nu^{+}\mathbf{e}^{-}), \tag{8.2}$$

where $\nu^{\pm} = \nu \pm (\mu\sigma)^{-1}$. Elsasser argues that MHD turbulence at large magnetic Reynolds numbers is as obvious as classical turbulence at large Reynolds numbers.

8.2. Initial evolution of MHD turbulence

Moffatt [1] imagined a thought experiment of sudden application of a magnetic field on a turbulent field. How will it evolve in an initial linear phase? In the framework of the quasilinear approximation (R_m is supposed to remain small), the linearized Navier–Stokes and induction equations can be written

$$\frac{\partial \mathbf{u}}{\partial t} = -\nabla p + Rm^{-1}S\,\mathrm{curl}\mathbf{b} \times \mathbf{B} + PmS^{-1}\nabla^2\mathbf{u},$$

$$\frac{\partial \mathbf{b}}{\partial t} = RmS^{-1}(\mathbf{B} \cdot \nabla)\mathbf{u} + S^{-1}\nabla^2\mathbf{b}.$$

In the spectral space, they become:

$$\frac{\partial \hat{\mathbf{u}}}{\partial t} = i\, Rm^{-1} S\, (\mathbf{B} \cdot \mathbf{k}) \hat{\mathbf{b}},$$

$$\frac{\partial \hat{\mathbf{b}}}{\partial t} = i\, Rm S^{-1} (\mathbf{B} \cdot \mathbf{k}) \hat{\mathbf{u}} - S^{-1} k^2 \hat{\mathbf{b}}.$$

Solutions are exponential $\exp(-\beta t)$, with

$$\beta = \frac{S^{-1}k^2}{2}\left[1 \pm \sqrt{1 - \zeta^2}\right],$$

where $\zeta = 2(\mathbf{B} \cdot \mathbf{k})S/k^2$. If $|\zeta| < 1$, they are monotonic damped solutions, if $|\zeta| > 1$ they are (damped) Alfvén waves. This can be seen in the following picture of the spectral space.

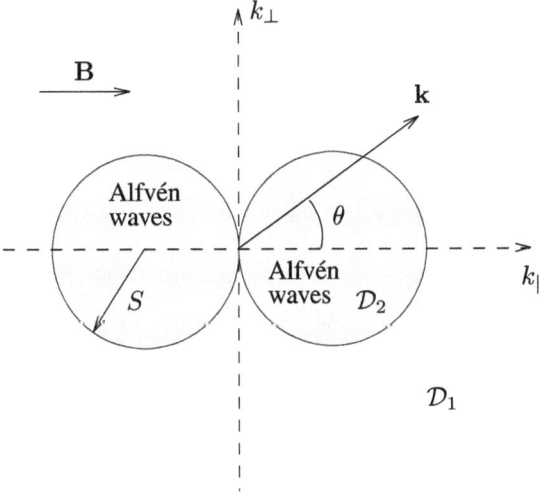

In $\mathcal{D}_1$, $|\zeta| \ll 1$ and by Taylor expansion $\beta \simeq (\mathbf{B} \cdot \mathbf{k})^2 S/k^2 = S\cos^2\theta$, inversely dimensional to $t_d = \frac{\rho}{\sigma B^2}$. This timescale is named Joule time or sometimes magnetic damping time.

In $\mathcal{D}_2$, $|\zeta| \gg 1$ and decay is nearly isotropic, of order $\beta \simeq \frac{S^{-1}k^2}{2}$, inversely dimensional to $t_\eta = l^2/\eta = St_d$.

In terms of energy decay of MHD turbulence in the linear phase, after integrating the evolution of the various turbulent modes initially present, kinetic energy decays like $t^{-1/2}$ when the Lundquist number is small. At large Lundquist numbers, The decay depends on the initial spectrum of turbulent energy: $t^{-5/2}$ for an initial spectrum of order k^4 at small wavenumbers.

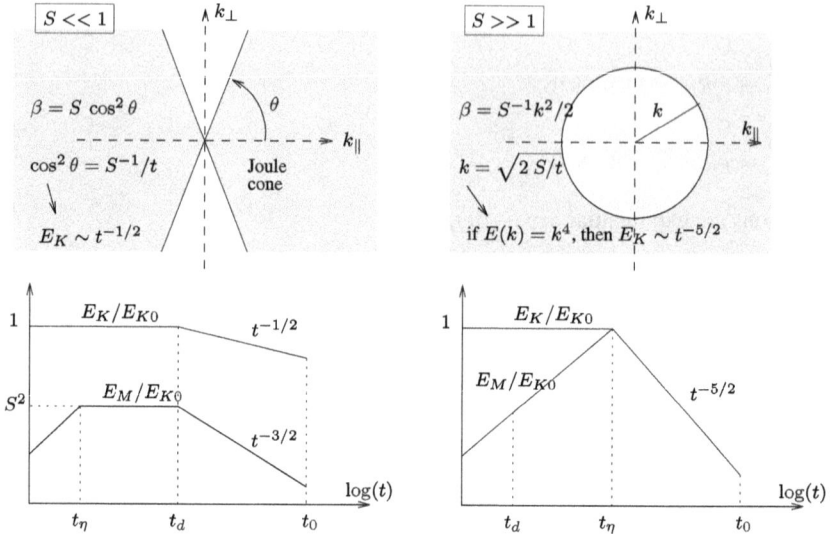

8.3. Strong MHD turbulence

After the initial linear phase, the regime of strong MHD turbulence (small Lundquist) is characterized heuristically by an equilibrium between non-linear inertial terms and Lorentz forces. This is represented by the dimensionless interaction parameter (or Stuart number), $N = \sigma B^2 l/(\rho u)$, where l is the integral length scale and u the scale of velocity fluctuations. However, as shown in [3], this interaction parameter has to be weighted by the degree of anisotropy of turbulence which tends to elongate the vortices in the direction of the applied magnetic field.

So two length scales must be considered, one in the direction of the magnetic field $l_\parallel$ and one in the perpendicular direction $l_\perp$. When evaluating Lorentz forces, the degree of anisotropy $l_\perp/l_\parallel$ plays a role at two levels. First, it reduces the electric current actually flowing in the liquid (in the extreme case of two-dimensional flows, there is no electric current), and secondly, anisotropy has a tendency to make Lorentz forces nearly curl-free. In the end, the correct ratio of Lorentz to inertial effects is represented by a "true" interaction parameter, which is assumed to be close to unity:

$$N_t = \frac{\sigma B^2 l_\perp}{\rho u} \left(\frac{l_\perp}{l_\parallel}\right)^2 \simeq 1.$$

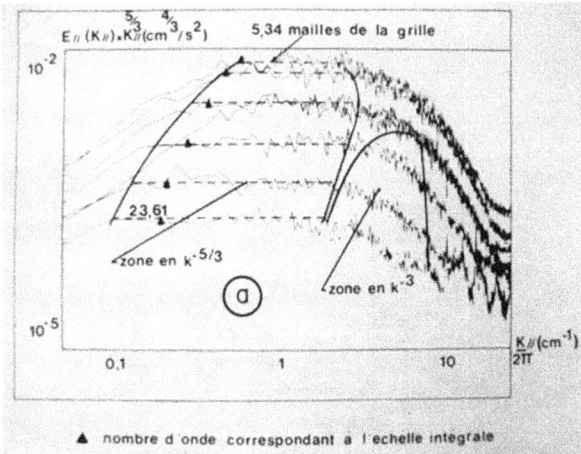

Capéran et al., J. de Mécanique, 1985

When this assumption $N_t \simeq 1$ is made for all scales of turbulence, and when anisotropy is supposed to be constant across those scales from the integral length-scale to the smallest scales, it follows immediately that the energy spectrum of strong MHD turbulence is proportional to k^{-3} at small scales. This is indeed clearly observed in MHD experiments.

8.4. Weak MHD turbulence

Weak MHD turbulence (low R_m, large Lundquist S) proves harder to model. The original heuristic model, based on the statistical effect of small (weak) wave inter-actions [5] leads to an energy spectrum proportional to $k^{-3/2}$ for kinetic energy, while observations (solar wind, interstellar medium) indicate a Kolmogorov-like $k^{-5/3}$ spectrum. Anisotropy again plays an important role and more elaborate models show that different scalings are possible [6].

8.5. Turbulence in Hartmann layers

Surprisingly, turbulence in Hartmann layers is not of MHD but of classical type [7]. It is shown that the interaction parameter is small in these layers and that the Lundquist number is small (for low P_m fluids). Although the Hartmann layer is a typical manifestation of MHD effects, the nature of turbulence within these layers is classical.

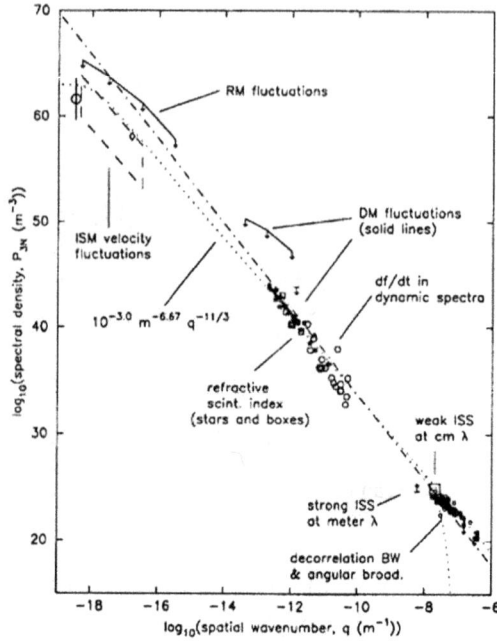

Armstrong et al., The Astroph. J., 1995

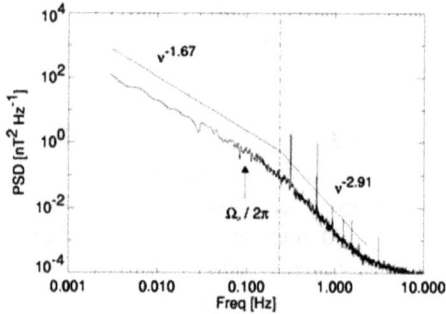

Leamon et al., JGR, 1998

8.6. A global picture

The picture below is an attempt to present different types of MHD turbulences (assuming low P_m) depending on the Lundquist and magnetic Reynolds numbers. DTS is a sodium experiment developed in Grenoble (LGIT) where a strong magnetic field is imposed by permanent magnets. Galalfven is a Galinstan experiment which has been run in a 13 Tesla magnetic field, which has been developed to study Alfvén waves.

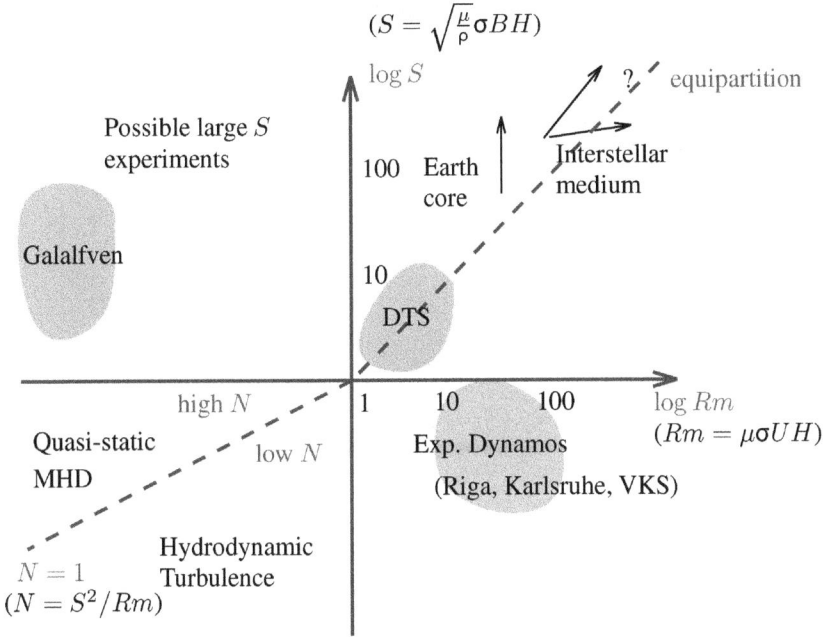

It should be noted that we do not have much theoretical knowledge regarding the $R_m > 1$ side of this diagram. Yet, this is where dynamos can exist and therefore it is crucial to investigate turbulence in this regime. In particular, it is interesting to investigate whether dynamos will be attracted towards an equipartition regime. If not, it remains to discover the selection principles governing saturation in a turbulent regime.

It should also be noticed that all experimental attempts are far from geo and astrophysically interesting regimes. In particular, all working dynamos had a Lundquist number smaller than one, while it must have a huge value in natural dynamos.

References

[1] H.K. Moffatt, Suppression of turbulence by a uniform magnetic field, Journal of Fluid Mechanics (1967).

[2] A. Alemany, R. Moreau, P.L. Sulem and U. Frisch, Influence of an external magnetic field on homogeneous MHD turbulence, J. de Mécanique **18**(2), 277–313 (1979).

[3] B. Sreenivasan and T. Alboussière, Evolution of a vortex in a magnetic field, Eur. J. Mech. B, Fluids **19**, 403–421 (2000).

[4] A. Pothérat and T. Alboussière, Small scales and anisotropy in low R_m magnetohydrodynamic turbulence, Phys. Fluids **15**(10), 3170–3180 (2003).

[5] R.H. Kraichnan, Inertial-range spectrum of hydromagnetic turbulence, Phys. Fluids **8**, 1385 (1965).

[6] S. Galtier, A. Pouquet and A. Mangeney, On spectral scaling laws for incompressible anisotropic magnetohydrodynamic turbulence, Physics of Plasmas **12**(9) (2005).

[7] T. Alboussière and R. J. Lingwood, A model for the turbulent Hartmann layer, Phys. Fluids **12**(6), 1535–1543 (2000).

Course 2

DYNAMO THEORY

Chris A. Jones

Department of Applied Mathematics, University of Leeds, Leeds LS2 9JT, UK

Ph. Cardin and L.F. Cugliandolo, eds.
Les Houches, Session LXXXVIII, 2007
Dynamos
© *2008 Published by Elsevier B.V.*

Contents

Introduction

These lectures were designed to give an understanding of the basic ideas of dynamo theory, as applied to the natural dynamos occurring in planets and stars. The level is appropriate for a graduate student with an undergraduate background in electromagnetic theory and fluid dynamics. There are a number of recent reviews which cover some of the material here. In particular, I have drawn extensively from the recent book *Mathematical Aspects of Natural Dynamos*, edited by Emmanuel Dormy and Andrew Soward (2007), and from the article on *Dynamo Theory* by Andrew Gilbert [24] which appeared in the Handbook of Mathematical Fluid Dynamics, edited by Susan Friedlander and Denis Serre (2003). Some older works, which are still very valuable sources of information about dynamo theory, are *Magnetic field generation in electrically conducting fluids* by Keith Moffatt [38] (1978), *Stretch, Twist, Fold: the Fast Dynamo* by Steve Childress and Andrew Gilbert [9] (1995), and the article by Paul Roberts on *Fundamentals of Dynamo Theory* [48] (1994).

1. Kinematic dynamo theory

1.1. Maxwell and pre-Maxwell equations

Maxwell's equations are the basis of electromagnetic theory, and so they are the foundation of dynamo theory. They are written (e.g. [12])

$$\nabla \times \mathbf{E} = -\frac{\partial \mathbf{B}}{\partial t}, \qquad \nabla \times \mathbf{B} = \mu \mathbf{j} + \frac{1}{c^2} \frac{\partial \mathbf{E}}{\partial t}, \qquad (1.1.1, 1.1.2)$$

$$\nabla \cdot \mathbf{B} = 0, \qquad \nabla \cdot \mathbf{E} = \frac{\rho_c}{\epsilon}. \qquad (1.1.3, 1.1.4)$$

Here $\mathbf{E}$ is the electric field, $\mathbf{B}$ the magnetic field, $\mathbf{j}$ is the current density, μ is the permeability. We use S.I. units throughout (metres, kilogrammes, seconds), and in these units $\mu = 4\pi \times 10^{-7}$ in free space. c is the speed of light, ρ_c is the charge density and ϵ is the dielectric constant. In free space $\epsilon = (\mu c^2)^{-1}$.

(1.1.1) is the differential form of Faraday's law of induction. In physical terms it says that if the magnetic field varies with time then an electric field is produced.

51

In an electrically conducting body, this electric field drives a current, which is the basis of dynamo action.

(1.1.3) says there are no magnetic monopoles. That is there is no particle from which magnetic field lines radiate. However, (1.1.4) says that there are electric monopoles from which electric field originates. These are electrons and protons.

Maxwell's equations are relativistically invariant, but in MHD we assume the fluid velocity is small compared to the speed of light. This allows us to discard the term $\frac{1}{c^2}\frac{\partial \mathbf{E}}{\partial t}$. If the typical length scale is L_* (size of planet, size of star, size of experiment) and the timescale on which $\mathbf{B}$ and $\mathbf{E}$ vary is T_*, then

$$\frac{|\mathbf{E}|}{L_*} \sim \frac{|\mathbf{B}|}{T_*}$$

from (1.1.1) so then

$$|\nabla \times \mathbf{B}| \sim \frac{|\mathbf{B}|}{L_*} \quad \text{and} \quad \frac{1}{c^2}\left|\frac{\partial \mathbf{E}}{\partial t}\right| \sim \frac{|\mathbf{B}|}{L_*}\frac{L_*^2}{c^2 T_*^2}.$$

So the term $\frac{1}{c^2}\frac{\partial \mathbf{E}}{\partial t} \ll \nabla \times \mathbf{B}$ provided $L_*^2/T_*^2 \ll c^2$. The dynamos we consider evolve slowly compared to the time taken for light to travel across the system. Even for galaxies this is true: light may take thousands of years to cross the galaxies, but the dynamo evolution time is millions of years.

The MHD equations are therefore

$$\nabla \times \mathbf{E} = -\frac{\partial \mathbf{B}}{\partial t}, \qquad \nabla \times \mathbf{B} = \mu \mathbf{j}, \qquad\qquad (1.1.5, 1.1.6)$$

$$\nabla \cdot \mathbf{B} = 0, \qquad \nabla \cdot \mathbf{E} = \frac{\rho_c}{\epsilon}. \qquad\qquad (1.1.7, 1.1.8)$$

Equation (1.1.6) is called Ampére's law, or the pre-Maxwell equation.

The final law required is Ohm's law, which relates current density to electric field. This law depends on the material, so has to be determined by measurements. In a material at rest the simple form

$$\mathbf{j} = \sigma \mathbf{E} \qquad\qquad (1.1.9)$$

is assumed, though as we see in Section 1.3 below, this changes in a moving frame. In some astrophysical situations, (1.1.9) no longer holds, and new effects, such as the Hall effect and ambipolar diffusion become significant.

1.2. Integral form of the MHD equations

1.2.1. Stokes' theorem

Stokes' theorem says that for any continuous and differentiable vector field **a** and simply connected surface S, enclosed by perimeter C,

$$\int_S (\nabla \times \mathbf{a}) \cdot d\mathbf{S} = \int_C \mathbf{a} \cdot d\mathbf{l}. \tag{1.2.1}$$

Ampére's law then becomes

$$\int_S \mu \mathbf{j} \cdot d\mathbf{S} = \int_C \mathbf{B} \cdot d\mathbf{l}. \tag{1.2.2}$$

If a current of uniform current density $\mathbf{j} = j\hat{\mathbf{z}}$ in cylindrical polars (s, ϕ, z) flows inside a tube of radius s, then the magnetic field generated is in the $\hat{\phi}$ direction, and has strength $B_\phi = \pi s^2 \mu j / 2\pi s = \mu j s / 2$. Since the z-component of $\nabla \times \mathbf{B}$ is $\frac{1}{s}\frac{\partial}{\partial s} s B_\phi = \mu j$ this is consistent with the differential form (1.1.6). It is convenient to think of the field **B** being created by the current.

1.2.2. Potential fields

If there is no current in a region of space, $\nabla \times \mathbf{B} = 0$ there, and so the magnetic field is a *potential* field. This means that $\mathbf{B} = \nabla V$, for some scalar potential V, and (1.1.7) gives

$$\nabla^2 V = 0, \tag{1.2.3}$$

which is Laplace's equation. If there are no currents anywhere and $V \to 0$ at infinity, the only solution of (1.2.3) is $V = 0$, so no currents anywhere means no field. In the Earth's core there are currents, but outside there is comparatively low conductivity, so outside the core we have approximately equation (1.2.3). In spherical geometry the general solution of (1.2.3) which decays at infinity can be written

$$V = \frac{a}{\mu} \sum_{l=1}^{\infty} \sum_{m=0}^{l} P_l^m \left(\frac{a}{r}\right)^{l+1} (g_l^m \cos m\phi + h_l^m \sin m\phi) \tag{1.2.4}$$

in spherical polar coordinates (r, θ, ϕ). Here the P_l^m are the Schmidt normalised associated Legendre functions, and if a is the radius of the Earth, the g_l^m and h_l^m are the Gauss coefficients of the Earth's magnetic field, listed in geomagnetic tables. Several different normalisations of the associated Legendre functions are

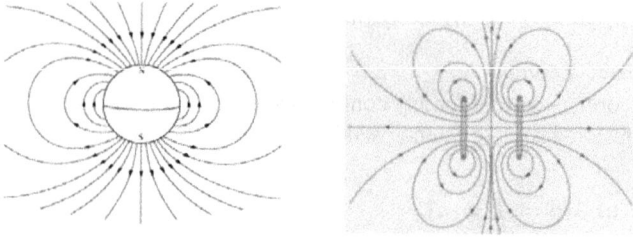

Fig. 1. Left: Axial dipole field. Right: axial quadrupole field.

given in the literature: see e.g. [37]. (1.2.4) can be written as the real part of a complex form, and the expression

$$P_l^m \left(\frac{a}{r}\right)^{l+1} e^{im\phi} = Y_l^m \left(\frac{a}{r}\right)^{l+1} (\theta, \phi) \tag{1.2.5}$$

defines the spherical harmonic Y_l^m. The $l = 1$, $m = 0$ component is the axial dipole component, and the $l = 1$, $m = 1$ component is called the equatorial dipole component, because it has the form of a rotated axial dipole field. The dipole component dominates at large r. If we are a long way away from a planet, star or experiment, all we see is the dipole component of the magnetic field. At the surface of the conducting region the field may be, probably is, a complicated superposition of spherical harmonics, and inside the conducting fluid it is not even a potential field. The $l = 2$, $m = 0$ component is called the axial quadrupole term. Note that it has a different parity about the equator from the axial dipole component.

1.2.3. Faraday's law
It is also useful to apply Stokes' theorem to Faraday's law, (1.1.5). Then we get

$$\int_C \mathbf{E} \cdot d\mathbf{l} = -\frac{\partial}{\partial t} \int_S \mathbf{B} \cdot d\mathbf{S}. \tag{1.2.6}$$

$\int_C \mathbf{E} \cdot d\mathbf{l}$ is called the e.m.f. around the circuit, and it drives a current which can be measured. So if the magnetic field inside a loop of wire changes, a current flows. Faraday originally deduced his law by moving permanent magnets near loops of wire and measuring the resulting currents.

1.3. Electromagnetic theory in a moving frame

The Maxwell equations are invariant under Lorentz transformations. The pre-Maxwell MHD equations are invariant under a Galilean transformation, which is simply a shift to a uniformly moving frame,

$$x_i' = x_i - u_i t, \quad t' = t, \quad u_i \text{ being a constant vector.} \tag{1.3.1}$$

In the moving frame, the electric, magnetic fields and current become

$$\mathbf{E}' = \mathbf{E} + \mathbf{u} \times \mathbf{B}, \qquad \mathbf{B}' = \mathbf{B}, \qquad \mathbf{j}' = \mathbf{j}. \tag{1.3.2, 1.3.3, 1.3.4}$$

To see where these come from, note that from Galilean invariance, the MHD equations in the moving frame (1.1.5)–(1.1.8) are

$$\nabla' \times \mathbf{E}' = -\frac{\partial \mathbf{B}'}{\partial t'}, \qquad \nabla' \times \mathbf{B}' = \mu \mathbf{j}', \tag{1.3.5, 1.3.6}$$

$$\nabla' \cdot \mathbf{B}' = 0, \qquad \nabla' \cdot \mathbf{E}' = \frac{\rho_c'}{\epsilon}, \tag{1.3.7, 1.3.8}$$

where $\nabla' = \frac{\partial}{\partial x_i'}$. The chain rule for transforming variables gives

$$\frac{\partial}{\partial x_i'} = \frac{\partial}{\partial x_i}, \qquad \frac{\partial}{\partial t'} = \frac{\partial x_i}{\partial t'} \frac{\partial}{\partial x_i} + \frac{\partial t}{\partial t'} \frac{\partial}{\partial t} \tag{1.3.9}$$

so using (1.3.1)

$$\frac{\partial}{\partial t'} = u_i \frac{\partial}{\partial x_i} + \frac{\partial}{\partial t}. \tag{1.3.10}$$

From (1.3.9), $\nabla' = \nabla$, so (1.3.6) and (1.3.7) are consistent with (1.3.3) and (1.3.4), but (1.3.10) means $\mathbf{E}'$ is not simply $\mathbf{E}$. If we substitute (1.3.2) into (1.3.5) we get

$$\nabla \times \mathbf{E} + \nabla \times (\mathbf{u} \times \mathbf{B}) = -\frac{\partial \mathbf{B}}{\partial t} - \mathbf{u} \cdot \nabla \mathbf{B}, \tag{1.3.11}$$

and since $\mathbf{u}$ is constant, using (1.1.7) gives $\nabla \times (\mathbf{u} \times \mathbf{B}) = -\mathbf{u} \cdot \nabla \mathbf{B}$ so (1.3.11) reduces to (1.1.5) as required. This shows that (1.3.2–1.3.5) are consistent with Maxwell's equations being invariant under a Galilean transformation. For a derivation of the transformations of the full Maxwell equations under the Lorentz transformation see [24]. In this case, $\mathbf{B}'$ is not the same as $\mathbf{B}$ as there is a correction of order u^2/c^2.

A surprising consequence of these transformation laws is that

$$\frac{\rho_c'}{\epsilon} = \frac{\rho_c}{\epsilon} + \nabla \cdot (\mathbf{u} \times \mathbf{B}) \tag{1.3.12}$$

so the charge density changes in the moving frame. $\nabla \cdot (\mathbf{u} \times \mathbf{B}) = -\mathbf{u} \cdot \nabla \times \mathbf{B} = -\mu(\mathbf{u} \cdot \mathbf{j})$, and since current is just moving charge, in a moving frame current can be equivalent to charge. By the same argument, the current should change in a moving frame, and it does, but only by a negligible amount if $u \ll c$ which is why in MHD we have (1.3.4).

Table 1

Electrical conductivity and magnetic diffusivity

	σ Sm^{-1}	η m^2 s^{-1}
Earth's core	4×10^5	2
Jupiter's core	10^5	8
Sodium	2.1×10^7	0.04
Gallium	6.8×10^6	0.12
Solar convection zone	10^3	10^3
Galaxy	10^{-11}	10^{17}

1.4. Ohm's law, induction equation and boundary conditions

We have already mentioned that Ohm's law in a stationary medium is given by (1.1.9), a simple proportionality between current and electric field. However, we now know that in a moving frame $\mathbf{E}$ must be replaced by $\mathbf{E} + \mathbf{u} \times \mathbf{B}$ while $\mathbf{j}$ stays the same. So in MHD Ohm's law is

$$\mathbf{j} = \sigma(\mathbf{E} + \mathbf{u} \times \mathbf{B}). \tag{1.4.1}$$

The SI unit of electrical conductivity is Siemens/metre. It is also useful to define the magnetic diffusivity

$$\eta = \frac{1}{\mu\sigma}, \tag{1.4.2}$$

which has dimensions metre^2/second. So poor conductors have large magnetic diffusivity η and the perfectly conducting limit is $\eta \to 0$.

1.4.1. Lorentz force

In a static medium, the force on an electron is $e\mathbf{E}$. If the electron moves with speed $\mathbf{u}$, the force is

$$\mathbf{F} = e(\mathbf{E} + \mathbf{u} \times \mathbf{B}). \tag{1.4.3}$$

Since current is due to the movement of charge, the force on a moving conductor is the Lorentz force

$$\mathbf{F} = \mathbf{j} \times \mathbf{B}. \tag{1.4.4}$$

Here $\mathbf{F}$ is actually the force per unit mass of conductor.

Ohm's law can be derived on the assumption that the electrons and ions whose movement gives the current are continually colliding with neutrals. This gives rise to a 'drag force' which balances the electric force. The drag force is proportional to **j** just as in viscous fluid a small particle experiences a drag proportional to its velocity **u**. Ohm's law therefore assumes that the ions and electrons are accelerated to their final speeds in a very short time.

1.4.2. Induction equation

Dividing Ohm's law (1.4.1) by σ and taking the curl eliminates the electric field to give

$$\nabla \times \left(\frac{\mathbf{j}}{\sigma} \right) = \nabla \times \mathbf{E} + \nabla \times (\mathbf{u} \times \mathbf{B}) = -\frac{\partial \mathbf{B}}{\partial t} + \nabla \times (\mathbf{u} \times \mathbf{B}), \qquad (1.4.5)$$

and using (1.1.6) to eliminate **j**,

$$\frac{\partial \mathbf{B}}{\partial t} = \nabla \times (\mathbf{u} \times \mathbf{B}) - \nabla \times \eta (\nabla \times \mathbf{B}), \qquad (1.4.6)$$

remembering (1.4.2). (1.4.6) is the induction equation, and is the fundamental equation of dynamo theory. If the conductivity is constant we can use the vector identity curl = grad div -del^2 and (1.1.7) to write the constant conductivity induction equation as

$$\frac{\partial \mathbf{B}}{\partial t} = \nabla \times (\mathbf{u} \times \mathbf{B}) + \eta \nabla^2 \mathbf{B}. \qquad (1.4.7)$$

An alternative form of the constant diffusivity induction equation is

$$\frac{\partial \mathbf{B}}{\partial t} + \mathbf{u} \cdot \nabla \mathbf{B} = \mathbf{B} \cdot \nabla \mathbf{u} + \eta \nabla^2 \mathbf{B}, \qquad (1.4.8)$$

where we have assumed incompressible flow, $\nabla \cdot \mathbf{u} = 0$ and (1.1.7).

1.4.3. Boundary conditions

We usually have to divide the domain in different regions, and apply boundary conditions between them. If one of the domains is perfectly conducting, it is possible to have surface charges and surface currents. Denoting [.] as the value just outside a surface S, **n** being the outward pointing normal, integrating (1.1.5)–(1.1.8) gives

$$[\mathbf{n} \cdot \mathbf{E}] = \frac{\rho_S}{\epsilon}, \qquad [\mathbf{n} \cdot \mathbf{B}] = 0, \qquad [\mathbf{n} \times \mathbf{B}] = \mu \mathbf{j}_S, \qquad [\mathbf{n} \times \mathbf{E}] = 0.$$

$$(1.4.9a,b,c,d)$$

Unless we have a perfect conductor involved, there are no surface currents, and (1.4.9b and c) imply **B** is continuous, provided μ is constant. This is all we need if the outside region is an **insulator**. If it is not, then $\nabla \cdot \mathbf{B} = 0$ implies the normal derivative of $\mathbf{n} \cdot \mathbf{B}$ is also continuous. However, the normal derivatives of the tangential components of **B** are not necessarily continuous. If we take $\mathbf{n} \times$ (1.4.1), Ohm's law,

$$\mathbf{n} \times \mathbf{j} = \sigma [\mathbf{n} \times \mathbf{E} + (\mathbf{n} \cdot \mathbf{B})\mathbf{u} - (\mathbf{n} \cdot \mathbf{u})\mathbf{B}]. \tag{1.4.10}$$

Now from (1.4.9d) $\mathbf{n} \times \mathbf{E}$ is continuous across the boundary, but σ may well not be. At a no-slip boundary, $\mathbf{u} = 0$ so then the ratio of the tangential current across the boundary is just the ratio of the conductivities. In general, the continuity conditions on the normal derivatives of **B** will involve the velocity at the boundary. However, if this is known, then the continuity of $\mathbf{n} \times \mathbf{E}$ across the boundary gives the required relations between the currents across the layer, and hence the normal derivatives of the tangential field components.

If the outside of the region is a static **perfect conductor**, it may have a trapped magnetic field which cannot change, but the usual assumption is there is no magnetic field inside the perfect conductor. Then assuming no normal flow across the boundary (1.4.9a) and (1.4.10) give

$$\mathbf{n} \cdot \mathbf{B} = 0, \qquad \mathbf{n} \times \mathbf{j} = 0, \tag{1.4.11}$$

in the fluid. This gives

$$B_z = \frac{\partial B_x}{\partial z} = \frac{\partial B_y}{\partial z} = 0, \tag{1.4.12}$$

at a Cartesian boundary $z = constant$, or

$$B_r = \frac{\partial (r B_\theta)}{\partial r} = \frac{\partial (r B_\phi)}{\partial r} = 0, \tag{1.4.13}$$

at a spherical boundary $r = constant$.

1.5. *Nature of the induction equation: Magnetic Reynolds number*

There are a number of important limits for the induction equation. If $\mathbf{u} = 0$, (1.4.7) reduces to the diffusion equation,

$$\frac{\partial \mathbf{B}}{\partial t} = \eta \nabla^2 \mathbf{B}. \tag{1.5.1}$$

If there is no fluid motion to maintain the dynamo, the field diffuses away. More precisely, if there is no field at infinity it diffuses away to zero, but if a conductor is immersed in a uniform field, the field inside the conductor eventually becomes uniform. How long does this diffusion process take? Suppose at time $t = 0$ $\mathbf{B} = (B_0 \sin ky + B_1, 0, 0)$ in Cartesian coordinates, so we have a uniform field of strength B_1 with a sinusoidal field of strength B_0 superimposed. Then the solution of (1.5.1) is $\mathbf{B} = (B_0 \sin ky \exp(-\eta k^2 t) + B_1, 0, 0)$. The sinusoidal part decays leaving the constant field. If we require $\mathbf{B} \to 0$ at infinity, we must have $B_1 = 0$, so then the whole field disappears at large time. The e-folding time is the time taken for the field amplitude to drop by a factor e, which is here $1/k^2\eta$. A field with half wavelength $L = \pi/k$, with $L = 1$ metre (large sodium experiment) and $\eta = 0.04$ will have an e-folding time of $1/(0.04\pi^2)$, about 2 seconds. Diffusion acts rather quickly in experiments! The radius of the Earth's core is about 3.5×10^6 metres, so now $k = \pi/3.5 \times 10^6 \sim \times 10^{-6}$. With $\eta = 2$, the e-folding time is about 5×10^{11} seconds or about 20,000 years! In large bodies like the Earth, the diffusion time is long, though not so long as the age of the Earth, so there must be motion in the Earth's core to maintain a dynamo. The Sun is much bigger than the Earth, and so the diffusion time is longer still. Indeed, some relatively short lived stars may not have a dynamo at all, the field being a 'fossil field' left over from the star formation process.

The opposite limit to the diffusion limit is the perfect conductor limit where $\eta = 0$. Then (1.4.6) becomes

$$\frac{\partial \mathbf{B}}{\partial t} = \nabla \times (\mathbf{u} \times \mathbf{B}). \tag{1.5.2}$$

This is the frozen flux limit, so called because the flux through any closed loop, that is the surface integral of $\mathbf{B}$ over the loop, remains fixed as the loop moves around with the fluid velocity (Alfvén's theorem). This means we can think of magnetic field as being frozen in the fluid. This is no longer true if there is diffusion, because diffusion allows field lines to slip through the fluid.

To measure the relative importance of the two terms $\nabla \times (\mathbf{u} \times \mathbf{B})$ and $\eta\nabla^2\mathbf{B}$ in (1.4.7) we need to non-dimensionalise the induction equation. We choose a typical length scale L_* which is the size of the object or region under consideration and a typical fluid velocity U_*. This may be an imposed velocity in some problems or may be the root mean square velocity in others. Then introduce scaled ˜ variables

$$t = \frac{L_*}{U_*}\tilde{t}, \qquad \mathbf{x} = L_*\tilde{\mathbf{x}}, \qquad \mathbf{u} = U_*\tilde{\mathbf{u}} \tag{1.5.3}$$

so that $\nabla = \tilde{\nabla}/L_*$, and (1.4.6) becomes

$$\frac{\partial \mathbf{B}}{\partial \tilde{t}} = \tilde{\nabla} \times (\tilde{\mathbf{u}} \times \mathbf{B}) + Rm^{-1}\tilde{\nabla}^2\mathbf{B}, \qquad Rm = \frac{U_* L_*}{\eta}, \tag{1.5.4, 1.5.5}$$

Rm being the dimensionless magnetic Reynolds number. Large Rm means induction dominates over diffusion, close to the perfect conductor limit, small Rm
means diffusion dominates over induction. In astrophysics and geophysics Rm
is almost always large, but in laboratory experiments it is usually small, though
values up to $\sim$50 can be reached in large liquid sodium facilities.

1.6. The kinematic dynamo problem

The kinematic dynamo problem is where the velocity $\mathbf{u}$ is a given function of
space and possibly time. The dynamic, or self-consistent, dynamo problem is
when $\mathbf{u}$ is solved for using the momentum equation, usually in the form of the
Navier-Stokes equation. The simpler kinematic dynamo problem is linear in $\mathbf{B}$.
The most commonly studied case is when $\mathbf{u}$ is a constant flow, that is $\mathbf{u}$ independent of time. Then we can look for solutions with

$$\mathbf{B} = \mathbf{B}_0(x, y, z)e^{pt}, \qquad \mathbf{B}_0 \to 0 \quad \text{as } \mathbf{x} \to \infty. \tag{1.6.1}$$

In general there are an infinite set of eigenmodes $\mathbf{B}_0$ each with a complex eigenvalue

$$p = \sigma + i\omega, \tag{1.6.2}$$

σ is the growth rate, and ω the frequency. Most of the eigenmodes are very
oscillatory, and are dominated by the diffusion term, and so have σ very negative.
These modes decay, but if there is one or more modes that have σ positive, we
have a dynamo. A random initial condition will have some component of the
growing modes, and these dominate at large time. These kinematic dynamos go
on growing for ever. In reality, the field affects the flow through the Lorentz force
in the equation of motion and changes $\mathbf{u}$ so the dynamo stops growing. This is
the nonlinear saturation process, which is beyond the scope of kinematic dynamo
theory.

If $\omega = 0$, the mode is a steady growth, so these are called steady dynamos. If
the mode has $\omega \neq 0$ (the more usual case) the growth is oscillatory, and we have
growing dynamo waves.

If the flow is periodic rather than constant, (1.4.7) is a linear equation with
periodic coefficients, so Floquet theory applies. Solutions have the form of a
periodic function multiplied by an exponential time dependence (the Floquet exponents), so the story is similar, though numerical calculation is significantly
harder.

Unfortunately, even the kinematic dynamo problem is quite hard. We discuss
two kinematic dynamos, the Ponomarenko dynamo and the G.O. Roberts dynamo
in detail in the next lecture.

1.7. Vector potential, toroidal and poloidal decomposition

1.7.1. Vector potential
Because $\nabla \cdot \mathbf{B} = 0$, we can write

$$\mathbf{B} = \nabla \times \mathbf{A}, \quad \nabla \cdot \mathbf{A} = 0, \tag{1.7.1}$$

where $\mathbf{A}$ is called the vector potential. The condition $\nabla \cdot \mathbf{A} = 0$ is called the Coulomb gauge, and is necessary to specify $\mathbf{A}$ because otherwise we could add on the grad of any scalar. The Biot–Savart integral means we can write $\mathbf{A}$ explicitly as

$$\mathbf{A}(\mathbf{x}) = \frac{1}{4\pi} \int \frac{\mathbf{y} - \mathbf{x}}{|\mathbf{y} - \mathbf{x}|^3} \times \mathbf{B}(\mathbf{y}) \, d^3 y, \tag{1.7.2}$$

the integral being over all space. (Exercise: show this is consistent with $\nabla \cdot \mathbf{A} = 0$.) The induction equation in terms of $\mathbf{A}$ is then

$$\frac{\partial \mathbf{A}}{\partial t} = (\mathbf{u} \times \nabla \times \mathbf{A}) + \eta \nabla^2 \mathbf{A} + \nabla \phi, \quad \nabla^2 \phi = \nabla \cdot (\mathbf{u} \times \mathbf{B}). \tag{1.7.3}$$

1.7.2. Toroidal-poloidal decomposition
Since the magnetic field has three components, but $\nabla \cdot \mathbf{B} = 0$, only two independent scalar fields are needed to specify $\mathbf{B}$. In spherical geometry, we can write

$$\mathbf{B} = \mathbf{B}_T + \mathbf{B}_P, \quad \mathbf{B}_T = \nabla \times T\mathbf{r}, \quad \mathbf{B}_P = \nabla \times \nabla \times P\mathbf{r}. \tag{1.7.4}$$

T and P are called the toroidal and poloidal components respectively. (Note some authors have the unit vector in the r-direction in place of $\mathbf{r}$ in the definition.) This expansion (1.7.4) is used in many numerical methods, having the advantage that then $\nabla \cdot \mathbf{B} = 0$ exactly. The radial component of the induction equation and its curl then give equations for the poloidal and toroidal components. If we define the 'angular momentum' operator

$$L^2 = -\frac{1}{\sin\theta} \frac{\partial}{\partial\theta} \sin\theta \frac{\partial}{\partial\theta} - \frac{1}{\sin^2\theta} \frac{\partial^2}{\partial\phi^2}, \tag{1.7.5}$$

then

$$L^2 P = \mathbf{r} \cdot \mathbf{B}, \quad L^2 T = \mathbf{r} \cdot \nabla \times \mathbf{B}. \tag{1.7.6}$$

Also the radial component of the diffusion term and its curl can be separated into purely poloidal and toroidal parts. The only coupling arises from the induction term. Note that these relations suggest that expanding P and T in spherical harmonics Y_l^m (see 1.2.4) is a good idea, because of the very simple property

$$L^2 Y_l^m = l(l+1) Y_l^m. \tag{1.7.7}$$

Toroidal poloidal decomposition can be useful in Cartesian coordinates too, but care is needed! You have to write

$$\mathbf{B} = \nabla \times g\hat{\mathbf{z}} + \nabla \times \nabla \times h\hat{\mathbf{z}} + b_x(z,t)\hat{\mathbf{x}} + b_y(z,t)\hat{\mathbf{y}}, \tag{1.7.8}$$

including the mean field terms b_x and b_y otherwise your expansion is not complete. [Problem: explain why in spherical polars any field can be expanded as (1.7.4) but in Cartesians you need to have these additional terms.]

1.7.3. Axisymmetric field decomposition

If the magnetic field and the flow are axisymmetric (a non-axisymmetric flow always creates a non-axisymmetric field, but a non-axisymmetric field can be created from an axisymmetric flow), a simpler decomposition is

$$\mathbf{B} = B\hat{\boldsymbol{\phi}} + \mathbf{B}_P = B\hat{\boldsymbol{\phi}} + \nabla \times A\hat{\boldsymbol{\phi}}, \quad \mathbf{u} = s\Omega\hat{\boldsymbol{\phi}} + \mathbf{u}_P = s\Omega\hat{\boldsymbol{\phi}} + \nabla \times \frac{\psi}{s}\hat{\boldsymbol{\phi}},$$

$$s = r\sin\theta. \tag{1.7.9}$$

The induction equation now becomes quite simple,

$$\frac{\partial A}{\partial t} + \frac{1}{s}(\mathbf{u}_P \cdot \nabla)(sA) = \eta\left(\nabla^2 - \frac{1}{s^2}\right)A, \tag{1.7.10}$$

$$\frac{\partial B}{\partial t} + s(\mathbf{u}_P \cdot \nabla)\left(\frac{B}{s}\right) = \eta\left(\nabla^2 - \frac{1}{s^2}\right)B + s\mathbf{B}_P \cdot \nabla\Omega. \tag{1.7.11}$$

This gives some important insight into the dynamo process. Both equations have a $(\mathbf{u}_P \cdot \nabla)$ advection term, which moves field around, and a $(\nabla^2 - \frac{1}{s^2})$ diffusion term, which cannot create field. The azimuthal field can be generated from poloidal field through the term $s\mathbf{B}_P \cdot \nabla\Omega$ term, provided there are gradients of angular velocity along the field lines. Poloidal field is stretched out by differential rotation $\nabla\Omega$ to generate azimuthal field. However, the poloidal field itself has no source term, so it will just decay unless we can find a way to sustain it. This requires some nonaxisymmetric terms to be present.

1.7.4. Symmetry

Y_l^m is symmetric about equator if $l - m$ is even, antisymmetric if $l - m$ is odd. Poloidal field with $P \sim Y_l^m$, $l - m$ odd, has B_r, B_ϕ antisymmetric and B_θ symmetric. Other way round if $l - m$ even. A toroidal field has T with $T \sim Y_l^m$, $l - m$ odd, also has B_r, B_ϕ antisymmetric and B_θ symmetric. Other way round if $l - m$ even.

In the Earth and Sun, dominant modes have P with $l - m$ odd and T with $l - m$ even.

1.7.5. Free decay modes

Seek solutions with $\mathbf{u} = 0$, the free decay modes. There are poloidal and toroidal decay modes,

$$\frac{\partial P}{\partial t} = \eta \nabla^2 P, \qquad \frac{\partial T}{\partial t} = \eta \nabla^2 T, \quad r < a, \tag{1.7.12}$$

$$\nabla^2 P = 0, \qquad T = 0, \quad r > a, \tag{1.7.13}$$

P, $\partial P / \partial r$ and T are continuous at $r = a$. The toroidal decay mode solution is

$$T = r^{-1/2} J_{l+\frac{1}{2}}(kr) Y_l^m e^{-\sigma t}, \qquad \sigma = \eta k^2 = \frac{\eta x_l^2}{a^2}, \tag{1.7.14}$$

x_l being the lowest zero of $J_{l+\frac{1}{2}}$. For $l = 1$, this is $x_1 = 4.493$. The e-folding time is $a^2 / \eta x_1^2$. Poloidal decay modes have

$$P = r^{-1/2} J_{l+\frac{1}{2}}(kr) Y_l^m e^{-\sigma t}, \qquad \sigma = \eta k^2, \quad r < a, \tag{1.7.15}$$

$$P = \frac{A}{r^{l+1}} Y_l^m e^{-\sigma t}, \quad r > a. \tag{1.7.16}$$

Matching at $r = a$ gives

$$\frac{\partial}{\partial r}(r^{-1/2} J_{l+\frac{1}{2}}) + (l+1) r^{-3/2} J_{l+\frac{1}{2}} = 0, \quad \text{at } r = a. \tag{1.7.17}$$

Using the Bessel function recurrence relations this gives just

$$J_{l-\frac{1}{2}}(ka) = 0. \tag{1.7.18}$$

For the dipole $l = 1$, the lowest zero is π, so the e-folding time is $a^2 / \eta \pi^2$. This is the least damped mode.

1.8. The anti-dynamo theorems

Theorem 1. *In Cartesian coordinates (x, y, z) no field independent of z which vanishes at infinity can be maintained by dynamo action. So its impossible to generate a 2D dynamo field.*

If the field is 2D, the flow must be 2D. In Cartesian geometry, the analogue of (1.7.9)–(1.7.11) is

$$\mathbf{B} = B\hat{\mathbf{z}} + \mathbf{B}_H = B\hat{\mathbf{z}} + \nabla \times A\hat{\mathbf{z}}, \quad \mathbf{u} = u_z\hat{\mathbf{z}} + \mathbf{u}_H = u_z\hat{\mathbf{z}} + \nabla \times \psi\hat{\mathbf{z}}, \quad (1.8.1)$$

$$\frac{\partial A}{\partial t} + (\mathbf{u}_H \cdot \nabla)(A) = \eta\nabla^2 A, \tag{1.8.2}$$

$$\frac{\partial B}{\partial t} + (\mathbf{u}_H \cdot \nabla)B = \eta\nabla^2 B + \mathbf{B}_H \cdot \nabla u_z. \tag{1.8.3}$$

Multiply (1.8.2) by A and integrate over the whole volume,

$$\frac{\partial}{\partial t} \int \frac{1}{2}A^2 \, dv + \int \nabla \cdot \frac{1}{2}\mathbf{u}A^2 \, dv = -\eta \int (\nabla A)^2 \, dv. \tag{1.8.4}$$

The divergence term vanishes, because it converts into a surface term which by assumption vanishes at infinity. The term on the right is negative definite, so the integral of A^2 continually decays. It will only stop decaying if A is constant, in which case there is no field. Once A has decayed to zero, $\mathbf{B}_H$ is zero, so there is no source term in (1.8.3). We can then apply the same argument to show B decays to zero. This shows that no nontrivial field 2D can be maintained as a steady (or oscillatory) dynamo.

Note that if A has very long wavelength components, it may take a very long time for A to decay to zero, and in that time B might grow quite large as a result of the driving by the last term in (1.8.3). But ultimately it must decay.

Theorem 2. *No dynamo can be maintained by a planar flow $(u_x(x, y, z, t),$ $u_y(x, y, z, t), 0)$ [58]. No restriction is placed on whether the field is 2D or not in this theorem.*

The point here is that the z-component of the field decays to zero. The z-component of (1.4.8) is

$$\frac{\partial B_z}{\partial t} + \mathbf{u} \cdot \nabla B_z = \eta\nabla^2 B_z, \tag{1.8.5}$$

because the $\mathbf{B} \cdot \nabla u_z$ is zero because u_z is zero. Multiplying (1.8.5) by B_z and integrating, again the advection term gives a surface integral vanishing at infinity, and so B_z decays. If $B_z = 0$, then $\frac{\partial B_x}{\partial x} + \frac{\partial B_y}{\partial y} = 0$ means we can write $B_x = \frac{\partial A}{\partial y}$, $B_y = -\frac{\partial A}{\partial x}$ for some A, and then the z-component of the curl of the induction equation gives

$$\frac{\partial}{\partial t} \nabla_H^2 A + \nabla_H^2 (\mathbf{u} \cdot \nabla A) = \eta \nabla_H^2 \nabla^2 A, \quad \nabla_H^2 = \frac{\partial^2}{\partial x^2} + \frac{\partial^2}{\partial y^2}. \tag{1.8.6}$$

If we take the Fourier transform of this in x and y, the ∇_H^2 operator leads to multiplication by $k_x^2 + k_y^2$ which then can be cancelled out, so (1.8.6) is just (1.8.2) again, which on multiplying through by A leads to the decay of A again. So the whole field decays if the flow is planar.

Theorem 3. *Cowling's theorem* [11]. *An axisymmetric magnetic field vanishing at infinity cannot be maintained by dynamo action.*

This is just the polar coordinate version of Theorem 1. We first show that (1.7.10) implies the decay of A, and then (1.7.11) has its source term removed, so it decays as well. Multiplying (1.7.10) by $s^2 A$ and integrating, and eliminating the divergence terms by converting them to surface integrals which vanish at infinity, we get

$$\frac{\partial}{\partial t} \int \frac{1}{2} s^2 A^2 \, dv = -\eta \int |\nabla(sA)|^2 \, dv \tag{1.8.7}$$

which shows that sA decays. Exercise: write this out fully, with divergence terms included! Note that $sA = constant$ would give diverging A at $s = 0$ which is not allowed. Once A has decayed, $\mathbf{B}_P = 0$ in (1.7.11), and now multiplying (1.7.11) by B/s^2 gives

$$\frac{\partial}{\partial t} \int \frac{1}{2} s^{-2} B^2 \, dv = -\eta \int \left| \nabla \left(\frac{B}{s} \right) \right|^2 dv \tag{1.8.8}$$

and since we don't allow B proportional to s, which doesn't vanish at infinity, this shows that B must decay also. So there can be no steady axisymmetric dynamo.

Note this theorem disallows axisymmetric $\mathbf{B}$ not axisymmetric $\mathbf{u}$. The Ponomarenko dynamo and the Dudley and James dynamos (see Section 2 below) are working dynamos with axisymmetric $\mathbf{u}$ but nonaxisymmetric $\mathbf{B}$.

Theorem 4. *A purely toroidal flow, that is one with* $\mathbf{u} = \nabla \times T\mathbf{r}$ *cannot maintain a dynamo* [4]. *Note that this means that there is no radial motion,* $u_r = 0$.

This is the polar coordinate version of Theorem 2. First we show the radial component of field decays, because

$$\frac{\partial}{\partial t}(\mathbf{r} \cdot \mathbf{B}) + \mathbf{u} \cdot \nabla(\mathbf{r} \cdot \mathbf{B}) = \eta \nabla^2(\mathbf{r} \cdot \mathbf{B}), \tag{1.8.9}$$

so multiplying through by $(\mathbf{r} \cdot \mathbf{B})$ and integrating does the job. Then a similar argument to that used to prove Theorem 2 shows that the toroidal field has no source term and so decays. For details see Gilbert (2003), p. 380. It is not necessary to assume either flow or field is axisymmetric for this theorem. Exercise: fill in the details of the toroidal flow theorem!

2. Working kinematic dynamos

2.1. Minimum Rm for dynamo action

If the magnetic Reynolds number Rm is too small, diffusion dominates over induction and no dynamo is possible. There are a number of ways in which a minimum Rm can be estimated. These are merely lower bounds on possible Rm. Just because a particular dynamo has Rm above these bounds is no guarantee it will work, but if Rm is below the bound it cannot possibly work.

2.1.1. Childress bound
Following Childress (1969) The magnetic energy equation is formed by taking the scalar product of the induction equation with $\mathbf{B}/\mu$,

$$\frac{\partial}{\partial t}\int \frac{\mathbf{B}^2}{2\mu}\, dv = \int \mathbf{j} \cdot (\mathbf{u} \times \mathbf{B})\, dv - \int \mu \eta \mathbf{j}^2\, dv. \tag{2.1.1}$$

Multiplying by μ and rearranging,

$$\mu \frac{\partial E_M}{\partial t} = -\eta \int |\nabla \times \mathbf{B}|^2\, dv + \int (\nabla \times \mathbf{B}) \cdot (\mathbf{u} \times \mathbf{B})\, dv. \tag{2.1.2}$$

The last term is a vector triple product, and since $\mathbf{a} \cdot \mathbf{b} \times \mathbf{c} \leq |\mathbf{a}||\mathbf{b}||\mathbf{c}|$ with equality only when all three vectors are perpendicular, then we have the inequality

$$\int (\nabla \times \mathbf{B}) \cdot (\mathbf{u} \times \mathbf{B})\, dv \leq u_{\max} \left(\int |\nabla \times \mathbf{B}|^2\, dv \right)^{1/2} \left(\int |\mathbf{B}|^2\, dv \right)^{1/2}. \tag{2.1.3}$$

Now a general result for divergence free fields confined in a sphere of radius a matching to a decaying potential outside is that

$$\int |\nabla \times \mathbf{B}|^2\, dv \geq \frac{\pi^2}{a^2}\int |\mathbf{B}|^2\, dv. \tag{2.1.4}$$

This can be proved by expanding **B** in poloidal and toroidal functions. (Exercise: see if you can do it!) So

$$\int (\nabla \times \mathbf{B}) \cdot (\mathbf{u} \times \mathbf{B}) \, dv \leq \frac{au_{max}}{\pi} \int |\nabla \times \mathbf{B}|^2 \, dv. \tag{2.1.5}$$

Putting this in (2.1.2) gives

$$\mu \frac{\partial E_M}{\partial t} \leq \left(\frac{au_{max}}{\pi} - \eta\right) \int |\nabla \times \mathbf{B}|^2 \, dv. \tag{2.1.6}$$

So a growing dynamo requires

$$Rm = \frac{au_{max}}{\eta} \geq \pi. \tag{2.1.7}$$

2.1.2. Backus bound
An alternative approach was provided by Backus [1],

$$\int (\nabla \times \mathbf{B}) \cdot (\mathbf{u} \times \mathbf{B}) \, dv = -\int B_i B_j \frac{\partial u_j}{\partial x_i} \leq e_{max} \int |\mathbf{B}|^2 \, dv, \tag{2.1.8}$$

where e_{max} is the maximum of the rate of strain tensor,

$$e_{ij} = \frac{1}{2}\left(\frac{\partial u_j}{\partial x_i} + \frac{\partial u_i}{\partial x_j}\right). \tag{2.1.9}$$

This gives using (2.1.4)

$$\mu \frac{\partial E_M}{\partial t} \leq \left(e_{max} - \frac{\eta \pi^2}{a^2}\right) \int |\mathbf{B}|^2 \, dv. \tag{2.1.10}$$

So a growing dynamo requires

$$Rm = \frac{a^2 e_{max}}{\eta} \geq \pi^2 \tag{2.1.11}$$

defining *Rm* in a slightly unusual way.

2.2. Faraday disc dynamos

2.2.1. Original Faraday disc dynamo (Fig. 2a)
Assume uniform magnetic field through the disc, $B\hat{\mathbf{z}}$. If no current flows through the meter,

$$\mathbf{j} = \sigma(\mathbf{E} + \mathbf{u} \times \mathbf{B}) = 0, \qquad \mathbf{u} = s\Omega\hat{\boldsymbol{\phi}} \tag{2.2.1}$$

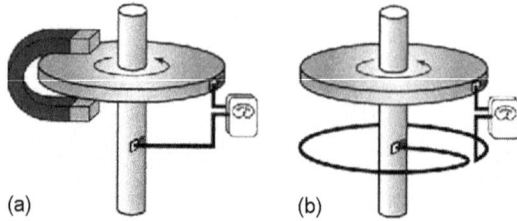

Fig. 2. (a) Original Faraday disc dynamo. Magnetic field supplied by permanent magnet. (b) Homopolar dynamo. Magnetic field now supplied by current flowing through loop of wire.

so $\mathbf{E} = -\mathbf{u} \times \mathbf{B} = -\Omega s B \hat{\mathbf{s}}$, and the voltage drop between axis and rim is

$$V = \int_0^a \mathbf{E} \cdot d\mathbf{l} = -\int_0^a (\mathbf{u} \times \mathbf{B}) \cdot d\mathbf{l} = \frac{1}{2} \Omega a^2 B. \qquad (2.2.2)$$

If the wire completing circuit has resistance R_W, and current I flows through wire, the voltage drop across the wire is $V_R = I R_W$. Suppose the current in the disc flows through a cross-section Σ, and $\mathbf{j} = j\hat{\mathbf{s}}$, then $I = \Sigma j$.

Now integrate (2.2.1) along the disc radius

$$\int_0^a \mathbf{j} \cdot d\mathbf{l} = \sigma \int_0^a \mathbf{E} \cdot d\mathbf{l} + \sigma \int_0^a (\mathbf{u} \times \mathbf{B}) \cdot d\mathbf{l}$$

so

$$\frac{Ia}{\Sigma} = -\sigma V_R + \frac{1}{2} \sigma \Omega a^2 B = -\sigma I R_W + \frac{1}{2} \sigma \Omega a^2 B, \qquad (2.2.3)$$

giving

$$I = \frac{\Omega B a^2}{2(R_D + R_W)}, \qquad R_D = \frac{a}{\Sigma \sigma}. \qquad (2.2.4)$$

R_D being the resistance of the disc. Rotate the disc faster, or get a bigger disc, to get more current. Since a is in units of meters and B is in units of Tesla, and 1 Tesla is a very big field, (strongest laboratory magnets are a few Tesla) the dynamo is not very efficient. Commercial dynamos have the field cutting through many turns of wire, thus multiplying the induction effect.

Note the electric field is in $-\hat{\mathbf{s}}$ direction, counter-acting $\mathbf{u} \times \mathbf{B}$, but the current is in the $+\hat{\mathbf{s}}$ direction.

2.2.2. Homopolar self-excited dynamo (Fig. 2b)

Now the field is generated by a current through the loop according to Ampère's law. The steady dynamo is similar to the original disc problem, but if the dynamo

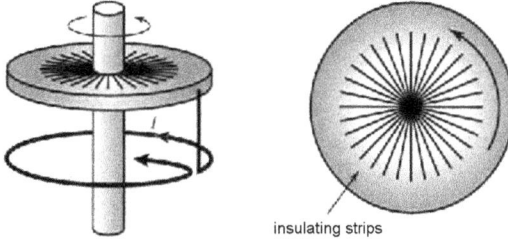

Fig. 3. Homopolar dynamo has segmented disc, to prevent azimuthal current in disc interior; from [39].

grows or decays B through disc will vary. We replace Ba^2 by Φ/π where Φ is the integral of B through the disc.

The dynamo is more interesting if we allow time-dependence. Now the field varies through the disc and the loop. According to Faraday's law, an e.m.f. around the rim of the disc is generated, so there is an azimuthal current as well.

Also, the flux through the wire loop changes, generating an additional e.m.f. there too.

2.2.3. Moffatt's segmented homopolar dynamo (Fig. 3)

The segmentation ensures separation into a radial current I_s and an azimuthal current I_ϕ around the rim.

I_s flows through the wire, and so produces magnetic flux through the disc. I_ϕ produces a magnetic flux through the wire, and its rate of change alters the e.m.f. round the wire loop.

2.2.4. Homopolar disc equations

If the current density through the wire is $\mathbf{j}$, the magnetic flux through the disc is from Biot–Savart

$$\Phi_D = \int_{disc} \mathbf{B}(\mathbf{x}) \cdot d\mathbf{S} = \int \left\{ \frac{\mu}{4\pi} \int \frac{\mathbf{j}(\mathbf{y})(\mathbf{x} - \mathbf{y})}{|\mathbf{x} - \mathbf{y}|^3} \, d\mathbf{y} \right\} \cdot d\mathbf{S}(\mathbf{x}). \qquad (2.2.5)$$

Rather than evaluate this we just write $\Phi_D = MI_s$, where M is called the mutual inductance. $\mathbf{j} = I_s \hat{\boldsymbol{\phi}}/\Sigma_W$ where Σ_W is wire cross-section, so M can be evaluated by doing the integral.

There are also magnetic fluxes through the wire due to the current through the wire, and fluxes through the disc, so

$$\Phi_D = MI_s + L_D I_\phi, \qquad \Phi_W = MI_\phi + L_W I_s, \qquad (2.2.6)$$

L_D, L_W being the self-inductance of the disc and wire.

Round the wire loop circuit, let the total resistance be R_W (includes resistance along radial path in disc). The sources of e.m.f. are the rotation of the disc and the changing flux through the wire loop.

$$R_W I_s = \frac{\Omega \Phi_D}{2\pi} - \frac{d\Phi_W}{dt} \qquad (2.2.7)$$

(minus sign in Faraday's law). If R_P is resistance round the perimeter,

$$R_P I_\phi = -\frac{d\Phi_D}{dt}, \qquad (2.2.8)$$

so we obtain a pair of coupled ODE's for Φ_W and Φ_D,

$$\frac{d\Phi_W}{dt} = -a_{11}\Phi_W + a_{12}\Phi_D, \qquad \frac{d\Phi_D}{dt} = -a_{21}\Phi_W + a_{22}\Phi_D, \qquad (2.2.9)$$

$$a_{11} = \frac{R_W L_D}{L_D L_W - M^2}, \qquad a_{12} = \frac{R_W M}{L_D L_W - M^2} + \frac{\Omega}{2\pi},$$

$$a_{21} = \frac{R_P M}{L_D L_W - M^2}, \qquad a_{22} = \frac{R_P L_W}{L_D L_W - M^2}. \qquad (2.2.10)$$

We seek solutions proportional to $\exp(pt)$, giving

$$p^2 + (a_{11} + a_{22})p + a_{11}a_{22} - a_{12}a_{21} = 0. \qquad (2.2.11)$$

Now $L_D L_W - M^2 > 0$, so $\Omega > 0$ guarantees two real roots, and one is positive if $a_{12}a_{21} > a_{11}a_{22}$, i.e.

$$\Omega > \frac{2\pi R_W}{M}. \qquad (2.2.12)$$

So its a growing dynamo if Ω is large enough. If Ω is very large,

$$p^2 \sim \frac{\Omega}{2\pi} \frac{R_P M}{L_D L_W - M^2}. \qquad (2.2.13)$$

So the dynamo requires resistance R_P, because the growth rate is small if R_P is small. The growth rate becomes zero if the resistance is small, which is a characteristic of a slow dynamo; see Section 4 below.

2.3. Ponomarenko dynamo

The Ponomarenko flow (Fig. 4) is given by

$$\mathbf{u} = s\Omega\hat{\phi} + U\hat{z}, \; s < a, \quad \mathbf{u} = 0, \; s > a,$$

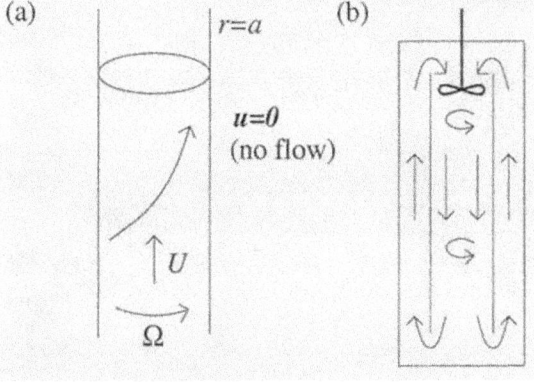

Fig. 4. (a) Sketch of the Ponomarenko flow, solid body screw motion inside cylinder $s = a$. (b) Riga dynamo experiment configuration; sketch from [24].

in polar coordinates (s, ϕ, z), so the flow has a discontinuity $s = a$. The flow has helicity,

$$H = \mathbf{u} \cdot \nabla \times \mathbf{u} = U \frac{1}{s} \frac{\partial}{\partial s} s^2 \Omega = 2U\Omega. \tag{2.3.1}$$

The discontinuity of $\mathbf{u}$ at $s = a$ provides strong shearing. The dynamo evades the cylindrical coordinate version of the planar motion anti-dynamo Theorem 2 through U, so $U = 0$ cannot give a dynamo. We define the magnetic Reynolds number as

$$Rm = \frac{a\sqrt{U^2 + a^2\Omega^2}}{\eta} \tag{2.3.2}$$

based on maximum velocity. We seek a nonaxisymmetric field of the form

$$\mathbf{B} = \mathbf{b}(s)\exp[(\sigma + i\omega)t + im\phi + ikz] \tag{2.3.3}$$

thus evading Cowling's antidynamo Theorem 3. The induction equation (1.4.7) is

$$\frac{\partial \mathbf{B}}{\partial t} + \mathbf{u} \cdot \nabla \mathbf{B} = \mathbf{B} \cdot \nabla \mathbf{u} + \eta \nabla^2 \mathbf{B}.$$

Using the definition of $\mathbf{u}$,

$$(\mathbf{u} \cdot \nabla)\mathbf{B} = (ikU + im\Omega)\mathbf{B} - \Omega B_\phi \hat{\mathbf{s}} + \Omega B_s \hat{\boldsymbol{\phi}}, \tag{2.3.4}$$

and

$$(\mathbf{B} \cdot \nabla)\mathbf{u} = -\Omega B_\phi \hat{\mathbf{s}} + \Omega B_s \hat{\boldsymbol{\phi}}, \qquad (2.3.5)$$

so

$$p^2 b_s = \Delta_m b_s - \frac{2im}{s^2} b_\phi, \qquad p^2 b_\phi = \Delta_m b_\phi + \frac{2im}{s^2} b_s, \qquad (2.3.6)$$

where

$$\Delta_m = \frac{1}{s}\frac{\partial}{\partial s}s\frac{\partial}{\partial s} - \frac{1}{s^2} - \frac{m^2}{s^2}. \qquad (2.3.7)$$

Inside, $s < a$, $p = p_i$,

$$\eta p_i^2 = \sigma + i\omega + im\Omega + ikU + \eta k^2. \qquad (2.3.8)$$

Outside, $s > a$, $p = p_e$,

$$\eta p_e^2 = \sigma + i\omega + \eta k^2. \qquad (2.3.9)$$

Defining $b_\pm = b_s \pm i b_\phi$,

$$p^2 b_\pm = \Delta_{m\pm1} b_\pm. \qquad (2.3.10)$$

Solutions of (2.3.10) that are finite at $r = 0$ and decay at infinity are

$$b_\pm = A_\pm \frac{I_{m\pm1}(p_i s)}{I_{m\pm1}(p_i a)}, \quad s < a, \qquad A_\pm \frac{K_{m\pm1}(p_e s)}{K_{m\pm1}(p_e a)}, \quad s > a, \qquad (2.3.11)$$

I_m and K_m are the modified Bessel functions (like sinh and cosh) that are zero at $= 0$ and zero as $s \to \infty$ respectively.

With this choice, the fields are continuous at $s = a$. One condition between the coefficients $A_\pm$ is set by $\nabla \cdot \mathbf{B} = 0$, and we also need E_z continuous (1.4.8d), so $\eta(\nabla \times \mathbf{B})_z + u_\phi B_s$ has to be continuous giving

$$\eta \left(\left.\frac{\partial b_\phi}{\partial s}\right|_{s\to a+} - \left.\frac{\partial b_\phi}{\partial s}\right|_{s\to a-} \right) = a\Omega b_s(a). \qquad (2.3.12)$$

Writing the jump as [.],

$$2\eta \left[\frac{\partial b_\pm}{\partial s} \right] = \pm i a\Omega (b_+(a) + b_-(a)). \qquad (2.3.13)$$

Defining

$$S_{\pm} = \frac{p_i I'_{m\pm1}(p_i a)}{I_{m\pm1}(p_i a)} - \frac{p_e K'_{m\pm1}(p_e a)}{K_{m\pm1}(p_e a)} \tag{2.3.14}$$

the dispersion relation is

$$2\eta S_+ S_- = ia\Omega(S_+ - S_-). \tag{2.3.15}$$

This needs a simple MATLAB code to sort it out.

We non-dimensionalise on a length scale a and a timescale a^2/η, so that the dimensionless parameters are the growth-rate $a^2 s/\eta$, the frequency $a^2 \omega/\eta$, the pitch of the spiral $\chi = U/a\Omega$, ka and m. The diffusion coefficient $\eta/a^2\Omega = (1+\chi^2)^{1/2} Rm^{-1}$. To find marginal stability we set $\sigma = 0$. For given χ, ka and m we adjust Rm and ω until the real and imaginary parts of $2\eta S_+ S_- - ia\Omega(S_+ - S_-) = 0$. We can use an iterative method such as Newton–Raphson iteration to do this automatically. Then we minimise Rm over m and ka to get the critical mode, and over χ to get the optimum pitch angle.

2.3.1. Ponomarenko dynamo results
When all this is done, we find $Rm_{crit} = 17.7221$, $ka_{crit} = -0.3875$, $m = 1$, $a^2\omega/\eta = -0.4103$ and $\chi = 1.3141$. The poloidal $\hat{\mathbf{z}}$ and toroidal $\hat{\boldsymbol{\phi}}$ components of the flow have similar magnitudes. This is a low value of Rm bearing in mind the lower bounds arguments in Section 2.1. The magnetic field is strongest near $s = a$ where it is generated by shear.

At large Rm, there is a significant simplification, because then $m\Omega + kU$ is small, so $p_e = p_i$ and the ηk^2 terms are small. Bessel functions have asymptotic simplifications at large argument which we can exploit. The fastest growing modes are given by

$$|m| = (6(1+\chi^{-2}))^{-3/4}\left(\frac{a^2\Omega}{2\eta}\right)^{1/2}, \qquad s = 6^{-3/2}\Omega(1+\chi^{-2})^{-1/2}. \tag{2.3.16}$$

A typical generated field is shown in Fig. 5.

2.3.2. Smooth Ponomarenko dynamo
Most fluids have viscosity, so the discontinuity in the Ponomarenko flow is not very realistic. At high Rm, the field is concentrated at the discontinuity. The high Rm analysis can be extended to the case

$$\mathbf{u} = s\Omega(s)\hat{\boldsymbol{\phi}} + U(s)\hat{\mathbf{z}},$$

where there is no discontinuity (see e.g. [24]). The magnetic field is then concentrated near the point $s = a$ where

$$m\Omega'(a) + kU'(a) = 0, \tag{2.3.17}$$

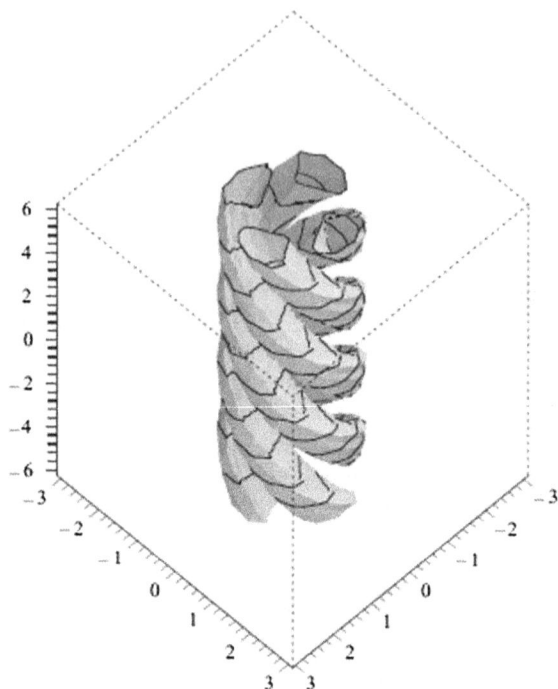

Fig. 5. Magnetic field for the Ponomarenko dynamo at large Rm. Surface of constant **B** shows spiralling field following flow spiral, and located near the discontinuity; from [24].

so the magnetic field is aligned with the shear at this radius. Not all choices of $\Omega(s)$ and $U(s)$ lead to dynamo action. A condition

$$a \left| \frac{\Omega''(a)}{\Omega'(a)} - \frac{U''(a)}{U'(a)} \right| < 4 \tag{2.3.18}$$

must hold for positive growth rates. A helical flow alone is not sufficient for dynamo action!

2.4. G.O. Roberts dynamo

The G.O. Roberts flow (Fig. 6) is two-dimensional, independent of z, but has a z-component.

$$\mathbf{u} = (\cos y, \sin x, \sin y + \cos x). \tag{2.4.1}$$

This avoids the planar antidynamo Theorem 2 through z velocity.

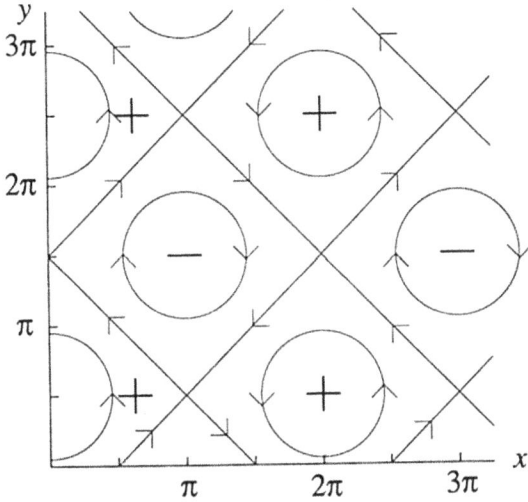

Fig. 6. The G.O. Roberts flow field at a section $z = 0$. The $+$ and $-$ denote the direction of flow in the z-direction, and there is no flow in the z-direction on the separatrices joining the stagnation points; from [45].

The Ponomarenko dynamo has a single roll, and the field at low Rm is on the scale of the roll, or even smaller at high Rm, so it models a small scale dynamo in which the length scale of the field is comparable with the length-scale of the flow. The G.O Roberts dynamo has a collection of rolls and the field can be coherent across many rolls, so the magnetic field can have a larger length-scale than the flow.

The Roberts flow is a special case of the ABC flows (named after Arno'ld, Beltrami and Childress)

$$\mathbf{u} = (C \sin z + B \cos y,\ A \sin x + C \cos z,\ B \sin y + A \cos x) \qquad (2.4.2)$$

with $A = B = 1$, $C = 0$. These flows have $\nabla \times \mathbf{u} = \mathbf{u}$, so vorticity = velocity. Clearly ABC flows have helicity.

The G.O. Roberts flow is integrable, and can be written in terms of a stream-function

$$\mathbf{u} = \left(\frac{\partial \psi}{\partial y},\ -\frac{\partial \psi}{\partial x},\ \psi \right), \qquad \psi = \sin y + \cos x. \qquad (2.4.3)$$

The generated magnetic field has to be z-dependent (anti-dynamo Theorem 1) so it can be taken to be of the form

$$\mathbf{B} = \mathbf{b}(x, y) \exp(pt + ikz), \qquad (2.4.4)$$

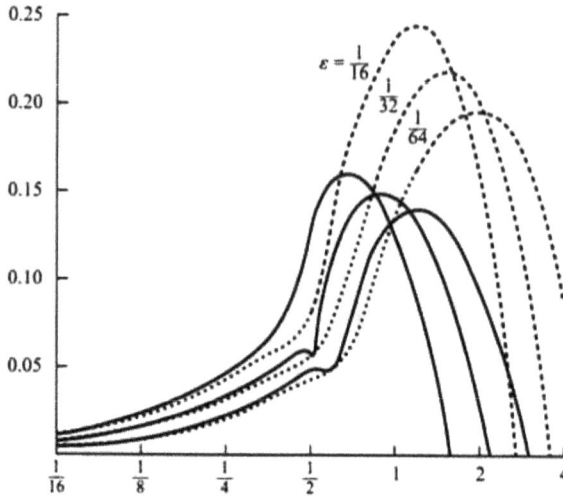

Fig. 7. Growth rate p as a function of z wavenumber, k for various $\epsilon = Rm^{-1}$. Solid lines: G.O. Roberts numerical results. Dashed lines, A.M. Soward's asymptotic large Rm theory; from [9].

where $\mathbf{b}(x, y)$ is periodic in x and y, but it has a mean part independent of x and y which spirals in the z-direction.

To solve the problem, Roberts inserted the form of $\mathbf{B}$ into the induction equation, using a double Fourier series expansion of $\mathbf{b}(x, y)$ truncated at a sufficiently large number of terms. The coefficients then form a linear matrix eigenvalue problem for p. The results are shown as the solid lines in Fig. 7. There is an optimum value of k, the wavenumber in the z-direction, which maximises the growth rate.

2.4.1. Large Rm G.O. Roberts dynamo

At large Rm the dynamo can be analysed in terms of the flows between the stagnation points, see Fig. 8. It is convenient to rotate Fig. 6 through 45°. The field generation occurs primarily in the flows along the separatrices between the stagnation points; see [50] for details. The asymptotic theory agrees qualitatively with the numerical results, as shown in Fig. 7. Since $p \to 0$ as $Rm \to \infty$, although it only decays logarithmically with Rm, this means the dynamo is slow, because a fast dynamo requires finite p in the limit $Rm \to \infty$.

2.4.2. Other periodic dynamos

G.O. Roberts also looked at

$$\mathbf{u} = (\sin 2y, \sin 2x, \sin(x + y)) \tag{2.4.5}$$

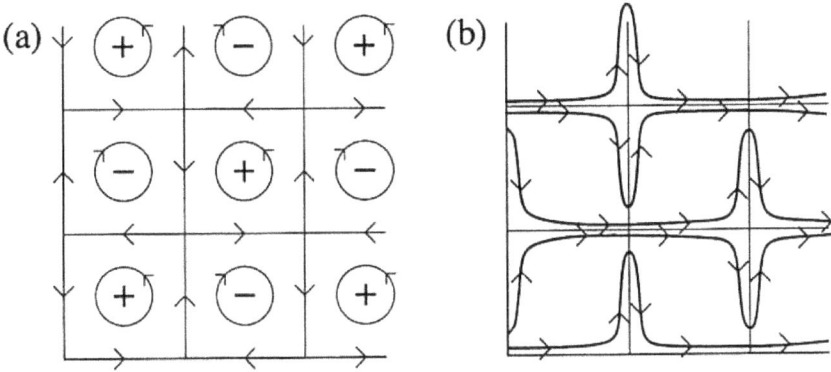

Fig. 8. Figure 6 rotated through 45°. At large Rm, generated field is expelled into boundary layers. This gives enhanced diffusion, leading to lower growth rates and ultimately to decay; from [24].

which has zero mean helicity. Nevertheless, dynamo action can occur! However the growth rates are much smaller than in the ABC case.

To quote H.K. Moffatt, 'Helicity is not essential for dynamo action, but it helps'.

2.5. Spherical dynamos

Following Bullard and Gellman, [4], the velocity for kinematic spherical dynamos can be written

$$\mathbf{u} = \sum_{l,m} \mathbf{t}_l^m + \mathbf{s}_l^m, \tag{2.5.1}$$

where $\mathbf{t}_l^m$ and $\mathbf{s}_l^m$ are the toroidal and poloidal components

$$\mathbf{t}_l^m = \nabla \times \hat{\mathbf{r}} t_l^m(r,t) Y_l^m(\theta,\phi), \qquad \mathbf{s}_l^m = \nabla \times \nabla \times \hat{\mathbf{r}} s_l^m(r,t) Y_l^m(\theta,\phi), \tag{2.5.2a,b}$$

where $-l \le m \le l$.

Bullard and Gellman used $\mathbf{u} = \epsilon \mathbf{t}_1^0 + \mathbf{s}_2^2$ with $t_1^0(r) = r^2(1-r)$, $s_2^2(r) = r^3(1-r)^2$. In their original calculations, they found dynamo action, but subsequent high resolution computations showed they were not dynamos. Warning: inadequate resolution can lead to bogus dynamos!

2.5.1. Dudley and James dynamos

Dudley and James looked at 3 models,

$$\mathbf{u} = \mathbf{t}_2^0 + \epsilon \mathbf{s}_2^0, \qquad \mathbf{u} = \mathbf{t}_1^0 + \epsilon \mathbf{s}_2^0, \qquad \mathbf{u} = \mathbf{t}_1^0 + \epsilon \mathbf{s}_1^0 \tag{2.5.3a,b,c}$$

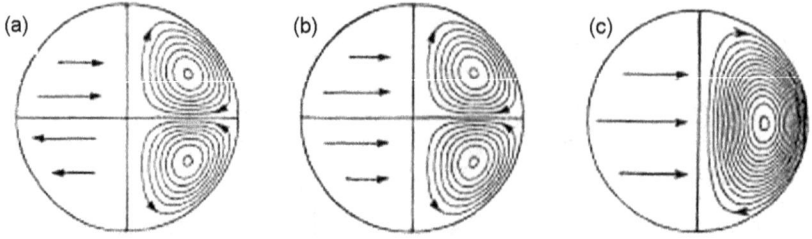

Fig. 9. The flow in the Dudley and James dynamos, [15].On the right the meridional flow, on the left the azimuthal flow direction is indicated.

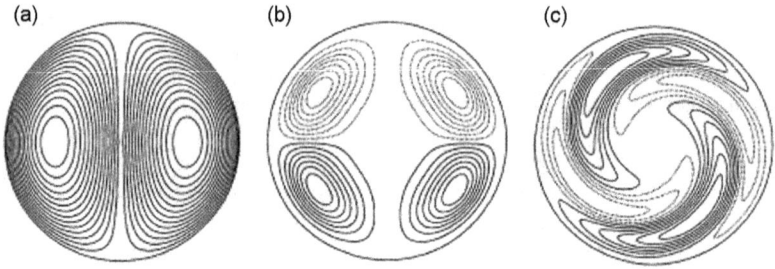

Fig. 10. The Kumar–Roberts fluid flow defined by (2.5.5). (a) Contours of u_ϕ controlled by ϵ_0, (b) streamlines of the meridional circulation controlled by ϵ_1, (c) streamlines of the convection rolls controlled by ϵ_2 and ϵ_3, [26].

with

$$t_1^0 = s_1^0 = r \sin \pi r, \qquad t_2^0 = s_2^0 = r^2 \sin \pi r. \qquad (2.5.4a,b)$$

All steady axisymmetric flows. The t components give azimuthal flow only, the s components give meridional flow.

All three models give dynamo action. Since the flow is axisymmetric, the field has $\exp im\phi$ dependence, and $m = 1$ is preferred. The three models studied in detail are (2.5.3a,b,c), sketched in Fig. 9, with (a) $\epsilon = 0.14$ has $Rm_{crit} \approx 54$ (steady). (b) $\epsilon = 0.13$ has $Rm_{crit} \approx 95$ (oscillatory). (c) $\epsilon = 0.17$ has $Rm_{crit} \approx 155$ (oscillatory).

In all cases, the toroidal and meridional flows are of similar magnitude. The field is basically an equatorial dipole, which in oscillatory cases rotates in time. The Dudley–James flow is probably the simplest spherical dynamo, but it doesn't look like convective flows, which are non-axisymmetric. The Kumar–Roberts flow sketched in Fig. 10 is more complex,

$$\mathbf{u} = \epsilon_0 \mathbf{t}_1^0 + \epsilon_1 \mathbf{s}_2^0 + \epsilon_2 \mathbf{s}_2^{2c} + \epsilon_3 \mathbf{s}_2^{2s}, \qquad (2.5.5)$$

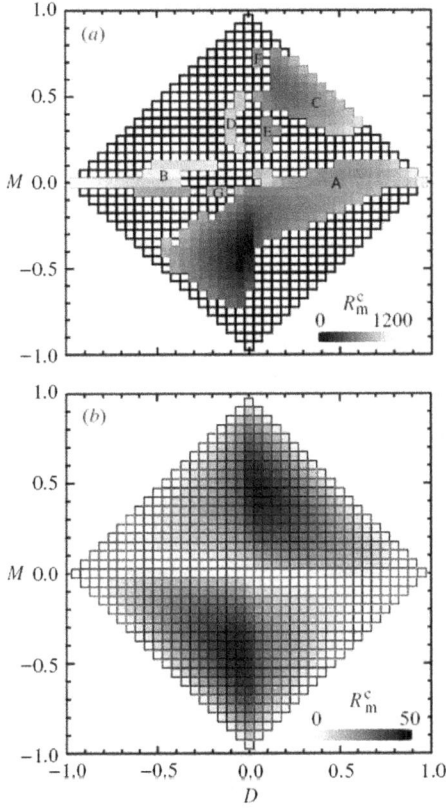

Fig. 11. The Love diamond. Upper diamond marks regions in $D - M$ space where steady dynamos occur. The letters A–F denote regions of growth. The lower diamond is helicity; [26].

where $2c$ means $\cos 2\phi$ and $2s$ means $\sin 2\phi$. The last two terms make the flow nonaxisymmetric, so more like a convective flow.

Gubbins et al. [26] studied these flows for a range of ϵ values. Various radial dependences were also considered. They define the relative energy in the flow as $D + M + C = 1$, where D is the differential rotation energy, the ω-effect, determined by ϵ_0, M is the energy of the meridional circulation, measured by ϵ_1, and C is convection energy from the other two terms. They then vary D and M to see which effects give dynamos at any Rm, see Fig. 11. Surprisingly, there are large areas in the diamond shaped domain where no dynamo occurs at any Rm. This is due to flux expulsion. As soon as a magnetic field tries to get going, it is expelled into the narrow regions between the convecting cells where it decays because of enhanced dissipation. This suggests that time-dependent flows of a convecting type might make better dynamos, because then the flow moves on before flux expulsion is established.

2.5.2. *Braginsky limit*

If ϵ_2 and ϵ_3 are small, the Kumar–Roberts flow is almost axisymmetric. Following Braginsky [3], we can seek fields which are almost axisymmetric. We can then do a perturbation expansion, axisymmetric quantities being large, non-axisymmetric quantities first order. Induction from the nonaxisymmetric quantities gives a mean part $\overline{\mathbf{u}' \times \mathbf{B}'}$ which is second order. At first sight, this doesn't seem to help sustain the leading order axisymmetric dynamo. However, it is the diffusion that makes the axisymmetric dynamo impossible. If we assume the diffusion is the same order as $\overline{\mathbf{u}' \times \mathbf{B}'}$ we can get a self-consistent solution. So we assume large Rm, and take Rm^{-1} as second order and balance the induction from the averaged nonaxisymmetric terms and the diffusion of the leading order axisymmetric field to get a working dynamo. This is Braginsky's 'nearly axisymmetric dynamo'

2.6. *More specimens from the dynamo zoo!*

2.6.1. *Gailitis dynamo*

The flow is in two axisymmetric ring vortices (Fig. 12). There is no toroidal flow. The field is of the form $\exp i\phi$. Two types of solution are found corresponding to different parities. (a) The lower ring generates a poloidal field $\mathbf{B}_1$ which permeates the upper ring, and is stretched by the flow to give a current F_2 in the upper ring. The corresponding field $\mathbf{B}_2$ generates the current in the lower ring. For details, see [17].

Gailitis analysed the dynamo using the Biot–Savart integral. This example shows that a purely poloidal flow can be a dynamo. Recall the antidynamo The-

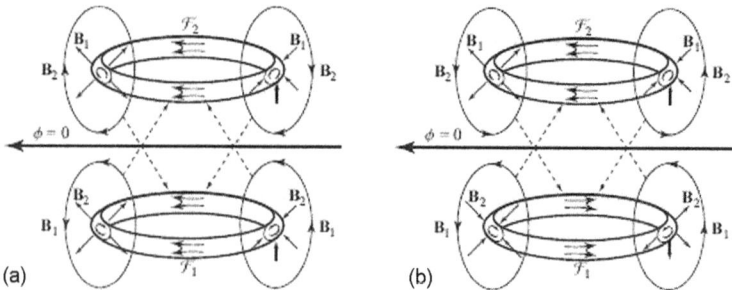

Fig. 12. The Gailitis dynamo; [19].

orem 4 that showed a purely toroidal dynamo cannot exist. The helicity is zero and the critical Rm is quite large, so this is not a particularly efficient dynamo.

2.6.2. *Herzenberg dynamo*
In the Herzenberg dynamo, the flow is a solid body rotation of spheres, with inclined rotation axes. The case with three such spheres is sketched in Fig. 13. The case with two spheres is sufficient to generate a magnetic field. This was an early model that demonstrated that dynamo action is possible despite Cowling's theorem.

2.6.3. *Lowes–Wilkinson dynamo experiment*
Lowes and Wilkinson constructed a laboratory dynamo (Fig. 14) based on the Herzenberg dynamo. The cylinders were copper, embedded in mercury. The cylinders were rotated by powerful motors, to achieve a high value of Rm. The

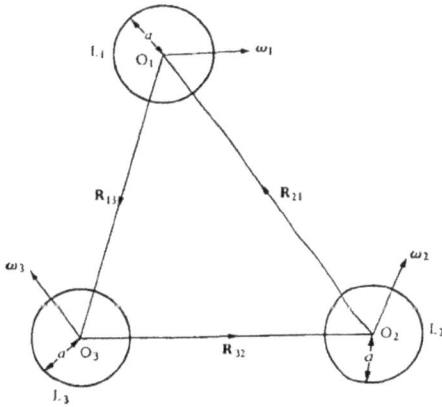

Fig. 13. The Herzenberg dynamo; [25].

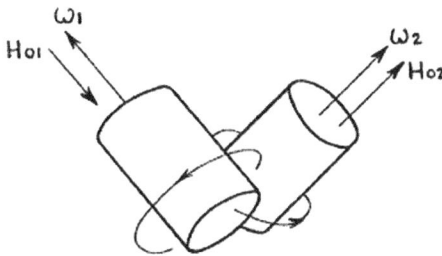

Fig. 14. The Lowes–Wilkinson dynamo experiment; [34].

experiment was successful, and a large field was generated. The Lorentz force generated by the magnetic field was often large enough to bring the motors to a stop, and hence blow them out! This illustrates the way that dynamo action can be limited by nonlinear effects inhibiting the flow, in this case the rotation rate of the cylinders. The field generated in the Lowes–Wilkinson experiment showed chaotic reversals.

3. Mean field dynamo theory

This subject divides into two areas

(i) The underlying theory of mean field dynamo theory, or MFDT, the conditions for its validity, its relationship to turbulence theory and its extension to include nonlinear effects.

(ii) The solutions of MFDT equations and the new types of dynamos they create: dynamo waves, $\alpha\omega$ dynamos and α^2 dynamos.

There is surprisingly little interaction between these two activities. Vastly more papers have been written on (ii), almost all accepting the MFDT equations as a useful model. There are, however, many important questions about (i) which have not yet been answered, so although MFDT is a useful source of ideas about dynamo behaviour, results dependent on it are not yet on firm basis.

3.1. Averaging the dynamo equations

The basic idea is to split the magnetic field and the flow into mean and fluctuating parts,

$$\mathbf{B} = \bar{\mathbf{B}} + \mathbf{B}', \qquad \mathbf{u} = \bar{\mathbf{u}} + \mathbf{u}' \qquad (3.1.1)$$

and apply the Reynolds averaging rules: assume a linear averaging process

$$\overline{\mathbf{B_1} + \mathbf{B_2}} = \bar{\mathbf{B}}_1 + \bar{\mathbf{B}}_2, \qquad \overline{\mathbf{u_1} + \mathbf{u_2}} = \bar{\mathbf{u}}_1 + \bar{\mathbf{u}}_2 \qquad (3.1.2)$$

and once its averaged it stays averaged, so

$$\bar{\bar{\mathbf{B}}} = \bar{\mathbf{B}}, \qquad \bar{\bar{\mathbf{u}}} = \bar{\mathbf{u}}. \qquad (3.1.3)$$

So averaging (3.1.1)

$$\overline{\mathbf{B}'} = \overline{\mathbf{u}'} = 0. \qquad (3.1.4)$$

Also, assume averaging commutes with differentiating, so

$$\frac{\overline{\partial \mathbf{B}}}{\partial t} = \frac{\partial}{\partial t}\overline{\mathbf{B}}, \qquad \overline{\nabla \mathbf{B}} = \nabla \overline{\mathbf{B}}. \tag{3.1.5}$$

Now we average the induction equation (1.4.7)

$$\frac{\overline{\partial \mathbf{B}}}{\partial t} = \overline{\nabla \times (\mathbf{u} \times \mathbf{B})} + \eta \overline{\nabla^2 \mathbf{B}}. \tag{3.1.6}$$

Using the Reynolds averaging rules,

$$\frac{\partial \overline{\mathbf{B}}}{\partial t} = \nabla \times \overline{(\mathbf{u} \times \mathbf{B})} + \eta \nabla^2 \overline{\mathbf{B}}. \tag{3.1.7}$$

The interesting term is $\overline{(\mathbf{u} \times \mathbf{B})}$.

$$\overline{\mathbf{u} \times \mathbf{B}} = \overline{(\bar{\mathbf{u}} + \mathbf{u}') \times (\bar{\mathbf{B}} + \mathbf{B}')} = \bar{\mathbf{u}} \times \bar{\mathbf{B}} + \overline{\bar{\mathbf{u}} \times \mathbf{B}'} + \overline{\mathbf{u}' \times \bar{\mathbf{B}}} + \overline{\mathbf{u}' \times \mathbf{B}'}. \tag{3.1.8}$$

3.1.1. Mean field induction equation
We can therefore write the induction equation as

$$\frac{\partial \overline{\mathbf{B}}}{\partial t} = \nabla \times (\bar{\mathbf{u}} \times \bar{\mathbf{B}}) + \nabla \times \mathcal{E} + \eta \nabla^2 \overline{\mathbf{B}}, \qquad \mathcal{E} = \overline{\mathbf{u}' \times \mathbf{B}'}. \tag{3.1.9}$$

$\mathcal{E}$ is called the mean e.m.f. and it is a new term in the induction equation. We usually think of the primed quantities as being small scale turbulent fluctuations, and this new term comes about because the average mean e.m.f. can be nonzero if the turbulence has suitable averaged properties.

No longer does Cowling's theorem apply! With this new term, we can have simple axisymmetric dynamos, a liberating experience. Not surprisingly, most authors have included this term in their dynamo work, though actually it can be hard to justify the new term in astrophysical applications.

3.1.2. Evaluation of $\overline{(\mathbf{u}' \times \mathbf{B}')}$
If we subtract the mean field equation (3.1.7) from the full equation (1.4.7),

$$\frac{\partial \mathbf{B}'}{\partial t} = \nabla \times (\bar{\mathbf{u}} \times \mathbf{B}') + \nabla \times (\mathbf{u}' \times \bar{\mathbf{B}}) + \nabla \times \mathcal{G} + \eta \nabla^2 \mathbf{B}',$$

$$\mathcal{G} = \mathbf{u}' \times \mathbf{B}' - \overline{\mathbf{u}' \times \mathbf{B}'}. \tag{3.1.10}$$

This is a linear equation for $\mathbf{B}'$, with a forcing term $\nabla \times (\mathbf{u}' \times \overline{\mathbf{B}})$. $\mathbf{B}'$ can therefore be thought of as the turbulent field generated by the turbulent $\mathbf{u}'$ acting on the mean $\overline{\mathbf{B}}$. We can therefore plausibly write

$$\mathcal{E}_i = a_{ij}\overline{\mathbf{B}}_j + b_{ijk}\frac{\partial \overline{\mathbf{B}}_j}{\partial x_k} + \cdots , \qquad (3.1.11)$$

where the tensors a_{ij} and b_{ijk} depend on $\mathbf{u}'$ and $\overline{\mathbf{u}}$.

We don't know $\mathbf{u}'$ and its unobservable, so we assume a_{ij} and b_{ijk} are simple isotropic tensors

$$a_{ij} = \alpha(\mathbf{x})\delta_{ij}, \qquad b_{ijk} = -\beta(\mathbf{x})\epsilon_{ijk}. \qquad (3.1.12)$$

We now have the mean field dynamo theory (MFDT) equations in their usual form,

$$\frac{\partial \overline{\mathbf{B}}}{\partial t} = \nabla \times (\overline{\mathbf{u}} \times \overline{\mathbf{B}}) + \nabla \times \alpha\overline{\mathbf{B}} - \nabla \times (\beta\nabla \times \overline{\mathbf{B}}) + \eta\nabla^2\overline{\mathbf{B}}. \qquad (3.1.13)$$

If β is constant, $\nabla \times (\beta\nabla \times \overline{\mathbf{B}}) = -\beta\nabla^2\overline{\mathbf{B}}$ so the β term acts like an enhanced diffusivity. Even if it isn't constant, we recall from (1.4.6) that the term has the same form as the molecular diffusion term.

We can now justify taking a large diffusion, choosing it to give agreement with observation.

3.2. Validity of MFDT

The α-effect does wonderful things, allowing simple dynamo solutions. But the argument given is very heuristic. There are two ways of trying to justify it, (i) examining the mathematical assumptions, (ii) trying to build a physical model.

3.2.1. The averaging process
For what sort of averaging are the Reynolds rules (3.1.1–3.1.5) valid?

A. Ensemble averages
If we had thousands of identical copies of the Sun, we could start them off with the same mean field, let them run and average all the results to get the ensemble average. Not very practical, but something similar is done in numerical weather forecasting to get a 'probability of rainfall' by running many different simulations.

B. Length scale separation

If the turbulence is small-scale and the mean field is large-scale, we can average over an intermediate length scale,

$$\overline{F}(\mathbf{x}, t) = \int F(\mathbf{x} + \boldsymbol{\xi}, t)\, g(\boldsymbol{\xi})\, d^3\xi, \qquad \int g(\boldsymbol{\xi})\, d^3\xi = 1. \tag{3.2.1}$$

We choose the weight function g to go to zero on the intermediate length scale, so the fluctuations average out but the mean field doesn't,

$$\int F'(\mathbf{x} + \boldsymbol{\xi}, t)\, g(\boldsymbol{\xi})\, d^3\xi = 0, \qquad \int \overline{F}(\mathbf{x} + \boldsymbol{\xi}, t)\, g(\boldsymbol{\xi})\, d^3\xi = \overline{F}. \tag{3.2.2}$$

For this to be strictly valid, the velocity spectrum must have a gap, i.e. all the energy is either in large or small scales. Otherwise the Reynolds rules don't work.

C. Time scale separation

We can do the same if the turbulence has a **short correlation time**, i.e. average over an intermediate timescale

$$\overline{F}(\mathbf{x}, t) = \int F(\mathbf{x}, t + \tau)\, g(\tau)\, d\tau, \qquad \int g(\tau)\, d\tau = 1. \tag{3.2.3}$$

D. Average over a coordinate

Braginsky [3] averaged over ϕ, so

$$\overline{F}(r, \theta, t) = \frac{1}{2\pi} \int F(r, \theta, \phi, t)\, d\phi. \tag{3.2.4}$$

This only applies to axisymmetric dynamo models, but it can be related to numerical simulations. However, Braginsky justified his 'almost axisymmetric dynamo' by assuming the non-axisymmetric components are small compared to the mean field. This is a fairly drastic assumption, and is not usually the case in numerically simulated dynamos.

3.2.2. Evaluation of $\overline{(\mathbf{u}' \times \mathbf{B}')}$, a closer look

Is it necessarily true that $\overline{\mathbf{B}} = 0$ means $\mathbf{B}' = 0$? We look again at (3.1.5),

$$\frac{\partial \mathbf{B}'}{\partial t} = \nabla \times (\overline{\mathbf{u}} \times \mathbf{B}') + \nabla \times (\mathbf{u}' \times \overline{\mathbf{B}}) + \nabla \times \mathcal{G} + \eta \nabla^2 \mathbf{B}',$$

$$\mathcal{G} = \mathbf{u}' \times \mathbf{B}' - \overline{\mathbf{u}' \times \mathbf{B}'}. \tag{3.2.5}$$

If there is no mean field or mean flow, we have the small-scale induction equation for the primed quantities, but this could be a dynamo! If so, we could have a non-zero $\mathcal{E}$ even when there is no mean field. So to justify having $\mathbf{B}'$ proportional to $\overline{\mathbf{B}}$ we need to assume the turbulent R_m is small.

Even if there is no small-scale dynamo, the solution of (3.1.10) is actually of the form

$$\mathcal{E}(\mathbf{x}, t) = \int \int K_{ij}(\mathbf{x}, t; \xi, \tau)\overline{B}_j(\mathbf{x} + \xi, t + \tau)\, d^3\xi d\tau \tag{3.2.6}$$

for some kernel K_{ij}. Only if the turbulence has a short correlation length and time compared to $\overline{\mathbf{B}}$, will $\mathcal{E}$ depend on the local $\overline{\mathbf{B}}$ as required for (3.1.11). If Braginsky averaging is adopted, this may not be true, or at least is an additional assumption.

If we have short correlation, then we can Taylor expand $\overline{\mathbf{B}}$ in (3.2.6),

$$B_j(\mathbf{x} + \xi) = B_j(\mathbf{x}) + \xi_k \partial_k B_j(\mathbf{x}) + \frac{1}{2}\xi_k\xi_m \partial_{km} B_j(\mathbf{x}) + \cdots \tag{3.2.7}$$

and since $|\xi|$ is small compared to the length scale of variation of $\overline{\mathbf{B}}$, the series converges rapidly, justifying the neglect of higher order terms, and so justifying (3.1.11). The time-derivative terms of $\overline{\mathbf{B}}$ can be removed using the mean field equation for $\overline{\mathbf{B}}$ [38].

3.3. Tensor representation of $\mathcal{E}$

Now we look more closely at (3.1.11)

$$\mathcal{E}_i = a_{ij}\overline{B}_j + b_{ijk}\frac{\partial \overline{B}_k}{\partial x_j} + \cdots. \tag{3.3.1}$$

a_{ij} *tensor*
Split this into a symmetric part $(a_{ij} + a_{ji})/2 = \alpha_{ij}$ and the antisymmetric part, $(a_{ij} - a_{ji})/2 = \epsilon_{ijk} A_k$. Then

$$\mathcal{E} = \alpha_{ij}\overline{B}_j + \mathbf{A} \times \overline{\mathbf{B}}. \tag{3.3.2}$$

We already have a $\overline{\mathbf{u}} \times \overline{\mathbf{B}}$ term, so the $\mathbf{A}$ term just modifies the mean velocity. The symmetric part will have three principal axes, and in general three different components along these axes, but isotropic turbulence gives

$$\alpha_{ij} = \alpha\delta_{ij}, \tag{3.3.3}$$

leading to the usual MFDT alpha-effect term.

β effect

The b_{ijk} tensor is treated by splitting $\partial_j \overline{\mathbf{B}}_k$ into symmetric, and antisymmetric, parts

$$\partial_j \overline{\mathbf{B}}_k = (\nabla \overline{\mathbf{B}})^s - \frac{1}{2} \epsilon_{jkm} (\nabla \times \overline{\mathbf{B}})_m. \tag{3.3.4}$$

The symmetric part is not believed to do much. To simplify the antisymmetric part, we decompose the 2nd rank tensor $b_{ijk} \epsilon_{jkm}$ into its symmetric and antisymmetric parts to get

$$\mathcal{E}_i = -\beta_{ij} (\nabla \times \overline{\mathbf{B}})_j - \delta \times (\nabla \times \overline{\mathbf{B}}). \tag{3.3.5}$$

The β-effect is in general an anisotropic eddy diffusion, usually taken as isotropic in applications. The δ-effect term has been discussed recently.

3.4. First order smoothing

The tensor approach is very general, but it gives lots of unknowns. Can we solve for $\mathbf{B}'$ in terms of $\mathbf{u}'$? With a short correlation length ℓ, the mean velocity term (which is constant over the short length scale) can be removed by working in moving frame. Then we have

$$\frac{\partial \mathbf{B}'}{\partial t} = (\overline{\mathbf{B}} \cdot \nabla) \mathbf{u}' + \nabla \times (\mathbf{u}' \times \mathbf{B}' - \overline{\mathbf{u}' \times \mathbf{B}'}) + \eta \nabla^2 \mathbf{B}', \tag{3.4.1}$$

$$O(B'/\tau) \qquad O(\overline{B} u'/\ell) \qquad O(B' u'/\ell) \qquad O(\eta B'/\ell^2), \tag{3.4.2}$$

where (3.4.2) gives the order of magnitude of the corresponding terms in (3.4.1). If the small-scale magnetic Reynolds number $u'\ell/\eta$ is small, the awkward curl term is negligible. This is the first order smoothing assumption, and gives

$$\frac{\partial \mathbf{B}'}{\partial t} = (\overline{\mathbf{B}} \cdot \nabla) \mathbf{u}' + \eta \nabla^2 \mathbf{B}'. \tag{3.4.3}$$

This implies $\mathbf{B}' \ll \overline{\mathbf{B}}$, which is probably not true in Sun. Now suppose the turbulence to be a random superposition of waves,

$$\mathbf{u}' = Re\{\mathbf{u} \exp i(\mathbf{k} \cdot \mathbf{x} - \omega t)\}. \tag{3.4.4}$$

Then using (3.4.3)

$$\mathbf{B}' = Re\left\{ \frac{i(\mathbf{k} \cdot \overline{\mathbf{B}})\mathbf{u}}{\eta k^2 - i\omega} \exp i(\mathbf{k} \cdot \mathbf{x} - \omega t) \right\}. \tag{3.4.5}$$

Now evaluate $\mathcal{E}$,

$$\overline{\mathbf{u}' \times \mathbf{B}'} = \frac{1}{2} \frac{i\eta k^2 (\mathbf{k} \cdot \overline{\mathbf{B}})}{\eta^2 k^4 + \omega^2} (\mathbf{u}^* \times \mathbf{u}), \tag{3.4.6}$$

where $*$ denotes complex conjugate, equivalent to

$$a_{ij} = \frac{1}{2} \frac{i\eta k^2}{\eta^2 k^4 + \omega^2} k_j \epsilon_{imn} u_m^* u_n. \tag{3.4.7}$$

3.4.1. Connection with helicity
If the turbulence has no preferred direction, i.e. it is isotropic,

$$\alpha_{ij} = \frac{1}{2} \frac{i\eta k^2}{\eta^2 k^4 + \omega^2} \delta_{ij} k_i \epsilon_{imn} u_m^* u_n. \tag{3.4.8}$$

Now consider the helicity

$$H = \overline{\mathbf{u}' \cdot \nabla \times \mathbf{u}'} = -\frac{1}{2} i\mathbf{k} \cdot (\mathbf{u}^* \times \mathbf{u}). \tag{3.4.9}$$

Taking the trace of (3.4.8) gives

$$\alpha = -\frac{1}{3} \frac{\eta k^2 H}{\eta^2 k^4 + \omega^2}. \tag{3.4.10}$$

This means that under first order smoothing, the mean e.m.f. is proportional to the helicity of the turbulence. Helical motion is 'Ponomarenko' type motion, left-handed or right-handed. Mirror-symmetric turbulence has zero helicity. Rotating convection has non-zero helicity in general.

3.4.2. Connection with G.O. Roberts dynamo
The G.O. Roberts dynamo had a flow which is an organised superposition of waves, see (2.4.1). In the case where the magnetic Reynolds number Rm based on the cell size length is small, we can do a two scale analysis (for details see [24]) in which the length scale of the mean field is large, $O(Rm^{-2})$, and the perturbed field is $O(Rm)$. $\mathbf{B}'$ is then given by the first order smoothing equations as a consequence of the expansion, and so $\overline{\mathbf{u}' \times \mathbf{B}'}$ can be evaluated using the same procedure as above.

The mean field then satisfies an equation on the large length scale X and a slow timescale $T = t\, Rm^3$

$$\frac{\partial \overline{\mathbf{B}}}{\partial T} = \nabla_X \times (\alpha_{ij} \overline{B}_j) + \nabla_X^2 \overline{\mathbf{B}}, \tag{3.4.11}$$

where $\alpha_{11} = \alpha_{22} = \alpha$, all other components being zero. (3.4.11) has growing solutions on the large length scale

$$\overline{\mathbf{B}} = (\pm \sin KZ, \cos KZ, 0) \qquad (3.4.12)$$

which is the helical form found in G.O. Roberts numerical solutions.

This analysis gives a more definite meaning to the mean field picture, but it also reveals a major weakness. The large scale modes grow alright, but only on the very slow $T = t Rm^3$ timescale. The modes where the mean field and fluctuating field have the same length scale (and where mean field theory is not valid) grow on a much faster timescale.

3.5. *Parker loop mechanism*

Mean field theory predicts an e.m.f. parallel to the mean magnetic field,

$$\frac{\partial \overline{\mathbf{B}}}{\partial t} = \nabla \times (\overline{\mathbf{u}} \times \overline{\mathbf{B}}) + \nabla \times \alpha \overline{\mathbf{B}} + \eta_T \nabla^2 \overline{\mathbf{B}}. \qquad (3.5.1)$$

This contrasts with $\overline{\mathbf{u}} \times \overline{\mathbf{B}}$ which is perpendicular to the mean field. With constant α, the α-effect predicts growth of field parallel to the current $\mu \nabla \times \mathbf{B}$. Recalling that the α-effect depends on helicity, we can picture this process as in Fig. 15. A rising twisting element of fluid brings up magnetic field. A loop of flux is created, which then twists due to helicity. The loop current is parallel to the original mean field. Poloidal field has been created out of azimuthal field.

Note that if there is too much twist, the current is in the opposite direction. First order smoothing assumes small twist.

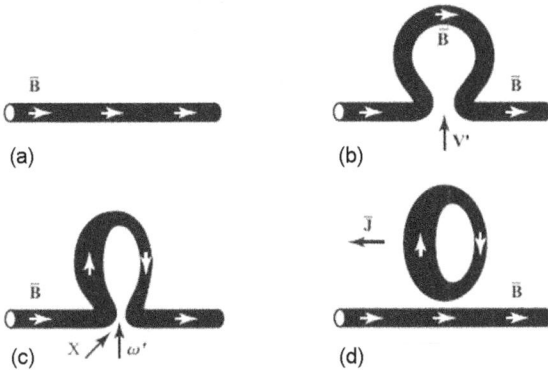

Fig. 15. Parker loop mechanism; from [48].

3.5.1. Joy's law

A sunspot pair is created when an azimuthal loop rises through solar photosphere. The vertical field impedes convection producing the cool dark spot. Joy's law says that sunspot pairs are systematically tilted, with the leading spot being nearer the equator. Assuming flux was created as azimuthal flux deep down, this suggests that loop has indeed twisted through a few degrees as it rose. This provides some evidence of the α-effect at work, and this idea is the basis of many solar dynamo models.

3.6. Axisymmetric mean field dynamos

The mean field dynamo equations with isotropic α are derived from (1.7.10) and (1.7.11) with the alpha-effect included,

$$\frac{\partial A}{\partial t} + \frac{1}{s}(\mathbf{u}_P \cdot \nabla)(sA) = \alpha B + \eta\left(\nabla^2 - \frac{1}{s^2}\right)A, \tag{3.6.1}$$

$$\frac{\partial B}{\partial t} + s(\mathbf{u}_P \cdot \nabla)\left(\frac{B}{s}\right) = \nabla \times \alpha \mathbf{B}_P + \eta\left(\nabla^2 - \frac{1}{s^2}\right)B + s\mathbf{B}_P \cdot \nabla\Omega. \tag{3.6.2}$$

The α-effect term is the source for generating poloidal field from azimuthal field, as envisaged by Parker [41], and Babcock and Leighton.

There are two ways of generating azimuthal field B from poloidal field $\mathbf{B}_P$: the α-effect or the ω-effect. If the α-effect dominates, the dynamo is called an α^2-dynamo. If the ω-effect dominates its an $\alpha\omega$ dynamo. We can also have $\alpha^2\omega$ dynamos where both mechanisms operate.

3.6.1. The Omega-effect

In Fig. 16, an initial loop of meridional field threads through the sphere. The inside of the sphere is rotating faster than the outside: so we have differential

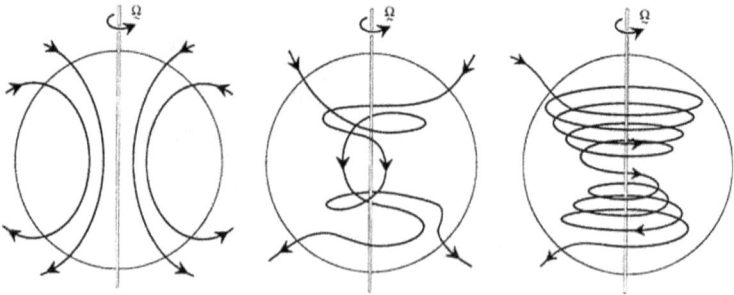

Fig. 16. An initial dipole field is sheared by differential rotation in the sphere to generate a large toroidal field.

rotation. The induction term $s\mathbf{B}_P \cdot \nabla\Omega$ generates azimuthal field by stretching. As we see in Fig. 16, opposite sign B_ϕ is generated on either side of the equator, as in the Sun.

3.6.2. Dynamo waves

The simplest analysis of dynamo waves uses Cartesian geometry, and we assume the waves are independent of y.

$$\mathbf{B} = (-\partial A/\partial z, B, \partial A/\partial x), \qquad \mathbf{u} = (-\partial\psi/\partial z, u_y, \partial\psi/\partial x), \tag{3.6.3}$$

$$\frac{\partial A}{\partial t} + \frac{\partial(\psi, A)}{\partial(x, z)} = \alpha B + \eta\nabla^2 A, \tag{3.6.4}$$

$$\frac{\partial B}{\partial t} + \frac{\partial(\psi, B)}{\partial(x, z)} = \frac{\partial(A, u_y)}{\partial(x, z)} - \nabla \cdot (\alpha\nabla A) + \eta\nabla^2 B. \tag{3.6.5}$$

Set $\psi = 0$, α constant, $u_y = U'z$, a constant shear, ignore the α term in the B equation ($\alpha\omega$ model) and set $A = \exp(\sigma t + i\mathbf{k} \cdot \mathbf{x})$. The dispersion relation is then

$$(\sigma + \eta k^2)^2 = ik_x\alpha U', \tag{3.6.6}$$

giving

$$\sigma = \frac{1 + i}{\sqrt{2}}(\alpha U'k_x)^{1/2} - \eta k^2 \tag{3.6.7}$$

with a suitable choice of signs. This gives growing dynamo waves if the $\alpha U'$ term overcomes diffusion.

If the wave is confined to a plane layer, $k_z = \pi/d$ gives the lowest critical mode, and there is a critical value of k_x for dynamo action. The dimensionless combination $D = \alpha U'd^3/\eta^2$ is called the dynamo number, and in confined geometry there is a critical D for onset.

Note fastest growing waves in unbounded geometry have $k_z = 0$, so they propagate perpendicular to the shear direction z. If $\alpha U' > 0$, a +ve k_x gives growing modes with $Im(\sigma) > 0$, so they propagate in the $-$ve x-direction. the direction of propagation depends on sign of $\alpha U'$.

3.6.3. α^2 dynamos

Now set $\psi = u_y = 0$, α constant, $A = \exp(\sigma t + i\mathbf{k} \cdot \mathbf{x})$ to get the dispersion relation

$$(\sigma + \eta k^2)^2 = \alpha^2 k^2, \tag{3.6.8}$$

$$\sigma = \pm \alpha k - \eta k^2 \qquad\qquad\qquad (3.6.9)$$

which means we can have growing modes with zero frequency. There are no dynamo waves in the α^2-dynamo, but a steady dynamo results. In bounded geometry there is a critical α for dynamo action.

3.7. Spherical $\alpha\omega$ dynamos

Figures 17 and 18 were obtained by integrating the spherical geometry mean field dynamo equations (3.6.1) and (3.6.2) with $\alpha = f(r)\cos\theta$ and $\mathbf{u} = s\omega(r)\hat{\boldsymbol{\phi}}$. Various choices of the functions $f(r)$ and $\omega(r)$ were considered in [47]. As expected from our simple analysis of plane dynamo waves, these $\alpha\omega$ dynamos give oscillatory solutions. Both dipolar and quadrupolar dynamos can occur. A dipolar solution is shown in Fig. 17, a quadrupolar one in Fig. 18. Dipolar dynamos generally onset before quadrupolar dynamos if $\alpha\omega' < 0$.

A brief summary of the numerical findings about spherical MFDT models is

(i) Generally, α^2 models give steady dynamos, $\alpha\omega$ dynamos give oscillatory solutions. But, some α distributions, particularly if there are positive and negative values in the same hemisphere, can give steady $\alpha\omega$ dynamos.

(ii) Meridional circulation, non-zero ψ, can also help to steady $\alpha\omega$ dynamos.

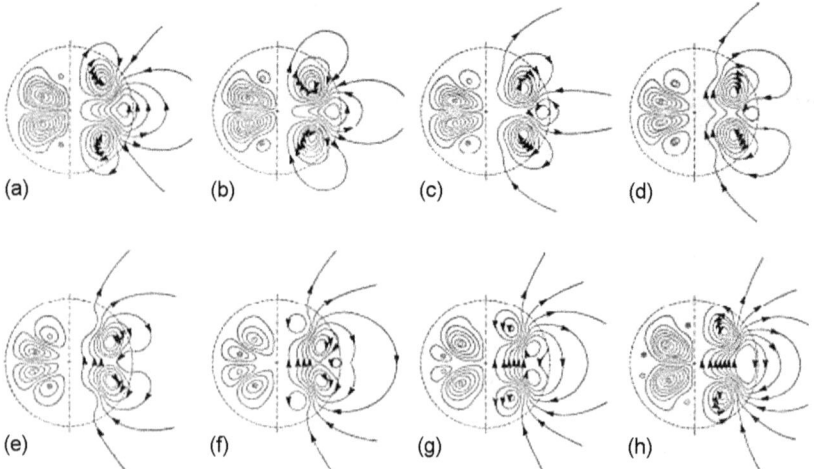

(a)   (b)   (c)   (d)

(e)   (f)   (g)   (h)

Fig. 17. Dipolar oscillatory solution of axisymmetric $\alpha\omega$-dynamo in a sphere. (a)–(h) goes through one period. Right meridional field, left azimuthal field. $\alpha = f(r)\cos\theta$, $\omega = \omega(r)$. B antisymmetric about equator, A symmetric; from [47].

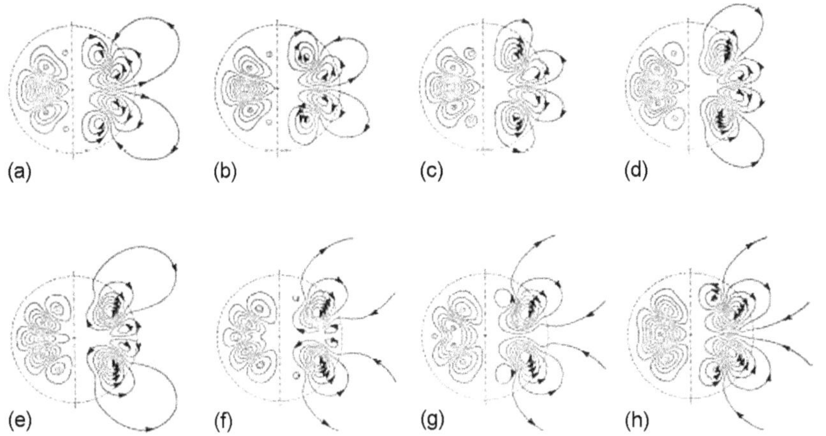

(a)　　　　(b)　　　　(c)　　　　(d)

(e)　　　　(f)　　　　(g)　　　　(h)

Fig. 18. As in Fig. 17, but a quadrupolar oscillatory solution of an axisymmetric $\alpha\omega$-dynamo in a sphere. B antisymmetric about equator, A symmetric; from [47].

(iii) Unfortunately, it seems that a vast range of different dynamo behaviour can be found depending on the spatial α distribution, even if α is restricted to the isotropic case. This is a major problem for modelling, as there is little prospect of determining α by observation. In practice, α-distributions are determined by adjusting until the numerical results agree with the observations, but such tuning of the parameters to fit the data is not a very satisfactory way to proceed.

4. Fast and slow dynamos

If the magnetic diffusion time is much longer than the turn-over time of the flow, the induction equation (1.4.8) can be written

$$\frac{\partial \mathbf{B}}{\partial t} + \mathbf{u} \cdot \nabla \mathbf{B} = \mathbf{B} \cdot \nabla \mathbf{u} + \epsilon \nabla^2 \mathbf{B},$$

where $\epsilon = Rm^{-1}$ is small. Time is scaled on the turnover time L/U, L being the length scale of the object, U a typical velocity.

For steady flow, a dynamo driven magnetic field grows exponentially, $\mathbf{B} \sim e^{\sigma t}$, and if $\gamma = Re(\sigma)$ the flow is a fast dynamo if

$$\lim_{\epsilon \to 0} \gamma(\epsilon) = \gamma_0 > 0.$$

The flow is a slow dynamo if

$$\lim_{\epsilon \to 0} \gamma(\epsilon) = \gamma_0 \leq 0.$$

Fast dynamos grow on the turnover time (months in the Sun) not the magnetic diffusion time (millions of years in the Sun). The solar magnetic cycle operates on a twenty-two year cycle, much shorter time than the diffusion time, so it must be a fast dynamo.

4.1. Magnetic helicity

Magnetic helicity is defined in terms of the vector potential $\mathbf{A}$, where $\mathbf{B} = \nabla \times \mathbf{A}$. The magnetic helicity,

$$H_m = \int \mathbf{A} \cdot \mathbf{B} \, dv. \tag{4.1.1}$$

Recalling the induction equation for $\mathbf{A}$ (1.7.3),

$$\frac{\partial \mathbf{A}}{\partial t} = \mathbf{u} \times (\nabla \times \mathbf{A}) + \eta \nabla^2 \mathbf{A} + \nabla \phi, \quad \nabla^2 \phi = \nabla \cdot (\mathbf{u} \times \mathbf{B})$$

and the induction equation for $\mathbf{B}$ (1.4.6),

$$\frac{\partial \mathbf{B}}{\partial t} = \nabla \times (\mathbf{u} \times \mathbf{B}) - \nabla \times \eta(\nabla \times \mathbf{B}),$$

multiplying the first by $\mathbf{B}$, the second by $\mathbf{A}$, and adding,

$$\frac{\partial}{\partial t} \mathbf{A} \cdot \mathbf{B} = \nabla \cdot (\phi \mathbf{B}) + \mathbf{A} \cdot \nabla \times (\mathbf{u} \times \mathbf{B}) - \mathbf{A} \cdot \nabla \times \eta(\nabla \times \mathbf{B}) - \eta(\nabla \times \mathbf{B}) \cdot \mathbf{B}. \tag{4.1.2}$$

By using standard vector identities we can write

$$\mathbf{A} \cdot \nabla \times (\mathbf{u} \times \mathbf{B}) = (\mathbf{u} \times \mathbf{B}) \cdot \nabla \times \mathbf{A} - \nabla \cdot \mathbf{A} \times (\mathbf{u} \times \mathbf{B})$$

$$= \nabla \cdot \mathbf{B}(\mathbf{A} \cdot \mathbf{u}) - \mathbf{u} \cdot \nabla(\mathbf{A} \cdot \mathbf{B}). \tag{4.1.3}$$

Also,

$$-\mathbf{A} \cdot \nabla \times \eta(\nabla \times \mathbf{B}) = \nabla \cdot (\mathbf{A} \times (\eta \nabla \times \mathbf{B})) - \eta \mathbf{B} \cdot \nabla \times \mathbf{B}, \tag{4.1.4}$$

so

$$\frac{D}{Dt} \mathbf{A} \cdot \mathbf{B} = \nabla \cdot [\mathbf{B}(\phi + \mathbf{u} \cdot \mathbf{A})] + \nabla \cdot \mathbf{A} \times (\eta \nabla \times \mathbf{B}) - 2\eta \mu \mathbf{B} \cdot \mathbf{j}. \tag{4.1.5}$$

The first divergence term can usually be eliminated by defining ϕ suitably. The second divergence term is resistive. If we integrate over the volume of the fluid region, and ignore the surface terms arising from the divergences in (4.1.5),

$$\frac{\partial}{\partial t} \int \mathbf{A} \cdot \mathbf{B} \, dv = -2\eta\mu \int \mathbf{B} \cdot \mathbf{j} \, dv, \tag{4.1.6}$$

so the total magnetic helicity is conserved if η is small (large Rm). $\mathbf{B} \cdot \mathbf{j}$ is called the current helicity.

4.2. The stretch-twist-fold dynamo

A loop of flux is first stretched to twice its length, reducing the cross-section by half. Alfvèn's theorem tells us that the integrated flux through the loop cannot change if diffusion is small, so since the area is halved, the field strength must double. Now twist the loop to get to (b), and then fold to get to (c). Apply small diffusion at X to reconnect. Since the two loops in (d) both have the same flux as in (a), because each has half the area and double the field strength, we have doubled the total flux. Repeating the process doubles the flux again, so we have exponential growth in this process.

The stretching phase of the process did work against the hoop stresses. The Lorentz force can be written as

$$\mathbf{j} \times \mathbf{B} = \frac{1}{\mu}(\mathbf{B} \cdot \nabla\mathbf{B}) - \frac{1}{2\mu}\nabla\mathbf{B}^2. \tag{4.2.1}$$

If $\mathbf{B} = B\hat{\phi}$,

$$\mathbf{j} \times \mathbf{B} = -\frac{B^2}{s\mu}\hat{\mathbf{s}} - \hat{\mathbf{s}}\frac{\partial}{\partial s}\frac{B^2}{2\mu} - \hat{\mathbf{z}}\frac{\partial}{\partial z}\frac{B^2}{2\mu} \tag{4.2.3}$$

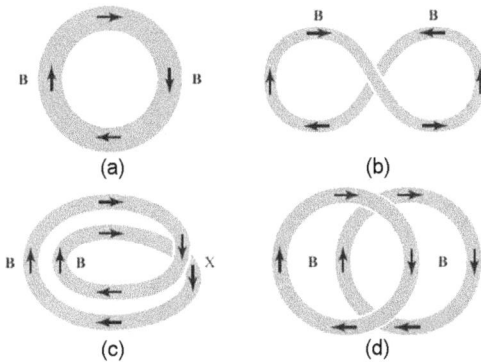

Fig. 19. Schematic diagram of the stretch twist fold dynamo; after [56].

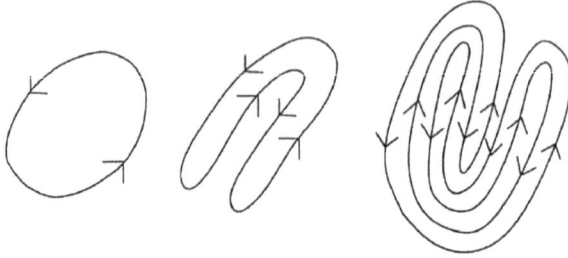

Fig. 20. Stretching and folding in 2D.

and the term $-B^2\hat{s}/s\mu$ is called the hoop stress. Energy conservation means that, because energy is needed to generate magnetic field, the fluid flows must be doing work against the magnetic forces. This work must come from an energy source, such as thermal convection or mechanical driving.

This is a fast process, because it happens on the fluid velocity turn over time, L/U. It does however, appeal to 'small diffusion' to reconnect in step (c) to (d). The hope is that this reconnection occurs over a very short length scale over which diffusion can act quickly, so the small diffusion does not slow the process down significantly.

4.2.1. Stretching and folding in 2D

Stretch-twist-fold is inherently 3D. Why can't we just stretch out field in 2D? We know that planar dynamos don't work from Theorem 2, but what is wrong with just 2D stretching and folding? It is clear from Fig. 20 that stretching in 2D generates more field just as it does in the stretch, twist and fold dynamo. The problem is that the fields generated have opposite sign. As stretching continues, opposite signed flux gets close together and diffusion acts fast over short distances. Diffusion wipes out field as fast as stretching generates it.

4.3. Baker's maps and stretch, fold, shear

These ideas can be put on a more formal basis using maps. The two stages of the baker's map shown in Fig. 21 are

$$(x, y) \to (2x, y/2), \quad 0 \le x \le \frac{1}{2},$$

$$(x, y) \to (2x - 1, (1 + y)/2), \quad \frac{1}{2} < x \le 1. \quad (4.3.1)$$

The baker's map is the basis of bread-making. This process has doubled the flux, but its a cheat, because how do you cut fluid in a continuous process? Now we

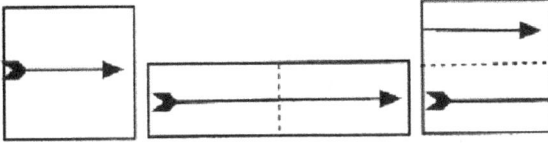

Fig. 21. The baker's map; from [24].

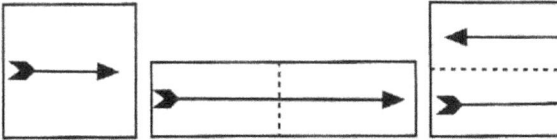

Fig. 22. The folded baker's map; from [24].

put the top slice the other way round as in Fig. 22.

$$(x, y) \rightarrow (2x, y/2), \quad 0 \leq x \leq \frac{1}{2},$$

$$(x, y) \rightarrow (2 - 2x, (1 - y)/2), \quad \frac{1}{2} < x \leq 1. \tag{4.3.2}$$

This less of a cheat, because we could fold the fluid round to achieve this. No cutting necessary. But its a 2D motion, so we have adjacent opposite signed field, and this process alone cannot give a dynamo.

4.3.1. Stretch-fold-shear, SFS

We can modify the folded baker's map of Fig. 22 by adding shear in the third dimension, as illustrated in Fig. 23. (a) is a stretch in the x-direction, stretching out the field in the x-direction. (b) is the folded baker's map. Note that when it is folded back, the grey region goes white because the x-component of field is now in the opposite direction. (c) shear in the z-direction. We now have all white field on back and front faces, all black in between. The shear step means field no longer cancels across z-planes.

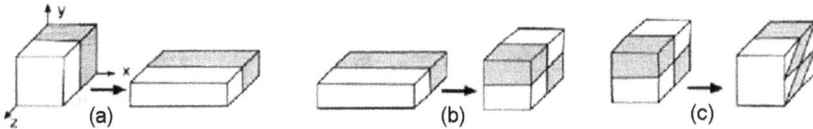

Fig. 23. Stretch fold shear. Grey regions have field in +ve x direction, white regions in −ve x direction; from [24].

C.A. Jones

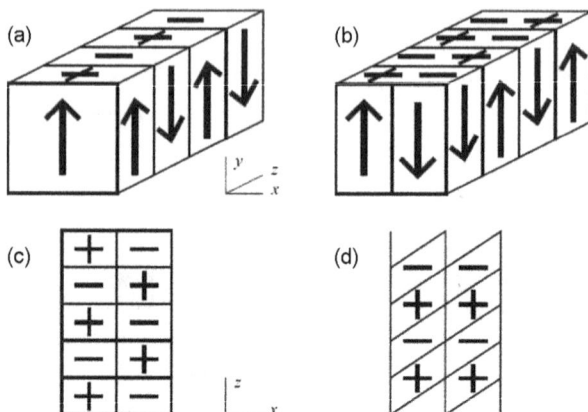

Fig. 24. Stretch fold shear. The magnetic field; from [24].

This stretch, fold and shear mechanism is illustrated again in Fig. 24. (a) Initial mean field is now taken in the y-direction. It is periodic in z, proportional to $\exp ikz$. (b) After the stretch in the y-direction and the fold in the x-direction. (c) is the same state as (b) but viewed from above. (d) After the shear. The fluid was given a shift in the z-direction by an amount proportional to x.

Note the bottom left hand corner in (c) has to be pulled down, and the bottom right pushed up, to keep the phase of (d) the same as (a). The net effect is that the final field configuration is close to that in (a), but the stretching means the field strength is increased.

4.3.2. Stretch-fold-shear in G.O. Roberts dynamo
In Fig. 8 the flow in the G.O. Roberts dynamo at large Rm was sketched. The flow is

$$\mathbf{u} = (\psi_y, -\psi_x, \sqrt{2}\psi), \quad \psi = \sin x \sin y. \tag{4.3.3}$$

Near the stagnation points the field is stretched out. This would lead to cancellation if there were no flow in the z-direction, but the z-flow provides the crucial shear, which is non-zero at the stagnation point because of the helicity. Thus a flow $u_z = \cos x \cos y$, which has zero net helicity, wouldn't give the required stretch-fold-shear mechanism.

We can consider the production of mean y-field from mean x-field in the Roberts cell. We take L to be half the wavelength in the z-direction. If the mean x-field is at $z = 0$, after shearing, there is mean negative y-field at $z = -L/2$, and mean positive field at $z = L/2$. Mean y-field is therefore $L/2$ out of phase with mean x-field, so the field turns as z increased. This spiralling of the mean field with z is a characteristic of the G.O Roberts dynamo.

4.4. ABC dynamos

The general ABC flow is

$$\mathbf{u} = (C \sin z + B \cos y, \; A \sin x + C \cos z, \; B \sin y + A \cos x). \qquad (4.4.1)$$

The G.O. Roberts flow was $A = B = 1, C = 0$, but now look at $A = B = C = 1$. This is not integrable, and the streamlines got by solving

$$\dot{x} = u_x, \qquad \dot{y} = u_y, \qquad \dot{z} = u_z \qquad (4.4.2)$$

are chaotic. Figure 25 shows a Poincaré section of the ABC flow, that is a trajectory is integrated forward in time, and each time the plane $z = 0$ is cut, the corresponding x, y values are plotted as a point. This flow is believed to be a fast dynamo. Because the chaotic regions which give stretching are small, it is quite difficult to show it is a fast dynamo numerically. However, in Fig. 26, where the growth rate γ has been computed numerically, it seems to be tending to 0.07 at large Rm.

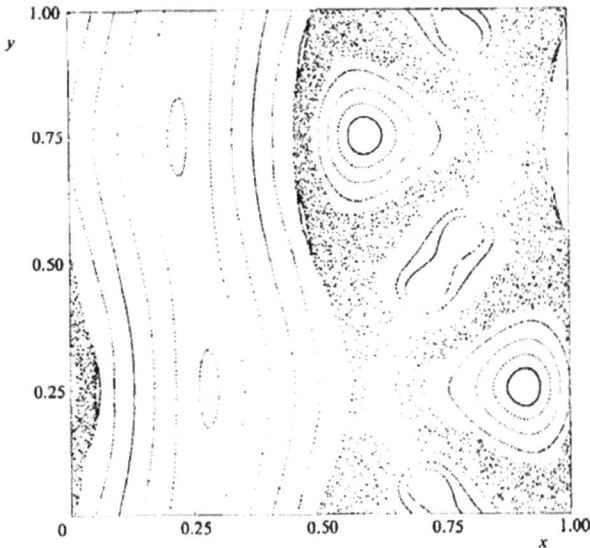

Fig. 25. Typical Poincaré section shows chaotic regions and ordered regions. Ordered regions called KAM regions, or KAM tori. 'Normal' in chaotic ODEs. The $ABC = 1$ flow is unusual in having rather large KAM regions and small chaotic regions; from [14].

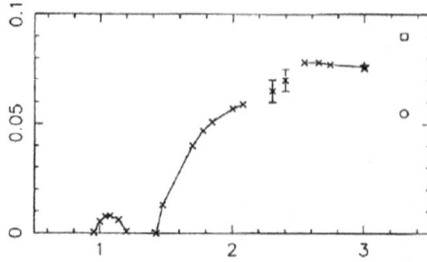

Fig. 26. Dynamo growth rate γ against $\log_{10} Rm$. The circle is the flux growth rate, and the square is h_{line}, which are defined in the text; from [20].

4.5. Stretching properties

Take a point $\mathbf{a}$ and a small vector $\mathbf{v}$ whose origin is at $\mathbf{a}$. Now integrate the particle path ODE's with initial conditions $\mathbf{x_1} = \mathbf{a}$ and $\mathbf{x_2} = \mathbf{a} + \mathbf{v}$, and monitor $d = |\mathbf{x_1} - \mathbf{x_2}|$. If stretching is occurring, d will grow exponentially.

The Lyapunov exponent is

$$\Lambda(\mathbf{a}) = \max_{\mathbf{v}} \lim_{t \to \infty} \sup \frac{\ln d}{t}. \tag{4.5.1}$$

The maximum Lyapunov exponent is found by taking the supremum over all $\mathbf{a}$. It can be computed, but it is expensive! A practical definition of chaos is that the Lyapunov exponent is positive. In a given chaotic region, Λ is usually the same for different trajectories in that region, but Λ is zero in KAM regions.

4.5.1. Line stretching
As a material curve is carried round by a chaotic flow, its length increases. If the length of the curve C is $M(C, t)$, then

$$h_{line} = \sup \lim_{t \to \infty} \frac{1}{t} \ln |M(C, t)|. \tag{4.5.2}$$

h_{line} is generally larger than Λ because the line gets twisted up, whereas the infinitesimal vectors $\mathbf{v}$ stay straight.

We might reasonably expect that fast dynamo growth rates will be a bit less than h_{line}, because diffusion acting over the short length scales will make the field grow a bit slower. This seems to be true of the known fast dynamos, and h_{line} is shown in Fig. 26.

4.5.2. Flux growth rate
Alfvèn's theorem says flux is conserved at large Rm so it is natural to define the flux growth rate. Choose a material surface S and define the flux through it as

$\Phi(S, t) = \int_S \mathbf{B} \cdot d\mathbf{S}$. Then the flux growth rate is

$$\Gamma(S, \epsilon) = \lim_{t \to \infty} \sup \frac{1}{t} \log |\Phi(s, t)|. \tag{4.5.3}$$

This may exist even at $\epsilon = 0$, infinite Rm. In a given chaotic region, Γ becomes independent of S and the initial choice of $\mathbf{B}$. Clearly, there is tricky mathematical analysis involved in these definitions, and proofs are complicated.

4.6. Time dependent flow fields

An alternative way of generating fast dynamos is rather than have a fully 3D steady flow field, like $A = B = C = 1$, stay 2D but have a time-dependent flow. Field still has $\exp ikz$ dependence, which makes the numerics a lot easier.

The Galloway–Proctor CP (circularly polarized) flow is an example of this approach:

$$\mathbf{u} = \nabla \times (\psi(x, y, t)\hat{\mathbf{z}}) + \gamma \psi(x, y, t)\hat{\mathbf{z}},$$

$$\psi = \sin(y + \sin t) + \cos(x + \cos t). \tag{4.6.1}$$

It is very like the G.O. Roberts flow, except the stagnation point pattern rotates round in a circle. Galloway and Proctor also defined an LP (linearly polarized) flow, with $\psi = \sin(y + \cos t) + \cos(x + \cos t)$. These flows are now non-integrable, and they have positive Lyapunov exponents. In Fig. 27 the Lyapunov exponents and the normal field B_z for the CP flow of the Galloway–Proctor dynamo are shown. As in the ABC flow there are regions of strong stretching where the Lyapunov exponent is strongly positive, and other regions where not

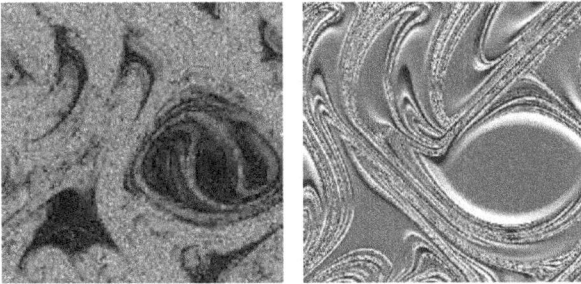

Fig. 27. Left: Lyapunov exponents: blue regions have little or no stretching, green/red has order one stretching. Right: snapshot of the normal field B_z. Note the good correlation between strong field and strong stretching; courtesy of D.W. Hughes and F. Cattaneo.

C.A. Jones

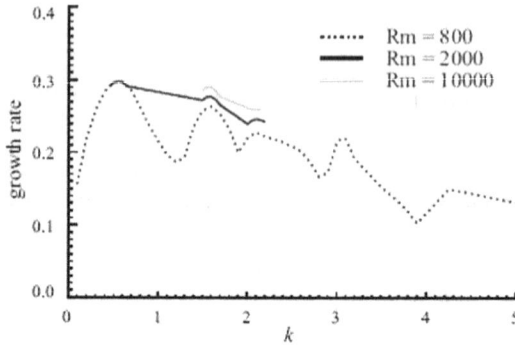

Fig. 28. Galloway–Proctor dynamo results for the CP flow. Growth rate against wavenumber k, $\exp ikz$, for $Rm = 800$, $Rm = 2,000$ and $Rm = 10,000$; from [21].

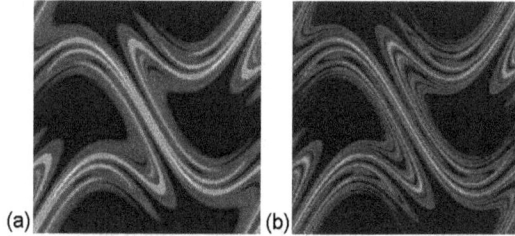

Fig. 29. Otani dynamo. (a) Eigenfunctions $b(x, y, t)$ at a snapshot in t. $\epsilon = Rm^{-1} = 5 \times 10^{-4}$. Actually, magnetic energy $|\mathbf{b}|^2$ is plotted. (b) As above with $\epsilon = Rm^{-1} = 5 \times 10^{-5}$ flow MW+ eigenfunctions; from [40].

much stretching occurs. It is clear from Fig. 27 that the stretching is crucial for generating the field.

In Fig. 28, the growth rate for the CP flow dynamo is shown as a function of k. This suggests that the optimum value of k does not change as $Rm \rightarrow \infty$, and the growth rate at the maximum k has attained its asymptotic value of about 0.3 by $Rm = 1000$.

Another time-dependent model, similar to the Galloway–Proctor model is Otani's modulated wave and flow dynamo, known as the MW+ flow dynamo. The flow is a little simpler than the Galloway–Proctor flow,

$$\mathbf{u} = 2\cos^2 2t\,(0,\,\sin x,\,\cos x) + 2\sin^2 t\,(\sin y,\,0,\,-\cos y). \tag{4.6.2}$$

The magnetic field has the form

$$\mathbf{B}(x, y, z, t) = \exp(ikz + \sigma t)\mathbf{b}(x, y, t), \tag{4.6.3}$$

where σ is the Floquet exponent and b is a 2π periodic function of time. There is rapid convergence of $Re(\sigma) = \gamma \approx 0.39$ with $k \approx 0.8$ as $Rm \rightarrow \infty$. The dynamo mechanism appears to be an SFS type.

5. Nonlinear dynamos

5.1. Basic ideas in nonlinear dynamos

The induction equation is linear in **B**, so it predicts dynamos that either decay or grow for ever. The field strength at which the dynamo stops growing is determined by terms nonlinear in **B**. The Lorentz force $\mathbf{j} \times \mathbf{B} = (\nabla \times \mathbf{B}) \times \mathbf{B}/\mu$ is the key nonlinear term.

Nonlinear dynamos therefore require analysis of the equation of motion. The dynamo stops growing when the Lorentz force changes the flow so that dynamo action is reduced. This process is called dynamo saturation. In the simplest models, Lorentz force only drives a large-scale flow, the dynamo being driven by an α-effect. More recently, models have been developed without an α-effect, in which the flow is driven by either a body force or by convection. An example of such a nonlinear dynamo is the plane layer dynamo driven by rotating convection at fixed Rayleigh number. In Fig. 30, the logarithm of the Magnetic Energy is plotted against time measured on the magnetic diffusion timescale. For time less than 0.5, the logarithm of the energy grows approximately linearly, which corresponds to the exponential growth we would expect with a kinematic dynamo. In this convection-driven model, the flow is chaotic, which accounts for why the en-

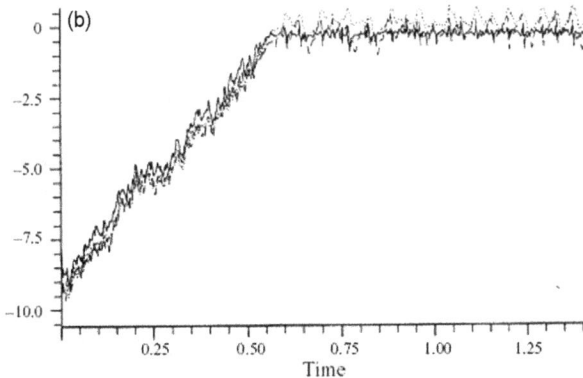

Fig. 30. A plane layer dynamo model is driven by rotating convection at fixed Rayleigh number. Log(Magnetic Energy) plotted against time on magnetic diffusion timescale; from [30].

ergy fluctuates over short times, but the overall exponential growth is quite clear. At about $t = 0.5$ the dynamo stops growing, i.e. it saturates. This happens when the field strength is large enough to change the flow significantly.

5.1.1. Dynamical regimes

Saturation is poorly understood, and is probably different in different dynamical regimes. Here we focus on three problems: (i) dynamo saturation in moderately rotating systems, e.g. the Sun. (ii) dynamo saturation in rapidly rotating systems, dominated by Coriolis force, e.g. the Earth's core, and (iii) dynamo saturation in experiments.

The essential difference between (i) and (ii) is whether the rotation rate is fast or slow compared to the flow turnover time. The Rossby number is defined by $U/L\Omega$, and is the dimensionless measure of this difference. In the interior of the Sun, Ro is about unity (it is larger near the surface). In the Earth's core it is $\sim 10^{-7}$, corresponding to very rapid rotation.

Dynamo experiments lie in a different regime again, because although they are strongly turbulent, Rm is only just above critical.

5.2. Stellar dynamo saturation mechanisms

Three different mechanisms of saturation have been proposed for stellar dynamos:

(i) omega-quenching, (ii) magnetic buoyancy, (iii) alpha-quenching.

Omega-quenching: in most solar dynamo models, differential rotation generates toroidal field from poloidal field. The Lorentz force acts to stop the differential rotation, because the tension in the field lines opposes the shear.
Magnetic buoyancy: a magnetic flux tube is lighter than its surroundings. Magnetic pressure in the tube means the gas pressure is reduced. From the gas law, this means the density is reduced, assuming thermal equilibrium. Flux tubes therefore float upwards, removing themselves from the active dynamo region.
Alpha-quenching: the magnetic field will stop the helical small-scale motions that create the mean field. We therefore expect the helicity to drop when the field strength is large, and thus dynamo action to cease. This is primarily a mean field dynamo mechanism. There is a large and controversial literature on alpha-quenching. Many models suggest that the alpha-effect should be quenched at relatively low field strengths, but nevertheless the Sun appears to achieve strong fields.

In convection-driven dynamos, the field can affect the stretching properties of the flow. Unfortunately, subtle changes in the flow pattern can radically alter stretching properties. The rate of creation of magnetic energy is through $\mathbf{u} \cdot \mathbf{j} \times \mathbf{B}$.

At large Rm, $\mathbf{u}$ and $\mathbf{B}$ are often nearly parallel. This means small changes in the angle between $\mathbf{u}$ and $\mathbf{j} \times \mathbf{B}$ can strongly affect field generation.

5.2.1. Modelling saturation mechanisms

We illustrate how these saturation mechanisms can be modelled in the context of a plane layer $\alpha\omega$ dynamo. From (3.6.4) and (3.6.5),

$$\frac{\partial A}{\partial t} = \alpha B + \eta \nabla^2 A, \qquad \frac{\partial B}{\partial t} = \frac{\partial A}{\partial x}\frac{\partial v}{\partial z} + \eta g \nabla^2 B, \qquad (5.2.1,2)$$

together with the y-component of the equation of motion,

$$\frac{\partial v}{\partial t} = \frac{1}{\mu\rho}\frac{\partial A}{\partial x}\frac{\partial B}{\partial z} + \nu \nabla^2 v + F. \qquad (5.2.3)$$

Here $B\hat{\mathbf{y}}$ is the toroidal field, $\nabla \times A\hat{\mathbf{y}}$ the poloidal field, and v is flow in $\hat{\mathbf{y}}$, so $\partial v/\partial z$ is differential rotation.

To model alpha-quenching and buoyancy set

$$\alpha = \frac{\alpha_0}{1 + \kappa B^2}, \qquad g = 1 + \lambda B^2. \qquad (5.2.4)$$

This is quite arbitrary, but it does model our physical expectations from saturation mechanisms (i) and (ii). As mentioned above, there has been much discussion about the magnitude of κ. The quenching rate shouldn't depend on sign of B, so the appearance of B^2 in the denominator is less controversial. The formula for g reflects magnetic buoyancy giving enhanced diffusion removing flux from the dynamo region.

5.2.2. A truncated system

These equations were originally solved by expanding in trig functions in x and Legendre polynomials in z and severely truncating the expansion so that only the first few terms are retained [57].

$$\dot{A} = 2D(1 + \kappa|B|^2)^{-1} - A, \qquad \dot{B} = i(1 + v_0)A - \frac{1}{2}iA^*v + (1 + \lambda|B|^2),$$

$$(5.2.5,6)$$

$$\dot{v}_0 = \frac{1}{2}i(A^*B - AB^*) - v_0 v_0, \qquad \dot{v} = -iAB - \nu v. \qquad (5.2.7,8)$$

Here A, B are complex coefficients of $\exp ikx$, because the system has dynamo waves. v_0 is the mean differential rotation, complex v the $\exp 2ikx$ part. We can explore various combinations of κ, λ, ν and v_0. D is the dynamo number, set above critical.

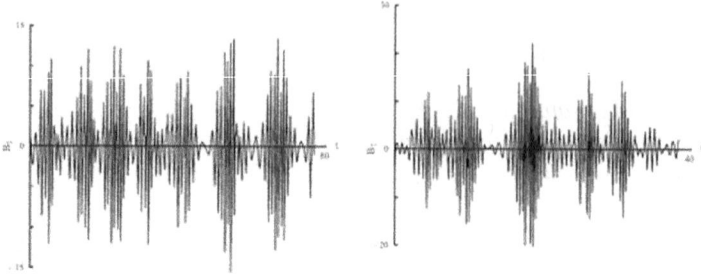

Fig. 31. Aperiodic oscillations of 6th order system (5.2.5–8) with $v_0 \to \infty$ for $D = 8$ and $D = 16$; from [57].

With ω-quenching switched off, stable periodic dynamo waves result. Stellar observations suggest $|B|$ increases with the convective velocity, and hence with D, and the frequency increases also. In the α-quenched model, the cycle period remains constant with D, though this result is somewhat model-dependent. With ω-quenching only, particularly when v dominates v_0, we can get chaotic behaviour (see Fig. 31), including 'Maunder minima'. The Maunder minimum was a period of about 70 years when there was reduced dynamo activity. The suggestion from this model, therefore, is that it was due to a reduced differential rotation, possibly at the bottom of the convection zone.

5.3. α-quenching

Writing the α-quenching formula as

$$\alpha = \frac{\alpha_0}{1 + B^2/\mathcal{B}^2} \tag{5.3.1}$$

highlights the fact that quenching occurs when $B \sim \mathcal{B}$. The value of $\mathcal{B}$ is more important than the exact functional form.

The simplest argument is to consider the force balance in turbulent convection. The buoyancy force $\rho g \alpha T' \hat{\mathbf{z}}$ balances the z-component of $\rho \mathbf{u} \cdot \nabla \mathbf{u} \sim \rho u^2/L$. Lorentz force is significant in this balance when

$$|(\nabla \times \mathbf{B}) \times \mathbf{B}/\mu| \sim B^2/L\mu \sim \rho u^2/L \tag{5.3.2}$$

$B^2/2\mu$ is the magnetic energy per unit volume, $\rho u^2/2$ is the kinetic energy per unit volume, so we expect energy equipartition, which determines $\mathcal{B}$ i.e. $\mathcal{B} = B_{eq} \sim (\rho\mu)^{1/2}u$. Then the Alfvèn speed and the flow speed are comparable.

5.3.1. α-quenching: small or large scale?
It is assumed in this argument that the length scale L is the same in the Lorentz force and the acceleration terms. Also, the acceleration term is really an eddy

diffusion term $\rho v_{turb} \nabla^2 \mathbf{u}$ with $v_{turb} \sim UL$. This still gives $\rho U^2 / L$, but again only by assuming the same L everywhere.

Numerical simulations suggest $\mathcal{B} \ll B_{eq}$, possibly even $\mathcal{B} \sim Rm^{-1/2} B_{eq}$. Why does (5.3.2) not hold in these simulations? The field is generated on short length scales, so the current is B/ℓ, and ℓ is small. Flux expulsion arguments suggest $\ell/L \sim Rm^{-1/2}$, which would imply saturation occurs at very low field strengths, because Rm is large in astrophysics.

5.3.2. α-quenching: magnetic helicity

If $\mathbf{a}$ and $\mathbf{b}$ are small-scale fluctuating components of the magnetic vector potential and magnetic field respectively, with a constant uniform mean field but no mean flow,

$$\frac{\partial \mathbf{a}}{\partial t} = \mathbf{u} \times (\nabla \times \overline{\mathbf{A}}) + \mathbf{u} \times (\nabla \times \mathbf{a}) + \eta \nabla^2 \mathbf{a} + \nabla \phi, \tag{5.3.3}$$

$$\frac{\partial \mathbf{b}}{\partial t} = \nabla \times (\mathbf{u} \times \overline{\mathbf{B}}) + \nabla \times (\mathbf{u} \times \mathbf{b}) - \nabla \times \eta (\nabla \times \mathbf{b}). \tag{5.3.4}$$

Multiply by $\mathbf{b}$ and $\mathbf{a}$ respectively and take the average

$$\frac{\partial}{\partial t} \overline{\mathbf{a} \cdot \mathbf{b}} = -2\overline{\mathbf{B}} \cdot \overline{\mathbf{u} \times \mathbf{b}} - 2\eta \overline{(\nabla \times \mathbf{b}) \cdot \mathbf{b}} + \text{divergence terms.} \tag{5.3.5}$$

Now $\overline{\mathbf{u} \times \mathbf{b}} = \alpha \overline{\mathbf{B}}$ from the definition of α, so in a steady state

$$\alpha = -\eta \mu \frac{\overline{\mathbf{j} \cdot \mathbf{b}}}{\overline{\mathbf{B}}^2}, \tag{5.3.6}$$

which is a small value of α since η is small. This argument ignores the divergence terms, and it has been suggested that possibly helicity leaks through boundary to give a larger value of α. However, at face value this argument supports the idea that α-quenching is controlled by the value of the small scale field rather than the large scale field, as supposed in (5.3.1).

5.3.3. β-quenching

If the Lorentz force limits α by reducing the helicity, we might also expect the turbulent diffusion to be reduced. This is called β-quenching. It is an open question whether it exists, and whether the $\mathcal{B}$ for α-quenching is the same for β-quenching.

Also, nonlinear β effects may be highly anisotropic, whereas in most applications to solar and stellar dynamos simple isotropic formulae are used. Both α and β quenching are the subject of ongoing numerical experiments.

5.4. Saturation in rapidly rotating systems

The motivation here is the dynamos in planetary cores. The fluid velocity is then very slow, so inertia is probably negligible, except on the smallest length scales. Viscosity is also small, so the dominant balance is between magnetic forces, buoyancy and Coriolis force. Archimedean for buoyancy gives the acronym MAC balance. Pressure is also important! Numerical simulations won't work without viscosity, but the aim is to reduce the viscous terms to as small a value as possible, which means getting the Ekman number $E = \nu/\Omega L^2 \ll 1$.

5.4.1. Busse rolls

In rapidly rotating convection, the convective rolls are tall and thin, see Fig. 32. At onset, they have thickness $E^{1/3}a$, the sphere radius being a. In the nonlinear regime, we still get thin columns though they are very time-dependent. The roll-width may then be controlled by the balance of inertial advection and vortex stretching, which leads to the Rhines length-scale, $(Ua/\Omega)^{1/2}$. As mentioned

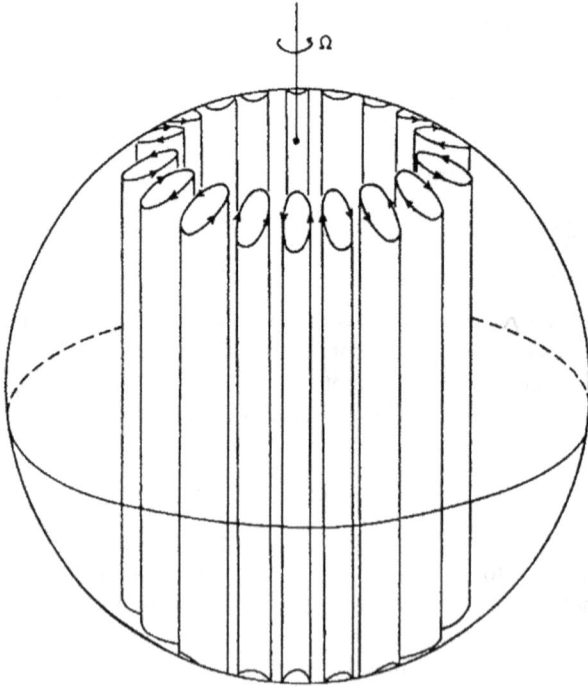

Fig. 32. Sketch of convection rolls in a rapidly rotating sphere; from [5].

above, because of the low velocity, the Rhines length-scale in planetary cores is still rather small, so this picture still has the convection occurring in tall thin columns.

5.4.2. J.B. Taylor's constraint

If we integrate the ϕ-component of the equation of motion over a co-axial cylinder S (see Fig. 33),

$$\frac{\partial}{\partial t}\int \rho u_\phi \, ds + \int 2\rho u_s \Omega \, ds = \int (\mathbf{j} \times \mathbf{B})_\phi \, ds - 2\pi s \frac{u_\phi (2E)^{1/2}}{(1-s^2)^{1/4}} \qquad (5.4.1)$$

ignoring any Reynolds stress. The last term comes from the drag at the boundary produced by Ekman suction. The Coriolis term is zero, because there is no net flow across the cylinder. The pressure term cancels out because the pressure at $\phi = 2\pi$ is the same as at $\phi = 0$. The buoyancy term has no ϕ-component.

On a short timescale, (5.4.1) predicts torsional oscillations, damped by the Ekman suction term. On longer timescales, after the oscillations have decayed

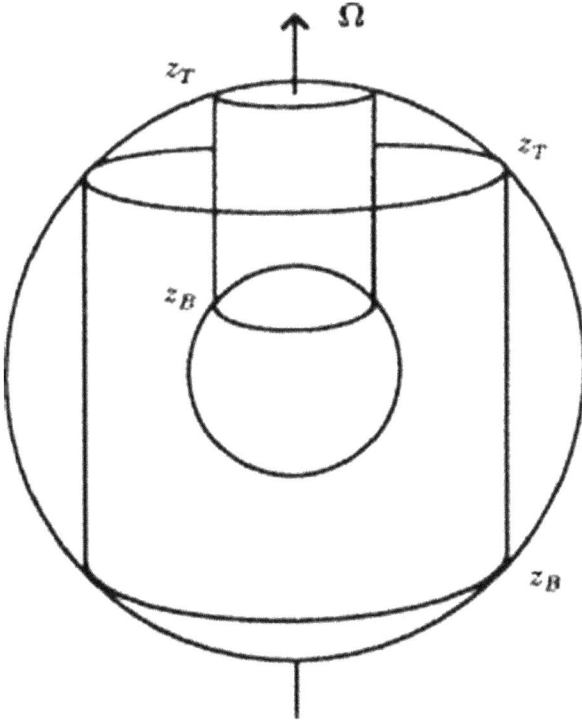

Fig. 33. Sketch of the co-axial cylinder used for Taylor's constraint.

C.A. Jones

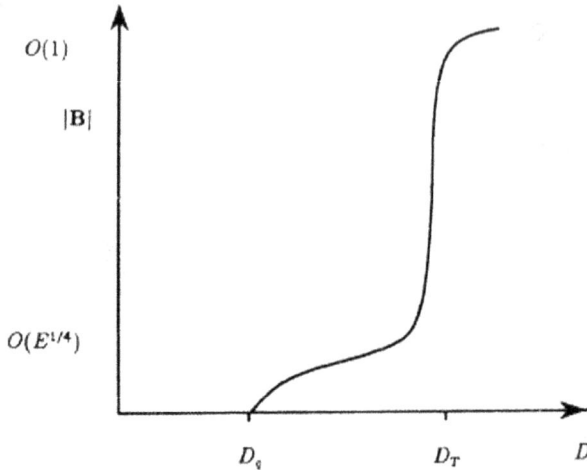

Fig. 34. Sketch of the field as a function of dynamo number, illustrating the Malkus–Proctor scenario; [28].

away,

$$\int (\mathbf{j} \times \mathbf{B})_\phi \, ds = 0, \tag{5.4.2}$$

which is called Taylor's constraint. We have ignored viscosity, since as mentioned above, it is very small in planetary cores.

So what happens if a dynamo generated field doesn't satisfy Taylor's constraint (5.4.2)? In general, the field distribution is given by the growing eigenfunctions of the dynamo problem, so there is no reason (5.4.2) will be satisfied. The torque on each cylinder then has to be balanced by viscosity (Ekman suction), so the field strength is small, $O(E^{1/4})$. However, Malkus and Proctor [35] suggested that in the nonlinear regime the field would adjust; the Lorentz force alters the large scale flow until the generated field satisfies Taylor's constraint. Then it is no longer necessary to balance the magnetic torque on the cylinder against small viscous terms, and the amplitude of the field rises to reach a new equilibrium as illustrated in Fig. 34. This behaviour is known as the Malkus–Proctor scenario. This problem was solved numerically [52] for an axisymmetric α^2-dynamo, and the Malkus–Proctor scenario was found to occur.

5.4.3. Elsasser number
The new field level reached when Taylor's constraint is satisfied has the Elsasser number $\Lambda \sim O(1)$, i.e.

$$\Lambda = \frac{B^2}{\mu \rho \Omega \eta} \sim 1. \tag{5.4.3}$$

This makes the Coriolis force $2\mathbf{\Omega} \times \mathbf{u}$ the same order as the Lorentz force, provided $U \sim \eta/L$, i.e. $Rm \sim 1$. Also, this is the field-strength level when the magnetic field breaks the Proudman–Taylor theorem and the flow is no longer columnar. The Proudman–Taylor theorem says that steady inviscid motion has $\mathbf{u}$ independent of z. This breaks down when the Lorentz force is significant, which happens at Elsasser number about unity.

5.5. Dynamo models and Taylor's constraint

Numerical dynamo models have viscosity for numerical reasons, and the field is always changing with time. Nevertheless at the lowest Ekman numbers attainable it appears that Taylor's constraint is becoming satisfied. The 'Taylorization'

$$\frac{\int (\mathbf{j} \times \mathbf{B})_\phi \, ds}{\int |(\mathbf{j} \times \mathbf{B})_\phi| \, ds} \tag{5.5.1}$$

is computed, and it does appear to go to zero as E is lowered. The field strength usually has Elsasser number order 1 in rapidly rotating dynamo models.

5.5.1. Equipartition in rapid rotation?

In non-rotating systems, we saw that magnetic and kinetic energy are expected to be approximately equal. In rapidly rotating systems, $B^2/\mu \sim \rho\Omega\eta$. $Rm = UL/\eta$ and $Ro = U/L\Omega$ so $B^2/\mu \sim \rho U^2/(Ro\,Rm)$. In planetary cores, $Ro \sim 10^{-7}$, $Rm \sim 10^3$, so the magnetic energy is much greater than the kinetic energy. There is therefore no equipartition; there is no fundamental law saying that the magnetic and kinetic energies are comparable in electrically conducting fluids. However, it can be difficult to get Ro small in computations, so in simulations there often is approximate equipartition.

5.5.2. Dissipation time

Christensen and Tilgner [10] argued that the time taken to dissipate field in a turbulent medium is

$$\tau_{diss} \sim \frac{\int \mathbf{B}^2 \, dv}{\int \eta (\nabla \times \mathbf{B})^2 \, dv} \sim \frac{L^2}{\eta Rm}, \tag{5.5.2}$$

where L is the integral length scale, the size of the container. This is an empirical law, derived from numerical and laboratory experiments, but it is consistent with the idea that the magnetic field is expelled by eddies into ropes of length $L Rm^{-1/2}$ where it is dissipated by diffusion.

In rapidly rotating convection-driven dynamos, the energy input (the rate of working of buoyancy, which is simply related to the heat flux passing through the system) is mainly dissipated as ohmic dissipation, so

$$\int \eta (\nabla \times \mathbf{B})^2 \, dv \sim \frac{g\alpha F \mu}{c_p},$$

(5.5.3)

where F is heat flux through the convecting system.

If we eliminate the current terms from (5.5.2) and (5.5.3) we get an expression for the magnetic energy in terms of F the heat flux, which leads to an estimate of the field. We obtain

$$B \sim \mu^{1/2} L^{1/2} \left(\frac{g\alpha F}{c_p} \right)^{1/2} \frac{1}{U^{1/2}}$$

which is entirely independent of diffusivities, including η. It does, however, require knowledge of the typical velocity. This is then a different estimate from the Elsasser number order one estimate which is dependent on η. Arguments for the Elsasser number being of order one are based on the flow having moderate Rm, and when this is the case the two estimates are not necessarily in conflict. It is also possible to estimate U in terms of the heat flux, using convection theory arguments. The resulting formula for B is then given in terms of the heat flux through the system: see [29] for details.

5.6. Dynamo saturation in experiments

In dynamo experiments, it is hard to get large Rm. Even if the container contains cubic metres of sodium and is stirred very vigorously, although $Re \gg 1$, typically Rm is only just above critical.

If we ignore the large Re, we can look at a slightly supercritical Ponomarenko dynamo to see at what level it saturates. The required flow is forced. Fauve & Petrelis [16] show that weakly nonlinear theory predicts

$$\mathbf{B}^2 \sim \frac{\rho \nu \eta \mu}{L^2} (Rm - Rm_{crit})$$

a balance between Lorentz force and viscosity. This is a rather small field strength, far from energy equipartition.

What happens at large Re? We expect the viscosity to be unimportant, because of the turbulent cascade. If we use a simple mixing length argument to replace the laminar viscosity ν by an eddy viscosity UL, setting

$$U \sim Rm_{crit} \frac{\eta}{L}$$

gives

$$\mathbf{B}^2 \sim \frac{\rho\eta^2\mu}{L^2} Rm_{crit}(Rm - Rm_{crit})$$

which gives reasonable agreement with experiments. This formula can also be obtained from dimensional arguments. We only get energy equipartition at large Rm.

6. Numerical methods for dynamos

The majority of numerical dynamo simulations are done with spectral methods. Cartesian geometry and spherical geometry are the most commonly used. We illustrate first with Cartesian Geometry. Spectral methods are commonly used for:

(i) Eigenvalue problems for kinematic dynamos.

(ii) Forced nonlinear dynamos: here an arbitrary body force drives the flow.

(iii) Convection driven dynamos.

6.1. The pseudo-spectral method for a convection-driven plane layer dynamo

We take a horizontal plane layer of electrically conducting fluid bounded between $z = \pm 0.5d$. Gravity and the rotation axis are usually in the $\hat{\mathbf{z}}$ direction, though sometimes the rotation axis is tilted. The layer is heated from below, usually with constant temperature on the boundaries.

We assume periodic boundary conditions in the x and y directions. Why do we not also assume periodicity in z? There is trouble with the 'elevator' mode of convection, independent of z, in which fluid elements simply rise and fall without moving sideways. This the preferred mode of convection if there is periodicity in z, and being two-dimensional, it does not give a dynamo (Theorem 2).

We can assume there is electrically insulating material outside the fluid layer, though a perfect conductor is another popular choice. The equation of motion, the induction equation and the temperature equation give

$$\frac{\partial \mathbf{u}}{\partial t} + (\mathbf{u} \cdot \nabla)\mathbf{u} + 2\Omega\hat{\mathbf{z}} \times \mathbf{u} = -\nabla p + \frac{1}{\rho}\mathbf{j} \times \mathbf{B} + g\alpha T\hat{\mathbf{z}} + \nu\nabla^2\mathbf{u}, \qquad (6.1.1)$$

$$\frac{\partial \mathbf{B}}{\partial t} = \nabla \times (\mathbf{u} \times \mathbf{B}) + \eta\nabla^2\mathbf{B}, \qquad (6.1.2)$$

$$\frac{\partial T}{\partial t} + \mathbf{u} \cdot \nabla T = \kappa \nabla^2 T + H, \tag{6.1.3}$$

H being the heat sources if any.

$$\nabla \cdot \mathbf{B} = 0, \qquad \nabla \cdot \mathbf{u} = 0. \tag{6.1.4a, 6.1.4b}$$

The temperature is fixed at ΔT at $z = -0.5$, 0 and $z = 0.5$. The boundaries are assumed to be either no-slip or stress-free.

6.1.1. Dimensionless plane layer equations
The unit of time is taken as d^2/η, length d, the unit of field $(\Omega \mu \eta \rho)^{1/2}$, the unit of temperature ΔT, and (6.1.1)–(6.1.3) become

$$E_m \left[\frac{\partial \mathbf{u}}{\partial t} + (\mathbf{u} \cdot \nabla)\mathbf{u} \right] + 2\hat{\mathbf{z}} \times \mathbf{u} = -\nabla p + (\nabla \times \mathbf{B}) \times \mathbf{B} + q\,RaT\hat{\mathbf{z}} + E\nabla^2\mathbf{u}, \tag{6.1.5}$$

$$\frac{\partial \mathbf{B}}{\partial t} = \nabla \times (\mathbf{u} \times \mathbf{B}) + \nabla^2 \mathbf{B}, \tag{6.1.6}$$

$$\frac{\partial T}{\partial t} + \mathbf{u} \cdot \nabla T = q\nabla^2 T, \tag{6.1.7}$$

$$E = \frac{\nu}{\Omega d^2}, \quad E_m = \frac{\eta}{\Omega d^2}, \quad Pm = \frac{\nu}{\eta}, \quad q = \frac{\kappa}{\eta} = \frac{Pm}{Pr}, \quad Ra = \frac{g\alpha \Delta T d}{\Omega \kappa}. \tag{6.1.8}$$

6.1.2. Toroidal-poloidal expansion
As outlined in Section 1.7 we decompose the velocity and the magnetic field into toroidal and poloidal parts

$$\mathbf{u} = \nabla \times e\hat{\mathbf{z}} + \nabla \times \nabla \times f\hat{\mathbf{z}} + U_x(z, t)\hat{\mathbf{x}} + U_y(z, t)\hat{\mathbf{y}}, \tag{6.1.9}$$

$$\mathbf{B} = \nabla \times g\hat{\mathbf{z}} + \nabla \times \nabla \times h\hat{\mathbf{z}} + b_x(z, t)\hat{\mathbf{x}} + b_y(z, t)\hat{\mathbf{y}}, \tag{6.1.10}$$

where $U_x(z, t)$, $U_y(z, t)$, $b_x(z, t)$, $b_y(z, t)$ are the mean parts. We have five scalar fields to solve for, e, f, g, h and T, as well as the mean parts. The divergences of $\mathbf{u}$ and $\mathbf{B}$ are automatically zero. The required equations are formed by taking the z-components of the curl and the double-curl of the momentum equation, and the z-component of the induction equation and its curl, and the temperature equation (6.1.8).

6.1.3. Toroidal-poloidal equations
We obtain

$$-E_m \frac{\partial}{\partial t} \nabla_H^2 e + \nabla_H^2 \frac{\partial f}{\partial z} + E \nabla^2 \nabla_H^2 e = F_1, \tag{6.1.11}$$

$$E_m \frac{\partial}{\partial t} \nabla_H^2 \nabla^2 f + \nabla_H^2 \frac{\partial e}{\partial z} - E \nabla^4 \nabla_H^2 f = F_2 - q Ra \nabla_H^2 \theta, \tag{6.1.12}$$

$$\nabla_H^2 \frac{\partial h}{\partial t} - \nabla^2 \nabla_H^2 h = -G_1, \qquad \nabla_H^2 \frac{\partial g}{\partial t} - \nabla^2 \nabla_H^2 g = -G_2, \tag{6.1.13}$$

$$F_1 = \hat{\mathbf{z}} \cdot \nabla \times (\mathbf{j} \times \mathbf{B}), \qquad F_2 = \hat{\mathbf{z}} \cdot \nabla \times \nabla \times (\mathbf{j} \times \mathbf{B}), \tag{6.1.14}$$

$$G_1 = \hat{\mathbf{z}} \cdot \nabla \times (\mathbf{u} \times \mathbf{B}), \qquad G_2 = \hat{\mathbf{z}} \cdot \nabla \times \nabla \times (\mathbf{u} \times \mathbf{B}), \tag{6.1.15}$$

$$E_m \frac{\partial}{\partial t} U_x - U_y = E \frac{\partial^2 U_x}{\partial z^2} + \hat{\mathbf{x}} \cdot < \mathbf{j} \times \mathbf{B} >, \tag{6.1.16}$$

$$\frac{\partial b_x}{\partial t} = \frac{\partial^2 b_x}{\partial z^2} + \hat{\mathbf{x}} \cdot < \nabla \times (\mathbf{u} \times \mathbf{B}) >, \tag{6.1.17}$$

and similarly for U_y and b_y. ∇_H^2 is the horizontal part of ∇^2.

6.1.4. Fourier decomposition
The scalar functions are expanded as

$$e = \sum_{l=-N_x+1}^{N_x} \sum_{m=-N_y+1}^{N_y} \sum_{n=1}^{N_z+2} e_{lmn} \exp i(l\alpha x + m\beta y) T_{n-1}(2z) \tag{6.1.18}$$

and similarly for the other five variables, except that the f expansion runs up to $N_z + 4$. The T_n are the Chebyshev polynomials. Chebyshev polynomials are preferred for no-slip boundaries. We could use $\sin n\pi(z + 0.5)$ for stress-free boundaries. Note that the e_{lmn} coefficients are complex functions of time, with $e_{l,m,n} = e^*_{-l,-m,n}$ since e is real. $\alpha = \beta = 2\pi$ gives the special case of a cubical periodic box.

6.1.5. Boundary conditions

Mechanical boundary conditions
No-slip conditions are

$$e = f = \frac{\partial f}{\partial z} = U_x = U_y = 0 \quad \text{on } z = \pm \frac{1}{2}. \tag{6.1.19}$$

$e = 0$ and $\partial f/\partial z = 0$ ensure that $u_x = u_y = 0$, and $f = 0$ implies $\nabla_H^2 f = u_z = 0$.

Magnetic boundary conditions

$j_z = 0$ implies $\nabla_H^2 g = 0$ outside fluid. So letting $a^2 = \alpha^2 l^2 + \beta^2 m^2$, then $-a^2 g = 0$ so $g = 0$ in the insulator.

The $l = m = 0$ modes are replaced by the mean fields b_x and b_y. Since $j_x = j_y = 0$ outside layer, $db_x/dz = db_y/dz = 0$, so b_x and b_y are constant in space fields outside the layer. If they are non-zero, this corresponds to a uniform externally imposed field. If we want self-excited dynamo action, we do not want to impose an external field, so we take $b_x = b_y = 0$ at the boundaries.

Since $\mathbf{j} = 0$ in insulators, $\hat{\mathbf{z}} \cdot \nabla \times \mathbf{j} = 0$. We have already shown $g = 0$ outside the layer, so $\hat{\mathbf{z}} \cdot \text{curl}^4 h \hat{\mathbf{z}} = 0$. From the vector identity curl curl = grad div - ∇^2, we deduce $\hat{\mathbf{z}} \cdot \text{curl}^2 A \hat{\mathbf{z}} = -\nabla_h^2 A$, so $-a^2 \nabla^2 h = 0$, so $\nabla^2 h = 0$ outside the layer.

The solution of $\nabla^2 h = 0$ outside the layer for each Fourier mode is

$$h = \exp i(l\alpha x + m\beta y) \exp -az, \quad z > 0.5, \tag{6.1.20}$$

$$h = \exp i(l\alpha x + m\beta y) \exp +az, \quad z < -0.5. \tag{6.1.21}$$

To make B_z continuous at the boundaries, $\nabla_H^2 h$ is continuous, so each component of h is continuous. Since B_x and B_y are continuous, so is j_z, so $j_z = 0$ at $z = \pm 0.5$ just inside the fluid, so $\nabla_H^2 g = 0$ so $g = 0$ at $z = \pm 0.5$.

Since $g = 0$ at boundaries, continuity of $B_x = \partial^2 h/\partial x \partial z$ and $B_y = \partial^2 h/\partial y \partial z$ implies continuity of $\partial h/\partial z$ as well as continuity of h. Outside we have the exact solution, so we deduce that

$$b_x = 0, \qquad b_y = 0, \qquad g_{lm} = 0, \qquad \frac{\partial h_{lm}}{\partial z} = \mp a h_{lm}, \qquad \text{on } z = \pm \frac{1}{2}$$
$$\tag{6.1.22}$$

are the required magnetic boundary conditions.

We see that the toroidal and poloidal expansions with Fourier decomposition gives very convenient boundary conditions, and this is a major advantage of the spectral method. The same ideas works in spherical geometry, and very simple boundary conditions result. In other geometries things are not so simple. If we take for example a finite cube rather than a periodic layer, there is no simple solution for the external field. We must either solve numerically $\nabla^2 h = 0$ outside the box, or write the boundary conditions in terms of surface integrals using Green's functions.

6.1.6. Collocation points

The remaining equations for the coefficients e_{lmn} and those for the other variables are derived by requiring that the equations (6.1.11)–(6.1.17) are solved exactly at collocation points. These collocation points in z are chosen as the N_z zeroes of

$T_{Nz}(2z)$. This bunches the points near the boundaries, where extra resolution is needed, and it allows the use of the Fast Fourier Transform. Since

$$T_n(x) = \cos n\theta, \quad x = \cos\theta \qquad (6.1.23)$$

the collocation points are uniformly spaced in θ. So evaluation of the expansions at the collocation points, and the inverse transform, are only $O(N_z \log_2 N_z)$ operations.

We also need collocation points in x and y. These are uniformly spaced. Sometimes 'de-aliasing' used, that is we take 3/2 as many mesh-points in x and y as needed, to improve accuracy; see [2] for details.

6.1.7. Pseudo-spectral method

The essence of pseudo-spectral method is that all differentiating is done in spectral space, and all multiplication is done in physical space. Linear terms and nonlinear terms are treated differently. The linear parts use an implicit Crank–Nicolson scheme, to avoid the very small time-steps' needed for explicit schemes for the diffusion equation. So

$$\frac{f_{lmn}^{t+1} T_{n-1}(2z_i) - f_{lmn}^{t} T_{n-1}(2z_i)}{\delta t}$$
$$= \frac{f_{lmn}^{t+1} (T_{n-1}''(2z_i) - a^2 T_{n-1}(2z_i)) + f_{lmn}^{t} (T_{n-1}''(2z_i) - a^2 T_{n-1}(2z_i))}{2} + \cdots .$$

$$(6.1.24)$$

For each l, m, this is $1 \leq i \leq N_z$ equations for the $1 \leq n \leq N_z$ unknowns for each of the five scalar fields.

We can invert the relevant matrices at the start of the calculation to get

$$f_{lmn}^{t+1} = M_{lmnn'} f_{lmn'}^{t} + N_{lmn}, \qquad (6.1.25)$$

where N denotes the nonlinear terms. Since the nonlinear terms couple different l and m together, we evaluate quantities such as $\hat{\mathbf{z}} \cdot \nabla \times (\mathbf{u} \times \mathbf{B})$ by evaluating components of $\mathbf{u}$ and $\mathbf{B}$ on the mesh, multiply together and FFT the result, including the Chebyshev transforms. Any curls are then done in Fourier space.

The nonlinear terms are evaluated at time t, and previous values $t-1$ are stored. A scheme such as Adams–Bashforth can be used, but there are advantages in using a predictor-corrector scheme. In these, the nonlinear terms are found at time t, then a predictor value of f^{t+1} is found. The nonlinear terms are evaluated again using these predicted values, and the result is averaged with the original nonlinear terms.

The advantage of the predictor-corrector method is that it allows time-step adjustment. If the difference between the predictor and corrector is unacceptably large, we reduce the timestep, but if it is tiny we can increase the timestep. Many different schemes have been used successfully. For example, the Coriolis and buoyancy terms are linear, but they couple different scalars together. Often it is more convenient to include them with the nonlinear terms, to reduce the size of matrices we have to invert at the start of the calculation. The slowest part of the calculation is usually evaluating the Fourier transforms. It is therefore crucial to do this efficiently with high quality FFT routines.

All dynamo calculations are demanding on computational resources. Multiprocessor clusters are therefore commonly used. It is then necessary to parallelize the code. This can be done by keeping all the y, z transforms on one processor, and storing each x-point on an individual processor. This does however require some all-to-all communication, so clusters with fast interconnects are essential for spectral methods. Gigabit Ethernet doesn't work so well. Using fast interconnects, pseudo-spectral codes can scale almost perfectly up to 1024 processors, that is the time taken on N processors is proportional to $1/N$.

6.2. Methods for kinematic dynamos

Similar methods can be used for solving kinematic dynamo problems. Typically, the problem is periodic in all 3 directions, so Fourier expansion in all three directions is used. If the velocity only involves periodic functions with a small number of Fourier components, e.g. for dynamos of G.O. Roberts type, then the coupling between magnetic field components is simple: a flow $\exp i\alpha x$ only couples a field $\exp im\alpha x$ to $m + 1$ and $m - 1$ and its conjugates.

The resulting matrices are therefore relatively sparse. For steady flows, eigenvalues can be found using inverse iteration or a related method such as Arnoldi iteration. Inverse iteration involves iteratively solving

$$(A - \sigma_0 \mathbf{I})\mathbf{x}_{n+1} = \mathbf{x}_n.$$

This ends up with $\mathbf{x}_n$ tending to the eigenvector with eigenvalue closest to σ_0.

For time-dependent velocities, a time-stepping method can be used to find the fastest growing mode. There are huge computational savings if $\mathbf{B}$ has only $\exp ikz$ dependence, which is why two-dimensional time-dependent velocity fields are very popular. The numerical problem is then just 2D, and high resolution can be achieved.

6.3. Hyperdiffusion

For a given set of parameters, how do we decide how many modes to expand in, i.e. how to choose the truncation levels N_x, N_y and N_z? We can look at the energy spectrum. Modes with l, m and n close to N_x, N_y and N_z should have amplitudes at least $\sim 10^{-3}$ of the amplitudes of the low order modes. Amplitudes drop off exponentially fast when the solution is properly converged, so this another useful test to check there is adequate resolution.

Fully resolved solutions are essential. If the truncation levels are set too low, all sorts of spurious behaviour can occur. Many of the early numerical dynamos were subsequently shown not to be dynamos at all when the truncation levels were increased. However, large N_x, N_y and N_z slows the code down: are there ways of reducing the number of modes without damaging the solution?

A frequently used device is hyperdiffusion, which replaces diffusion terms such as $\nabla^2 f$ with a formula such as $(\nabla^2 - \lambda^2 \nabla^4) f$. Here λ is small, so the higher order derivatives only switch in for high order modes. Hyperdiffusion can be used on velocity, temperature or magnetic field. Because $m^4 \alpha^4$ gets larger much faster than $m^2 \alpha^2$ as $m \to \infty$, the higher order terms get damped out quickly, making the energy spectrum look much nicer. In practice, hyperdiffusive schemes $\nabla^2 \to \nabla^2 - \lambda^2 \nabla_H^4$ are used, because this doesn't raise the order of the system in the z-direction. If you do raise the order in z, extra boundary conditions are required and it is not obvious what they should be. The same issue arises in spherical geometry, so that in spherical codes if hyperdiffusion is used it is applied only in the θ and ϕ directions. Using ∇_H^4 then introduces artificial anisotropy, diffusion in x and y being larger than in z. Potentially this can change the nature of the solution. In spherical geometry, hyperdiffusion seems to enhance the toroidal field compared with the poloidal field, so hyperdiffusive dynamos are rather different to non-hyperdiffusive dynamos. Also, dynamo action can depend on small-scale behaviour, which may be wiped out by the hyperdiffusion.

6.4. LES models

Since it is very expensive to have high resolution, Large Eddy Simulation methods are being developed. The idea is that we model the very small scale behaviour by extra terms in the equations rather than by including all the modes that represent them explicitly as in DNS, Direct Numerical Simulation. Currently there are many models under consideration, e.g. the LANS-α model which has been applied to dynamo theory. It is too early to say how successful they will be.

6.4.1. Similarity model
One of the more promising LES models is the similarity model e.g. [7] and references therein. Here the equations are filtered at a length-scale Δ, with a Gaussian filter the most popular choice,

$$\bar{\mathbf{u}}(\mathbf{x}, t) = \int G_\Delta(\mathbf{x} - \mathbf{r}) \mathbf{u}(\mathbf{r}, t) d\mathbf{r}, \quad G_\Delta(\mathbf{x}) = \sqrt{\frac{6}{\pi}} \frac{1}{\Delta} \exp\left(-\frac{6\mathbf{x}^2}{\Delta^2}\right). \quad (6.4.1)$$

Then terms such as

$$\tau^B = \overline{\mathbf{uB}} - \overline{\mathbf{u}}\,\overline{\mathbf{B}} - (\overline{\mathbf{Bu}} - \overline{\mathbf{B}}\,\overline{\mathbf{u}}) \tag{6.4.2}$$

appear in the induction equation, and similar terms appear in the momentum equation and the heat equation. How do we represent these terms?

The idea of the similarity method is to introduce two filtering operations, $\overline{\mathbf{u}}$ at scale Δ and $\tilde{\mathbf{u}}$ at $\lambda\Delta$, $\lambda > 1$. We write

$$\tau_{sim}^B = C_{ind}(\widetilde{\overline{\mathbf{uB}}} - \widetilde{\overline{\mathbf{u}}}\,\widetilde{\overline{\mathbf{B}}} - (\widetilde{\overline{\mathbf{B}}}\,\widetilde{\overline{\mathbf{u}}} - \widetilde{\overline{\mathbf{B}}}\,\widetilde{\overline{\mathbf{u}}})) \approx \overline{\mathbf{uB}} - \overline{\mathbf{u}}\,\overline{\mathbf{B}} - (\overline{\mathbf{Bu}} - \overline{\mathbf{B}}\,\overline{\mathbf{u}}) \tag{6.4.3}$$

with similar expressions for the small-scale heat flux, Lorentz and Reynolds stresses. Since we do the simulation on the $\overline{\mathbf{u}}$ variables, we can numerically evaluate terms like $\widetilde{\overline{\mathbf{uB}}}$. We then assume similarity, i.e. the small-scale terms are a constant multiple of the computed term. This assumes that the small-scale behaviour at the highest order modes in the simulation, between scales $\overline{\mathbf{u}}$ and $\tilde{\mathbf{u}}$, behaves similarly to the behaviour of the scales we can't compute below the scale $\overline{\mathbf{u}}$.

6.4.2. Dynamical similarity model

The similarity model can give sensible results, even with an 'ad hoc' choice of similarity constant C_{ind}. The idea has been developed [7] by taking a third filtering scale, which allows the small-scale stresses to be found at two different length-scales. We can then get an estimate for C_{ind} at each timestep by the similarity argument. This is an attractive philosophy, but it can get a bit expensive, as a great many filtering operations have to be done at each timestep. Also, it won't work if there is 'new physics' operating at a length scale below our smallest filter level.

6.5. Finite volume methods

The spectral method is used by most researchers working on numerical dynamos, but finite volume method codes have also been developed. The domain is divided into discrete volumes. Often fully implicit schemes are used to advance the variables in each volume. Sometimes just one variable value is used in each cell, more often a low order polynomial representation is used in each cell as in spline methods.

Differentiation is less accurate in finite volume methods, so it is avoided as much as possible. This means that the primitive equations are used rather than the toroidal-poloidal decomposition. The variables are then $\mathbf{u}$, $\mathbf{B}$, T and p, eight scalar fields. It is necessary to correct the magnetic field at each step to ensure $\nabla \cdot \mathbf{B} = 0$, similarly for $\mathbf{u}$. The pressure is computed by taking the divergence of the momentum equation and solving the resulting Poisson equation. The great advantage of finite volume methods is there are only nearest neighbour interactions, so they can be parallelized in a natural way. This means they can be used on distributed memory clusters without fast interconnects. Also we can use fully implicit methods, solving the equations for $\mathbf{u}^{t+1}$ etc. by iterating at each timestep.

The disadvantage is that solving the pressure Poisson equation and the other equations typically takes many iterations at each timestep. Another problem with finite volume methods is the magnetic boundary conditions. Matching on to a decaying solution outside the conducting region is easy in the spectral formulation, but is much less easy with local methods. Often finite volume papers use perfectly conducting boundaries, which are local. Iskakov and Dormy 2005 [31] give a Green's function method for applying boundary conditions for finite volume/difference methods. Nevertheless, finite volume methods are beginning to compete with spectral methods, particularly at extreme parameter values.

6.6. Spherical geometry: spectral methods

Dynamos in spherical shells are also often treated by a toroidal and poloidal expansion in terms of the radial vector as in (1.7.4), and no additional mean field component is required. We expand the poloidal and toroidal scalars into spherical harmonics,

$$Y_{lm} = P_l^{|m|} \cos \theta \exp(im\phi). \tag{6.6.1}$$

The field outside matches onto a decaying field $r^{-l-1} Y_{lm}$, again giving simple boundary conditions. The radial dependence can be treated either by a Chebyshev expansion in polynomials in r, or by finite differences (often 4th order). These schemes are usually parallelised by putting each radial grid-point on a node (or several points per node).

The main problem is the lack of an efficient FFT for the associated Legendre functions. FFT's do exist, but they only compete with the 'slow' method for $N > 256$ or even 512. So the spherical spectral codes are significantly slower than plane layer codes. There is a particular problem for large radius ratio cases: i.e. when $r_{inner}/r_{outer} > 0.75$. Then we need large resolution in latitude, so a large number of spherical harmonics required, and the lack of an effective FFT in the θ direction is really limiting.

6.6.1. Spherical geometry: finite volume/element methods
Additional problems are found in finite volume and finite difference methods because of pole/axis singularities. If the grid is defined on equal latitude and longitude intervals, then points crowd together near the poles, where high resolution

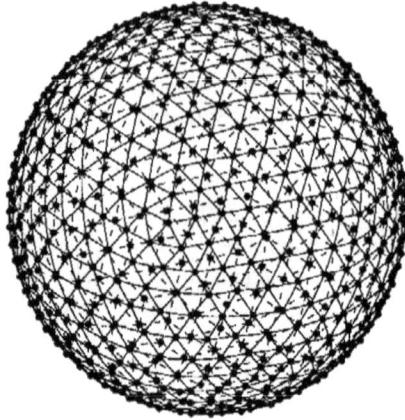

Fig. 35. Finite element grid used by K. Zhang for spherical convection problems.

is not required. This is not only wasteful, it can lead to numerical problems, because very small timesteps are required if the condition that fluid is not advected
further than one grid spacing in one timestep is to be respected. Codes generally
go unstable if this condition is not met. However, spectral methods are less efficient also, so potentially these methods could compete in spherical geometry.
There are some interesting ways of distributing the grid points or finite elements
to avoid the pole singularity problem, for example the grid shown in Fig. 35.

7. Convection driven plane layer dynamos

7.1. Childress–Soward dynamo

An infinite horizontal plane layer of conducting fluid bounded between $z = \pm 0.5$
is considered, so this is the Bénard layer configuration. Gravity and the rotation
axis are in the $\hat{\mathbf{z}}$ direction, and the layer is heated from below.

In the linear approximation, using equations (6.1.11), (6.1.12) and (6.1.7),
rotating convection rolls in a plane layer satisfy

$$\frac{\partial f}{\partial z} + E \nabla^2 e = 0, \tag{7.1.1}$$

$$-\frac{\partial e}{\partial z} + E \nabla^4 f = q \, Ra\theta, \tag{7.1.2}$$

$$u_z + q \nabla^2 \theta = 0, \qquad u_z = -\nabla_H^2 f \tag{7.1.3}$$

since stationary modes with frequency $\omega = 0$ are preferred in this geometry at Prandtl number 1 or greater. We look for a solution

$$u_z = \cos \pi z \exp i\mathbf{k} \cdot \mathbf{x}, \qquad \mathbf{k} = (k_x, k_y, 0), \qquad k^2 = k_x^2 + k_y^2. \qquad (7.1.4)$$

This corresponds to a single convection roll, but in general we can add together any number of such rolls in our linear approximation. For a single roll,

$$f = 1/k^2 \cos \pi z \exp i\mathbf{k} \cdot \mathbf{x}, \qquad e = -\pi/Ek^2(\pi^2 + k^2) \sin \pi z \exp i\mathbf{k} \cdot \mathbf{x},$$
$$(7.1.5, 7.1.6)$$

$$\theta = [1/q(\pi^2 + k^2)] \cos \pi z \exp i\mathbf{k} \cdot \mathbf{x}, \qquad (7.1.7)$$

$$Ra = \frac{E^2(\pi^2 + k^2)^3 + \pi^2}{Ek^2}, \qquad (7.1.8)$$

and minimising Ra over k and taking the limit $E \to 0$ gives

$$k \sim (\pi/E\sqrt{2})^{1/3}. \qquad (7.1.9)$$

Since E is small, k is large, so these are tall thin rolls. The helicity of the roll is

$$H = \mathbf{u} \cdot \nabla \times \mathbf{u} = \frac{\pi}{2Ek^2} \sin 2\pi z. \qquad (7.1.10)$$

Note that it is nonzero. Rotation is needed because the helicity vanishes at large E, the non rotating limit. Also note that the helicity has opposite sign about the midplane.

Childress and Soward [8] investigated whether this flow with non-zero helicity can generate a magnetic field, assuming any field generated is too weak to affect the flow. We use first order smoothing to find $\mathbf{B}'$ from a horizontal mean field $\overline{\mathbf{B}}$, so we solve

$$-\nabla^2 \mathbf{B}' = \overline{\mathbf{B}} \cdot \nabla \mathbf{u}, \qquad \mathcal{E} = \overline{\mathbf{u} \times \mathbf{B}'}. \qquad (7.1.11)$$

If our roll is aligned with y-axis, so $k_y = 0$ we get

$$\mathcal{E}_x = -\frac{\pi \overline{B}_x \sin 2\pi z}{2Ek^4}, \qquad \mathcal{E}_y = \frac{\pi \overline{B}_x \sin 2\pi z}{k^2} \qquad (7.1.12)$$

to leading order. We insert this into

$$\frac{\partial \overline{\mathbf{B}}}{\partial t} = \nabla \times \mathcal{E} + \nabla^2 \overline{\mathbf{B}} \qquad (7.1.13)$$

to see whether the mean field grows or decays. This single roll doesn't give a dynamo, as we might expect from the planar antidynamo theorem, Theorem 2 of Section 1.8. However, we can evaluate $\mathcal{E}$ for two rolls inclined to each other, and for $\mathcal{E}$ we just get the sum of the contributions from each roll. We then find growing solutions of (7.1.13), so we have shown that in some regimes, small-scale convection can lead to a growing large scale field.

7.1.1. Weak field—strong field branches

Adding a uniform horizontal magnetic field to a rotating convecting layer can reduce the critical Rayleigh number, because the constraint imposed by the Proudman–Taylor constraint is relaxed by the Lorentz force. So the dynamo is expected to be subcritical, that is in the presence of a magnetic field, convection and therefore a dynamo can exist when nonmagnetic convection is stable.

The weak field branch found by Childress and Soward exists close to critical, and its amplitude is controlled by viscosity: in the strong field branch the magnetic field amplitude is controlled by the Lorentz–Coriolis balance.

7.2. Numerical simulations

7.2.1. Meneguzzi and Pouquet results

Meneguzzi and Pouquet [36] investigated nonrotating and mildly rotating convection driven dynamos. The Rossby number was about 1, and they had stress-free, perfectly conducting boundaries. $Ra \sim 100 Ra_{crit}$. $q = Pm/Pr \sim 10$. They found dynamo action only for $q = \kappa/\eta > 5$, corresponding to $Rm > Re$. They only considered aspect ratio order 1, i.e. the period of the solution was similar to the layer depth. This means there are only a few rolls in the periodic box. The magnetic field was smaller scale than the velocity field, but there was some mean field. The magnetic energy was typically 0.05 times the kinetic energy.

7.2.2. St. Pierre's dynamo

St. Pierre [53] had stress-free, perfectly conducting boundaries. He considered the rapidly rotating case, with $E \sim 10^{-5}$. The Rayleigh number was a few times critical. He chose $Pr = 1$, $Pm = 2$, and an aspect ratio about 1. He found dynamo action, but not much mean field. He expanded the field and the flow as

$$\sim \sum_{l=-N_x+1}^{N_x} \sum_{m=-N_y+1}^{N_y} \sum_{n=1}^{N_z+2} e_{lmn} \exp i(l\alpha x + m\beta y)\sin n\pi z \qquad (7.2.1)$$

and the coefficients with $l = m = 0$ were small. The magnetic energy was much larger than the kinetic energy, and the Elsasser number was of order one, so it was a strong field dynamo. He looked for evidence of subcriticality, but didn't really find it.

7.2.3. Jones and Roberts results

Jones and Roberts [30] also looked at the rapidly rotating problem with E in the range $10^{-3} \geq E \geq 2 \times 10^{-4}$. No-slip, insulating boundaries were used. This work used the large Pr limit, so the inertial terms are ignored, as the motivation was the geodynamo. They considered $q = 1$ and $q = 5$, with an aspect ratio 1, i.e. a cubical, periodic box. Ra was set about 10 times critical.

If we integrate the x-component of the induction equation over a y-z square,

$$\frac{\partial}{\partial t} \int_S (\hat{\mathbf{x}} \cdot \mathbf{B}) dy dz = - \int_C (\nabla \times \mathbf{B}) \cdot \mathbf{dl} \qquad (7.2.2)$$

because by Stokes' theorem the induction term is zero. With perfectly conducting boundaries, the RHS is zero, too, so the **total** mean x-flux is zero. Similarly with y. It is possible to have mean field, with perfectly conducting boundary conditions but the z average must be zero.

They found dynamo action, with Elsasser number order 1, and a much larger mean field than St. Pierre found. The field was quite symmetric about the equator (quadrupole field). The convection is chaotic and time-dependent, but there is a broad spatial structure. A snapshot of the flow is given in Fig. 36. The flow resembled a 2D roll whose horizontal axis rotates around the z axis, repeated periodically. In Fig. 36, the right panel shows the fluid is rising in the middle of the box and falling at the edges. The Coriolis force then drives a flow parallel to the roll structure as shown in the left panel. The Kuppers–Lortz instability predicts that the roll axis will rotate about the z-axis at Rayleigh number just above critical, and this rotation persists to quite high Ra. The mean magnetic field was mainly lined up with rotation axis, so that rotated too. In Fig. 37, the x and z components of the magnetic field are shown, for the same snapshot as in Fig. 36.

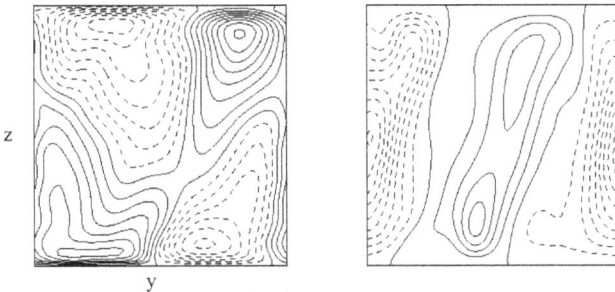

Fig. 36. Snapshot of flow in a rotating plane layer dynamo. Left u_x in y-z plane. Right u_z in y-z plane. Snapshot across the roll structure; from [30].

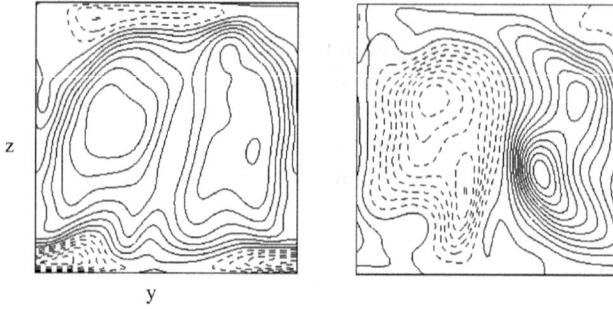

Fig. 37. Snapshot of the magnetic field in a rotating plane layer dynamo. Left B_x in y-z plane. Right B_z in y-z plane. Snapshot across the roll structure; from [30].

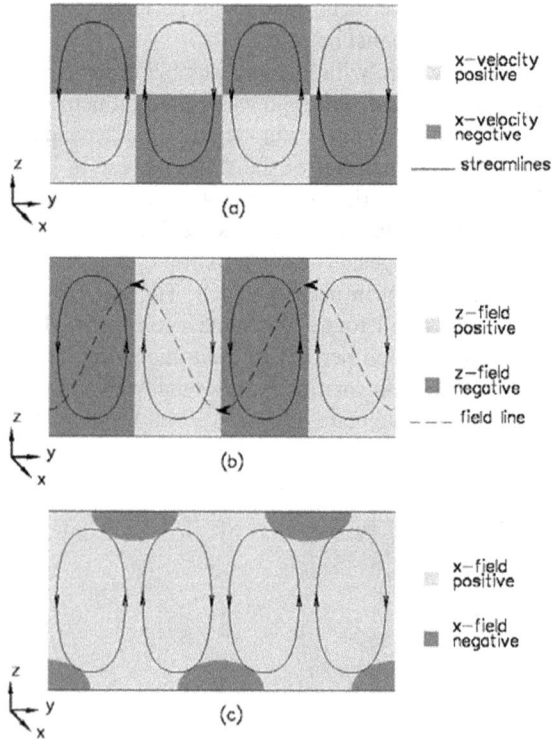

Fig. 38. Cartoon of the field generation process in the Jones–Roberts dynamo. (a) Initial y-field stretched by roll to give z-field in (b). Then rotate roll so this x-field becomes new y field; from [30].

The processes generating the field can be described using a cartoon based on the snapshots of the flow and field, shown in Fig. 38. We start in (a) with convection rolls with their axis aligned with the x-axis, and suppose there is an initial field in the y-direction. The rolls stretch this y-field in the z-direction, creating new B_z field. This B_z field is oriented oppositely in the clockwise and anticlockwise neighbouring rolls as in (b) in Fig. 38. Now the x-velocity shown in (a) stretches out this B_z, tilting it in the x-direction. The shear $\partial u_x / \partial z$ is oppositely directed in the clockwise and anticlockwise rolls, so since the B_z also alternates in adjacent rolls, the B_x created has the same sign everywhere as in (c). Actually, if the initial field in the y-direction is just stretched and sheared without diffusion, there must be regions where the B_x field has the opposite sign, and there are dark regions in (c), but these are located near the insulating boundary and so diffuse away. So the net effect is to generate mean x-field from mean y-field. If the roll stayed fixed, the y-field would eventually disappear, consistent with Theorem 2 that planar flows cannot be dynamos. However, the Kuppers–Lortz instability turns the roll round, so the x-field becomes equivalent to the y-field, and in the next phase of the cycle mean y-field is created from the x-field. As expected the mean field continually rotates about the z-axis, following the roll-axis.

7.2.4. Rotvig and Jones results

Rotvig and Jones, [49] used the same model same as above, except they had the rotation vector tilted at $45°$ to the gravity in the z-direction, so it is now $\Omega = \Omega(0, -\sin\theta, \cos\theta)$. This corresponds to a localised region in spherical geometry in the midlatitudes. The aim was to stop the Kuppers–Lortz rotation of the rolls, because in the geodynamo the constraints of the geometry mean that rotating rolls as in the Jones–Roberts dynamo are unlikely to occur. The Ekman number was set very low at $E \sim 10^{-5}$. The Rayleigh number was kept at a few times critical, so Ra/Ra_{crit} was fixed, and they varied E to look at how quantities such as the flow speed and the field scaled with E.

The Elsasser number found was well above unity, so there is a strong field dynamo. A mean field exists, but is now significantly less than the spatially fluctuating field, so this is a significant difference from the case when rotation and gravity are parallel. In Fig. 39 a snapshot of u_x and B_z is shown in the $y - z$ plane. Note the Proudman–Taylor effect: the velocity has much less variation in the direction of the rotation, which is at $45°$ to the axes, than across the rotation vector. The same is true for the other components of velocity. The magnetic field is on a smaller scale than the velocity field, probably because $q = \kappa/\eta = 4$ in these runs. Good evidence was found that Taylor's constraint is becoming satisfied at low E. In this geometry, we define $h = \hat{\mathbf{z}} \cdot \nabla \times (\mathbf{j} \times \mathbf{B})$, and then Taylor's constraint is $\langle h \rangle_{ra} = 0$, for all x and y, where $\langle\rangle_{ra}$ means average along

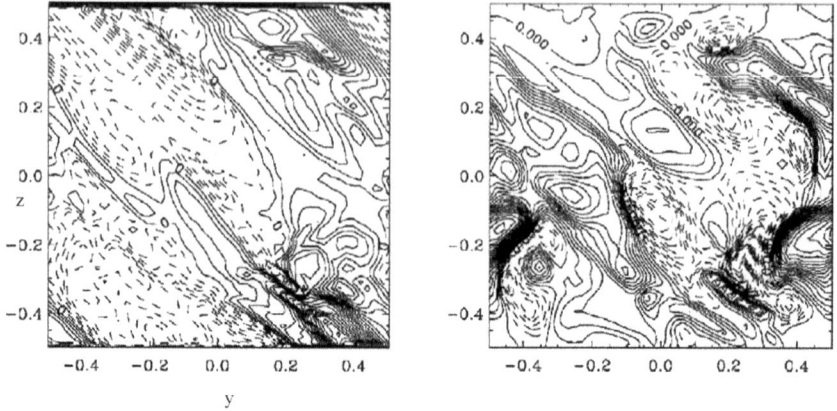

Fig. 39. Left u_x in y-z plane. Right B_z in y-z plane; from [49].

the rotation axis. So

$$Tay = \frac{\langle |\langle h \rangle_{r.a.}|^2 \rangle_{xy}}{\langle \langle |h|^2 \rangle_{r.a.} \rangle_{xy}}$$

measures how close we are to satisfying the constraint. The results showed that $Tay < 0.01$ at the smallest values of E attainable. This is clear evidence that low E dynamos do satisfy Taylor's constraint, even though the velocity field and the magnetic field are strongly time-dependent, and spatially quite complex.

They also examined the force balance between Coriolis, Lorentz, pressure, buoyancy and viscous forces. The viscous force was found to be negligible except in very thin Ekman layers at the boundaries. The other forces were roughly in balance, though the Lorentz force is rather patchy, being stronger than the other forces in localised patches, but weaker in the bulk of the fluid. This supports the idea that field is generated in localised stretching events, and the field is limited by the Lorentz force stopping the stretching.

7.2.5. Stellmach and Hansen model

Stellmach and Hansen [54] investigated a model very similar to the St. Pierre model, but at lower values of E. They used stress-free, perfectly conducting boundaries as did St. Pierre. Their numerical scheme was a fully implicit finite volume method rather than the spectral method used by all the other researchers discussed in this section. The rotation axis was taken parallel to gravity. Very low Ekman number was achieved, $E = 5 \times 10^{-6}$. They looked at aspect ratio unity, i.e. a cubical periodic box. Note that at this very small E, a lot of tall

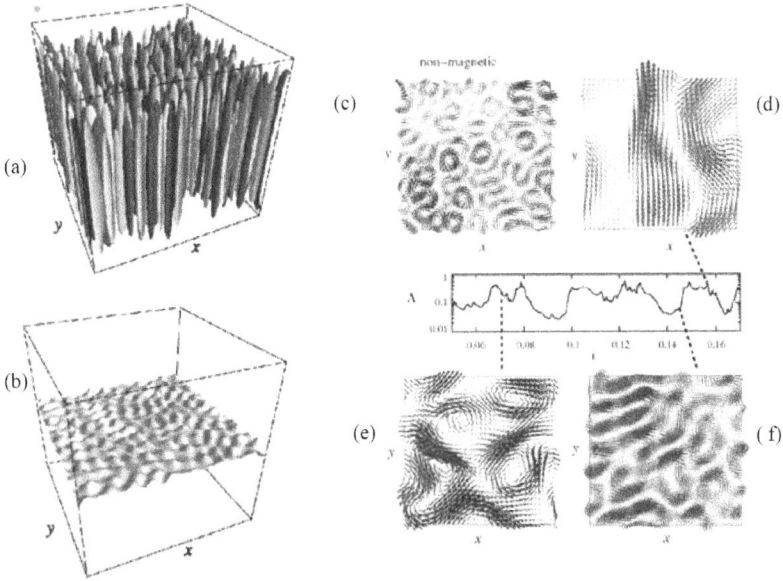

Fig. 40. Left: Convection with no magnetic field, (a) vertical velocity, u_z, (b) temperature profile at $z = 0$. Right: velocity field at upper boundary at different times, according to magnetic field strength. $E = 5 \times 10^{-7}$, $Pr = q = 1$. (c) non-magnetic, (d) time of strong field, (e) time of moderate field, (f) time of weak field; from [54].

thin cells fit in to a cubical box, see Fig. 40(a,b), so unlike the Jones Roberts model, there are many cells within one period. They fixed the Rayleigh number at only a few times critical, obtaining a magnetic Reynolds number $Rm \sim 100$. Some runs were performed with high Pr, but most runs had $Pr = q = 1$. A notable feature of their results is that the magnetic field appears to be influencing the convection strongly. In Fig. 40(a,b), we see that with no magnetic field there are many small scale rolls. However, when the field is created, the roll size is significantly increased. As the field strength waxes and wanes during the dynamo run, the size of the columns varies in step with the field. Thus in Fig. 40(d), when the field is strong, the velocity is large scale, but in Fig. 40(f), at a time when the field is weak, the columns are significantly thinner. This effect is predicted by magnetoconvection, but there is less evidence of it in spherical geometry, except in the polar regions. In the Stellmach–Hansen model, the coherent small scale rolls generate a substantial mean field, as shown in Fig. 41. The direction of the mean field also rotates as z increases, similar to the behaviour predicted by G.O. Roberts and the Childress–Soward analysis. They also found some evidence (not conclusive) of subcritical behaviour. These simulations are probably the

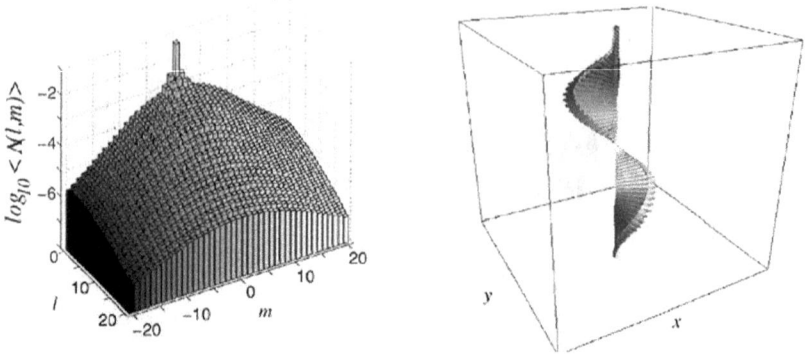

Fig. 41. Left: magnetic power spectrum over Fourier modes l and m. $E = 5 \times 10^{-6}$. Note the strong peak at $l = m = 0$, the mean field.Right: Plot of horizontally averaged mean field as a function of z. Note the zero average in accord with perfect conductor boundaries. Also note spiralling structure as in G.O. Roberts spatially periodic flow dynamo; from [54].

closest approach to the original Childress–Soward scenario that have yet been achieved numerically.

7.2.6. Cattaneo and Hughes 2006 model

Cattaneo and Hughes [6] also used perfectly conducting, stress-free boundaries as in the Stellmach–Hansen model. However, they focussed on large aspect ratio, using only slowly rotating or non-rotating cases. Because of the large aspect ratio, there are many cells in the box at all times, but no tall thin cells. This allowed them to explore the issue of whether a mean field can develop when all the individual convection cells have no phase coherence.

An optimised spectral method used, and the runs mostly had $Pr = 1$ and $Pm = 5$. The motivation was more for astrophysical convection rather than Earth's core convection, as in the Sun the Rossby number is of order unity. The focus was on whether there is a mean field and whether it can be described by an alpha effect. Somewhat remarkably, they found that the ratio of mean field magnetic energy to total magnetic energy is typically less than 0.001. This means that the fluctuating field completely dominates the mean field. This happens despite there being a significant amount of helicity in the convection, which does have a large scale component. This result is the opposite of the first order smoothing case, where fluctuating field is assumed small compared to mean field. The problem is that the small scale magnetic Reynolds number is not that small in these convection driven dynamos, so that the helicity twists any initial field not through a small angle but through a random large angle, preventing any coherent alpha-effect developing. They found the same result, no significant mean field,

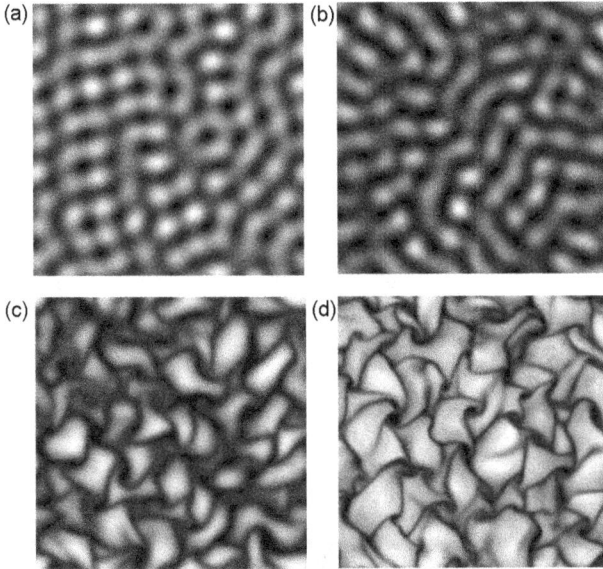

Fig. 42. $E = 13 \times 10^{-3}$. Temperature fluctuations near upper boundary. (a)–(d) series with increasing Ra. (a) is $Ra = 6.2 \times 10^4$, near critical. (d) has $Ra = 5 \times 10^5$. Note transition from ordered to disordered pattern; from [6].

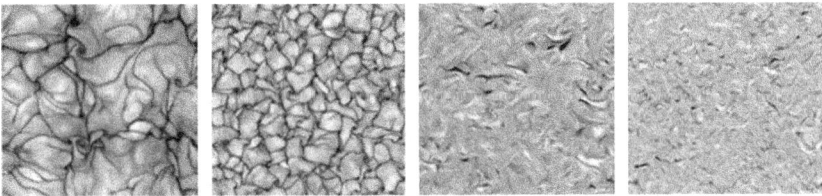

Fig. 43. From left to right: temperature near boundary of nonrotating case: temperature in rotating case: magnetic field nonrotating case: magnetic field rotating case. Note the decrease in cell size with rotation. Also note the lack of any mean field in any case! from [6].

even in the mildly rotating case. It is interesting to speculate whether this would remain true in the rapidly rotating case.

Summary

• Everyone found dynamo action over a wide range of parameters. This is surprising as 'most' steady flows are not kinematic dynamos. Time-dependent chaotic flows generally seem to be dynamos at high enough Rm, consistent with fast dynamo stretching ideas.

- Most researchers didn't find dynamos at low $q = Pm/Pr$. This is a numerical problem, because to get a dynamo at all you must have a reasonably large Rm and at low Pm that means Re is very large. This means there are very small scales in flow. Very recent work suggests there is dynamo action at low Pm: it is just very hard to compute.
- An emerging issue is whether the dynamo generates mean field or not. Most workers only got a small mean field, but much small scale dynamo action giving small scale fields.
- Jones and Roberts found a large mean field, but there was a large scale coherent flow in their problem.
- With many cells in flow, Cattaneo and Hughes found negligible mean field unlike Stellmach and Hansen. Possibly the difference is in the larger Ra used by Cattaneo and Hughes. It is also possible that the tall thin cells in the Stellmach–Hansen model give local Rm small, thus validating first order smoothing, and leading to a mean field larger than the small scale field.
- First order smoothing is clearly not a valid assumption for dynamos of the Cattaneo and Hughes type, when Rm over an individual cell is large. Then each cell acts as its own dynamo, and different regions never get into phase. The development of phase coherence of the magnetic field, that is how each small scale element of generated field builds up to give a large-scale coherent field, is clearly something that needs exploring!

Acknowledgement

I am indebted to Wietze Herreman, for reading through the lecture notes and making many helpful suggestions to improve them.

References

[1] G. Backus, A class of self-sustaining dissipative spherical dynamos, Ann. Phys. **4**, 372–447 (1958).

[2] J.P. Boyd, *Chebyshev and Fourier Spectral Methods*, Dover, 2001.

[3] S.I. Braginsky, Nearly axisymmetric model of the hydromagnetic dynamo of the Earth, Geomag. Aeron. **15**, 122–128 (1975).

[4] E.C. Bullard and H. Gellman, Homogeneous dynamos and terrestrial magnetism, Phil. Trans. R. Soc. Lond. A **247**, 213–278 (1954).

[5] F.H. Busse, Thermal instabilities in rapidly rotating systems, J. Fluid Mech. **44**, 441–460 (1970).

[6] F. Cattaneo and D.W. Hughes, Dynamo action in a rotating convective layer. J. Fluid Mech. **553**, 401–418 (2006).

[7] Q.N. Chen and C.A. Jones, Similarity and dynamic similarity models for large-eddy simulations of a rotating convection-driven dynamo, Geophys. J. Int. **172**, 103–114 (2008).

[8] S. Childress and A.M. Soward, Convection-driven hydromagnetic dynamo. Phys. Rev. Lett. **29**, 837–839 (1972).

[9] S. Childress and A.D. Gilbert, *Stretch, Twist, Fold: The Fast Dynamo*, Lecture Notes in Physics: Monographs. Springer, 1995.

[10] U.R. Christensen and A.Tilgner, Power requirement of the geodynamo from ohmic losses in numerical and laboratory dynamos, Nature **429**, 169–171 (2004).

[11] T.G. Cowling, The magnetic field of sunspots, Mon. Not. R. Astr. Soc. **94**, 39–48 (1934).

[12] P.A. Davidson, *An Introduction to Magnetohydrodynamics*, Cambridge University Press, U.K., 2001.

[13] B. Desjardins and E. Dormy, in Chapter 1 of *Mathematical Aspects of Natural Dynamos*, eds. E. Dormy and A.M. Soward, CRC Press, 2007.

[14] T. Dombré, U. Frisch, J.M. Greene, M. Hénon, A. Mehr and A.M. Soward, Chaotic streamlines in ABC flows. J. Fluid Mech. **167**, 353–391 (1986).

[15] M.L. Dudley and R.W. James, Time-dependent kinematic dynamos with stationary flows, Proc. R. Soc. Lond. A **425**, 407–429 (1989).

[16] S. Fauve and F. Petrelis, in Chapter 2 of *Mathematical Aspects of Natural Dynamos*, eds. E. Dormy and A.M. Soward, CRC Press, 2007.

[17] D.R. Fearn, Hydromagnetic flow in planetary cores, Rep. Prog. Phys. **61**, 175–235 (1998).

[18] D.R. Fearn, P.H. Roberts and A.M. Soward, in: *Energy, Stability and Convection*, eds. G.P. Galdi and B. Straughan, Longmans, 1988.

[19] A. Gailitis, Self-excitation of a magnetic field by a pair of annular vortices, Magnetohydrodynamics **6**, 14–17 (1970).

[20] D.J. Galloway and U. Frisch, A note on the stability of a family of space-periodic Beltrami flows, J. Fluid Mech. **180**, 557–564 (1986).

[21] D.J. Galloway and M.R.E. Proctor, Numerical calculations of fast dynamos for smooth velocity fields with realistic diffusion, Nature **356**, 691–693 (1992).

[22] A.D. Gilbert, Fast dynamo action in the Ponomarenko dynamo, Geophys. Astrophys. Fluid Dynam. **44**, 214–258 (1988).

[23] A.D. Gilbert, in Chapter 1 of *Mathematical Aspects of Natural Dynamos*, eds. E. Dormy and A.M. Soward, CRC Press, 2007.

[24] A.D. Gilbert, Dynamo theory, in: *Handbook of Mathematical Fluid Dynamics*, eds. S. Friedlander and D. Serre, vol. 2, Elsevier Science BV, 2003, pp. 355–441.

[25] A. Herzenberg, Geomagnetic dynamos, Phil. Trans R. Soc. Lond. A **250**, 543–583 (1958).

[26] D. Gubbins, C.N. Barber, S. Gibbons and J.J. Love, Proc. Roy. Soc. Lond. A **456**, 1333–1353 (1999).

[27] D.W. Hughes, in Chapter 2 of *Mathematical Aspects of Natural Dynamos*, eds. E. Dormy and A.M. Soward, CRC Press, 2007.

[28] C.A. Jones. Dynamo models, in: *Advances in Solar System Magnetohydrodynamics*, eds. E.R. Priest and A.W. Hood, Cambridge University Press.

[29] C.A. Jones, Thermal and Compositional Convection in the Outer Core, in the *Treatise of Geophysics*, Core dynamics volume, Volume editor, P. Olson, Editor-in-Chief, G. Schubert, Elsevier, 2007, pp. 131–185.

[30] C.A. Jones and P.H. Roberts. Convection-driven dynamos in a rotating plane layer, J. Fluid Mech. **404**, 311–343 (2000).

[31] A. Iskakov and E Dormy, On the magnetic boundary conditions for non-spectral dynamo simulations, Geophys. Astrophys. Fluid Dynam. **99**, 481–492 (2005).

[32] C.A. Jones, N.O. Weiss, and F. Cattaneo, Nonlinear dynamos: a complex generalization of the Lorenz equations, Physica D **14**, 161–176 (1985).

[33] F. Krause and K.-H. Rädler, *Mean Field Magnetohydrodynamics and Dynamo Theory*, Pergamon Press, New-York, 1980.

[34] F.J. Lowes and I. Wilkinson, Geomagnetic dynamo: an improved laboratory model, Nature **219**, 717–718 (1968).

[35] W.V.R. Malkus and M.R.E. Proctor, The macrodynamics of alpha-effect dynamos in rotating fluids, J. Fluid Mech. **67**, 417–443 (1975).

[36] M. Meneguzzi and A. Pouquet, Turbulent dynamos driven by convection, J. Fluid Mech. **205**, 297–318 (1989).

[37] R.T. Merrill, M.W. McElhinny and P.L. McFadden, *The Magnetic Field of the Earth*, Academic Press, 1996.

[38] H.K. Moffatt, *Magnetic Field Generation in Electrically Conducting Fluids*, Cambridge University Press, 1978.

[39] H.K. Moffatt, A self-consistent treatment of simple dynamo systems, Geophys. Astrophys. Fluid Dynam. **14**, 147–166 (1979).

[40] N.F. Otani, A fast kinematic dynamo in two-dimensional time-dependent flows, J. Fluid Mech. **253**, 327–340 (1993).

[41] E.N. Parker, Hydromagnetic dynamo models, Astrophys. J. **122**, 293–314 (1955).

[42] Yu. B. Ponomarenko, On the theory of hydromagnetic dynamo, J. Appl. Mech. Tech. Phys. **14**, 775 (1973).

[43] M.R.E. Proctor, in Chapter 1 of *Mathematical Aspects of Natural Dynamos*, eds. E. Dormy and A.M. Soward, CRC Press, 2007.

[44] G.O. Roberts, Spatially periodic dynamos, Phil. Trans. R. Soc. Lond. **A 266**, 535–558 (1970).

[45] G.O. Roberts, Dynamo action of fluid motions with two-dimensional periodicity, Phil. Trans. R. Soc. Lond. **A 271**, 411–454 (1972).

[46] P.H. Roberts and A.M. Soward, Dynamo Theory, Ann. Rev. Fluid Mech. **24**, 459–512 (1992).

[47] P.H. Roberts, Kinematic Dynamo Models, Phil. Trans. R. Soc. Lond. **A 272**, 663–703 (1972).

[48] P.H. Roberts, Fundamentals of dynamo theory, in: *Lectures on Solar and Planetary Dynamos*, eds. M.R.E. Proctor and A.D. Gilbert, Cambridge University Press, 1994, pp. 1–58.

[49] J. Rotvig and C.A. Jones, Rotating convection-driven dynamos at low Ekman number, Phys. Rev. E **66**, 056308:1–15 (2002).

[50] A.M. Soward, Fast dynamo action in a steady flow. J. Fluid Mech. **180**, 267–295 (1987).

[51] A.M. Soward, *Fast Dynamos*, in Lectures on Solar and planetary dynamos, eds. M.R.E. Proctor, A.D. Gilbert, Cambridge University Press, 1994, pp. 181–217.

[52] A.M. Soward and C.A. Jones, Alpha-squared dynamos and Taylor's constraint, Geophys. Astrophys. Fluid Dynamics **27**, 87–122 (1983).

[53] M.G. St. Pierre, The strong field branch of the Childress–Soward dynamo, in: *Theory of Solar and Planetary Dynamos*, eds. M.R.E. Proctor et al., 1993, pp. 295–302.

[54] S. Stellmach and U. Hansen, Cartesian convection driven dynamos at Low Ekman number, Phys. Rev E. **70**, 056312:1–16 (2004).

[55] J.B. Taylor, The magnetohydrodynamics of a rotating fluid and the Earth's dynamo problem, Proc. R. Soc. Lond. A **274**, 274–283 (1963).

[56] S.I. Vainshtein and Ya.B. Zeldovich. Origin of magnetic fields in astrophysics, Usp. Fiz. Nauk **106** 431–457. [English translation: Sov. Phys. Usp. **15**, 159–172] (1972). Proc. R. Soc. Lond. A **274**, 274–283 (1963).

[57] N.O. Weiss, F. Cattaneo and C.A. Jones, Periodic and aperiodic dynamo waves, Geophys. Astrophys. Fluid Dyn. **30**, 305–341 (1984).

[58] Ya.B. Zeldovich, The magnetic field in the two-dimensional motion of a conducting turbulent fluid, Sov. Phys. JETP **4**, 460–462 (1957).

Course 3

PLANETARY MAGNETISM

Peter Olson

Johns Hopkins University, Baltimore, MD, USA

Ph. Cardin and L.F. Cugliandolo, eds.
Les Houches, Session LXXXVIII, 2007
Dynamos

Contents

1. Planetary dynamos introduction

1.1. Planetary magnetic fields

Nearly all large solar system objects have magnetic fields of their own, as do some of the smaller objects. The magnetic planets include Jupiter, Saturn, Earth, Uranus, Neptune, probably Mercury, and Ganymede (Connerney, 2007). Why is planetary magnetism ubiquitous? The general answer is that only three ingredients are necessary for magnetic field generation via a self-sustaining dynamo. These are:

1. A large volume of electrically conducting fluid in the planet interior.

2. An energy supply to stir the fluid, either convection, tidal dissipation, or some other energy source.

3. Planetary rotation, to organize the fluid motion in a way that provides the necessary positive feedbacks for sustained magnetic field generation to occur.

If these ingredients are present then a self-sustaining dynamo is possible. Among the planets in our solar system, however, we find a substantial variety of planetary magnetic field structures. Figure 1 shows that the five largest planetary magnetic fields, which originate in Jupiter, Saturn, Earth, Uranus, and Neptune, all have a different dipole orientation with respect to their rotation axes. The magnitude (intensity) and dipolarity of the fields also varies greatly. Why is there such a diversity of magnetic fields? This question is a lot harder to answer!

1.1.1. Definition of a planetary magnetic field

Matter in the neighborhood of a planet experiences a force $\mathbf{F}$ proportional to its charge q, given by:

$$\mathbf{F} = \mathbf{F}_E + \mathbf{F}_L = q(\mathbf{E} + \mathbf{u} \times \mathbf{B}) \tag{1.1}$$

The force $\mathbf{F}$ consists of two parts, an electrostatic part $\mathbf{F}_E$ and an electrodynamic part $\mathbf{F}_L$ called the Lorentz force, which is proportional to the velocity of the matter relative to the planetary field. In (1.1), $\mathbf{E}$ is the planetary electric field, $\mathbf{u}$

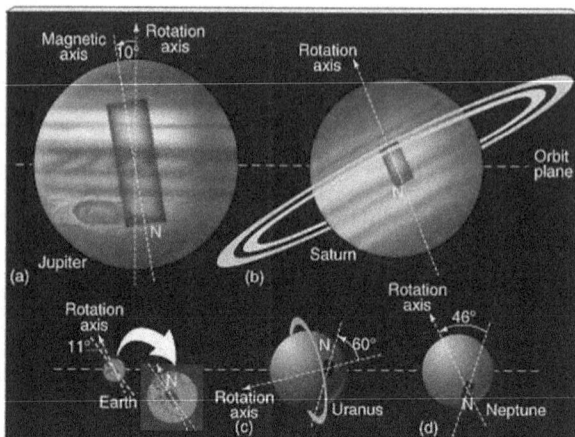

Fig. 1. Orientations of the best-fitting dipole fields of Jupiter, Saturn, Earth, Uranus, and Neptune. The field orientation with respect to the rotation axis and the orbital plane is indicated by a hypothetical bar magnet in the planetary interior.

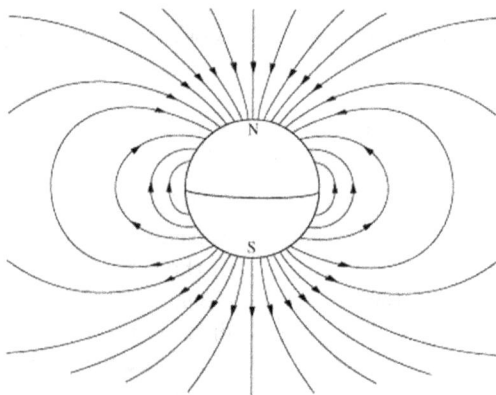

Fig. 2. Depiction of a dipole field where **B** is everywhere tangent to the magnetic field lines and $|B|$ is inversely proportional to the spacing between the lines.

is the relative velocity of the matter, and **B** is the planetary magnetic field. The units of **B** are therefore (Woan, 2000)

$$|B| = |Fq^{-1}u^{-1}| = N\ A^{-1}\ m^{-1} = Tesla \tag{1.2}$$

The simplest possible external planetary magnetic field structure is the dipole field depicted in Fig. 2, where the magnetic field lines are curves that are everywhere tangent to the field vector **B**, and the intensity of the field $|B|$ is inversely proportional to the spacing of the field lines.

1.1.2. *What can be learned from planetary magnetic fields?*

Magnetic fields can be used to probe planetary interiors and as recorders of their history. In addition, their generation involves fundamental physical processes, worthy of study in their own right (Stevenson, 2003). Self-sustaining magnetic fields originate deep in planetary interiors, where there is an electrically conducting fluid. This requirement places important constraints on the composition, state, and thermal regime of magnetic planets. The structure and temporal change of an external magnetic field can provide additional information on the planets interior structure, such as the depth to the dynamo-producing region and possibly its thickness. In some cases crustal magnetization can preserve the past magnetic and tectonic history of a planet, possibly including information on the evolution of its atmosphere and surface environments. Finally, self-sustaining planetary dynamos are natural laboratories for complex, non-linear magnetohydrodynamic and multi-physics processes, involving rotational fluid dynamics, turbulence, convection, and the interaction of these phenomena with self-generated magnetic fields.

1.1.3. *Internal sources for planetary magnetic fields*

There are three general types of internal sources for planetary magnetic fields. These are:

1. Permanent magnetization. Only ferromagnetism is important here, and it is limited to temperatures above the Curie temperature, around $800K$ in iron-titanium oxides. Fields produced this way are steady, generally weak, and spatially heterogeneous, given that the magnetization is shallow.

2. Magnetotelluric internal currents, induced by an externally-applied magnetic field. Such currents only require an electrical conductor, which could be solid as well as fluid.

3. Free internal electric currents. Internal electric currents are associated with a magnetic field according to Ampere's Law:

$$\mu\mathbf{J} = \nabla \times \mathbf{B} \tag{1.3}$$

where μ is the magnetic permeability and $\mathbf{J}$ is the current density. Free currents require a mechanism to maintain them against the effects Ohmic decay, that is, they require a dynamo mechanism.

1.1.4. Rationale for the self-sustaining fluid dynamo mechanism for planetary magnetism

All of the magnetic sources listed are capable of producing a planetary magnetic field. The challenge is to infer the nature of the source, using the character of the external field and our knowledge of the planet structure. For the major planets, the dynamo mechanism is the most viable explanation, for the following combination of reasons:

• All large solar system objects have (or had) global-scale magnetic fields, possibly excepting Venus.

• We know that planetary magnetic fields predominantly have an internal origin; fields of external origin are generally a small percentage of the respective internal fields.

• Permanent magnetization is a weak, near-surface phenomenon because planetary interior temperatures are generally too high for ferromagnetism at great depth.

• Planetary magnetic fields change rapidly in time. This is not consistent with permanent magnetization as the primary source of the field.

• The magnetic free decay time is much less than the planet age. This means that a regeneration mechanism is required.

• Electrically-conducting fluids, necessary for the dynamo mechanism, are in fact commonplace in the solar system.

1.1.5. Planet interior classification

To compare the dynamo processes in each object, it is useful to classify the planets and large satellites into the following four groups (see Fig. 3). The known properties of all the objects in our solar system can be found in general planetary science texts and monographs, such as De Pater and Lissuaer (2001) and Lodders and Fegley (1998).

1. Terrestrial planets: Mercury, Venus, Earth, Moon, Mars, Io. These consist of high-temperature condensates, with mostly solid Mg, Fe oxide and silicate mantles, plus Fe-rich, partially molten metallic cores. Mantle heat transport occurs by subsolidus convection (solid-state, buoyantly-driven creep). Their cores have remained partly liquid because the melting point is depressed by addition of lighter alloying elements such as S, O, Si, etc.

2. Gas giants: Jupiter, Saturn, and possibly hot Jupiter extrasolar planets (such as HD179949; see Shkolnik and Walker, 2003, for example). Their composition is mainly H and He, and they possibly have small rocky central cores. Pressure-induced metallization transforms H, and possibly He, into electrical conductors. The energy source for their dynamos is probably convection, associated with heat released by on-going gravitational contraction.

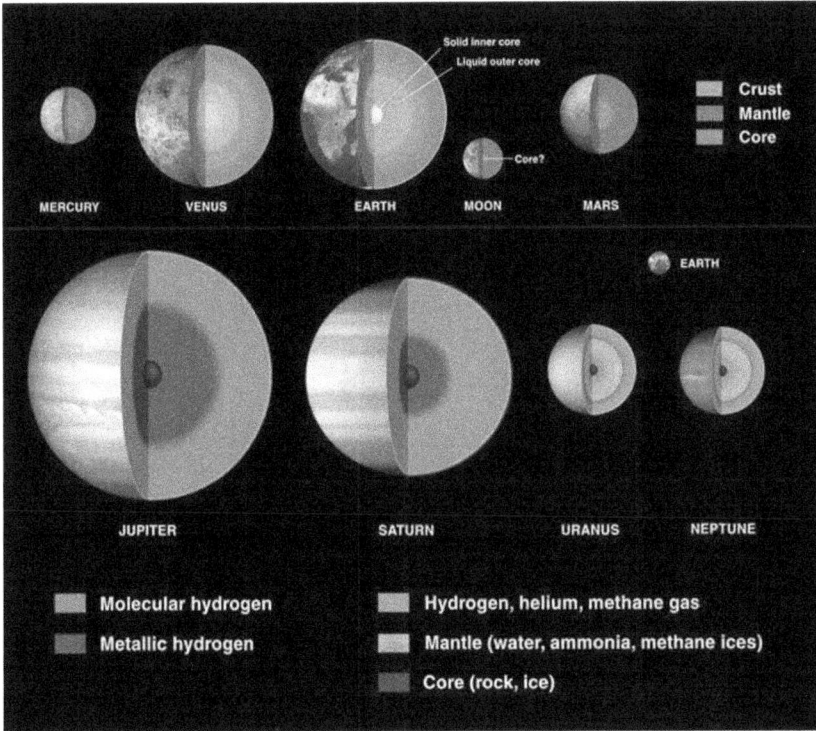

Fig. 3. Gross internal radial structure and compositions of the planets and the moon. Credit NASA.

3. Ice giants: Uranus, Neptune. Their composition includes a H-rich envelope surrounding a conducting H_2O, NH_3, CH_4 mantle, which is probably the dynamo-generating region.

4. Icy planets and large satellites: Ganymede, Callisto, Titan, Triton, Pluto, Europa. This set consists of weakly conducting H_2O ice-rock mixtures, some with metallic cores. These objects are marginal for sustained dynamo action by virtue of their small size, as they may not contain enough high conductivity material, or the conducting material may have already solidified.

1.2. How to make a self-sustaining planetary dynamo

The following is a thought experiment, intended to illustrate the necessary ingredients for a self-sustaining fluid dynamo. Our purpose here is to identify the important variables, physical parameters, and necessary conditions for a self-

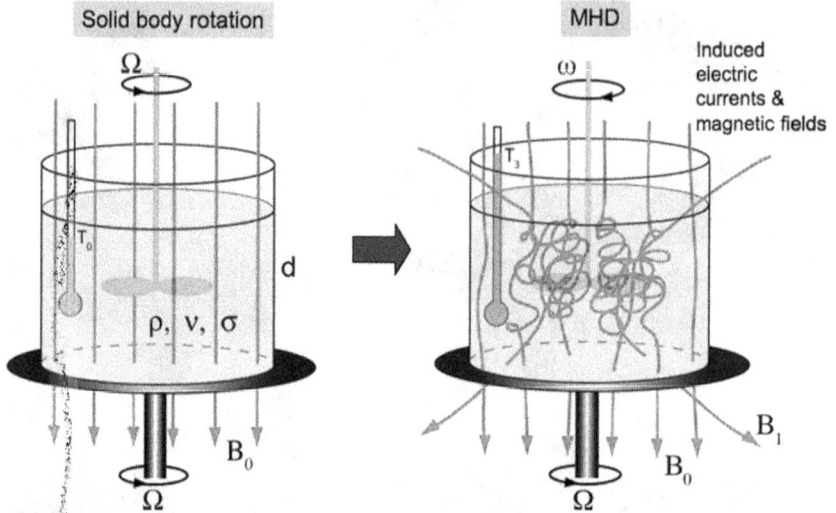

Fig. 4. A hypothetical fluid dynamo experiment. Left: Conducting fluid in solid body rotation in a uniform external magnetic field. Right: Same fluid stirred, with induced electric currents and magnetic fields.

sustaining planetary dynamo. For a fuller discussion of mhd experiments, see Davidson (2001) or Cardin and Olson (2007).

Suppose a cylindrical volume of electrically conducting fluid is placed on a rotating platform, as shown in Fig. 4. The fluid is subjected to a uniform external magnetic field B_0 aligned with the rotation axis (note: if the external field is uniform, co-rotation of the field and fluid is not at issue; if it is not uniform, we suppose the field co-rotates with the platform). A propeller is inserted in the fluid, capable of rotating independently of the tank, and a thermometer is inserted to measure temperature. The motion of the propeller is the energy (power) source for the dynamo in this experiment.

Relevant physical parameters include the rotation rate (angular velocity) Ω, the fluid depth d, its density ρ, kinematic viscosity ν, electrical conductivity σ, the strength of the imposed field B_0, and the propeller speed ω measured relative to the background rotation rate. The basic magneto-static state (left side of Fig. 4) consists of a solid body rotation produced by rotating the platform and the propeller at the same rate for a long time, until the fluid is spun up. The thermometer reads the lab temperature T_0, the fluid is at rest in the rotating coordinate system, and the magnetic field is unchanged.

The magnetohydrodynamic state is shown on the right side of Fig. 4. It is produced by rotating the propeller differentially with respect to the fluid, induc-

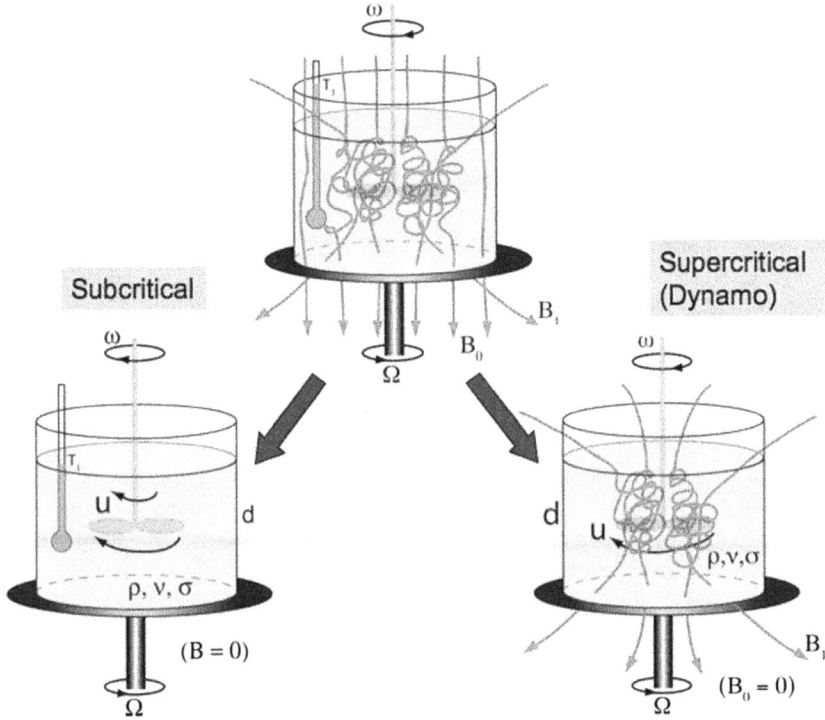

Fig. 5. Remove external field and wait. If the fluid motion is subcritical for dynamo action, the internal magnetic field decays away (bottom left); if it is supercritical for dynamo action, the internal field persists (bottom right).

ing fluid motion. The fluid motion in turn induces electric currents in the fluid, which induce magnetic fields of their own. These induced fields are shown in the figure as complex field lines within the fluid and as an induced field B_1 outside the fluid. Because of viscous dissipation by the fluid motion and Joule heating by the electric currents, heat is produced in the fluid and the temperature rises to T_1, as indicted on the thermometer in the figure.

The next step starts from the magnetohydrodynamic state, shown at the top of Fig. 5. We switch off the externally-applied field B_0 and wait for a while. The system will evolve into either one of two general regimes, labeled subcritical and supercritical (dynamo) in the bottom left and bottom right of Fig. 5. In the subcritical regime the induced magnetic field B_1 decays away, leaving only the fluid motion driven by the propeller. There is still heating from viscous dissipation, but this is less than the heating in the previous situation, so the temperature T_2 is correspondingly lower. In the supercritical regime, the induced field B_1 remains, although perhaps with a different form than it had before B_0 was removed. This regime can be called a "self-sustaining" dynamo if the internally-induced field remains for many magnetic diffusion times without further help from an external field.

The threshold condition that separates these two regimes is called the "critical state". It depends on the magnitude of a dimensionless parameter, the magnetic Reynolds number

$$Rm = \mu_0 \sigma u d = \frac{ud}{\eta} \qquad (1.4)$$

where μ_0 is free-space magnetic permeability and u is the characteristic fluid velocity. The second definition of Rm involves $\eta = 1/\mu_0\sigma$, the *magnetic diffusivity* of the fluid. Theory, lab experiments, and numerical models show that the value of the critical magnetic Reynolds number depends on the geometry of the fluid and on the way the motion is induced (mechanically versus buoyantly, for example) and on the rotation rate. In many rotating flows, the critical magnetic Reynolds number is about $Rm_{critical} \simeq 40$, although this is just a typical (not a precise) value.

If $Rm > Rm_{critical}$, the induced field strengthens and ultimately equilibrates, either as a steady field, an oscillatory field, or a field with chaotic time behavior. The geometry of the induced field depends on the physical parameters of the dynamo. Its strength also depends on the dynamo parameters. However, it is sometimes found, and it is often argued, that in rotating fluid dynamos the field equilibrates at strength given by

$$B_{1\ equilibrium} \approx \sqrt{\rho\Omega/\sigma} \qquad (1.5)$$

1.2.1. Energy flow

It is instructive to contrast the energy flow in an ordinary hydrodynamic system with the energy flow in a self-sustaining fluid dynamo. Figure 6 depicts the energy flow in hydrodynamic systems powered by mechanical, convective, and inertial forcings, respectively. Mechanical forcing involves an external power source (such as the propeller in the Fig. 4) which adds kinetic energy to the fluid motion, which is then converted to heat via viscous dissipation. A convective system converts the gravitational potential energy of adverse density gradients into kinetic energy via a convective instability, which again is dissipated as heat. Inertial systems convert the rotational potential energy into kinetic energy through an inertial instability.

Figure 7 depicts the energy flow in a dynamo powered by convective and inertial forcings. The first parts of these paths are the same as the hydrodynamic

Hydrodynamic Energy Flow Paths

Mechanical   Convective   Inertial

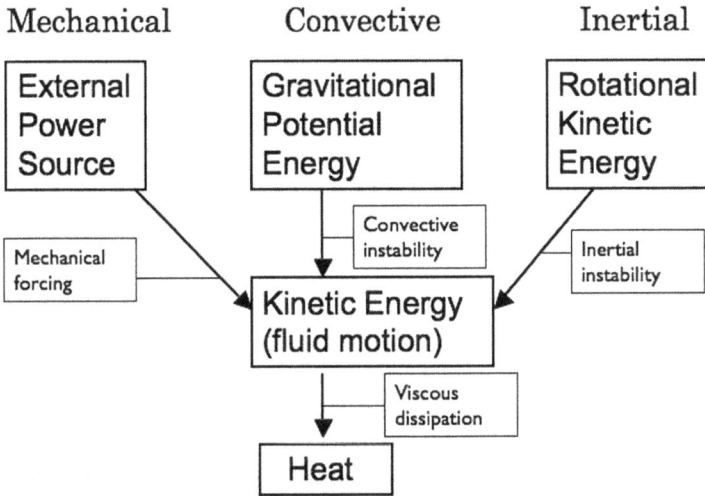

Fig. 6. Energy flow paths in a hydrodynamic system driven by mechanical, convective, and inertial forcings.

Planetary Dynamo Energy Flow Paths

Electromagnetic   Convective   Inertial

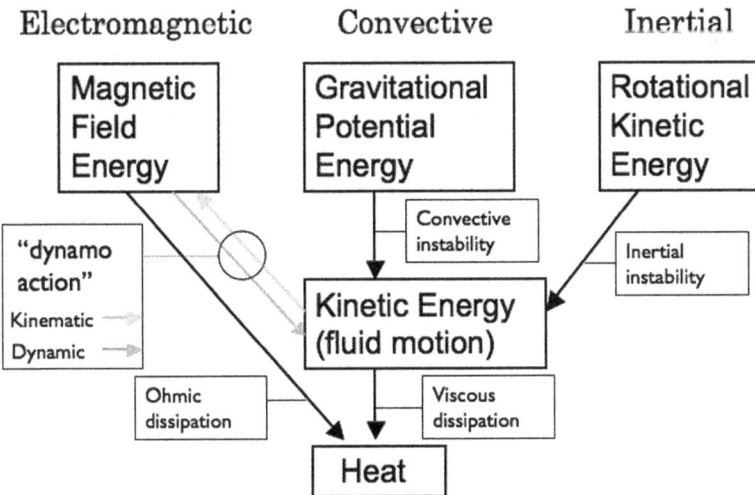

Fig. 7. Energy flow paths in a dynamo driven by convective and inertial forcings.

paths: Convective systems convert the gravitational potential energy due to adverse density gradients to kinetic energy via a convective instability. Inertial systems convert the rotational potential energy into kinetic energy through an inertial instability. An additional source of energy in the dynamo, not found in the hydrodynamic system, is the conversion of kinetic energy into electromagnetic energy, shown in Fig. 7 as the "kinematic" step in dynamo action. Magnetic energy is converted back to kinetic energy, labeled as the "dynamic" step in Fig. 7. Magnetic energy is also dissipated by Joule heating (Ohmic dissipation).

1.2.2. Experimental parameters

What physical and dimensionless parameters define a planetary dynamo? Here we must distinguish between input (i.e., control) parameters and output (i.e., result) parameters. Also, we will continue to use mechanical forcing as in the thought experiment, and later introduce convective forcing. The input physical parameters include external (structural) parameters such as the typical length scale d, planetary rotation rate Ω, and differential rotation ω, as well as material properties such as the fluid density ρ, viscosity ν, conductivity σ, permeability μ_0, and thermal diffusivity κ. The output physical parameters are typically the magnetic field B and velocity field u. Within these 10 physical parameters there are four fundamental units: mass m, length L, time t, and charge q. The Buckingham Pi Theorem (see Barenblatt, 1996) then dictates that the number of independent dimensionless parameters needed to fully describe the behavior of this system is $n = 10 - 4 = 6$.

Characteristic time scales   There are formal methods for finding $n = 6$ independent dimensionless parameters (Bridgemann, 1931), but a less formal method that also provides useful insight is to simply list the characteristic time scales defined by the physical parameters of the dynamo, and form ratios of them. In this case the physical parameters fall into three groups: diffusive, dynamic, and forcing.
   Diffusive:

$$t_\nu = \frac{d^2}{\nu}, \qquad t_\kappa = \frac{d^2}{\kappa}, \qquad t_\eta = \frac{d^2}{\eta} = \mu_0 \sigma d^2 \tag{1.6}$$

where t_ν is the viscous diffusion time scale, t_κ is the thermal diffusion time scale and t_η is the magnetic diffusion time scale.
   Dynamic:

$$t_\Omega = \Omega^{-1}, \qquad t_u = \frac{d}{u}, \qquad t_A = (\rho\mu_0)^{1/2}\frac{d}{B} \tag{1.7}$$

where t_Ω is the time scale for rotation, t_u is the circulation time, and t_A is the time scale based on the Alfven wave propagation speed.

Forcing (mechanical or buoyant):

$$t_\omega = \omega^{-1} \quad \text{or} \quad t_F = d^{2/3} F^{-1/3} \tag{1.8}$$

where t_ω is the mechanical forcing frequency, and we have written the time scale of convective forcing t_F based on the buoyancy flux F. There are many ways to represent the effects of convection, however, the buoyancy flux is a particularly useful parameter to describe the actual strength of the convective forcing, either thermal or chemical. In terms of the buoyancy $g' = \rho' g / \rho_0$ and the vertical (or radial) velocity of the fluid w, we have $F = \langle g' w \rangle$ where the brackets $\langle \rangle$ denote horizontal (or spherical) surface average. In thermal convection for example, the convective heat flux q' is related to the buoyancy flux as $F = \alpha g q' / \rho C_p$, where α, g, ρ, and C_p are thermal expansivity, gravity, density, and specific heat, respectively.

Dimensionless parameters   We can now form 6 independent dimensionless parameters by taking ratios of the 7 characteristic time constants. There are an infinite number of possibilities here, but the following are the most commonly used set. Control parameters defined in terms of material properties and kinematics include the Prandtl number,

$$Pr = \frac{t_\kappa}{t_\nu} = \frac{\nu}{\kappa} \tag{1.9}$$

magnetic Prandtl number,

$$Pm = \frac{t_\eta}{t_\nu} = \frac{\nu}{\eta} \tag{1.10}$$

and the Ekman number,

$$E = \frac{t_\nu}{t_\Omega} = \frac{\nu}{\Omega d^2} \tag{1.11}$$

The Prandtl numbers are ratios of the three diffusivities. In most planetary dynamos, $Pr \simeq 0.1$ and $Pm \simeq 10^{-5}$–10^{-6}. The Ekman number measures the relative importance of viscous to Coriolis effects, and is typically very small in planetary dynamos, in the range $E = 10^{-7}$–10^{-15}. Control parameters that define the forcing include the frequency ratio ϵ for mechanical forcing,

$$\epsilon = \frac{t_\Omega}{t_\omega} = \frac{\omega}{\Omega} \tag{1.12}$$

or alternatively, the Rayleigh number Ra for convective forcing,

$$Ra = \frac{t_v^2 t_\kappa}{t_F^3} = \frac{Fd^4}{\nu^2 \kappa} \tag{1.13}$$

Just as there are many different ways to define convective forcing, there are many different definitions of the Rayleigh number. The one used here is based on the buoyancy flux F. For thermal convection, the Rayleigh number is sometimes based on the heat flux q, i.e. $Ra = \alpha g q d^4 / K \kappa \nu$, where α and K are thermal expansivity and conductivity, or the superadiabatic temperature gradient ΔT, i.e. $Ra = \alpha g \Delta T d^3 / \kappa / \nu$ the rate of secular cooling, or ϵ, the density of radioactive heat sources, in which case $Ra = \alpha g \epsilon d^5 / K \kappa \nu$. For compositional convection, the chemical mass flux, compositional gradient, or rate of change of composition are used to define Ra, as we shall show later.

Response parameters include the magnetic Reynolds number,

$$Rm = \frac{t_\eta}{t_u} = \frac{ud}{\eta} \tag{1.14}$$

the ratio of magnetic diffusion time to the circulation time, and the Elsasser number,

$$\Lambda = \frac{t_\eta t_\Omega}{t_A^2} = \frac{\sigma B^2}{\rho \Omega} \tag{1.15}$$

which measures the dynamo magnetic field strength. Note that $\Lambda = 1$ corresponds to the estimate of saturated magnetic field strength in the dynamo thought experiment.

1.3. Implications for planetary dynamos

As discussed in the context of our hypothetical fluid dynamo, there is a critical Rm above which dynamo action is possible, with $Rm_{critical} \simeq 40$. Laboratory experimental fluid dynamos seek to achieve this large Rm using large fluid velocities u (i.e., small t_u). In contrast, planetary dynamos achieve large Rm by virtue of large length scales d (i.e., large t_η). Generally this condition can be met with rather small (1–10 mm/s) planetary fluid velocities. More problematic is the question of the field intensity in a planetary dynamo. It is often assumed that planetary dynamos equilibrate at $\Lambda \sim 1$, which implies $B_{equilibrium}$ is independent of dynamo energy sources. This is an attractive result, because it is usually difficult to quantify the energy source in planetary dynamos, but is it a good assumption?

Formal solutions to the planetary dynamo problem consist of determining functional relationships of the form: $\Lambda = f(Pr, Pm, E, Ra)$ and $Rm = f(Pr, Pm, E, Ra)$. Similar relationships can be constructed for the external and internal magnetic field and the fluid velocity and other dynamo properties of interest. In practical terms, this means that we should know the following items for each planet:

- External magnetic field.
- Internal structure.
- In-situ physical properties.
- Internal energy sources.
- Evolutionary history of the magnetic field and the planet itself.

In addition, we would like to have a good theoretical or numerical model that incorporates all of the above effects. Its easy to see why planetary dynamos are multifaceted processes, and it is no wonder that formal solutions of this type have not been found!

Since planetary dynamos involve so many inter-related parameters and physical effects, we should consider possible shortcuts in applying dynamo theory and dynamo models to them. Here are a list of some of the possible simplifying approaches that have proven to be useful:

- Similarity theories: assume that some of the dimensionless parameters have constant values in planetary dynamos (e.g., the $\Lambda = constant$ assumption).
- Asymptotic theories: simplify dynamo theory by assuming the dynamo equations are independent of certain physical parameters (such as fluid viscosity).
- Energy arguments: use energy and entropy balances to constrain planetary dynamos (i.e., ignore dynamics and magnetic induction processes).
- Kinematic dynamos: analyze only the magnetic induction due to an assumed velocity field (e.g., ignore the full dynamics and energetics).
- Scaling laws: extrapolate dynamo models or lab experiments to planetary conditions using scaling laws written in terms of the dimensionless parameters.
- Direct application of numerical or experimental dynamos: use physical insight to interpret planetary dynamo behavior in light of simpler numerical and experimental dynamos, without scaling.

References

Barenblatt, G.I. 1996. *Scaling, Self-sililarity, and Intermediate Asymptotics*, Cambridge University Press.

Bridgemann, P.W. 1931. *Dimensional Analysis*, Yale University Press.

Cardin, P. and Olson, P. 2007. Experiments on core dynamics, in: *Treatise on Geophysics*, vol. 8, ch. 11, Elsevier, B.V., pp. 319–344.

Connerney, J.E.P. 2007. Planetary magnetism, in: *Treatise on Geophysics*, vol. 10, ch. 7, Elsevier B.V., pp. 243–280.

Davidson, P.A. 2001. *An Introduction to Magnetohydrodynamics*, Cambridge University Press.

De Pater, I. and Lissauer, J.J. 2001. *Planetary Sciences*, Cambridge University Press.

Lodders, K. and Fegley, B. 1998. *The Planetary Scientist's Companion*, Oxford University Press.

E. Shkolnik, G., Walker, AH 2003. Evidence for Planet-induced chromospheric activity on HD 179949, ApJ **597**, 1092–1096.

Stevenson, D.J. 2003. Planetary magnetic fields, Earth Planet. Sci. Lett. **208**, 1–11.

Woan, G. 2000. *The Cambridge Handbook of Physical Formulas*, Cambridge University Press.

2. Planetary dynamos, a short tour

2.1. Planetary magnetic fields: comparison parameters

Detailed magnetic field models exist for only some of the planets, while in other cases, only the largest scale field structure is known. In addition, it is understood that many aspects of planetary magnetic fields are transient, so that spacecraft measurements, which represent snapshots in time, may not capture the true time-average state of the field. For these reasons, comparisons of planetary magnetic fields often is based on a few of their largest scale properties (Connerney, 2007). An additional complication for planetary dynamo studies is that comparison of some magnetic field parameters (such as rms magnetic field strength) should properly be made at the boundary of the dynamo region, rather than at the planetary surface. However, determining the location of the outer boundary of the dynamo region requires structural and compositional information about the planetary interior that is often poorly constrained or lacking.

The standard representation of an external planetary magnetic field is as a potential field. The theoretical basis for describing planetary magnetic fields this way is given in Backus et al. (1996). Assuming the absence of local electric currents, the magnetic field outside the dynamo region satisfies $\mathbf{B} = -\nabla\Psi$, in which the magnetic potential Ψ has the following definition:

$$\Psi = r_s \sum_{n=1}^{\infty} \sum_{m=0}^{n} \left(\frac{r_s}{r}\right)^{n+1} P_n^m(\cos\theta)(g_n^m \cos m\phi + h_n^m \sin m\phi) \tag{2.1}$$

where (r, θ, ϕ) are planet-centric spherical coordinates (radius, colatitude, and east longitude, respectively), P_n^m are the Schmidt-normalized associated Legendre polynomials, r_s is the surface radius, and g_n^m and h_n^m are the Gauss coefficients of degree n and order m.

Two important large-scale properties of the field are the dipole moment vector $\mathbf{M}$ and dipole tilt angle θ. Physically, a centered dipole field can be produced by either a small current loop at the center of the planet or by a uniformly magnetized sphere. The dipole moment can be expressed in terms of the Gauss coefficients of degree $n = 1$ in the above spherical harmonic expansion of the field as

$$\mathbf{M} \equiv (M_x \hat{\mathbf{x}} + M_y \hat{\mathbf{y}} + M_z \hat{\mathbf{z}}) = \frac{4\pi r_s^3}{\mu_0}(g_1^1 \hat{\mathbf{x}} + h_1^1 \hat{\mathbf{y}} + g_1^0 \hat{\mathbf{z}}) \tag{2.2}$$

where $(\hat{\mathbf{x}}, \hat{\mathbf{y}}, \hat{\mathbf{z}})$ are Cartesian unit vectors with origin at planet's center (z is the polar axis, x and y are axes in the equatorial plane, with x through $0°$ longitude and y through $90°$ East longitude) and (M_x, M_y, M_z) are the dipole moment components along these axes. The magnitude of the dipole moment is

$$M = \frac{4\pi r_s^3}{\mu_0} g_1 \tag{2.3}$$

where

$$g_1 = \sqrt{(g_1^1)^2 + (h_1^1)^2 + (g_1^0)^2} \tag{2.4}$$

The units of the dipole moment are $|M| = $ Am^2; physically, this can be thought of as the product of the electric current times the area of the equivalent electric circuit. Table 1 gives the dipole moment (magnitudes) of the various magnetic planets in units of Z Am^2 ($Z \equiv 10^{21}$). Much of this variation simply reflects the large differences in the volumes of the dynamo-producing regions. However, some of the variation is due to differences in the intrinsic magnetic field strengths; it is these differences that dynamo theory seeks to explain.

Although the dipole moment is just the leading order term in an infinite series expansion of the external magnetic field, it plays an important role in planetary dynamos, for several reasons. On the Earth, Jupiter, and Saturn, the dipole field dominates the other harmonics. In addition, the dipole field represents the largest

Table 1

Planetary dipole moments

Planet	Dipole moment M (Z Am^2)
Jupiter	1.46×10^6
Earth	7.8×10^1
Saturn	4.3×10^4
Uranus	3.7×10^3
Neptune	2.0×10^3
Ganymede	$\sim 4 \times 10^{-1}$
Mercury	4.2×10^{-2}
Venus	$< 6 \times 10^{-2}$

scale current system in the planetary interior. This can be seen by writing the dipole moment vector in terms of the electric current density $\mathbf{J}$ or alternatively the magnetic field $\mathbf{B}$ inside the planet as

$$\mathbf{M} = \frac{1}{2} \int (\mathbf{r} \times \mathbf{J}) dV = \frac{3}{2\mu_0} \int \mathbf{B} dV \qquad (2.5)$$

where dV denotes volume integration and $\mathbf{r}$ is the radial position vector. The dipole moment is one-half times the first moment of the current density, and is proportional to the volume integral of the internal field.

The dipole tilt, measured relative to the planet rotation axis, is given by

$$\theta = \cos^{-1}\left(-\frac{m_z}{m}\right) = \cos^{-1}\left(-\frac{g_1^0}{g_1}\right) \qquad (2.6)$$

and is illustrated in Fig. 8. The minus signs in (2.6) conform to the convention in geomagnetism, in which the geomagnetic tilt is the angle between the rotation axis and the vector $-\mathbf{m}$. This convention is not used for all the other planets. On Jupiter for example, the dipole tilt angle is usually defined without the minus sign. Figure 8 shows the field lines of a tilted dipole.

A wide range of dipole tilts are found in the solar system. Saturn's field has (essentially) zero dipole tilt, the dipole tilts on Earth and Jupiter are 10.8 and 9.6 degrees, respectively, while Uranus and Neptune have large dipole tilts, although it is known that their fields are not dipole-dominant. It is likely that the

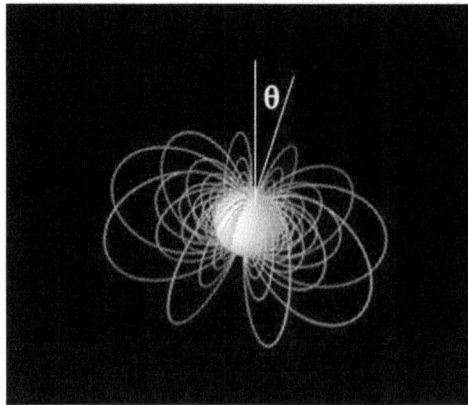

Fig. 8. Dipole field lines outside of dynamo region. Dipole tilt θ measured with respect to the rotation axis.

dipole tilt of all planets change with time. Geomagnetic and paleomagnetic measurements reveal that Earth's dipole tilt changes continuously with time (Merrill et al., 1996), and there is evidence for dipole tilt changes on Jupiter as well.

2.1.1. *Internal magnetic field intensity*

The internal magnetic energy density and the rms magnetic field intensity can be calculated at an internal radius r using the Gauss coefficients, assuming there are no intervening electric current sources, i.e., that $\mathbf{J} = 0$, between r_s and r. In this situation, the mean-square intensity over the surface of radius r is given by

$$B_{rms}^2(r) = \sum_{n=1}^{\infty} \left(\frac{r_s}{r}\right)^{2n+4} R_n \tag{2.7}$$

where

$$R_n = (n+1) \sum_{m=0}^{n} \left[(g_n^m)^2 + (h_n^m)^2\right] \tag{2.8}$$

Formulas (2.7)–(2.8) define the so-called the Mauersberger–Lowes Spectrum (Backus et al., 1996). This spectrum gives the distribution of magnetic energy density (or the rms field intensity) as a function of radius and harmonic degree n. Mauersberger–Lowes spectra allow the surface-measured field to be downward continued inside the planet, but not inside the dynamo region, because $\mathbf{J} \neq 0$ there.

Table 2 compares the surface rms magnetic field intensities of the major bodies in the solar system. What do these surface field intensities imply about the dynamo field strengths inside the dynamo region? First, it is clear that Jupiter's dynamo is intrinsically stronger than the other planets. Second, the dynamo field intensities in Earth, Saturn, Uranus, and Neptune may not be vastly different from one another. Superficially, Mercury seems to have a very weak internal field compared to the other dynamo planets.

2.2. *Sketches of individual planetary dynamos*

2.3. *Mercury*

In 1974 and 1975, the Mariner 10 spacecraft discovered that Mercury has a global magnetic field. The Mariner 10 mission observations were taken in two passes through the magnetosphere, the first and third flybys of Mercury, respectively. Maximum field intensities measured were 98 nT on the first flyby (3/29/1974), and 400 nT on the third flyby (3/16/1975). It was found that the field decreases

Table 2

Nominal planetary surface magnetic field intensities, in nanoTesla.
Probable magnetic field sources are indicted

Planet	Surface field, rms (nT)	Source
Jupiter	580,000	Active dynamo
Earth	42,000	Active dynamo
Saturn	31,000	Active dynamo
Uranus	~32,000	Active dynamo
Neptune	~18,000	Active dynamo
Ganymede	2,500	Active dynamo?
Io	<1,000	Jovian Induced?
Mercury	~450	Active dynamo?
Europa	~100	Jovian induced
Venus	<10	No dynamo
Callisto	~5	Jovian induced?
Mars	>10,000 in patches	Extinct dynamo
Moon	>100 in patches	Extinct dynamo?

like $1/r^3$, indicating a significant dipolar magnetic field. For comparison the Interplanetary Field in this region is ~20 nT (De Pater and Lissauer, 2001). The
inferred rms field strength at Mercury's surface is only about 1% that of the geomagnetic field at the Earth's surface. The limited spatial extent of Mariner observations restricts our ability to separate the dipole and quadropole terms, or to resolve higher order moments. Most field models give $g_1^0 \simeq$ 330–350 nT, reversed
with respect to the present-day geomagnetic field polarity, however, estimates of
this parameter for Mercury vary by factor of ~2. The northern hemisphere magnetic field is due to be mapped by the NASA Messenger spacecraft starting in
2011 and later by the ESA Bepi-Colombo mission.

The argument that Mercury has a partially liquid core comes from its orbital
dynamics and moments of inertia. Mercury is in a 3:2 spin-orbit resonance: the
planet rotates 3 times about its spin axis every 2 orbits, as shown in Fig. 10. In addition, solar gravitational torques acting on Mercury's equatorial bulge produce
librations, including a longitude oscillation $\phi(t)$ shown in Fig. 11. Radar measurements (Margot et al., 2007) indicate an oscillatory libration with a 36 arcsec
amplitude at the 88 day orbital period. The longitude of libration ϕ depends on
the difference in equatorial moments of inertia A, B, polar moment of inertia C,
and orbit eccentricity e according to:

$$\phi(t) \sim \frac{(B-A)}{C} f(e) \sin(\omega t) \tag{2.9}$$

The coefficient $(B - A)$ is determined using Mariner 10 gravity data: $C_{22} = (B - A)/4Mr_s^2 \sim 10^{-5}$. Large ϕ requires a small value of C, which implies

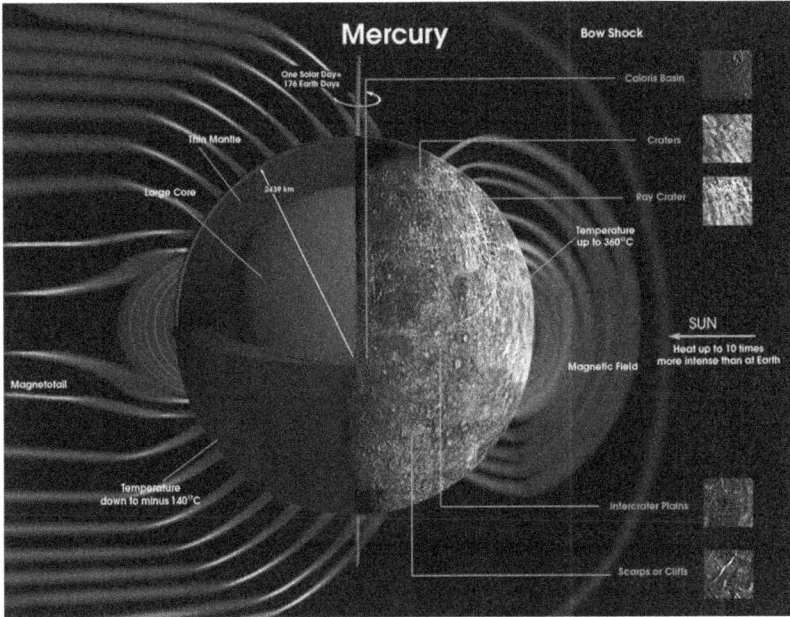

Fig. 9. Mercury's magnetic field structure. Credit: ESA.

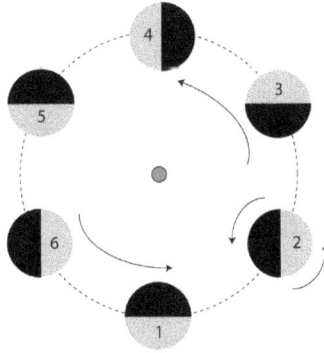

Fig. 10. Mercury's 3:2 spin-orbit resonance.

that only the solid mantle contributes to C, not the core. Decoupling between the mantle and core would be expected with a liquid outer core.

Proposals for the origin of Mercury's field include an internally generated but rather weak dynamo, remanent crustal magnetization, and thermo-electric

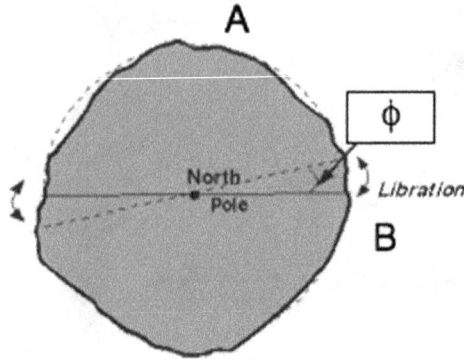

Fig. 11. Schematic of Mercury's orbital libration.

currents (Stevenson, 2003). Most investigators now favor the weak dynamo hypothesis. A uniformly magnetized crust with constant thickness does not work as a source, because its external field would be identically zero, a consequence of Runcorn's theorem. To produce an external field, the crustal magnetization would have to be non-uniform and distributed somehow to produce a global dipolar field, an unlikely situation. The crust of Mercury does not appear to have such global heterogeneity, although it is possible there are some variations in the depth of the Curie isotherm due to heterogeneity in the near-surface thermal regime. The thermo-electric current hypothesis is rather speculative, as it requires large-scale temperature variations along Mercury's core-mantle boundary.

Dynamo models seek to explain why Mercury's field is so weak compared to Earth's, despite its overall similar mode of generation. Interestingly, two very different dynamo models have been proposed, one assuming dynamo action in a thin shell near the outer part of the core (Stanley et al., 2005; Takahashi and Matsushima, 2006)), the other assuming dynamo action deep within the core (Christensen, 2006). Both predict weaker external fields than the Earth, but for different reasons. Most thermal evolution models for Mercury (Schubert et al., 1988) predict the heat flow from the core is small, $q_{cmb} \sim 5$ mW/m^2, which would imply subadiabatic, i.e., stable thermal stratification in the core. The same models predict low absolute core temperatures, implying either a solid iron core or a liquid iron alloy core with a substantially depressed melting temperature relative to pure iron.

Summary of properties:

- Estimated surface field $\sim$300–500 nT.
- Best-fitting dipole tilt $\theta = 7°$.
- Slow rotation period of 88 days. Spin-orbit resonance of 3:2. See Fig. 10.

• High mean density $(5,440 \text{ kg/m}^3)$ indicates a high percentage of metal, around 70% by mass.
• Fe–FeS core would occupy about 75% of the planet's radius and 42% of its volume.
• Small Elsasser number implied by surface field: $\Lambda = \sigma B^2 / \rho \Omega \sim 10^{-4}$.
• Is this a weak field dynamo?
• Small size implies rapid planetary cooling. How has a dynamo lasted so long?

2.4. Venus

Venus is similar to Earth in terms of its size, mean density, and bulk composition. Unlike Earth, however, it has no measurable magnetic field of its own. Apart from exotic possibilities (such as a "hidden" dynamo that has zero external field) we are forced to conclude that it lacks an active dynamo. Because of its elevated surface temperature, there is little or no crustal magnetization, so we can't even say if Venus ever had an active dynamo. There are several proposed explanations of the lack of a dynamo on Venus. The high surface temperature suggests that perhaps the internal heat flux is too low to support convection (Schubert et al., 1997), and that its core is stably stratified, with fluid velocities that are subcritical for dynamo action. Another proposal is that Venus lacks a growing solid inner core, and is therefore missing an important energy source for its dynamo (Nimmo and Stevennson, 2002). Previously it was thought that Venus' rotation was too slow for dynamo action, but this notion is contradicted by successful numerical dynamos which rotate even more slowly!!

Summary of properties:
• No measured external magnetic field.
• Lacks evidence of crustal magnetization.
• Probably no dynamo, although some have suggested a dynamo without an external field.
• Elevated surface temperature from runaway greenhouse atmosphere.
• Liquid metallic core probable.
• 60 day rotation period—not too slow for a dynamo.

2.5. Earth

Earth's magnetic field is by far the best known planetary field, in terms of both spatial and temporal resolution. Here we summarize just its basic properties (which are fully described elsewhere in this Volume and in several general references; see Merrill et al., 1996; Gubbins and Berverra, 2007), in order to provide comparison with the other planets discussed in this chapter.

The present-day geomagnetic core field (up to spherical harmonic degree 14) is about 85% dipolar at the surface, and about 60% dipolar at the cmb (core-

P. Olson

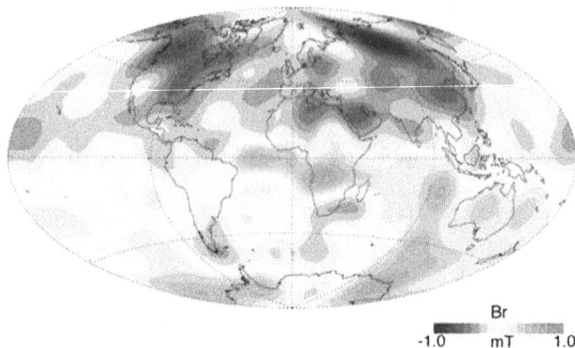

Fig. 12. Radial magnetic field at the cmb from the 2000 Oersted core field model.

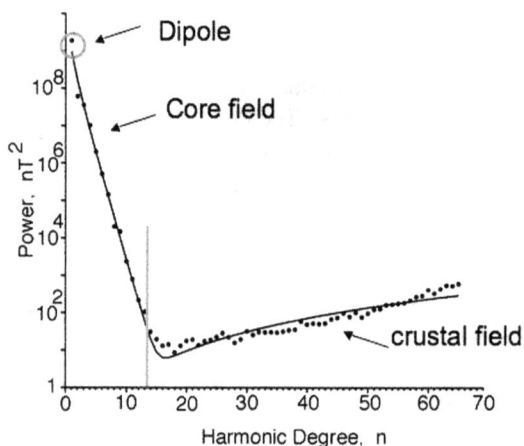

Fig. 13. Earth's surface magnetic field spectrum obtained by Oersted satellite for epoch 2000.

mantle boundary). The map in Fig. 12 shows contours of the radial component of the geomagnetic field on the cmb at epoch 2000 from the Oersted satellite, resolved to spherical harmonic degree $n = 14$. At higher harmonics the filtering effects of crustal magnetization obscure the core field. The axial dipole moment comes mostly from the four high latitude flux spots arrayed more-or-less symmetrically with respect to the equator. The rms field is about 0.4 mT on the cmb. Reversed flux spots are concentrated in the southern hemisphere.

The history of the geomagnetic field is unique in offering a perspective on the evolution of a planetary dynamo, including variations on a wide range of time scales. The geomagnetic dipole moment has been decreasing monotonically since its first measurement by Gauss just prior to 1840. As shown in Fig. 14, the average rate of decay is about 6% per century. For comparison, the free Ohmic

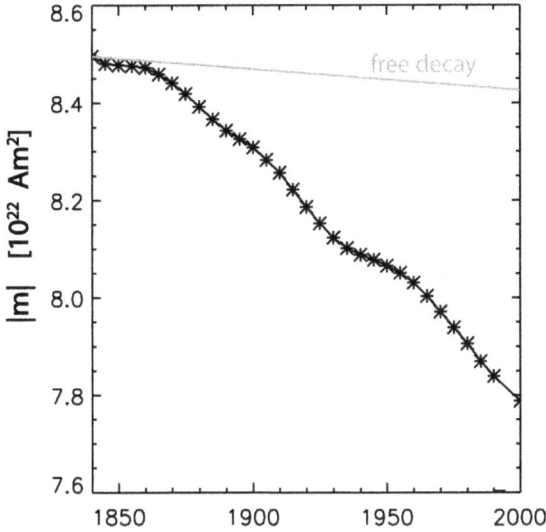

Fig. 14. Earth's dipole intensity decay (after Jackson et al., 2000).

decay (the decay due to electrical resistance alone, without any effects of fluid motion) of the dipole moment in a sphere with radius r_c and uniform electrical conductivity σ is given by

$$\mathbf{M} = \sum_{p=1}^{\infty} \mathbf{M}_p \exp(-t/\tau_p) \tag{2.10}$$

where p is the radial mode number, $\mathbf{M}_p$ is its amplitude at time $t = 0$, and $\tau_p = \mu_0 \sigma r_c^2/(p\pi)^2$. Using $r_c = 3480$ km for the core radius and $\sigma = 4 \times 10^5$ Sm^{-1} (corresponding to a magnetic diffusivity of $\eta = 1/\mu_0\sigma = 2$ m^2 s^{-1}) for the electrical conductivity of the core (Poirier, 1994; 2000) gives $\tau_1 = 19{,}600$ years for the free decay time of the fundamental mode and $\tau_p = 19{,}600/p^2$ for the free decay time of the pth dipole mode. The historical dipole decay is therefore about 12 times its free decay rate, indicating that the geodynamo is actively extracting energy from the dipole part of the geomagnetic field. As shown in Fig. 15, this decrease has coincided with the development of reversed magnetic flux regions on the cmb.

Figure 16 shows the historic path of the north geomagnetic pole, the projection of the negative end of the geocentric dipole moment vector onto the surface. Both the tilt and the longitude of the dipole axis vary with time, as shown. Figure 17 shows the variation of the Virtual Axial Dipole Moment (VADM) over the past 2

REVERSED MAGNETIC FLUX ON THE CMB

RADIAL FIELD AT 1900   REVERSED FLUX SPOTS

RADIAL FIELD AT 2000   SECULAR VARIATION

-1.0   mT   1.0   -10.   nT/yr   10.

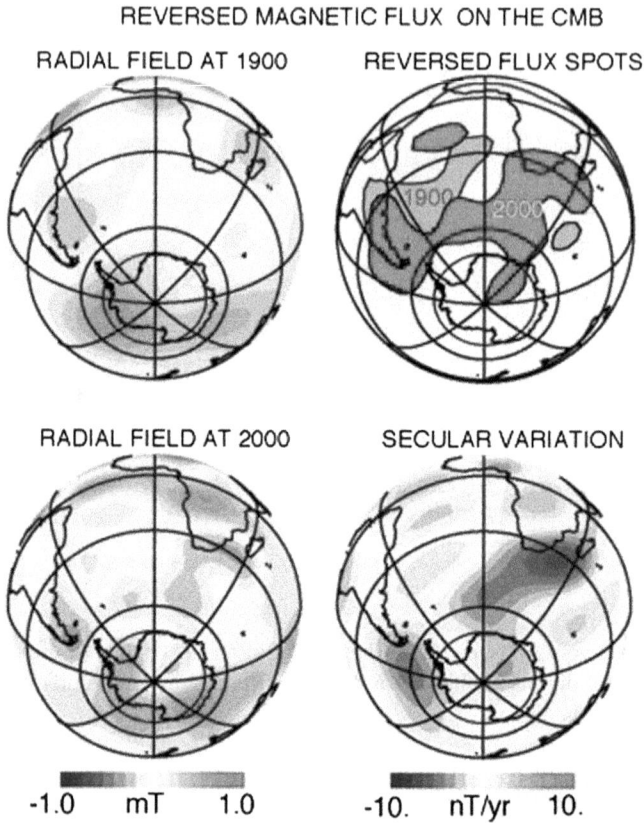

Fig. 15. Historical changes in the southern hemisphere of the core field. Left: Radial magnetic field in 2000 and 1900. Right: Reversed flux and secular variation.

Ma, as determined from paleomagnetic measurements. Magnetic polarity chrons, shown in black and white above the plot, tend to have boundaries where the VADM decayed rapidly for a short time.

Summary of properties:

- $B_{rms,surface} \sim 40,000$ nT.
- Surface field $\sim$85% dipolar.
- Core-Mantle boundary field $\sim$60% dipolar.
- $M = 78$ Z Am^2.
- Dipole tilt $\theta = 10.8°$.
- The only active planetary dynamo with a well-constrained internal structure.
- Wide spectrum of magnetic variability in time, including polarity reversals and excursions.

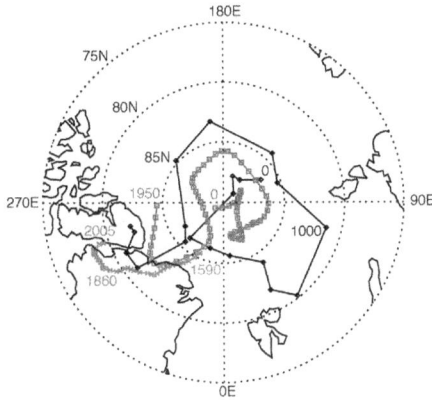

Fig. 16. Earth's north geomagnetic dipole path. Green is 2000 yr global field model by Korte and Constable (2006). Historical path is from Jackson et al. (2000). Black is archeomagnetic pole positions from Merrill et al. (1996).

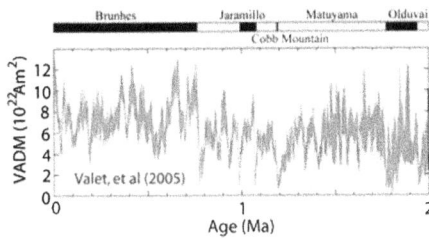

Fig. 17. Virtual axial dipole moment over the past 2 Ma. Magnetic polarity chrons are shown above the plot (Valet et al., 2005).

2.6. Mars

Following numerous unsuccessful attempts, Mariner 4 succeeded in measuring a magnetic field around Mars in 1965. However, that field was indistinguishable from the interplanetary field. The first unambiguous measurements indicating an internal field were made during the Mars Global Surveyor mission in 1997. During its mapping orbit (a near circular polar orbit), no evidence of a present-day dynamo-type field was seen, however very large crustal anomalies were measured, > 1000 nT at spacecraft altitudes. The largest anomalies are limited to the southern hemisphere, as shown in Fig. 18. Smaller anomalies are scattered elsewhere, including the northern hemisphere. Globally, there is a lack of strong anomalies in and around the largest impact basins. The broad-scale anti-correlation between large impact basins and the largest crustal anomalies

P. Olson

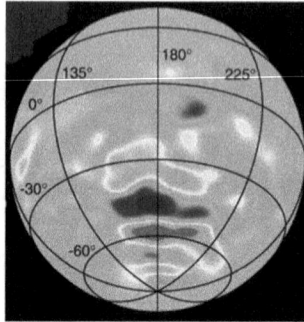

Fig. 18. North-south gradient of the radial magnetic field in the southern hemisphere of Mars, from Connerney et al., 2004.

suggests that the inducing field was turned off prior to the end of heavy bombardment, that is, prior to about ~3.8 Ga.

Meteorites from Mars offer some tantalizing clues about its early magnetic history. Martian meteorites are often magnetized and sometimes show a preferred direction of magnetization. Shergotty, Nahkla, Chassigny (SNC) and similar ones have ages ranging from 180 Ma–1.8 Ga and suggest a paleofield intensity of 0.5–5.0 μT based on their magnetization (Weiss et al., 2002). The oldest Martian meteorite, Allen Hills ALH84001, was famously interpreted as indicating life formerly on Mars, but this interpretation has since fallen into disrepute. Gas inclusions contain relatively unfractionated atmosphere, as opposed to Mars's present atmosphere, suggesting that an early magnetic field was strong enough to protect the atmosphere from stripping by solar wind that has led to the present-day fractionation. Our current knowledge of the Martian core is limited. From the moment of inertia $I = 0.365Mr^2$ and mean density $\rho = 3{,}930$ kg/m^3 we infer a radius of the core $r_c \sim 1{,}300$–$1{,}500$ km. The core-to-surface ratio is therefore $r_c/r_s \sim 0.5$–0.6, assuming an Earth-like structure and bulk composition (Schubert et al., 2001). One big question is whether the Martian core is still liquid, which could be answered by placing seismometers on the Martian surface.

Assuming that Mars had an ancient dynamo, then a determination of its onset and demise is important for several reasons. For example, it may preserve information on how terrestrial planetary dynamos initiated. Its demise may be related to changes in core-mantle interaction, possibly correlated with changes in Martian hotspot volcanism at the surface. Lastly, the loss of Martian dynamo might possibly have precipitated surface environment changes that affected its habitability for living organisms.

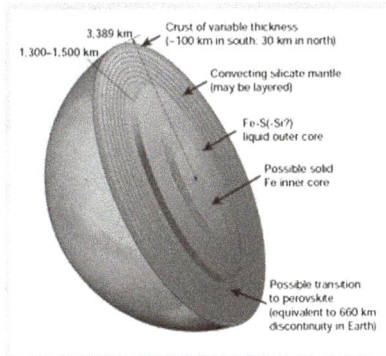

Fig. 19. A model of the internal structure of Mars. Credit: D. Stevenson, Nature.

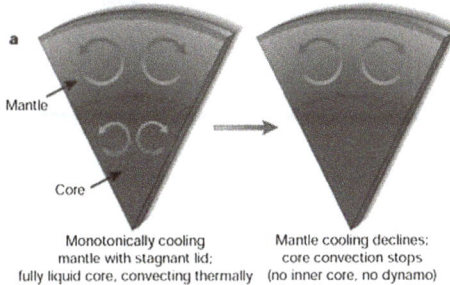

Fig. 20. Demise theory I for the Martian dynamo. Credit: D. Stevenson, Nature.

2.6.1. Why did the Martian dynamo die?

Here there are lots of speculative possibilities. The following is a summary of these, as reported in Stevenson (2001).

Demise Theory I supposes monotonic planetary cooling through a static lid lithosphere with a convecting mantle. The mantle cooling rate declines with time. Eventually the core stratifies and the dynamo dies, as illustrated in Fig. 20.

Demise Theory II supposes enough monotonic cooling of the core to freeze the outer liquid core and grow the solid inner core. This freezing then consumes too much of the convectively-active region, and the convection becomes subcritical for dynamo action, as illustrated in Fig. 21.

Demise Theory III assumes that early Mars had substantial mantle plate tectonics and/or mantle plumes, which cooled the mantle enough to allow for convection in the outer core and an active dynamo. Subsequently the plume and the subduction activity diminishes, core heat loss diminishes, the core stratifies, and the dynamo dies due to a lack of convection. This progression is illustrated in Fig. 22.

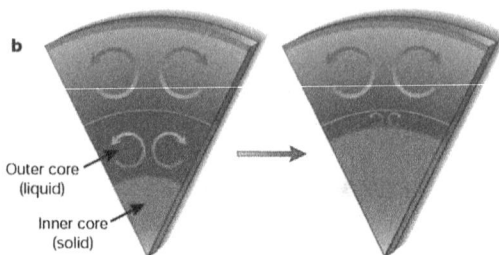

Fig. 21. Demise theory II for the Martian dynamo. Credit: D. Stevenson, Nature.

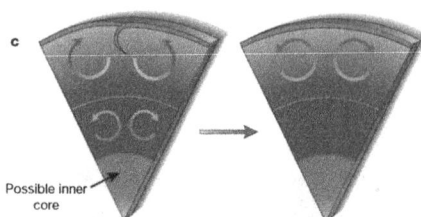

Fig. 22. Demise theory III for the Martian dynamo. Credit: D. Stevenson, Nature.

2.6.2. Big questions

There are many first-order questions surrounding the ancient Martian dynamo, such as:

- Why are Martian crustal anomalies so intense?
- What is the major magnetic carrier?
- Why are largest anomalies limited to southern hemisphere?
- What caused the demise of the Martian magnetic field?
- Did the loss of its magnetic field cause loss of a habitable surface environment?

2.7. Jupiter

Jupiter's magnetic field was originally detected by Burke and Franklin (1955), when in the process of calibrating their radio telescope, they discovered a strong radio source coming from the planet. Jupiter's field was confirmed and first mapped during the Pioneer 10 mission flyby in 1973. Since then it has been mapped by Pioneer 11, Voyager 1 and 2, and most recently in 1992 by Ulysses. The Galileo and Cassini missions provided some additional flyby measurements.

Jupiter's field is tilted at about 9.6°, compared to the tilt of Earth's dipole, which is now about 10.8°. The sign of the Jovian field is reversed from Earth's,

Table 3
Gauss coefficients for Jupiter and Saturn in mT

n	m	Jupiter[a]		Saturn[b]	
		g_n^m	h_n^m	g_n^m	g_n^m
1	0	4.208	0.	0.2144	0.
1	1	−0.660	0.261	−0.0014	0.0014
2	0	−0.034	0.	0.0188	0.
2	1	−0.759	−0.294	−0.0052	−0.0043
2	2	0.483	0.107	0.0050	−0.0004
3	0	. . .	0.	. . .	0.
3	1	. . .	. . .	. . .	. . .
3	2	0.263	0.695	. . .	. . .
3	3	−0.069	−0.247	. . .	. . .

[a] from Connerney et al. (1982).
[b] from Acuña et al. (1983).

and its dipole moment is by far the largest among the planets—nearly 20000 times larger than the Earth's. The surface dipolarity of Jupiter's field (the ratio of dipole to total field) is comparable to the Earth's, and there is evidence for secular variation from the change of dipole tilt between the Voyager and Galileo missions. Table 3 gives the Gauss coefficients of a model of Jupiter's field derived from Voyager measurements.

Jupiter interior is mainly comprised of hydrogen and helium, in proportions that are similar to the sun. Beneath the cloud layers that are commonly seen in images of the planet is a "mantle" of fluid molecular hydrogen and helium, again with a mixing ratio (composition) similar to the sun (Guillot, 1999). At a depth interval starting around $0.8R_J$ and pressures near 150 GPa, there is a transformation of the molecular hydrogen into an electrically conducting metallic fluid. It is now understood that this transition is not a discontinuity as was first thought, but is distributed over a substantial pressure interval. As the electrical conductivity increases, the magnetic diffusivity decreases slowly with depth reaching perhaps $\eta = 10$–$100 \text{ m}^2 \text{ s}^{-1}$ in the metallic region (Nellis, 2000). In spite of the gradual nature of this transition, it is generally thought that dynamo processes similar to those that generate the geodynamo on Earth also are responsible for generating Jupiter's field (Jones, 2003). In particular, it is very likely that thermal convection is occurring in the metallic region. Jupiter radiates about twice as much energy as it receives from the sun (Pirraglia, 1984), and thermal convection is the only mechanism that transfers heat efficiently enough from the deep interior to account for this. The internal energy probably comes from its continued gravitational collapse (Boss, 2002). However, there are some significant differences with the Earth that may be important for understanding the Jovian

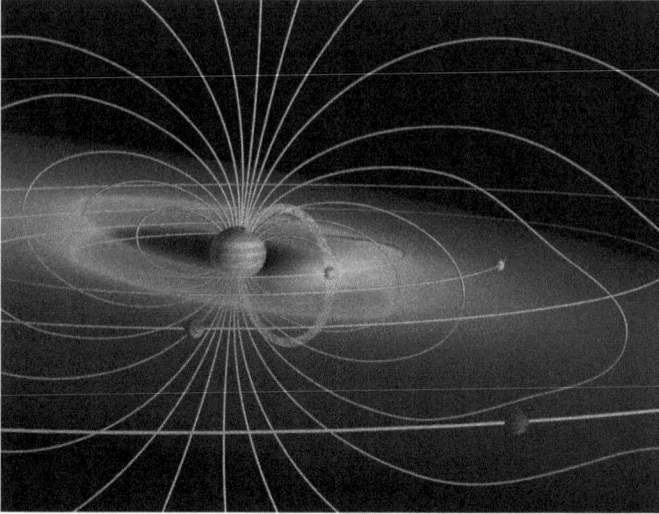

Fig. 23. Graphic of Jupiter's external field interacting with satellites. (Credit: John Spencer.)

dynamo, particularly the depth-dependent electrical conductivity and the effects of compression.

2.7.1. Atmosphere
- Absorption $\sim 5 \times 10^{17}$ W
- Emission $\sim 8.4 \times 10^{17}$ W
- Intrinsic $\sim 3.4 \times 10^{17}$ W
- Temperatures: 168 K at 1 bar; 428 at 22 bar
- Zonal winds 100 m/s to 22 bar (at least)
- Mixing ratios (element mass/total mass) Y[He], X[H]
- Jupiter: Y/(X + Y) 0.24
- Proto solar nebula: Y $\sim$0.275
- Jupiter's atmosphere is slightly He-depleted (via He-precipitation?)
- Polar regions: intrinsic emission > absorption

2.7.2. Deep interior
- 87–97% light elements (H+He)
- 3–13% heavy elements (solar = 2%)
- T $\sim$20,000 K; P $\sim$0.1–1 TPa fluid interior
- Dense liquid metallic H plasma interior state
- Residual contraction rate 3 cm/yr
- Buoyancy flux $\sim 10^{-11}$ m^2 s^{-3}

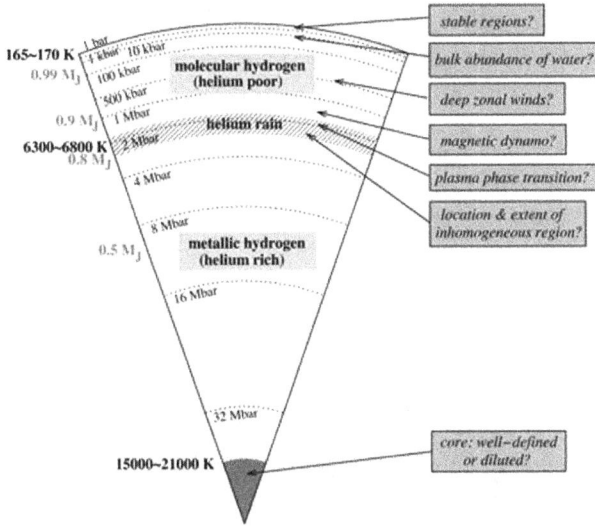

Fig. 24. Model of the interior of Jupiter.

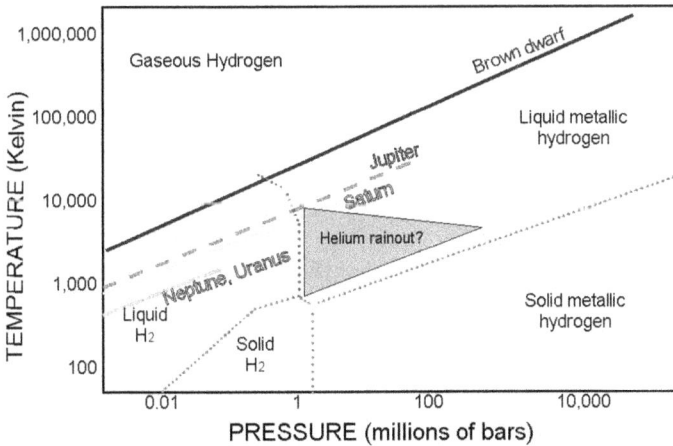

Fig. 25. Hydrogen phase diagram.

2.7.3. Big questions

There are many questions about the Jovian dynamo, some of which may be answered by the upcoming Juno spacecraft mission. These include:

• What is the form of convection in Jupiter's electrically conducting interior?

• Does the similarity between the Jovian and Geomagnetic external field structures signify similar dynamo mechanisms?

• How important are compressibility and gradients in electrical conductivity?

- Do the zonal atmospheric winds couple with the Jovian dynamo?
- Why is Saturn's dynamo far more axisymmetric than Jupiter's?

2.8. Saturn

Following the measurements of Jupiter's field, it was widely thought that Saturn's field would be very similar. The first flyby of the planet in 1979 by the Pioneer 11 mission and later Voyager 1 and 2 missions proved otherwise. Like Jupiter, Saturn's field is reversed in polarity compared to the Earth's and is also relatively strong, about 0.2 mT near the poles, although its dipole moment is only 1/30 that of Jupiter. But unlike Jupiter and the Earth, Saturn's dipole has little or no tilt with respect to its rotational axis (field models constrain it to be less than 1°). In fact, from the flybys of the planet, the field can be adequately described by the sum of an axial dipole, an axial quadrupole and an axial octupole field.

The interior of Saturn is thought to have a composition broadly similar to Jupiter's, with molecular hydrogen being transformed into metallic hydrogen at a depth corresponding to 40–50% of its surface radius. This is where the dynamo thought to be generated. Jupiter's metallic hydrogen region is about twice as deep as Saturn's, and this difference in conducting fluid volume may account for Jupiter's much stronger field.

The near axisymmetry of Saturn's field has been the subject of much confusing speculation, and it is a source of ambiguity in determining Saturn's rate of rotation. If Saturn's field were exactly axisymmetric and steady, it would violate Cowling's Theorem, which prohibits steady axisymmetric dynamos. However, the dynamo process may entail only small deviations from axisymmetry and steadiness, so there is no real contradiction in this case. Even so, it is clear that some physical process in Saturn's interior leads to greater magnetic symmetry than found in other planets. There is no widely accepted reason as to why Saturn's field has this axisymmetric structure. One idea is that the non-axisymmetric field is shielded by a layer of conducting fluid rotating differentially with respect to the rest of the planet (Stevenson, 1982). It has been speculated that induction effects in this layer block out the asymmetrical part of the field, preventing it from reaching the surface. However this concept is yet to be substantiated by models (Love, 2000). The interior dynamics of Saturn may differ from Jupiter in important respects. Since the interior of Saturn is a lower pressure, lower temperature environment compared to Jupiter's interior, it is possible that helium, which is dissolved in the metallic hydrogen, may exsolve and "rain out" from the dynamo region (Fortney and Hubbard, 2003).

2.9. Uranus and Neptune

Voyager II flew past Uranus and Neptune in 1986 and 1989, respectively, providing enough magnetic measurements to construct low-order global field models

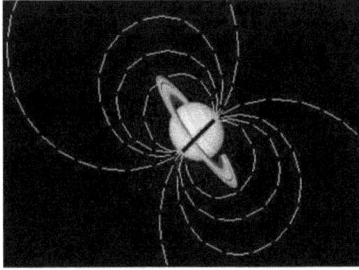

Fig. 26. Graphic of Saturn's external magnetic field.

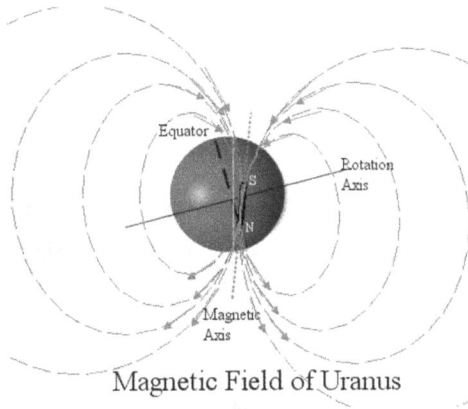

Fig. 27. Uranus' external magnetic field.

for both planets. To the surprise of many, the two planets have magnetic fields that deviate from the axisymmetry expected on the basis of their large size and relatively rapid rates of rotation. Furthermore, in neither case is the field even remotely dipole dominant. Figures 27 and 28 show the external field structure of the two planets.

In another paper in this Volume, S. Stanley describes what is known about the interior structure and physical properties of both planets, and reviews existing dynamo theories for why their magnetic fields differ so starkly from Jupiter and Saturn on one hand, and the Earth on the other.

2.10. Ganymede

One of the most important results of the Galileo spacecraft mission was the discovery that Ganymede possesses its own global magnetic field, and that a self-sustaining dynamo seems to be the most viable explanation (Kivelson et al.,

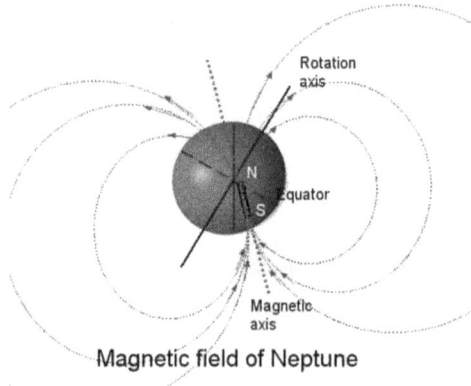

Magnetic field of Neptune

Fig. 28. Neptune's external magnetic field.

2002). Ganymede is the largest satellite in the solar system, with a central core of nearly 700 km in radius, assuming it is composed of iron. This is adequately large for a self-sustaining dynamo provided the core is in fact liquid. Nominal thermal history considerations applied to an object of its size would indicate the core should have long since solidified, unless it possesses some anomalous heat source, or unless its melting temperature is markedly reduced compared to pure iron (Hauck et al., 2006). Another interesting possibility, suggested by the near alignment of Ganymede's field with the Jovian field, is that the Jovian field is assisting Ganymede's dynamo (Sarson et al., 1997).

2.11. Comparison of the magnetic state of the Galilean satellites

• *I*o It is inconclusive whether Io has an intrinsic magnetic field of its own, or whether its field is entirely induced by Jupiter's field.
• *E*uropa The evidence indicates that Europa has an induced magnetic field. One interesting possibility is that the induction takes place in a slightly conducting, subsurface ocean.
• Ganymede As discussed above, Ganymede seems to have both internally and externally- induced magnetic fields. The internal field is strong enough to stand-off Jupiter's main field and create mini-magnetosphere. Ganymede is the most centrally-condensed solid planetary body in the solar system, with a moment of inertia coefficient $I/MR^2 = 0.3105$. The ready explanation is a large iron-rich core. Experimental determination of eutectic melting temperatures in the Fe–FeS system show a decrease in melting temperature with increasing pressure over the range of pressures relevant to Ganymede (up to 14 GPa), suggesting the possibility of iron "snowing" from shallow levels in core to deeper levels (Hauck et al., 2006).

Table 4

Estimated planetary core properties

Planet	Rotation rate	R_{core} (km)	ρ_{core}	Electrical conductivity	Core viscosity	Magnetic Moment (Z Am^2)
Mercury	1.24e-6	1900	10e3	6e5	1e-6	0.03
Earth	7.29e-5	3485	11e3	6e5	1e-6	78
Jupiter	1.76e-4	60000	1.8e3	2e4	1e-6	1.4e6
Saturn	1.64e-4	35000	1.8e3	2e4	1e-6	4.5e4
Uranus	1.01e-4	17000	2e3	1e4	1e-6	3.9e3
Neptune	1.09e-4	17000	2e3	1e4	1e-6	1.8e3

Table 5

External planetary properties

Planet	Distance (AU)	Rotation Period	Orbital Period	Axis Tilt	Radius (x R_E)	Mass (x M_E)	Dipole-Moment (x M_{DE})	Dipole Tilt	Moons
Sun	0.000	25 days	N/A	7.25°	109	332,000	Varies	Varied	N/A
Mercury	0.387	58.7 days	88 days	~0°	0.382	0.055	~4.7x10^{-4}	~10°	0
Venus	0.723	243 days	225 days	-2°	0.949	0.815	N/A	N/A	0
Earth	1.000	23.9 hrs	365 days	23.5°	1.000	1.000	1.000	11.5°	1
Mars	1.520	24.6 hrs	1.88 yrs	24°	0.533	0.107	?	?	2
Jupiter	5.200	9.92 hrs	11.9 yrs	3.1°	11.200	318	20,000	9.6°	28
Saturn	9.540	10.7 hrs	29.5 yrs	29°	9.450	95.200	540	0.7°	18
Uranus	19.200	-17.29hrs	84.0 yrs	-82.1°	4.100	14.600	48	59°	17
Neptune	30.100	16.11 hrs	165 yrs	28.8°	3.880	17.200	26	47°	8
Pluto	39.400	6.39 days	248 yrs	>50°?	~0.24	0.002	?	?	1

- Callisto Callisto evidently has only an induced field. It is less differentiated than Ganymede, with $I/MR^2 = 0.3549$, still large enough to imply a denser core of some unknown composition.

2.12. The ancient Lunar dynamo

Lunar samples collected during the Apollo missions show that the surface rocks on the Moon were magnetized by a relatively strong magnetic field. However, due to the estimated cooling rate of the moon and evidence that the rocks after

Table 6

Additional planetary properties

Body	Mass (kg)	Avg. Radius (km)	Avg. Density (Mg/m³)	Moment of inertia factor (I/MR²)	Internal composition
Sun	1.991×10^{30}	695,950	1.410	0.006	By mass H 78.5%, He19.7%, O 0.86%, C 0.4%, Fe 0.14%, others 0.54%.
Mercury	3.181×10^{23}	2,433	5.431		Large Fe core.
Venus	4.883×10^{24}	6,053	5.256		Similar to Earth. Convecting mantle, core.
Earth	5.979×10^{24}	6,371	5.519	0.3308	Crust/mantle/core.
Moon	7.354×10^{22}	1,738	3.342	0.392	Thick lithosphere, olivine/pyroxene mantle, tiny core.
Mars	6.418×10^{23}	3,380	3.907	0.366	Thick lithosphere, olivine mantle, moderate core.
Jupiter	1.901×10^{27}	69,758	1.337	0.254	Hydrogen → metallic form when compressed. Ice/rock core.
Saturn	5.684×10^{26}	58,219	0.688	0.210	Hydrogen, less He than Jupiter? Sizeable ice/rock core.
Uranus	8.682×10^{25}	23,470	1.603	0.225	Either mostly ices, or hydrogen/helium/rock, or combination of both.
Neptune	1.027×10^{26}	22,716	2.272	0.23	Either mostly ices, or hydrogen/helium/rock, or combination of both.
Pluto	$1.0 ? \times 10^{24}$	5,700	1.65 ?		Ice covered.

the last great bombardment show no signs of a magnetic field, a possible Lunar dynamo is estimated to have lasted for only a few million years, and went extinct before 3.5 Ga. The moon's core is quite small, and the evidence for its liquidity is limited (Williams et al., 2001). However, a transient dynamo could possibly have been sustained during a short-lived, early episode of mantle convection (Stegman et al., 2003).

2.13. Outstanding questions in planetary dynamo theory

- Why is Mercury's (dynamo) field so weak?
- Why does Venus not have any measurable magnetic field?
- Why does Earth's field reverse its polarity?
- How did the geodynamo work prior to inner core formation?

- What killed the Martian (and Lunar) dynamos?
- What keeps Ganymede's dynamo alive today?
- Why is Saturn's field so axisymmetric while Jupiter's is more Earth-like?
- Why are the fields of Uranus and Neptune both non-dipolar?
- Are there planetary magnetic field scaling laws?

References

Acuna, M.H., Connerney, J.E., Wasilewski, P. et al. 2001. Magnetic field of Mars: 2001 Summary of results from the aerobraking and mapping orbits, J. Geophys. Res. Planets **106**, 23403–23417.

Backus, G., Parker, R. and Constable, C. 1996. *Foundations of Geomagnetism*, Cambridge University Press.

Boss, A.P. 2002. Formation of gas and ice giant planets, Earth Planet. Sci. Lett. **202**, 513–523.

Burke, Franklin, 1955.

Christensen, U.R. 2006. A deep dynamo generating Mercury's magnetic field, Nature **444**, 1056–1058.

Connerney, J.E.P., Acuöa, M.H., Ness, N.F., Spohn, T. and Schubert, G. 2004. Mars crustal magnetism, Space Science Reviews, vol. 111, issue 1, pp. 1–32, doi: 10.1023/B:SPAC.0000032719.40094.1d.

Gubbins, D., Herrero-Bervera, E., eds. 2007. *Encyclopedia of Geomagnetism and Paleomagnetism*, Springer.

Guillot, T., 1999. Interiors of giant planets in side and outside the solar system, Science **286**, 72–77.

Merrill, R.T., McElhinny, M.W. and McFadden, P.L. 1996. *The Magnetic Field of the Earth*, Academic Press.

Margot, et al. 2007. Science **316**, 710.

Nimmo, F. and Stevenson, D.J. 2002. Why does Venus lack a magnetic field? Geology **30**(11), 987–990.

Fortney, J.J. and Hubbard, W.B. 2003. Phase separation in giant planets: inhomogeneous evolution of Saturn, Icarus **164**, 228–243.

Hauck, S.A., Aurnou, J.M. and Dombard, A.J. 2006. Sulfur's impact on core evolution and magnetic field generation on Ganymede, J. Geophys. Res. **111**, E09998.

Jones, C.A. 2003. Dynamos in planets, in: Thompson, M.J. and Christensen-Daalsgard, J., eds., *Stellar Astrophysical Fluid Dynamics*, Cambridge University Press, pp. 159–176.

Kivelson, M.G., Khurana, K.K. and Volwerk, M. 2002. The permanent and inductive magnetic moments of Ganymede, Icarus **157**(2), 507.

Love, J.J. 2000. Dynamo action and the nearly axisymmetric magnetic field of Saturn, Geophys. Res. Lett. **27**, 2888–2892.

Nellis, W.J. 2000. Metallization of fluid hydrogen at 140 GPa (1.4 bar): implications for Jupiter, Planet. Space Sci. **48**, 671–766.

Pirraglia, J.A. 1984. Meridional energy balance of Jupiter, Icarus **59**, 169–176.

Sarson, G.R., Jones, C.A., Zhang, K. and Schubert, G. 1997. Magnetoconvection dynamos and the magnetic field of Io and Ganymede, Science **276**, 1106–1108.

Schubert, G., Ross, M.N., Stevenson, D.J. and Spohn, T. 1988. Mercury's thermal history and the generation of its magnetic field, in: F. Vilas et al., eds., Mercury, University of Arizona Press, Tucson, AZ, pp. 429–460.

Schubert, G., Solamotov, V.S., Tackley, P.J. and Turcotte, D.L. 1997. Mantle convection and the thermal evolution of Venus, in: S.W. Bougher et al., eds., Venus II, University of Arizona Press, Tucson, AZ, pp. 1245–1287.

Schubert, G., Turcotte, D.L. and Olson, P. 2001. *Mantle Convection in the Earth and Planets*, Cambridge University Press.

Stanley, S., Bloxham, J., Hutchinson, W.E. and Zuber, M.T. 2005, Thin shell dynamos consistent with Mercury's weak observed magnetic field, Earth Planet Sci. Lett. **234**, 27–38.

Stegman, D.R., Jellinik, M., Zatman, S.A., et al. 2003. Lunar core dynamo driven by thermo-chemical mantle convection, Nature **421**, 143–145.

Stevenson, D.J. 2003. Planetary magnetic fields, Earth Planet. Sci. Lett. **208**, 1–11.

Stevenson, D.J. 2001. Mars' core and magnetism, Nature **412**, 214–219.

Stevenson, D.J. 1982. Reducing the non-axisymmetry of a planetary dynamo and an application to Saturn, Geophys. Astrophys. Fluid Dyn. **21**, 113–127.

Takahashi F. and Matsushima, M. 2006. Dipolar and non-dipolar dynamos in a thin shell geometry with implications for the magnetic field of Mercury, Geophys. Res. Lett. **33**, L10202.

Weiss, B.P., Hojatollah, V., Beudenbachere, F.J., Kirschvink, J.L., Steward, S.T. and Shuster, D.L. 2002. Records of an ancient Martian magnetic field in ALH84001, Earth Planet. Sci. Lett. **201**, 449–463.

Williams, J.G., Boggs, D.H., Yoder, C.F. et al. 2001. Lunar rotational dissipation in solid body and molten core, J. Geophys. Res. Planet. **106**, 27933–27968.

3. Deep Earth structure

3.1. The structure of the Earth's core

The Earth is the natural laboratory for planetary dynamo studies. In addition to knowing far more about the geomagnetic field than the magnetic fields of the other planets, we have much more precise information on Earth's interior structure, composition, and temperature, compared to the other planets.

Seismic waves provide the most quantitative information on Earth's deep interior. Observations of seismic body waves provide information on the distribution of compression and shear velocity throughout the core and the mantle. Reviews of seismic waves in the core and lower mantle can be found in Romanowicz and Dziewonski (2007), particularly the chapters on the core (Sauriau, 2007) and lower mantle (Lay, 2007). Here we show how the radial profile of density in the core can be retrieved from these wave speeds. Density and seismic velocity are consistent with an iron core slightly diluted with some lighter elements. Seismic waves also indicate that the liquid outer core is very nearly homogeneous (that is, well-mixed), whereas the two solid regions that bound the outer core—the lowermost mantle and the solid inner core—are laterally heterogeneous. Seismic and thermodynamics properties of the core (see Stacey, 1992) also indicate that the adiabatic thermal gradient in the outer core is large and the flow of heat from the core is also large.

The propagation speeds of the compression (P) and shear (S) body waves are given by

$$V_p = \sqrt{(K + 4\mu/3)/\rho}, \qquad V_s = \sqrt{\mu/\rho} \qquad (3.1)$$

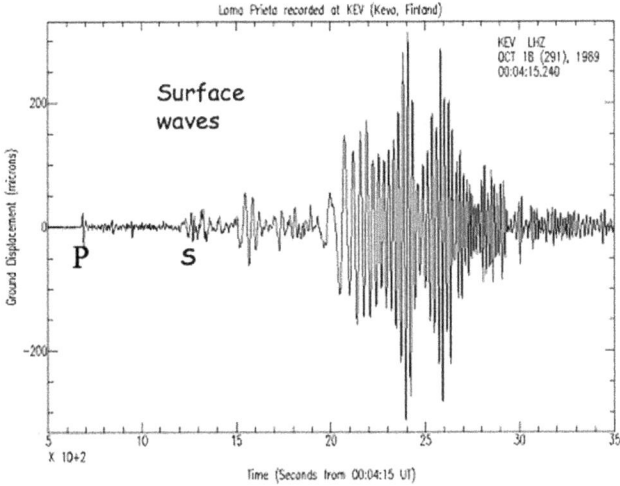

Fig. 29. Seismogram showing surface vertical displacement versus time from the 1979 Loma Prieta earthquake recorded in Kevo, Finland, with P and S body waves and surface waves labeled.

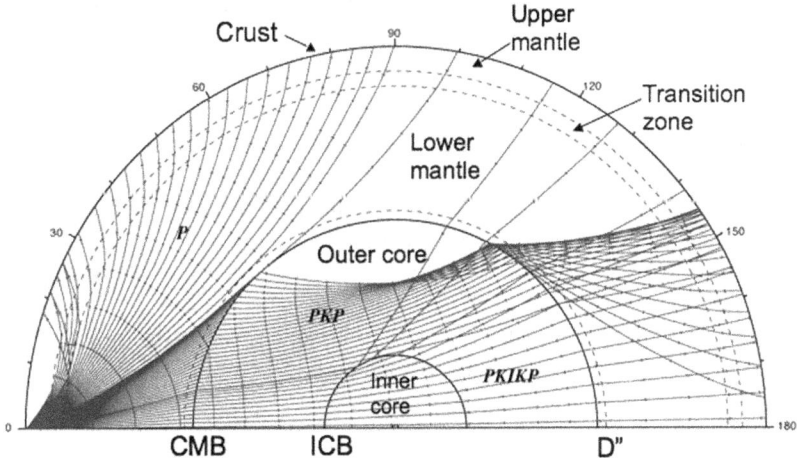

Fig. 30. Ray paths for compression waves in the deep Earth. Nomenclature: mantle shear S, mantle compression P, outer core compression K, inner core compression I. Credit: B. Kennett.

where ρ is density, K is incompressibility (bulk modulus), and μ is rigidity (shear modulus).

Figure 30 shows Snell's Law ray paths of seismic compression waves from a shallow earthquake source in a great circle section of the Earth. The major radial

P. Olson

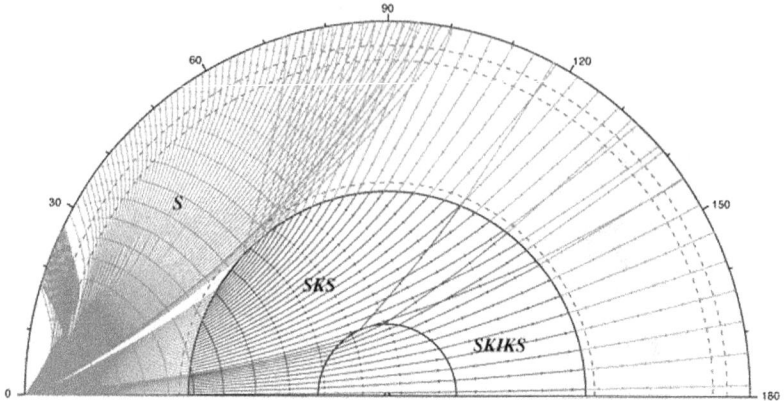

Fig. 31. Ray paths for shear waves in the deep Earth. Credit: B. Kennett.

subdivisions of the Earth are shown, the primary boundaries denoted by solid lines, the secondary boundaries denoted by dashed lines. Closely spaced ray paths indicate high wave energy density. Ray turning points are regions where the constraints on the velocity are good.

3.1.1. Interpretations
• Mantle P waves refract downward entering the core, creating shadow zone and poor sampling of the outermost core.
• Mantle P reflects and diffracts on the core-mantle boundary (cmb), providing good sampling of the D″ layer at the mantle base.
• Core K reflects and transmits at inner core boundary (icb), providing good sampling there.

Figure 31 shows Snell's Law ray paths of seismic shear waves from the same shallow source, in the same radially symmetric Earth model.

3.1.2. Interpretations
• Mantle S waves convert to core K (compression) waves at the core-mantle boundary with a velocity increase, providing good sampling of the outer core. Lots of mantle S reflects at the cmb, providing good imaging of D″ structure.
• Absence of outer core shear wave transmission demonstrates zero rigidity (fluidity) of the outer core.

Figure 32 shows ray paths of the compression wave core interface reflections. Note the abundance of inner core boundary reflections, which are vital for imaging the structure near that boundary.

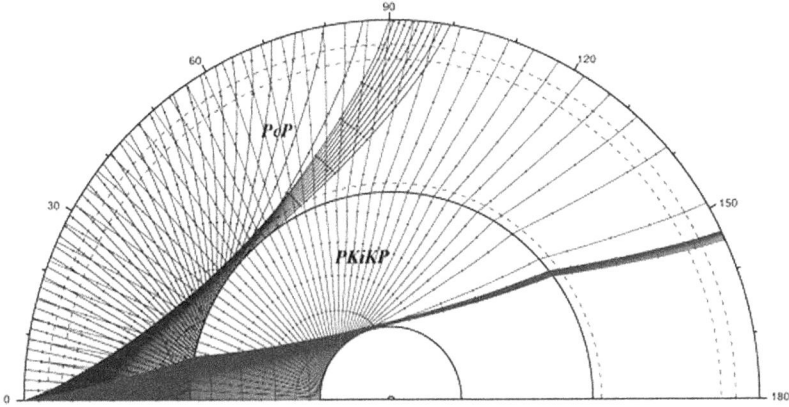

Fig. 32. Core interface reflections. Credit: B. Kennett.

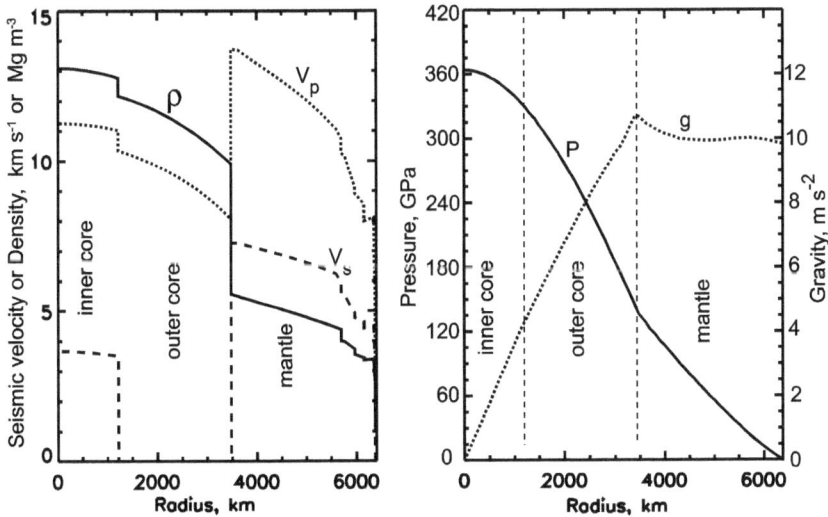

Fig. 33. PREM model of the Earth.

3.1.3. PREM radial Earth model

The highly successful Earth model PREM (Dziewonski and Anderson, 1981) shown in Fig. 33 is based on seismic wave travel times and free oscillation frequencies. In PREM the cmb radius is $r_c = 3480$ km and the hydrostatic pressure $P = 136$ GPa there. The icb radius is $r_i = 1221$ km and $P = 330$ GPa there.

The inner core has a small rigidity (Dziewonski and Gilbert, 1971) and it is several percent denser than the outer core (Shearer and Masters, 1990; Cao and Romanowicz, 2004). The PREM density and P wave profiles in the outer core are smooth, and gravity increases almost linearly with radius.

The Adams–Williamson equation of state is commonly used to interpret core density variations. In its simple form, this equation of state is applicable to "well-mixed" layers with uniform composition and uniform entropy, where the radial density gradient is isentropic (or, adiabatic). In such layers, the adiabatic density gradient can be written

$$\left(\frac{d\rho}{dr}\right)_{ad} = \left(\frac{\partial\rho}{\partial p}\right)_S \frac{dp}{dr} = \frac{1}{\phi}\frac{dp}{dr} \tag{3.2}$$

where

$$\phi = \frac{K}{\rho} = V_P^2 - \frac{4}{3}V_S^2 \tag{3.3}$$

is called the seismic parameter (Stacey, 1992). Here again K denotes the bulk modulus. Along with (3.2) we add the hydrostatic equation for the pressure

$$\frac{dp}{dr} = -\rho g \tag{3.4}$$

and Poisson's equation for gravity

$$\frac{d}{dr}(r^2 g) = 4\pi G \rho r^2 \tag{3.5}$$

(G is the gravitational constant) giving three first order differential equations in radius for the three unknowns: ρ, g and P. Gravity and pressure are continuous, but density jumps occur at interfaces. These density jumps are constrained by the planet mass and moment of inertia, and also, in the case of the Earth, by seismic wave reflection and transmission that constrain the depth of internal interfaces.

Figure 34 is a comparison of Adams–Williamson density profiles with the observed core density variation. The top figure shows the range of Adams–Williamson profiles consistent with a suite of core P-wave profiles. The lower figure compares the Adams–Williamson (A-W) density profiles with the observed density of the liquid outer core, including error bar uncertainties. Within the uncertainties the core follows an A-W profile, although the error bars permits some deviations from homogeneity. This comparison shows that:

• The outer core density profile is about 0.5% uncertain.
• The inner core density profile is about 2% uncertain.
• The icb density jump is about 0.7–0.8 Mg/m^3.

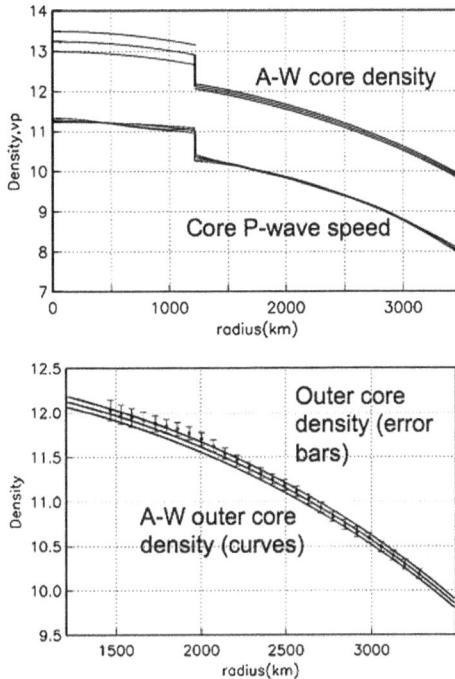

Fig. 34. Comparison of Adams–Williamson density profiles with seismic earth models (from Masters, 2007).

- Seismic models reject a strongly stratified core, but allow weak stratification.
- Overall, the outer core appears quite well-mixed.

Although the outer core appears to be rather well-mixed, it is important to emphasize that even small departures from A-W stratification are dynamically significant. For example, consider the forces on a small outer core fluid parcel displaced vertically from its equilibrium position, as shown in Fig. 35. A parcel initially at location *a* will move along the adiabatic density curve to position *b* if the displacement is "fast", i.e., involves no heat transfer and conserves entropy. Suppose the actual density profile is sub-adiabatic. Then the parcel will be denser than its surroundings by the amount indicated by the broken line in the figure. In the Earth's gravity field the buoyancy force on the parcel will be downward, i.e., restoring. Conversely, if the actual density profile is super-adiabatic, the displaced parcel will be less dense than its surroundings and the buoyancy force on it will be upward, i.e., accelerating.

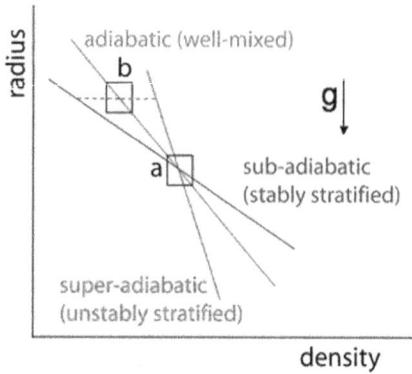

Fig. 35. The density of a parcel displaced isentropically radially upward from a to b, compared with three *in-situ* density profiles: super-adiabatic, adiabatic, and sub-adiabatic.

In an inviscid fluid (not subject to rotational or magnetic effects) the displacement of the parcel $z = r - a$ is governed by a balance between its inertia and buoyancy, which leads to the following equation of motion

$$\frac{d^2z}{dt^2} + N^2 z = 0 \tag{3.6}$$

where N is the so-called buoyancy frequency, defined by

$$N^2 = -\frac{g}{\rho}\left(\frac{d\rho}{dr} + \frac{\rho g}{\phi}\right) = -g\alpha\tau \tag{3.7}$$

The bracketed terms in the definition of N^2 are the A-W balance. The second definition of N^2 involves τ, the difference between the actual temperature gradient and the adiabatic temperature gradient (the temperature gradient that corresponds to adiabatic density stratification). Here

$$\tau = \left(\frac{dT}{dr}\right)_{ad} - \frac{dT}{dr} \tag{3.8}$$

The solution to (3.6) is oscillatory for $N^2 > 0$ and the motion consists of horizontally propagating internal gravity waves. Accordingly, this situation corresponds to stable density stratification. For $N^2 < 0$ the solutions are exponentials in time, so this stratification is therefore unstable or "convective".

In the outer core, the bracketed terms in (3.7) *nearly but not exactly cancel*, thus it is possible that $N^2 > 0$ at some depths. The comparisons in Fig. 34 constrain N to be less than about 10^{-3} rad/sec, but this would still be a dynamically-significant amount of stratification. Because of convective mixing, N^2 can never be large and negative in the outer core. Most geodynamo models assume that $N^2 \simeq 0$, but this is really just a simplification!

3.2. The adiabatic temperature gradient in the core

The adiabatic gradient, the radial temperature gradient that corresponds to a well-mixed (isentropic) hydrostatically-compressed fluid, is an important reference for planetary dynamos. To a good approximation, the temperature gradient in convection-driven dynamos is adiabatic except in boundary layer regions, which are typically very thin compared to the whole fluid depth. The adiabatic thermal gradient can be expressed several ways (Anderson, 2007):

$$\left(\frac{dT}{dr}\right)_{ad} = \left(\frac{\partial T}{\partial p}\right)_S \frac{dp}{dr} = -\frac{\alpha g T}{C_p} = -\frac{\gamma g T}{\phi} \tag{3.9}$$

where

$$\frac{C_p}{\alpha g} = H_T; \quad \gamma \simeq 1.4 \tag{3.10}$$

is the temperature scale height (again, C_p is specific heat and α is thermal expansivity). In the Earth's outer core $H_T \simeq 7000$ km. The second definition is in terms of the dimensionless Grueneisen parameter $\gamma = \alpha/C_p \partial P/\partial \rho$. The Grueneisen parameter is a weak function of temperature and pressure (Stacey and Davis, 2004), but a nominal value in the core is $\gamma \simeq 1.4$. Using the above values we get

- $\left(\frac{dT}{dr}\right)_{ad} \sim -0.9$ K/km near the cmb.
- $\left(\frac{dT}{dr}\right)_{ad} \sim -0.3$ K/km near the icb.
- Integration of the adiabatic gradient gives $\Delta T_{ad} \sim 1200$ K for the adiabatic temperature increase with depth through the outer core.

The parameter $Di = (r_c - r_i)/H_T$ is called the *dissipation number* (sometimes defined as $Di = \gamma (r_c - r_i)/H_T$). The dissipation number measures the importance of compressibility and also viscous dissipation in the fluid energy balance. For the Earth's outer core $Di \simeq 0.3$, which indicates that compression plays some role in convection, although not a dominant role. We note that $Di \gg 1$ in the giant planet atmospheres, and $Di \sim 1$ in giant planet dynamo regions, so that compressibility effects are significant in the dynamics of Jupiter and Saturn.

3.3. Heat conduction and the core adiabat

The steep adiabatic gradient implies a large conductive heat flow in planetary dynamos, particularly the terrestrial dynamos with iron-rich cores. To quantify this, consider Fourier's Law of heat conduction,

$$\mathbf{q} = -k\nabla\mathbf{T} \qquad (3.11)$$

where $\mathbf{q}$ is the heat flow vector (units: $W\,m^{-2}$) and k is the thermal conductivity of the fluid. The thermal conductivity of the outer core is estimated to be around $45 \pm 5\ W\,m^{-1}\,K^{-1}$. For adiabatic stratification

$$q_{ad} = -k\left(\frac{dT}{dr}\right)_{ad} = +\frac{k\alpha g T}{C_p} \qquad (3.12)$$

The total heat flow at the core-mantle boundary can be written as

$$Q_{cmb} = 4\pi r_c^2 \overline{q_{cmb}} \qquad (3.13)$$

where the overbar denotes spherical surface average. Assuming $\overline{q_{cmb}} = q_{ad}$ yields

$$Q_{cmb,ad} = 4\pi r_c^2 k\alpha g T_{cmb}/C_p \qquad (3.14)$$

for the total heat loss at the cmb from conduction down the outer core adiabat. Using $\alpha \simeq 1 \times 10^{-5}\ K^{-1}$ and $T_{cmb} \simeq 4000\ K$ gives $Q_{cmb,ad} \sim 4$–6 TW (1 TW $= 10^{12}W$). This is to be compared with the total surface heat loss of the Earth, $Q_{surface} = 40$–45 TW. The total core heat loss also contains a superadiabatic part, so we can write

$$Q_{cmb,total} = Q_{cmb,ad} + Q_{cmb,superad} \qquad (3.15)$$

The superadiabatic part is *very poorly constrained*, but various estimates range from 0–6 TW. This gives $Q_{cmb,total} = 4$–12 TW for the total heat loss from the core.

3.4. Adiabatic cooling of the core

We can use the above results to estimate the rate of cooling of the core (Labrosse et al., 1997). For simplicity, we assume the inner core plays no role in the heat balance (as we shall later see, this is not actually the case), and furthermore we will ignore radioactive heat sources. The heat equation is then

$$\rho C_p \frac{dT}{dt} = k\nabla^2 T \qquad (3.16)$$

and can be integrated over the volume of the core V_c to give

$$\rho C_p \dot{T}_c V_c = -Q_{cmb} \tag{3.17}$$

where $\dot{T}_c$ is the rate of change of the volume average core temperature and Q_{cmb} is the core heat loss defined in (3.13). Combining (3.17) with (3.13) gives

$$\overline{q_{cmb}} \simeq -\rho C_p r_c \dot{T}_c / 3 \tag{3.18}$$

Now assume the stratification is adiabatic. Using (3.12), (3.18) becomes

$$\frac{\dot{T}_c}{T_c} = -\frac{3\kappa}{H_T r_c} \tag{3.19}$$

where κ is thermal diffusivity. Using the previous parameter values, this gives an adiabatic cooling rate in the range -60 to -90 K/Gyr. The presence of radioactive heat sources and the solidification of the inner core will reduce this cooling rate, but this serves as a useful reference rate. Also, note that the cooling rate is inversely proportional to the core radius, so that small planetary cores cool faster than large ones by this mechanism.

3.5. Heterogeneity in the lower mantle

Lateral (i.e., non-radial) seismic heterogeneity has never been confirmed in the outer core. This is entirely consistent with it being an iron-rich fluid, in which its very low viscosity (Section 4.10) implies too much convective mixing to support detectable lateral gradients in any property connected to its density. However the solid lower mantle and inner core have large lateral heterogeneity in seismic properties, in part because of lateral variations in temperature. Lateral variations in lower mantle temperature can affect the geodynamo (Section 7.5).

Figure 36 shows shear wave rms amplitudes versus depth through the mantle at selected depths from four seismic tomography models. Lateral heterogeneity is large in the upper mantle, small in the mid-mantle and large again the D″-layer at the base of the mantle. The rms variations are consistent with a thick thermal and possibly a chemical boundary layer interpretation of the D″ region.

3.6. Dynamical interpretations of D″

The pattern of the D″ layer heterogeneity is dominated by long-wavelength structures at spherical harmonic degree 2. It features high seismic velocities in a ring below the Pacific and low seismic velocities beneath Africa and southcentral Pacific. These low velocity structures are sometimes referred to as mantle superplumes, on the interpretation that they represent low density, presumably upwelling mantle material. Conversely, the high velocity structures in D″ are

P. Olson

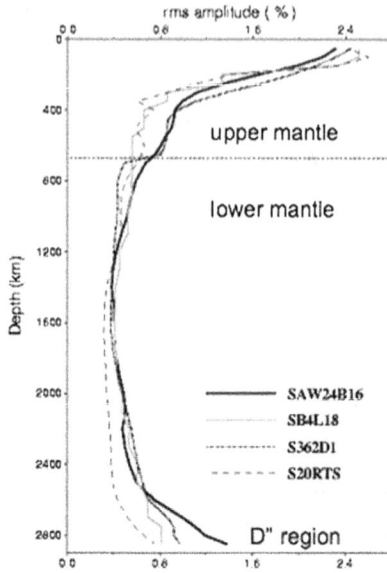

Fig. 36. Seismic heterogeneity amplitude in the mantle. Percent rms shear velocity variations vs. depth from four seismic tomography models.

usually interpreted as having anomalously high density and are therefore down-wellings, possibly the remnants of subducted lithosphere.

The increase in heterogeneity at all scales in the D'' includes thermal effects of large scale mantle flow, smaller scale instabilities in the D'' layer, chemical heterogeneity, possible melts (represented by the ultra-low velocity zone near the cmb) and also the post-perovskite phase change. The temperature gradient in D'' is highly superadiabatic, decreasing by almost 1000 K in 100–200 km above the cmb. Where mantle temperatures are low in D'' (representing mantle down-wellings, possibly including former lithosphere slab material) the heat flow at the cmb is expected to be higher than average. Conversely, where mantle tempera-tures in D'' high, the cmb heat flow expected to be lower than average. A more speculative interpretation calls for infiltration of core metals into D''. This would elevate the electrical conductivity of D'', allowing for better electromagnetic cou-pling between the core and mantle.

3.7. The surprising inner core

The inner core is known to be seismically anisotropic, with compression waves traveling a few percent faster along polar paths, compared to equatorial paths.

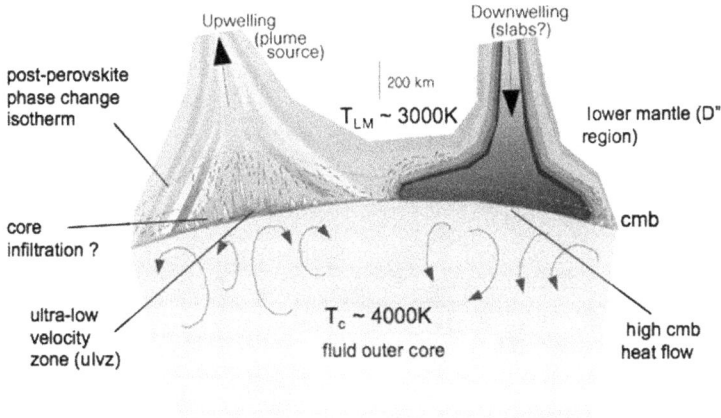

Fig. 37. Dynamical interpretations of seismic structure in the D″ region. Credit: E. Garnero.

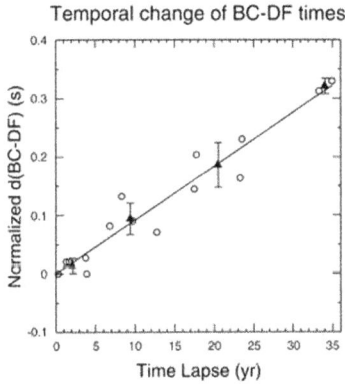

Fig. 38. Temporal change of travel times across the inner core (from J. Zhang et al., 2005).

Associated with this anisotropy are some interesting dynamical effects. There is seismic evidence that the structure of the solid inner core changes on decade time scales, as seen from fixed points on the Earth's surface. The primary evidence consists of changes in the relative travel time of compression waves entering the inner core versus waves with similar paths that do not enter the inner core. The travel time difference between these phases from doublet earthquakes (earthquakes with nearly identical focal points, occurring years apart) increases almost linearly with time lapse between events along polar (i.e., N-S) ray paths, as shown in Fig. 38.

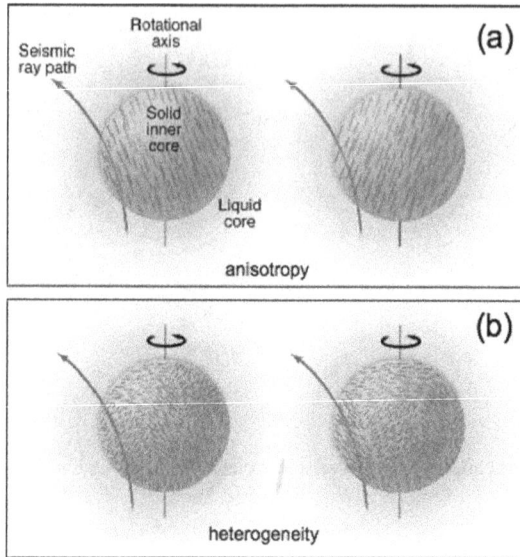

Fig. 39. Dynamical interpretations of the secular change of inner core seismic travel times.

Figure 39 shows two alternative interpretations of this secular change in travel times. Interpretation (*a*) assumes the inner core has uniform seismic anisotropy with fabric inclined with respect to the rotation axis, and rotates differentially with respect to the crust and mantle. Modeling indicates the anomalous rotation is retrograde (westward). Interpretation (*b*) is similar, but assumes the anisotropy is heterogeneous. The inferred anomalous rotation is about -0.2 deg/yr. However, other seismic data, most notably free oscillation frequencies, do not require anomalous inner core rotation. Seismic evidence also reveals that the western hemisphere of the inner core has higher compression wave speeds in its outer 100–200 km that the eastern hemisphere. This lateral heterogeneity is called the inner core hemispheric dichotomy. There is evidence for this hemispheric dichotomy in other seismic properties, including inner core anisotropy and attenuation.

3.8. Summary of deep Earth structure

According to the seismic evidence, we can infer that

• The outer core is (very nearly) laterally homogeneous, and is close to being adiabatic, i.e., well-mixed. The uncertainties in the radial density variations do not preclude a weak stratification, however.

• The inner core is anisotropic and laterally heterogeneous, particularly in its outer few hundred kilometers.

• The core-mantle boundary is structurally complex and laterally heterogeneous on the mantle side, in the D''-layer.

• The inner core boundary has a density jump near 0.7–0.8 Mg/m^3 and is laterally heterogeneous on the inner core side.

• The inner core is possibly super-rotating, but slowly, about 0.2 deg/yr or less.

• Core-mantle boundary heat flow is large, in the range 4–12 TW.

• The adiabatic thermal gradient in the core is large, 0.3–0.9 K/km.

3.9. Cores in other planets

The core structures of the other planets are inferred, rather than measured directly. Planetary core sizes can be estimated using mass and moment of inertia constraints, plus models of the planet bulk composition and experimentally-determined equations of state of the assumed constituents. Iron cores are assumed for the other terrestrial planets, however, except for Mercury, the state of the core (liquid vs. solid) is inferred from thermal modeling (Sohl and Schubert, 2007). The small central cores of Jupiter and Saturn probably consist of high-pressure oxides and post-silicates; their dynamos are generated in layers rich in liquid metallic hydrogen between their atmosphere and core (see Guillot, 2005; Guillot and Gautier, 2007). Figure 40 shows some standard radial structure models of the interiors of the other planets, scaled to the size of the Earth.

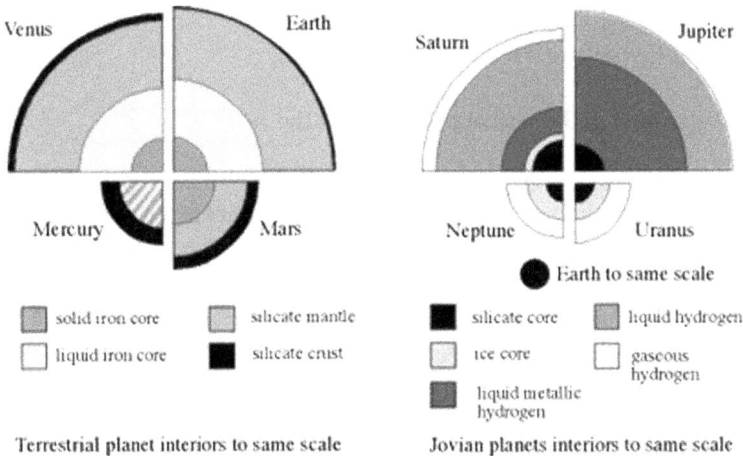

Fig. 40. Comparison of planetary interior structures, including relative core sizes.

194   *P. Olson*

References

Anderson, O.L. 2007. Core, adiabatic gradient, in: *Encyclopedia of Geomagnetism and Paleomagnetism*, D. Gubbins and E. Herroero-Bervera, eds., Springer.

Cao, A. and Romanowicz, B. 2004. Constraints on density and shear velocity contrasts at the inner core boundary. Geophys. J. Intl. **157**, 1–6.

Dziewonski, A.M. and Anderson, D.L. 1981. Preliminary reference Earth model, Phys. Earth Planet. Inter. **25**, 297–356.

Dziewonski, A.M. and Gilbert, F., 1971. Solidity of the inner core of the Earth inferred from normal model observations, Nature **234**, 465–466.

Guillot, T. and Gautier, D. 2007. Giant Planets, in: *Treatise on Geophysics*, vol. 10, ch. 13, Elsevier Press.

Guillot, T. 2005, The interiors of giant planets: models and outstanding questions. Annu. Rev. Earth Planet. Sci. **33**, 493–530.

Labrosse, S., Poirier, J. and LeMouel J.L. 1997. On cooling of the Earth's core, Phys. Earth. Planet. Inter. **99**, 1–17.

Lay, T. 2007. Deep Earth structure—lower mantle and D″, in: *Treatise on Geophysics*, vol. 1, ch. 19, Elsevier, B.V.

Masters, G. and Gubbins, D. 2003. On the resolution of density within the Earth, Phys. Earth. Planet. Inter. **140**, 159–167.

Masters, G. 2007. Density models using the Adams–Williamson Equation, in: *Encyclopedia of Geomagnetism and Paleomagnetism*, D. Gubbins and E. Herrero-Bervera, eds., Springer, pp. 83–84.

Romanowicz, B. and Dziewonski, A.M., eds. 2007. Seismology and structure of the Earth, *Treatise on Geophysics*, vol. 1, Elsevier, B.V. 858 pp.

Shearer, P. and Masters, G. 1990. The density and shear velocity contrast at the inner core boundary, Geophys. J. Int. **10**, 491–498.

Sohl, F. and Schubert, G. 2007. Interior structure, composition and mineralogy of the terrestrial planets, in: *Treatise on Geophysics*, vol. 10, ch. 2, Elsevier, B.V.

Souriau, A. 2007. Deep Earth structure—The Earth's cores, in: *Treatise on Geophysics*, vol. 1, ch. 19, Elsevier B.V.

Stacey, F. 1992. *Physics of the Earth*, 3rd ed., Brookfield Press.

Stacey, F.D. and Davis, P.M. 2004. High pressure equations of state with applications to the lower mantle and core, Phys. Earth Planet. Int. **124**, 153–162.

Zhang J. et al. 2005. Science **309**, 1357–1360.

4. Composition and physical properties of the core

4.1. Evidence from meteorites

Meteorites constrain the overall abundance of elements in the terrestrial planets and their metallic cores. There are three main categories of meteorites: irons, stony-irons, and stony (Wasson, 1985). Stony meteorites are further divided into chondrites and achondrites. Among the near-Earth meteorites, roughly 84% are chondrites, 8% are achondrites, 7% are irons, and 1% are stony-irons. Figure 41 shows examples of the three different types of meteorites. Iron meteorites are composed primarily of Kamacite and Taenite, both iron-nickel alloys. Most iron meteorites have nickel contents of 6–10% . and commonly display Widmanstätten patterns (see Fig. 41) due to regular geometric intergrowth of Kamacite and Taenite. Kamacite contains α-iron, which is the low temperature structure with a

Fig. 41. Upper left: Chondrules in a carbonaceous chondrite meteorite. Upper right: Fe–Ni iron meteorite. Lower: Pallasite meteorite, a mixture of silicate and iron.

body-centered cubic (BCC) structure. Taenite contains iron in the higher temperature, lower pressure form γ-iron, with a face-centered cubic (FCC) structure. Finally, there is the ϵ-phase of iron with an hexagonal closest packed crystal structure with AB stacking and is the high temperature, high pressure phase. The lattice structures of these three types of iron are shown in Fig. 43.

Chondrules are spherical bodies that form by rapid cooling of silicate melt, and typically consist of olivine or pyroxene crystals embedded in a glassy or microcrystalline matrix. Radioactive dating of chondrite meteorites indicates that they are all about the same age as Earth, roughly 4.56 Ga. Carbonaceous chondrites are a special class of chondrites that contain more than 0.2% carbon. In terms of chemical composition, carbonaceous chondrites are the most similar of all the chondrites to the Sun. Their composition is thought to be similar to planetary embryos in the early solar system. Finally, the stony-iron meteorites are a mix of iron-nickel alloy and silicates. The silicates are sometimes distributed in these meteorites as if they crystallized in a small gravity field, without segregating, as would be the case in a small planetesimal.

Figure 42 shows a comparison of two models of the Earth's composition and the composition of the core, derived from solar abundances and meteorite compo-

wt%	Si-bearing		O-bearing	
	Earth	core	Earth	core
Fe %	32.0	85.5	32.9	88.3
O %	29.7	0	30.7	3
Si %	16.1	6	14.2	0
Ni %	1.82	5.2	1.87	5.4
S %	0.64	1.9	0.64	1.9
Cr %	0.47	0.9	0.47	0.9
P %	0.07	0.20	0.07	0.20
C %	0.07	0.20	0.07	0.20
H %	0.03	0.06	0.03	0.06
mean atomic #	23.5		23.2	
atomic proportions				
Fe	0.768		0.783	
O	0.000		0.093	
Si	0.107		0.000	
Ni	0.044		0.045	
S	0.030		0.029	
Cr	0.009		0.009	
P	0.003		0.003	
C	0.008		0.008	
H	0.030		0.029	
total	1.000		1.000	

Fig. 42. Elemental abundance models for the whole Earth and the core, from McDonough, 2005.

sitions by McDonough (2005). Iron is clearly the dominant element in the core, comprising about 86% of its mass. Nickel is second, with both models in Fig. 42 containing about 5% Ni. These two elements, together with minor amounts of Cr, constitute the *heavy element* (HE) of the core bulk composition. The most abundant light elements (LE) include Si, O, and S.

4.2. The phase diagram of iron

The phase diagram of iron is shown in Fig. 43 over the pressure range of the core. The high pressure portion of this diagram is based on data from shock wave and diamond cell measurements, and remains controversial (Boehler, 1996; Stixrude and Cohen, 1997; Vocaldo, 2007). Most (but not all) evidence indicates that ϵ-Fe is the stable phase for the core. The crystal structure of the three phases are shown in Fig. 43. The melting curve for ϵ-Fe increases rapidly with pressure, reaching about 6000 K at the pressure corresponding to Earth's center,

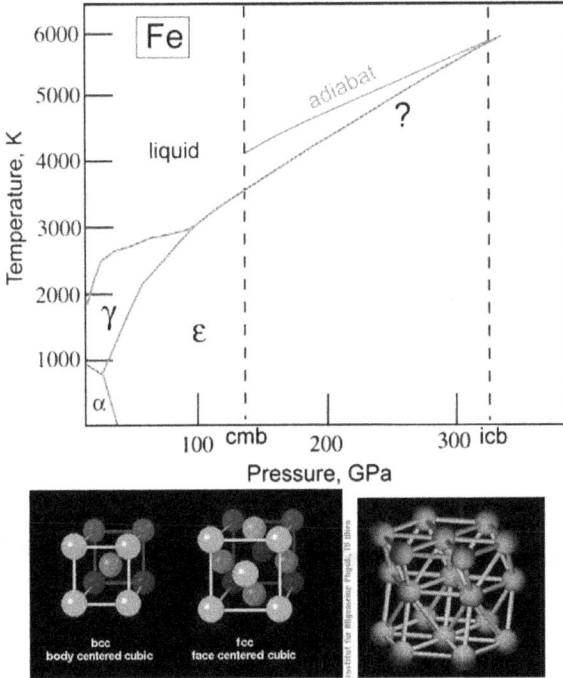

Fig. 43. Top: Phase diagram of iron in the core. Lower: structures of the various phases of Fe. From left to right: α (body centered cubic), γ (face centered cubic), ϵ, hexagonal closest packing.

and around 5800 K at the icb pressure (Nguyen and Holmes, 2004). Uncertainties on these two melting temperatures are large, around 800 K and possibly larger.

Assuming the icb corresponds to the melting point, an adiabat drawn through the melting curve at the icb approximates the temperature distribution through the convecting outer core. It predicts that the cmb temperature just above 4000 K. However, the actual temperature distribution in the core may be somewhat lower than this, due to the presence of light elements (Alfe et al., 2003). The presence of light alloys depresses the melting point of the core relative to pure iron, by an amounts that depends on the light element composition.

4.3. Light elements in the core

Comparison of the density of solid ϵ-iron extrapolated to core pressures (shown in Fig. 44) indicates it is nearly 10% more dense than the outer core. Even with

P. Olson

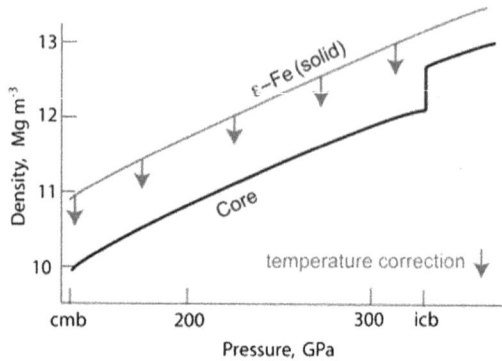

Fig. 44. Comparison of ϵ-Fe density with Earth's core density. Temperature correction is shown with arrows.

the temperature and melting effects applied, ϵ-iron is still 7–8% denser than the outer core, and about 2% denser than the inner core. Adding Ni to Fe makes little difference to the density (so the Ni content of the core is not well constrained this way). Addition of about 10% by mass of the cosmochemically plausible light elements S, Si, and O removes the outer core density discrepancy (Poirier, 1994). An approximately 3% addition of some mix of these light elements to solid Fe–Ni removes the density discrepancy with the inner core. Thus the inner and outer core have different concentrations of the light elements: the outer core is enriched in light elements relative to the inner core. Comparison of the outer core and iron bulk moduli (elastic incompressibility) yields a similar conclusion. Some high pressure experimental studies indicate other light elements may be soluble in liquid iron, such as H and He, but the abundance of these is too low to much affect the core density.

4.4. Thermal significance of light elements in the core

Light elements reduce the melting temperature of iron alloys and therefore reduce the melting temperature of the outer core. This melting point reduction is quite large at low pressure in iron-sulfur alloys. Figure 45 shows the phase diagram of Fe S at $P = 1$ atm pressure. Fe–FeS shows a solid solution for low S concentrations and a eutectic composition near 30% S, with a melting temperature of 988 K, compared to the 1538 K melting temperature of pure Fe. The outer core light element (LE) concentration would indicate $\sim$250 K melting point reduction. However the effect of pressure on the phase diagram is not well constrained (Alfe et al., 2003), so extrapolation to core pressures is problematic. The melting

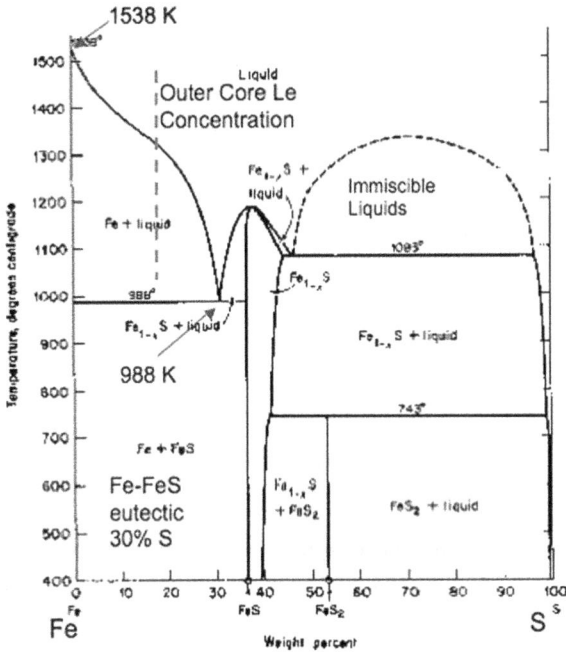

Fig. 45. Fe–S phase diagram at $P = 1$ atm.

point reduction at the icb could be as much as 1000 K, although it could also be much smaller than this.

The Fe–S phase diagram (Fig. 45) shows another complication—regions of limited solubility, with possibly immiscible liquids. In fact the Fe–FeO system shows immisicibility at low pressure, which is reputed to go away with increasing pressure. Even so, there is the possibility that some core liquids are not soluble, and would separate into layers.

4.5. Constraints on the deep Earth geotherm

Deep Earth geotherm constraints include the melting temperature of the core material at the icb pressure, and adiabatic extrapolation of upper mantle temperatures through the lower mantle (Boehler, 1996). Figure 46 shows a whole-Earth geotherm that satisfies these constraints. The mismatch between lower mantle and outer core adiabats is accommodated by a large thermal boundary layer in D″-region at the base of the mantle, where the temperature increases 800–1000 K

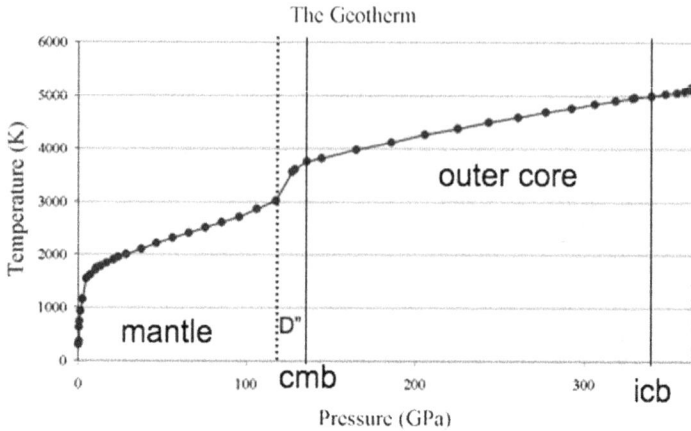

Fig. 46. Estimated deep Earth geotherm.

over about 200 km. Thus the potential temperature of the core (the temperature extrapolated adiabatically to surface pressure) is higher than the potential temperature of the mantle, and the core superheat (the potential temperature relative to the surface melting temperature) is also high, ~ 1000 K. Estimated core temperatures are T(icb)$\sim$5000 K; T(cmb)$\sim$3600–4000 K, and the thermal gradient in D$''$ shown in Fig. 46 is consistent with a total cmb heat flow $Q_{cmb} = 4$–12 TW.

4.6. Thermal boundary conditions at the cmb

We next consider the thermal regime in the neighborhood of the cmb. Both the temperature and the normal component of the heat flux vector must be continuous at the cmb. Temperature and heat flux continuity, assuming the cmb is solid and impermeable, can be written

$$T_c = T_m \quad \& \quad k_c \left(\frac{\partial T}{\partial r}\right)_c = k_m \left(\frac{\partial T}{\partial r}\right)_m \tag{4.1}$$

where the subscripts c and m denote the core and mantle side of the cmb.

Because the lower mantle is solid (or, at least it is mostly solid) and the outer core is a low viscosity fluid, transport velocities are much smaller in the mantle than in the outer core and lateral thermal (and density) gradients are also much larger than on the core side. This leads to the following two inequalities:

$$\delta_m \gg \delta_c \quad \& \quad |\nabla T_H|_{D''} \gg |\nabla T_H|_c \tag{4.2}$$

where δ_m and δ_c are the thermal boundary layer thicknesses on the mantle and core side of the cmb, respectively, and the subscript H denotes horizontal derivatives. Because of these inequalities, the thermal interaction between the core and mantle is highly asymmetric. The mantle "sees" the cmb as an isothermal boundary, and "sees" the outer core as an isothermal reservoir:

$$T_{mantle}(r = r_c) = T_{cmb} \qquad (4.3)$$

On the other hand, the core "sees" the cmb as a boundary on which the heat flow is imposed:

$$k_c \left(\frac{\partial T}{\partial r}\right)_{cmb}^{core} = -q_{cmb}^{mantle}(\theta, \phi) \qquad (4.4)$$

The variable heat flow $q_{cmb}^{mantle}(\theta, \phi)$ that the core "sees" is calculated from the temperature and thermal conductivity in the mantle D''-layer.

Assuming the cmb is an impermeable boundary for mass transfer (as is usually assumed for the geodynamo) then the normal flux of all chemical constituents vanishes there. For the light elements in the core with concentration χ, this condition gives

$$\left(\frac{\partial \chi}{\partial r}\right)_{cmb}^{core} = 0 \qquad (4.5)$$

The conditions described above for the core-mantle boundary region are physically similar to the situation in the neighborhood of the ocean floor, where the outer core plays a role analogous to the abyssal ocean and the D''-layer plays a role analogous to the oceanic crust. Based on this similarity, it has been proposed that many of the smaller-scale dynamical processes that characterize the water-rock interactions near the sea floor, such as hydrothermal circulation, sedimentation and sediment transport, have counterparts in the core-mantle boundary region. The possible influences that these smaller-scale processes might have on the geodynamo are not well understood.

4.7. Thermal and chemical conditions near the inner core boundary

Thermal and chemical conditions near the icb are shown in Fig. 47. The icb radius r_i is the intersection of the liquidus temperature of the core material T_L and the geotherm T_o. As the core cools, T_o decreases and r_i moves to larger radius. The inner core growth rate $\dot{r}_i$ is given by

$$\dot{r}_i = -\rho g \left(\frac{dT_L}{dp} - \frac{dT_{ad}}{dp}\right)^{-1} \dot{T}_o \qquad (4.6)$$

where is the core cooling rate.

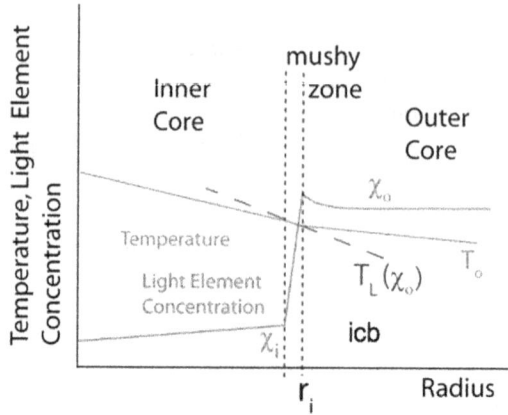

Fig. 47. Temperature and light element concentration profiles versus radius near the icb. T_o is the temperature profile, χ is the light element concentration, and T_L is the melting (liquidus) curve.

We have seen that light elements are less concentrated in the inner core than in the outer core, which is interpreted as light element rejection from the solidifying inner core. The rejection process occurs via porous flow separation in the two-phase region (mushy zone) shown in Fig. 47. Since the mushy zone is quite thin (it is estimated to be only a few tens of meters deep or less, because of compaction), the phase separation should be regarded as a microscopic process for the geodynamo. The macroscopic representation of this process is based on light element diffusion. Balancing the diffusive light element flux with the light element rejection at the icb gives

$$D\left(\frac{\partial \chi}{\partial r}\right)_{icb} = (\chi_o - \chi_i)\,\dot{r}_i \qquad (4.7)$$

where D is the light element diffusivity.

The continuity of heat flow includes heat conducted down the inner core adiabat, plus latent heat released as the inner core solidifies. Accordingly, the heat flux into the outer core at the icb is

$$k\left(\frac{\partial T}{\partial r}\right)_{icb} = -\rho L \dot{r}_i - \frac{k\alpha g_i T_i}{C_p} \qquad (4.8)$$

where L is the latent heat of fusion of the inner core material.

Figure 47 indicates that the outer core is enriched in light elements relative to the inner core by an amount $\chi_o - \chi_i$ and is therefore intrinsically less dense. In

Fig. 48. Three proposed mechanisms for producing inner core anisotropy, from Sumita and Bergman, 2007.

addition, the outer core is liquid, so its density is less than the inner core because of the volume change due to melting. At the icb pressure, it is estimated that the density change on melting is about $\Delta\rho_{melting} \sim -210$ kg/m^3 for iron. Recall that, from seismic observations, the density increase across the icb (going from the outer to inner core) is $\Delta\rho_i \sim 800$ kg/m^3. Thus the intrinsic (compositional) density difference across the icb is approximately $\Delta\rho_\chi \sim 580$ kg/m^3.

Note that the fluxes represented by (4.7) and (4.8) both contribute positive buoyancy flux at the icb, and are both linearly proportional to the inner core growth rate. The buoyancy flux just above the icb F_i due to their combined action is approximately

$$F_i = g_i \frac{\Delta\rho_\chi}{\rho_o}\dot{r}_i + \alpha g_i \frac{L}{C_p}\dot{r}_i \qquad (4.9)$$

4.8. Causes inner core anisotropy

There are many different proposals for the origin of inner core anisotropy (Sumita and Bergman, 2007; Dehant et al., 2003). As shown in Fig. 48, dynamical mechanisms include: subsolidus convection in the inner core with lattice preferred orientation; nonuniform growth in the equatorial plane and subsequent relaxation toward hydrostatic equilibrium with lattice preferred orientation; nonhydrostatic Maxwell stresses applied to the inner core by toroidal magnetic field and stress-induced grain re-orientation; grain alignment by isotropic heat flow.

4.9. Radioactivity in the core

The concentration of radioactive heat sources in the core is another long-standing controversial issue (Gessmann and Wood, 2002; Rama Murthy et al., 2003). Radioactive potassium K40 is likely the most important heat producing isotope for the core. Sulfur is an often-cited light element candidate for the outer core, and experiments indicate that K may be soluble in S at outer core pressures (Rama Murthy et al., 2003). Interpretations of these experiments suggest a possible K-content of the outer core of 0–400 ppm., which would amount to a present-day core heat production of 0–0.7 TW.

Significant core radioactive heat producing isotopes would have several effects on the geodynamo and the thermal evolution of the core. First, it would provide another power source for the geodynamo, in addition to secular cooling of the core and light element segregation. It also would slow the secular cooling of the core, especially in the past when the radioisotope was more abundant. For the same reason, its presence would slow the growth of the inner core. However, the actual abundance of K in the core remains unknown.

4.10. Transport properties in the core

4.10.1. Viscosity
The viscosity of the core is estimated from lab measurements in liquid iron alloys, although at much lower pressure than the core. The dynamic viscosity of liquid iron at surface pressure is less than $1 \times 10^{-6} \, m^2 \, s^{-1}$. Laboratory determinations indicate a low viscosity for liquid iron up to about 15 GPa, of the order $\nu \sim 2 \times 10^{-6} \, m^2 \, s^{-1}$ (Dobson et al., 2000; de Wijs et al., 1998). Extrapolating the lab measurements to the pressures of the core, the outer core viscosity would still be of order $10^{-5} \, m^2 \, s^{-1}$. However, some geodetic measurements suggest far higher outer core viscosity, in the range $\nu \sim 10^{+6} \, m^2 \, s^{-1}$, although this interpretation is far more controversial (Volcado, 2007). Indirect estimates of the inner core (subsolidus) viscosity are in the range $\nu \sim 10^{12} \, m^2 \, s^{-1}$.

4.10.2. Thermal and electrical conductivity
Thermal and electrical conductivity are linked via the Wiedemann-Franz relationship for metals

$$k = \mathcal{L}\sigma T \qquad (4.10)$$

where $\mathcal{L} = 2.44 \times 10^{-8} E\Omega \, K^{-1}$. Conductivity of iron is measured in shock wave experiments, and the effects of light elements on conductivity are then applied as a correction. For Fe–FeS–FeO core at $P = 136$ GPa is $k \simeq 45 \, W \, m^{-1} \, K^{-1}$ and $\sigma \simeq 5 \times 10^5 \, Sm^{-1}$ (Stacey, 2007) and the electrical conductivity is in the range $\sigma = 3\text{–}5 \times 10^5 \, Sm^{-1}$ (Boness et al., 1986; Poirier, 1988). The uncertainty on each of these parameters is at least $\pm 30\%$.

4.11. Reference tables

Property	Notation	Units	Value
Core-Mantle Radius (mean)	r_o	m	3.480×10^6
Inner Core Radius (mean)	r_i	m	1.22×10^6
Outer Core Thickness	d	m	2.26×10^6
Core-Mantle Boundary Area	A_o	m^2	1.52×10^{14}
Inner Core Boundary Area	A_o	m^2	1.87×10^{13}
Core-Mantle Boundary Ellipticity	ε_o	nd	2.5×10^{-3}
Core Volume	V	m^3	1.77×10^{20}
Inner Core Volume	V_i	m^3	7.6×10^{18}
Outer Core Volume	V_o	m^3	1.70×10^{20}
Core Moment of Inertia	I	$kg\ m^2$	9.2×10^{36}
Outer Core Mass	M_o	kg	1.835×10^{24}
Inner Core Mass	M_i	kg	9.68×10^{22}
Core Density (mean)	ρ	$kg\ m^{-3}$	1.09×10^4
Inner Core Density (mean)	ρ_i	$kg\ m^{-3}$	1.29×10^4
Core-Mantle Boundary Gravity	g_o	$m\ s^{-2}$	10.68
Inner Core Boundary Gravity	g_i	$m\ s^{-2}$	4.40
Core-Mantle Boundary Pressure	P_o	GPa	136
Inner Core Boundary Pressure	P_i	GPa	323
P-wave velocity below CMB	v_p	$km\ s^{-1}$	8.07
P-wave velocity above ICB	v_p	$km\ s^{-1}$	10.36
Poisson Ratio, Inner Core	v_P	nd	0.44
Angular Velocity of Rotation	Ω	$rad\ s^{-1}$	7.292×10^{-5}
Free Core Nutation Period	T_c	s	3.71×10^7
Magnetic Dipole Moment	m_d	$A\ m^2$	7.8×10^{22}
Magnetic Dipole Tilt	θ_d	deg	10.8
Magnetic Intensity, CMB (rms)	B_{CMB}	mT	0.42
Magnetic Dipole Intensity, CMB (rms)	B_{CMB}^{dip}	mT	0.263

Fig. 49. Known core properties.

Property	Notation	Units	Range
Core-Mantle Boundary Temperature	T_o	K	4000 ± 400
Inner Core Boundary Temperature	T_i	K	5500 ± 600
Outer Core Temperature (mean)	T	K	4500 ± 500
Adiabatic Temperature Gradient, CMB	$-dT/dr_{ad}$	$K\ km^{-1}$	0.3 ± 0.1
Adiabatic Temperature Gradient, ICB	$-dT/dr_{ad}$	$K\ km^{-1}$	0.9 ± 0.3
Density Jump, ICB	$\Delta\rho_i$	$kg\ m^{-3}$	700 ± 100
Light Element Concentration, Outer Core	C_o	%	10 ± 2
Light Element Concentration, Inner Core	C_i	%	3 ± 2
Thermal Expansivity	α	K^{-1}	$1.4 \pm 0.5 \times 10^{-5}$
Light Element Expansivity	α_C	nd	0.7 ± 0.3
Specific Heat	C_p	$J\ kg^{-1} K^{-1}$	850 ± 20
Latent Heat, Crystallization	L	$J\ kg^{-1}$	$1 \pm 0.5 \times 10^6$
Thermal Conductivity	k	$W\ m^{-1} K^{-1}$	45 ± 10
Thermal Diffusivity	κ	$m^2\ s^{-1}$	$5 \pm 3 \times 10^{-6}$
Compositional Diffusivity	D	$m^2\ s^{-1}$	$1 \times 10^{-9 \pm 2}$
Kinematic Viscosity, Outer Core	ν	$m^2\ s^{-1}$	$1 \times 10^{-5 \pm 2}$
Kinematic Viscosity, Inner Core	ν_i	$m^2\ s^{-1}$	$1 \times 10^{10 \pm 3}$
Electrical Conductivity	σ	$A^2\ kg^{-1} m^{-3} s^3$	$5 \pm 2 \times 10^5$
Magnetic Diffusivity	λ	$m^2\ s^{-1}$	1.5 ± 0.5

Fig. 50. Thermodynamic and transport properties.

Property	Notation	Units	Approximate Range
Total Heat Flux, CMB	Q	TW	8 ± 4
Adiabatic Heat Flux, CMB	Q_{ad}	TW	3 ± 1
Buoyancy Flux, Outer Core	F	$m^2\,s^{-3}$	$\sim 10^{-13}$
Core Age	t_c	yr	4.4×10^9
Inner Core Age	t_i	yr	$1 - 2 \times 10^9$
Inner Core Growth Rate	dr_i/dt	$m\,yr^{-1}$	$1 - 2 \times 10^{-3}$
Viscous Diffusion Time	t_ν	yr	$\sim 10^{10}$
Thermal Diffusion Time	t_κ	yr	$\sim 10^{11}$
Magnetic Diffusion Time	t_λ	yr	$\sim 10^5$
Dipole Diffusion Time	t_{dip}	yr	2×10^4
Circulation Time	d/U	yr	$100 - 300$

Fig. 51. Dynamic properties. Define buoyancy flux as $F = \frac{\alpha g q'}{\rho C_p}$.

Parameter	Notation	Definition	Approximate Value
Radius Ratio	r^*	r_i/r_o	0.35
Prandtl Number	Pr	ν/κ	$0.1 - 0.5$
Magnetic Prandtl Number	Pm	ν/λ	$\sim 10^{-5}$
Roberts Number	q	ν/κ	$\sim 10^{-6}$
Ekman Number	E	$\nu/\Omega d^2$	$\sim 10^{-14}$
Magnetic Ekman Number	E_λ	$\lambda/\Omega d^2$	$\sim 10^{-9}$
Thermal Ekman Number	E_κ	$\kappa/\Omega d^2$	$\sim 10^{-15}$
Rayleigh Number, ΔT	Ra_T	$\alpha g \Delta T d^3 / \kappa \nu$	$\sim 10^{20}$
Rayleigh Number, q	Ra_q	$\alpha g q d^4 / k \kappa \nu$	$\sim 10^{21}$
Dissipation Number	Di	$\alpha g d / C_p$	0.3
Elsasser Number	Λ	$\sigma B_{rms}^2/\rho\Omega$	~ 1
Elsasser Number, CMB	Λ_{CMB}	$\sigma B_{CMB}^2/\rho\Omega$	~ 0.05
Reynolds Number	Re	Ud/ν	$\sim 10^7$
Magnetic Reynolds Number	Rm	Ud/λ	$300 - 600$
Peclet Number	Pe	Ud/κ	$\sim 10^7$

Fig. 52. Dimensionless parameters for the core (nominal values).

References

General Core-Mantle Interaction

Jones, C., Soward, A. and Zhang, K., eds. 2003. *Earth's Core and Lower Mantle*, Taylor and Francis.

Dehant, V., Creager, K., Karato, S. and Zatman, S. 2003. *Earth's Core, Dynamics, Structure, Rotation*, AGU Geodynamics Series 31.

Core Temperatures

Boehler, R. 1996. Melting temperature of the Earth's mantle and core: Earth's thermal structure, Annu. Rev. Earth Planet Sci. **24**, 15–40.

Nguyen, J.H. and Holmes, N.C. 2004. Melting of iron at the physical conditions of the Earth's core, Nature **427**, 339–342.

Core Composition

Vocaldo, L. 2007. Mineralogy of the Earth—The Earth's core: Iron and Iron Alloys, in: *Treatise on Geophysics*, vol. 2, ch. 5, Elsevier Press.

Alfe, D., Gillan, M.J. and Price, G.D. 2003. Thermodynamics from first principles temperature and composition of the Earth's core. Min. Mag. **67**, 113–123.

McDonough, W. 2005. Chemical composition of the core, in: *Treatise on Geochemistry*, Elsevier B.V.

Poirier, J.P. 1994. Light elements in Earth's outer core: a critical review, Phys. Earth Planet. Inter. **85**, 319–337.

Stixrude, L., Waserman, E. and Cohen, R.E. 1997. Composition and temperature of the Earth's inner core, J. Geophys. Res. **102**, 24729–24739.

Radioactive Heat Sources

Gessmann, C. and Wood, B. 2002. Potassium in the earth's core? Earth Planet. Sci. Lett. **200**, 63–78.

Rama Murthy, V., von Westrenen, W. and Fei, Y. 2003. Experimental evidence that potassium is a substantial radioactive heat source in planetary cores, Nature **423**, 163–165.

Transport Properties

Poirier, J.P. 1988. Transport properties of liquid metals and viscosity of the Earth's core, Geophys. J. **92**, 99–105.

Dobson, D.P., Crichtom, W.A., Vocaldo, L. et al. 2000. In situ measurements of viscosity of liquids in the Fe–FeS system at high pressures and temperatures, Amer. Mineralogist **85**, 1838–1842.

de Wijs, G.A. et al. 1998. The viscosity of liquid iron at the physical conditions of the Earth's core, Nature **392**, 805–807.

Stacey, F.D. 2007. Core properties, physical, in: *Encyclopedia of Geomagnetism and Paleomagnetism*, D. Gubbins and E. Herrero-Bervara, eds., pp. 91–94.

Boness, D.A., Brown, J.M. and McMahan, A.K. 1986. The electronic thermodynamics of iron under Earth core conditions, Phys. Earth Planet Inter. **42**, 227–240.

Inner Core Dynamics

Sumita I. and Bergman, M.I. 2007. Inner-core dynamics, in: *Treatise on Geophysics*, vol. 8, ch. 10, Elsevier B.V.

Meteorites

Wasson, J.T. 1985. *Meteorites: Their Record of Early Solar-system History*, WH Freeman, NY, 274 pp.

5. Energetics of the core and the geodynamo

5.1. Growth of the solid inner core

Our estimates of the core geotherm and the core superheat both indicate substantial heat flow from the core to the mantle. The concentrations of radioactive K40 that have been proposed would not generate enough heat to balance this loss, so the core is inevitably cooling (Buffett, 2003). We have already seen that the cooling rate of the core due to heat conducted down its adiabat amounts to -60 to -90 K/Gyr. Of course this calculation ignored the growth of the inner core, which contributes latent heat to the energy balance of the whole core. But even with latent heat, our inference of heat flow at the cmb dictates that the average

P. Olson

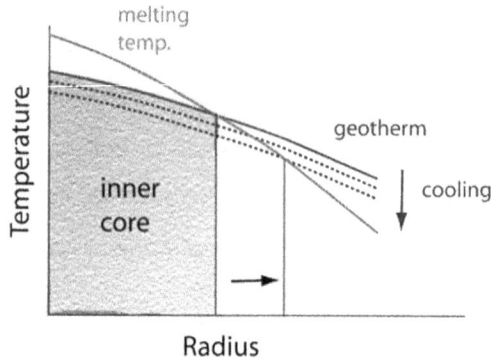

Fig. 53. Evolution of the geotherm and inner core boundary radius during core cooling. Credit: B. Buffett.

temperature of the core is decreasing and the inner core is growing by solidification. As shown in Fig. 53, the location of the icb at any given time is determined by the intersection of the melting curve and the geotherm. As the core cools and the geotherm evolves downward in the figure, the icb moves radially outward. The rate of inner core growth is therefore proportional to the cooling rate of the outer core.

5.2. Chemical evolution of the core

Chemical evolution accompanies the growth of the solid inner core. Figure 54 shows the evolution of the inner core and outer core compositions during secular cooling of the core and inner core growth (Buffett, 2000), depicted in an idealized binary Fe–Feχ phase diagram. As the temperature of the core decreases, the inner and outer evolve from the solid toward the dashed lines. The liquidus corresponds to the icb, the two-phase region represents a possible mush layer. Light elements tend to be rejected from the solid, much like dissolved salts are rejected as sea ice crystallizes from ocean water. Note that the outer core light element concentration and the concentration of light elements in the inner core region just below the icb both increase with time. Overall, the density of the outer core is reduced, with release of gravitational potential energy.

5.3. Gravitational dynamo mechanism

The chemical evolution of the core entails a reduction in the gravitational potential energy of the core, and this is a big source of kinetic energy for the geodynamo. The dynamics of this process is sketched schematically in Fig. 55. Prior to

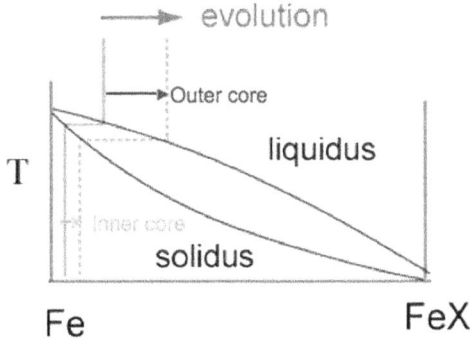

Fig. 54. Binary alloy model of the evolution of the core involving Fe and Fe alloyed with a light element X, possibly O, Si, or S. Credit: B. Buffett.

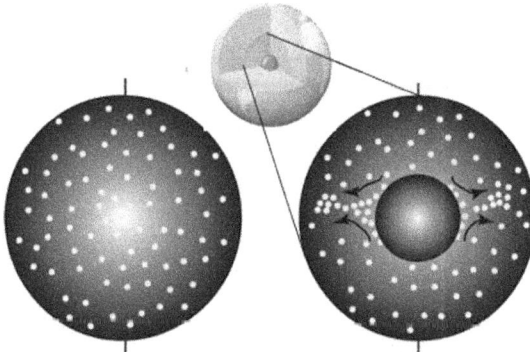

Fig. 55. Illustration of the gravitational dynamo mechanism. Left: light (colored) elements are uniformly mixed in the core prior to inner core nucleation. Right: light elements are rejected from the solidifying inner core, producing compositional convection. In this illustration, it is assumed that the solid inner core contains no light elements.

inner core nucleation, the light elements are supposed to be mixed uniformly in the entirely liquid core, as shown on the left. As the core temperature decreases and inner core solidification occurs, a concentrated solution rich in light elements rejected from the inner core forms above the inner core boundary. This layer is less dense than the outer core at pressure of the inner core, owing to its high light element concentration. Gravitational (convective) instability of this layer produces chemical convection in the liquid outer core, as shown on the right. In this mechanism, the basic energy source is the earth's gravitational potential (Loper, 1978). The power supplied to drive the geodynamo is proportional to the rate of inner core growth, which in turn is controlled by heat flow at the core-mantle boundary. Because the mantle dictates the heat flow at the core-mantle boundary, it is sometimes said that "the mantle controls the geodynamo".

5.4. Light element transport in the outer core

Here we derive an equation governing the concentration of light elements χ in the outer core, subject to inner core growth. Following Jones (2007), we begin with a transport-diffusion equation

$$\frac{\partial \chi}{\partial t} + \mathbf{u} \cdot \nabla \chi = D\nabla^2 \chi \tag{5.1}$$

where D is the mass diffusivity of the light elements in the iron-rich outer core. A light element transport equation for the outer core can be obtained by representing χ as the sum of a volume average which increases slowly with time $\chi_o(t)$, plus a spatially-variable perturbation χ':

$$\chi = \chi_o(t) + \chi' \tag{5.2}$$

with χ_o defined by

$$\chi_o(t) = V_o^{-1} \int_{V_o} \chi \, dV \tag{5.3}$$

Substituting (5.2) into (5.1) yields

$$\frac{\partial \chi'}{\partial t} + \mathbf{u} \cdot \nabla \chi' = D\nabla^2 \chi' - \frac{d\chi_o}{dt} \tag{5.4}$$

To a good approximation, we can assume that χ_o increases so slowly compared to the time scale of dynamo regeneration that both χ_o and $d\chi_o/dt$ can be considered constants over the course of a dynamo simulation. In this situation, $d\chi_o/dt$ acts mathematically as a "sink" term in (5.4), drawing light elements from the icb into the outer core. Equation (5.4) can be scaled using d as the length scale, d^2/ν as the time scale (ν is kinematic viscosity), and assuming that the light element perturbation scales like

$$\chi' = \frac{d^2 \dot{\chi}_o}{\nu} \chi^* \tag{5.5}$$

Then the dimensionless light element transport equation becomes

$$\frac{\partial \chi^*}{\partial t^*} + \mathbf{u}^* \cdot \nabla^* \chi^* = S\nabla^{*2}\chi^* - 1 \tag{5.6}$$

where

$$S = \frac{D}{\nu} \tag{5.7}$$

is the mass-to-momentum diffusivity ratio (yet another diffusivity ratio we need to consider!!).

5.5. Light element dynamics

For now, we ignore thermal buoyancy and consider only the buoyancy associated with the light elements. A simplified, linear equation of state can be written for the outer core density

$$\rho = \rho_o + \rho' = \rho_o(1 - \beta\chi') \tag{5.8}$$

in which

$$\beta = -\frac{1}{\rho}\frac{\partial\rho}{\partial\chi} \tag{5.9}$$

denotes the density change per unit change of the light element concentration. Since the densities of the candidate light elements are much less than the density of the outer core, we have $\beta = O(1)$. Here ρ' denotes the density perturbation due to light elements and ρ_o is the outer core mean density. The buoyancy of the fluid (considering only chemical buoyancy, not thermal buoyancy) can now be written as

$$g' = -\frac{\rho' g_o}{\rho_o} = \frac{\beta g_o d^2 \dot{\chi}_o}{\nu}\chi^* \tag{5.10}$$

where the second equality uses (5.5). If we non-dimensionalize g' by multiplying it by d^3 and dividing it by the product of the mass and viscous diffusivities, we get

$$\frac{g' d^3}{D\nu} = \frac{\beta g_o d^5 \dot{\chi}_o}{D\nu^2}\chi^* \tag{5.11}$$

The coefficient in (5.11) defines the Rayleigh number for light element convection:

$$Ra_\chi = \frac{\beta g_o d^5 \dot{\chi}_o}{D\nu^2} \tag{5.12}$$

We can re-write Ra_χ in terms of the inner core growth rate. For the extreme situation of no light elements in the inner core, conservation of light elements in the outer core implies

$$\overline{\chi_o \rho_o \dot{V}_o} = 0 \qquad (5.13)$$

where the overbar denotes that the time derivative is to be applied to the triple product, and V_o is the outer core volume. Applying the dilute solution approximation (corresponding to the limit $\chi_o \ll 1$) to the above equation yields

$$\dot{\chi}_o \simeq -\chi_o \frac{\dot{V}_o}{V_o} \simeq \frac{3\chi_o r_i^2}{r_c^3} \dot{r}_i \qquad (5.14)$$

where we have used the Boussinesq relation between inner and outer core volume changes, i.e. $\dot{V}_i = -\dot{V}_o$ and have ignored terms of order V_i/V_o. Substituting equation (5.14) into (5.12) gives

$$Ra_\chi = \frac{3\beta g_o d^5 \chi_o r_i^2}{D\nu^2 r_c^3} \dot{r}_i \qquad (5.15)$$

To estimate the magnitude of this Rayleigh number for the core, recall that the growth rate of the inner core can be expressed in terms of the core cooling rate as

$$\dot{r}_i = -\rho g \left(\frac{dT_L}{dp} - \frac{dT_{ad}}{dp} \right)^{-1} \dot{T}_o \qquad (5.16)$$

Using our previous estimate for the core cooling rate yields

$$\dot{r}_i \sim 10^{-3} \, myr^{-1} \qquad (5.17)$$

which is consistent the inner core growth rate predicted by more elaborate core evolution models (Gubbins et al., 2003; Labrosse, 2003). Depending on the highly uncertain mass diffusivity, (5.15) then gives

$$Ra_\chi \sim 10^{28-30} \qquad (5.18)$$

a very large number!!!

5.6. Geodynamo efficiency

In this part we derive equations governing the energetics of the geodynamo. We obtain a simplified energy balance, starting from the Boussinesq equations of motion, but including both compositional and thermal buoyancy. Although this energy balance is not exact (it ignores compression, among other things) its derivation is more transparent than if we were to start with the complete energy and entropy equations, and the results of this simpler analysis include the most important effects in dynamo energetics, including expressions for the dynamo efficiency. Later we outline the derivation of the full energy balance for the geo-dynamo and discuss its implications.

We begin with a linearized equation of state including both temperature and composition. Let

$$\rho = \rho_o + \rho' \tag{5.19}$$

where

$$\rho' = -\rho_o(\alpha T' + \beta \chi') \tag{5.20}$$

where ρ' is the *co-density* (sometimes denoted by C). The Navier–Stokes equation then becomes:

$$\frac{\partial \mathbf{u}}{\partial t} + (\mathbf{u} \cdot \nabla)\mathbf{u} + 2\Omega \times \mathbf{u} = -\frac{1}{\rho_o}\nabla P' + \mathbf{g}\frac{\rho'}{\rho_o} + \nu\nabla^2\mathbf{u} + \frac{1}{\rho_o}\mathbf{F}_L \tag{5.21}$$

where

$$\begin{array}{rcl}
\mathbf{u} & = & \text{Fluid velocity} \\
P' & = & \text{Perturbation pressure} \\
\rho' & = & \text{Co-density} \\
\mathbf{g} & = & \text{Gravity acceleration} \\
\Omega & = & \text{Rotation angular velocity} \\
\mathbf{F}_L & = & \text{Lorentz Force}
\end{array}$$

Four additional equations are necessary to describe Boussinesq convection in the core. These are the induction equation:

$$\frac{\partial \mathbf{B}}{\partial t} = \nabla \times (\mathbf{u} \times \mathbf{B}) + \eta\nabla^2\mathbf{B} \tag{5.22}$$

the continuity equation

$$\nabla \cdot \mathbf{u} = 0, \tag{5.23}$$

the heat equation

$$\frac{\partial T}{\partial t} + (\mathbf{u} \cdot \nabla)T = \kappa\nabla^2 T \tag{5.24}$$

and the mass transport equation for light elements

$$\frac{\partial \chi}{\partial t} + (\mathbf{u} \cdot \nabla)\chi = D\nabla^2\chi \tag{5.25}$$

If we assert that, because of turbulent mixing process the effective mass and thermal diffusivities in the outer core are equal (this is the so-called Prandtl hypothesis), we then have $D = \kappa$ and the two transport equations can be combined into a single transport equation for the co-density

$$\frac{\partial C}{\partial t} + (\mathbf{u} \cdot \nabla)C = \kappa \nabla^2 C \tag{5.26}$$

We now construct a mechanical energy equation by taking the inner product of the velocity with the terms in the Navier–Stokes equation (5.21) and integrating over the outer core volume. The first term on the left hand side becomes the temporal change in kinetic energy:

$$\int \rho_0 \mathbf{u} \cdot \frac{\partial \mathbf{u}}{\partial t} dV = \rho_0 \int \frac{\partial}{\partial t}\left(\frac{u^2}{2}\right) dV = \frac{\partial}{\partial t} E_k \tag{5.27}$$

The second term on the left hand side vanishes on volume integration. The third term on the left hand side vanishes because the velocity $\mathbf{u}$ is perpendicular to the Coriolis acceleration (i.e., the Coriolis acceleration does no work). The first term on the right hand side can be rewritten:

$$\int_V \mathbf{u} \cdot \nabla P dV = \int_V \nabla \cdot (\mathbf{u}P) dV - \int_V P(\nabla \cdot \mathbf{u}) dV \tag{5.28}$$

Using the incompressibility condition (continuity equation) and Stokes' theorem, this becomes

$$\int_V \mathbf{u} \cdot \nabla P dV = \oint_S (\mathbf{u}P) \cdot \mathbf{ds} \tag{5.29}$$

The area integral is zero if the velocity at the cmb is perpendicular to the area vector, assuming the cmb does not deform. Here we assume that the cmb radius remains constant, consistent with the Boussinesq approximation. In the fully compressible formulation, this term contributes to the energetics.

The second term on the right hand side of this equation represents the buoyancy production of kinetic energy. Only the radial velocity appears in this term because $\mathbf{u} \cdot \mathbf{r} = u_r$. The buoyancy production $\mathcal{B}$ can be written as the sum of thermal and compositional parts, as

$$\begin{aligned} \mathcal{B} &= \int (gCu_r)dV = -\rho_0\left(\int (g\alpha T'u_r)dV + \int (g\beta\chi'u_r)dV\right) \\ &= \mathcal{B}_T + \mathcal{B}_\chi \end{aligned} \tag{5.30}$$

The third term on the right hand side is the viscous heating

$$\Phi_v = \rho_0 v \int \mathbf{u} \cdot \nabla^2 \mathbf{u}\, dV \tag{5.31}$$

The last term on the right hand side is the work done against the Lorentz (magnetic) force. Using vector identities, it can be shown to be equal to:

$$\Phi_m = \int_V \frac{1}{\sigma} \mathbf{J}^2 dV \tag{5.32}$$

To obtain an electromagnetic energy equation, the left side of Ohm's law is dotted by $\mathbf{J}$ and the right side by the equivalent (Ampere's law) $\frac{1}{\mu_0} \nabla \times \mathbf{B}$. Using vector identities and applying Faraday's law to eliminate $\nabla \times \mathbf{E}$ gives

$$\frac{1}{\sigma} \mathbf{J}^2 = -\frac{1}{\mu_0} \nabla \cdot (\mathbf{E} \times \mathbf{B}) - \frac{\partial}{\partial t} \left(\frac{B^2}{2\mu_0} \right) - \mathbf{u} \cdot \mathbf{F_L} \tag{5.33}$$

where $\mathbf{F_L}$ is the Lorentz force. Integration of this equation over the whole space accounts for the total variation in the magnetic energy. The first term on the right hand side is zero because both $\mathbf{E}$ and $\mathbf{B}$ decrease faster than r^{-2}, and thus the integral vanishes at infinity. $\mathbf{J}$ and $\mathbf{u}$ are zero outside of the conducting sphere where there are no sources. Therefore, the expression for the change in magnetic energy is obtained by integration of the right hand side over the volume of the conducting sphere only, and yields

$$\frac{\partial}{\partial t} \int_\infty \frac{B^2}{2\mu_0} dV = - \int_V \mathbf{u} \cdot \mathbf{F_I} dV - \int_V \frac{1}{\sigma} \mathbf{J}^2 dV \tag{5.34}$$

The term on the left hand side is the change in magnetic energy. The first term on the right hand side is the work done by the fluid against the Lorentz force. The second term on the right hand side is due to Joule heating, and is called the Ohmic dissipation.

Adding the above mechanical and electromagnetic energy equations yields the Boussinesq form of the energy equation for the geodynamo:

$$\frac{\partial}{\partial t} (E_k + E_m) = \mathcal{B} - \Phi_m - \Phi_v \tag{5.35}$$

where E_k is the kinetic energy, E_m is the magnetic energy, $\mathcal{B}$ is the buoyancy production (thermal and compositional combined), Φ_v is the viscous dissipation and Φ_m is the Ohmic heating. Note that the two terms of work done by the Lorentz force (one from the Navier–Stokes equation and the other from Ohm's law) cancel each other. For convection in the outer core, it is usually assumed that $\Phi_m \gg \Phi_v$.

For thermal convection driven by secular cooling of the core, we can volume integrate the heat equation to obtain

$$\mathcal{B}_T \simeq \frac{r_c}{5H_T}(Q_{cmb} - Q_{ad}) \tag{5.36}$$

Recall that $H_T = C_p/\alpha g_c$ (C_p is the heat capacity, α is the thermal expansivity coefficient and g_c is the gravitational acceleration at the outer core), Q_{cmb} is the heat flux at the cmb and Q_{ad} is the adiabatic heat flux there.

For compositional convection due to the growth of the inner core, integration of the χ-equation (5.25) yields

$$\mathcal{B}_\chi \simeq \frac{r_c g_c \chi_0 \Delta\rho}{5\rho_0}\dot{m}_i \tag{5.37}$$

where $\dot{m}_i$ is the rate of increase of the inner core mass and $\Delta\rho = (\rho_0 - \rho_{LE})$ is the density difference between the outer core and the light elements. As the inner core crystallizes, there is an additional contribution to buoyancy production from latent heat release, which adds a term like $r_c \dot{m}_i L/2H_T$ to (5.37). For an equilibrium dynamo, (5.35) reduces to the following simple-looking balance:

$$\mathcal{B}_T + \mathcal{B}_\chi = \Phi_m + \Phi_v \tag{5.38}$$

Formulas (5.36), (5.37), and (5.38) can be applied to the geodynamo and other terrestrial-type convective dynamos to calculate what is called the *dynamo efficiency*. The dynamo efficiency ϵ is usually defined as the ratio of total dissipation to total heat loss. For dynamos driven by thermal convection, where the convective heat flow is used, (5.36) and (5.38) yield

$$\epsilon_T \simeq \frac{r_c}{5H_T}\frac{Q_{cmb} - Q_{ad}}{Q_{cmb}} \sim 0.05\text{--}0.1 \tag{5.39}$$

The efficiency factor for compositional convection can be calculated using (5.37) and (5.38), and relating the heat flow to the rate of inner core growth using formulas from previous sections. For the case of compositional convection the efficiency factor is somewhat higher:

$$\epsilon_\chi \sim 0.1\text{--}0.2 \tag{5.40}$$

because the buoyancy originates deeper in the core than buoyancy derived from secular cooling. In either case, the dynamo efficiency factor is rather low, so if the dissipation is high, as is generally thought, a lot of heat must be extracted from the core to maintain the geodynamo. In a later section we review estimates of the Ohmic dissipation in the core, and show that it is indeed rather high. This has implications for the present-day heat budget of the core, and also for its rate of thermal evolution.

5.7. Inner core formation

5.7.1. Energy considerations

We can estimate the energy available to drive the geodynamo that accompanies the formation of the inner core using the following simple model. Let T^-, ρ^-, V^-, etc. denote the temperature, density, volume, etc of the purely liquid core, prior to inner core nucleation, while T, ρ, V, etc. denote present-day variables. Assume that at some specific time in the past the core begins to differentiate into a solid inner core and a liquid outer core. The change in energy content of the core since that time due to inner core formation can be written

$$\Delta E = \Delta E_T + \Delta E_L + \Delta U_e + \Delta U_i \tag{5.41}$$

where the terms on the right hand side represent contributions from changes in sensible heat, latent heat release, and external and internal work, respectively. Here we ignore the effects of radioactive heat production. Using typical values for the core, the change in energy from each term can be estimated in Joules. For the sensible heat

$$\Delta E_T = \int_c \rho^- C_p^- T^- dV^- - \int_c \rho C_p T dV \simeq 20 \times 10^{28} \text{ J} \tag{5.42}$$

where C_p is the heat capacity. For the latent heat due to the solidification of the inner core

$$\Delta E_L = L m_i \simeq 8 \times 10^{28} \text{ J} \tag{5.43}$$

where m_i is the present-day mass of the inner core and L is the energy released by solidification per unit mass. ΔU_e is the change in "external" work, mostly due to the change in pressure, and is given by

$$\Delta U_e = \int_c P^- dV^- - \int_c P dV \simeq 10 \times 10^{28} \text{ J} \tag{5.44}$$

while ΔU_i is the change in "internal" work due to the change in the density distribution:

$$\Delta U_i \simeq \frac{3}{10} (gr)_c V_i (\rho_i - \rho^-) \simeq 5 \times 10^{28} \text{ J} \tag{5.45}$$

Summing the above contributions gives the net change in energy related to inner core formation

$$\Delta E \simeq (20 + 8 + 10 + 5) \times 10^{28} \text{ J} \simeq 43 \times 10^{28} \text{ J} \tag{5.46}$$

5.7.2. Implications

Even without meaningful amounts of radioactive heat sources in the core, the energy available since inner core nucleation is substantial, but because the heat loss from the core is large, it may be that the inner core is far younger than the Earth $\tau_e \simeq 4.56$ Ga, and possibly far younger than the geodynamo. Recall that the total heat flow at the cmb is estimated at $Q_{cmb} \simeq 4$–12 TW. The ratio of ΔE to total core heat flow defines an evolutionary time scale, a "model age" for the inner core:

$$\tau_i = \frac{\Delta E}{Q_{cmb}} \qquad (5.47)$$

Based on the limits of our estimate of core heat loss, two distinct regimes are possible:

1. If we assume $Q_{cmb} = 4$ TW, then $\tau_i \simeq 3.5$ Ga. The inner core is relatively old and has grown slowly. The geodynamo is driven by co-existing compositional and thermal convection. The heat flux through the cmb is relatively small, so the Ohmic dissipation by the geomagnetic field must also be rather small.

2. If we assume $Q_{cmb} = 12$ TW, then $\tau_i \simeq 1.2$ Ga. The inner core is relatively young and is growing rapidly. The core's evolution was dominated by thermal convection when the inner core was small, then by compositional convection as the area of the icb increased. The heat flux through the cmb is large enough to support a rather dissipative dynamo. However, this model has problems in the deep past, before inner core nucleation.

5.8. Dynamo energetics

More comprehensive analyses of dynamo energetics makes use of the first and second laws of thermodynamics for a self-gravitating, compressible, dissipative fluid. There are many different formulations of these balances for the geodynamo in the literature (e.g. Verhoogen, 1979; Labrosse et al., 2001; Gubbins et al., 2003; Roberts et al., 2003; Nimmo et al., 2004) Here we outline one such derivation, following Nimmo (2007). The first law of thermodynamics provides an energy balance that places constraints on the core heat loss, as shown above for inner core growth energetics. However, this balance does not involve the magnetic field. In order to bring the magnetic field into the picture, the second law of thermodynamics (entropy balance) must be added. Figure 56 identifies the most important contributions to the energy and entropy balances for the geodynamo.

We begin with the first law of thermodynamics in rate form, volume-integrated over the whole core:

$$Q_{cmb} = Q_R + Q_C + Q_G + Q_L + Q_P \qquad (5.48)$$

Full Core Energy Balance -- Major Players

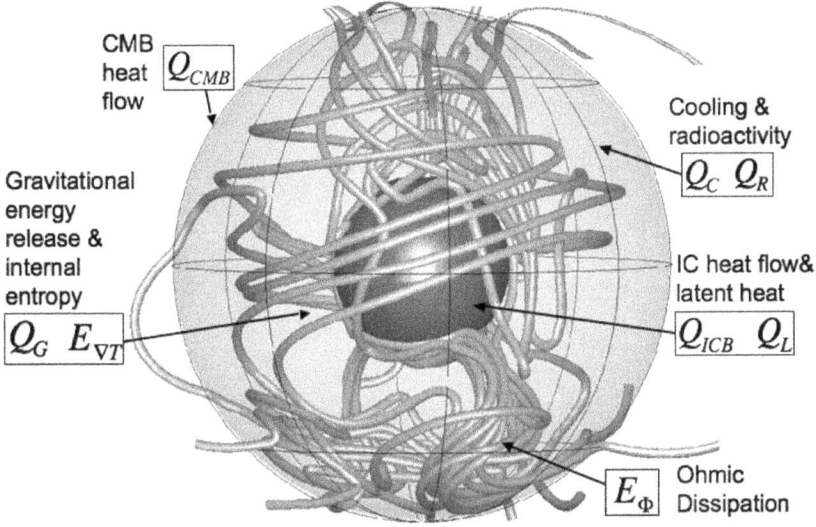

Fig. 56. Important core energy sources (Q) and entropy sources (E) for the geodynamo.

where

$$Q_G = \int \rho \psi \frac{\Delta \rho}{\rho_o} \frac{d\chi}{dt} dV \tag{5.49}$$

is the gravitational energy,

$$Q_L = 4\pi r_i^2 \rho_i L \frac{dr_i}{dt} \tag{5.50}$$

is the energy from latent heat release,

$$Q_P = \int \alpha T \frac{dP}{dt} dV \tag{5.51}$$

is the pressure-work energy,

$$Q_R = \int \rho h \, dV \tag{5.52}$$

is energy from radioactive sources with density h, and

$$Q_C = -\int \rho C_p \frac{dT}{dt} dV \tag{5.53}$$

is the energy from secular cooling (sensible heat loss).

Similarly, we can form an entropy production balance starting from the rate form of the second law of thermodynamics. *Note: In equations (5.54) to (5.66) we use E to denote entropy change per unit time.* The entropy balance is

$$E_C + E_L + E_P + E_G + E_R = E_k + E_\Phi \tag{5.54}$$

in which

$$E_C = \int \rho C_p (T^{-1} - T_c^{-1}) \frac{dT}{dt} dV \tag{5.55}$$

is the entropy production from secular cooling,

$$E_L = -\frac{4\pi r_i^2 L(T_i - T_c)}{(dT_m/dP - dT/dP)g T_c} \tag{5.56}$$

is the entropy production from latent heat,

$$E_R = -\int \rho h (T^{-1} - T_c^{-1}) dV \tag{5.57}$$

is entropy production from radioactive heating,

$$E_P = \int \alpha T (T^{-1} - T_c^{-1}) \frac{dP}{dt} dV \tag{5.58}$$

is entropy production from pressure work,

$$E_G = \frac{Q_g}{T_c} \tag{5.59}$$

is entropy production from chemical differentiation,

$$E_k = \int k \left(\frac{\nabla T}{T}\right)^2 dV \tag{5.60}$$

is entropy production from heat conduction, and

$$E_\Phi = \int \frac{\phi}{T} dV \tag{5.61}$$

is entropy production from dissipation, where

$$\phi = \phi_m + \phi_v \simeq \phi_m = \sigma^{-1} \mathbf{J}^2 \tag{5.62}$$

where ϕ is the dissipation per unit volume.

These balances can be re-written by factoring out the rate of change of the core-mantle boundary temperature $\dot{T}_c$:

$$Q_C + Q_L + Q_G = Q_T^* \dot{T}_c \tag{5.63}$$

and

$$E_C + E_L + E_G - E_k = E_T^* \dot{T}_c \tag{5.64}$$

where we have neglected the small pressure term. We also define

$$T_R = \frac{Q_R}{E_R} \tag{5.65}$$

and combine (5.63) and (5.64) to get

$$Q_{cmb} = Q_R\left(1 - \frac{Q_T^*}{T_R E_T^*}\right) + \frac{Q_T^*}{E_T^*}(E_\Phi + E_k) \tag{5.66}$$

With these equations, we can now calculate thermal evolution model of the core and the geodynamo. Table 7 gives the size of the various terms in the energy and entropy balances for two different assumed concentrations K40, according to Nimmo (2007), along with calculated rates of cooling, inner core growth, and inner core age for each case. These were calculated using a nominal estimate of the dissipation, $\Phi = 1$ TW, and a cmb heat flow of 9 TW. Note that the latent heat contributes heavily to the energy balance, but the compositional differentiation term contributes most heavily to the entropy balance. Also, the inner core grows fast in both cases, being less than 1 Ga in age in both cases.

Table 7
Energy Q and entropy E terms from Nimmo (2007). Notes: s = C; g = G in our notation

	$K = 0$				$K = 300$ ppm			
	Q		E		Q		E	
	TW	%	MW/K	%	TW	%	MW/K	%
Q_s, E_s	2.2	25	73	8	1.7	19	56	7
Q_L, E_L	4.2	47	268	28	3.3	37	205	26
Q_g, E_g	2.5	28	618	64	1.9	21	474	59
Q_R, E_R	0	0	0	0	2.1	24	65	8
Q_k, E_k	4.9	–	−162	–	4.9	–	−162	–
Q_H, E_H	0	–	−219	–	0	–	−168	–
$Q_{cmb}, \Delta E$	9.0	–	537	–	9.0	–	431	–
dT_c/dt (K/dyr)			−37				−30	
dr_i/dt (km/dyr)			788				605	
Ic age (Myr)			590				780	

P. Olson

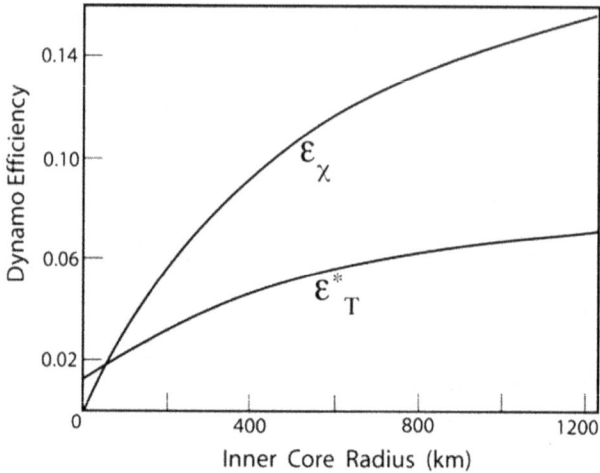

Fig. 57. Nominal dynamo efficiency versus inner core radius for thermal and compositional convection.

5.9. Dynamo efficiency vs inner core radius

The dynamo efficiency factors depend on the inner core radius and therefore have changed during evolution of the core, as shown in Fig. 57. Compositional convection ϵ_χ has overall higher efficiency than thermal convection ϵ_T^*. Efficiency in general depends on the nature of the buoyancy source and its location, with "deep" buoyancy sources having intrinsically higher efficiency than "shallow" or volumetrically-distributed sources, and compositional buoyancy having intrinsically higher efficiency compared to thermal.

5.10. Estimating the Ohmic dissipation in the core

We have been able to place geophysical constraints on Q_{cmb}, but what about the Ohmic dissipation Φ in the core? This is harder, for several reasons. First, the magnetization of crustal rocks confounds our estimate of the high wavenumber part of spectrum of the core field, so we don't know the short wavelength (i.e., high spherical harmonic degree) part of the field on the cmb. This is a real problem for estimating Ohmic dissipation. Because harmonics of the squared electric current density and the power in the magnetic field are related by $j_n^2 \sim n(n+1)B_n^2 \sim n(n+1)R_n$, the dissipation spectrum contains proportionally more energy at high harmonics compared to the power spectrum of the magnetic field. Second, we have few constraints on the toroidal part of the magnetic field inside the core, and so we have almost no information on the poloidal part of the electric current density. For these reasons, severe assumptions about the internal magnetic field structure must be invoked in order to estimate the Ohmic dissipation in the core. Several methods of doing this have been proposed, and all have obvious limitations.

5.10.1. Ohmic dissipation estimated from the core field structure

Figure 13 shows the power spectrum of the core field at the Earth's surface. The knee around $n = 14$ is usually taken to be the edge of the visible core field. Some assumptions about the continuation of the core field spectrum must be made in order to extrapolate the core field to higher degrees. Usually the power spectrum R_n is assumed to extrapolate to large n like

$$R_n = R_o e^{-bn} \tag{5.67}$$

Large b amounts to truncation; small b assumes a broad-band core field. The present-day core field on the cmb can be fit to (5.67) with $b \simeq 0.1$. Even this extrapolation is not enough, however, because this applies to the cmb field, not the rms interior field, and it also does not take the toroidal field into account. So, further constraints must be invoked. Some of these are:

Minimum dissipation. Calculate the Ohmic dissipation for the smoothest possible internal field structure that matches the observed cmb field. This approach gives $\Phi \sim 45$ MW, which is unrealistically low.

Free decay. Calculate the Ohmic dissipation for a freely-decaying field that matches the observed cmb field. This approach gives $\Phi \sim 2$ GW, which is better, but is still unrealistically low. In addition, we already know from the geomagnetic secular variation that free decay is not a good model for the behavior of the present-day core field.

Isotropic extrapolation. Extrapolate the core field spectrum using the best-fitting b-value and assume an isotropic electric current structure inside the core to calculate the Joule heating. This gives $\Phi \sim 100$ GW for $b \simeq 0.1$, which is more realistic. However, this method is very sensitive to the b-value, which is poorly constrained.

5.10.2. Ohmic dissipation estimated from numerical dynamos

Figure 58 shows the 3D structure of a numerical dynamo driven by thermal convection heated from below. The Ekman number is 6×10^{-3} and the Rayleigh number is about 5 times critical. The external field (*a*) is dipole dominant and the convection is columnar and quasi-geostrophic, as shown in (*b*) by the 3D distribution of axial vorticity. The Ohmic dissipation in (*d*) is seen to be concentrated near the inner core where the buoyancy originates. In addition, the Ohmic dissipation has a high wavenumber content, characterized by short length scales.

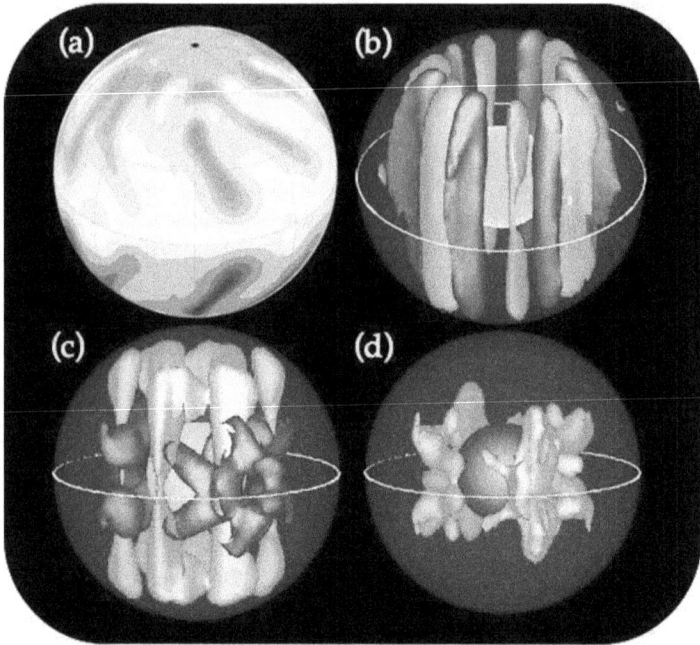

Fig. 58. Numerical dynamo model images. (a) cmb radial magnetic field. (b) Axial vorticity. (c) Kinetic (yellow) and magnetic (blue) energy density. (d) Ohmic dissipation. In all images the equator is shown in white, and the inner core is a solid color.

Table 8

Some dynamo model estimates of Ohmic dissipation in the core

	Dissipation	Method
High:		
Labrosse (2003)	1.5–3 TW	assumed toroidal field
Gubbins et al. (2003)	2.2–3.6 TW	assumed toroidal field
Intermediate:		
Roberts, Jones & Calderwood (2003)	1–2 TW	dynamo models & scaling
Nimmo (2007)	0.5–2 TW	assumed range
Low:		
Buffett & Bloxham (2002)	0.1–0.5 TW	scaled dynamo model
Christensen & Tilgner (2004)	0.2–0.45 TW	experimental comparison

Because of these properties, Ohmic dissipation in the core is very difficult to constrain with numerical dynamo model results. Table 8 lists several studies in which the Ohmic dissipation in the core has been estimated from numerical dynamo model results. There is a wide range of values, from a minimum near 0.1 TW to over 3 TW. Some of the range is due to differences in dynamo models.

Models with more toroidal flow generate larger toroidal fields, with typically higher dissipation. There are also differences in the ways the dynamo models were scaled to core conditions. Even so, the spread is indicative of the true uncertainty in this parameter.

5.11. A simple evolution model of the core

For the present-day geodynamo, let us assume that we can write

$$\Phi = \epsilon Q_{cmb} \tag{5.68}$$

and further assume that $Q_{cmb} = 9$ TW, and $\epsilon = 0.12$, midway between the efficiencies for thermal and purely compositional convection. Then the dissipation is $\Phi \sim 1$ TW and the entropy production from this dissipation is $E_\Phi \approx 200$ MW K^{-1}. To estimate the age of the inner core, we repeat the procedure from the last lecture, forming the ratio of the available energy during inner core growth to the heat flow at the cmb,

$$\tau_{ic} \approx \frac{\Delta E}{Q_{cmb}} = \frac{43 \times 10^{28} \text{ J}}{9 \times 10^{12} \text{ W}} \approx 1.5 \text{ Ga} \tag{5.69}$$

Note that the age of the inner core is inversely proportional to Q_{cmb}, and accordingly, the model age of the inner core is only about 1.5 Ga. The next question is: how do these values compare with more elaborate thermal evolution models?

5.12. Detailed thermal evolution modeling

More elaborate models of the evolution of the core are derived from the rate form of the energy and entropy balances. Usually these evolution equations are arranged so that the cmb temperature or the average core temperature is the dependent variable, and take the following general form:

$$\frac{dT_c}{dt} = \frac{Q_{cmb} - Q_R}{Q_T^*} \tag{5.70}$$

where the Q-factors have their previous definitions. The corresponding energy changes are then given by

$$(E_C, E_G \ldots) = \int (Q_C, Q_G \ldots) \, dt \tag{5.71}$$

There are two general classes of these evolution models in the literature: isolated core models, in which the core evolves separate from the mantle, and coupled core-mantle models, in which the core and mantle are coupled and co-evolve. Within each of these broad classes, the models differ based on which parameters are assumed to be constant in time. Some models assume the cmb heat flow is invariant, others assume the total dissipation is invariant, still others (the coupled models in particular) only assume particular values for the dynamo efficiency factors.

Isolated Core Evolution Models These solve (5.70) for $T_c(t)$, assuming either fixed Q_{cmb} or fixed entropy production E_ϕ The inner core radius is determined from the intersection of the core geotherm and the melting curve. These models just involve ordinary differential (time-stepping) equations, so they can be solved backward as well as forward in time.

Coupled Core-Mantle Models These models drive the evolution equation (5.70) with $Q_{cmb}(t)$ derived from a mantle thermal evolution model. For these, $T_c(t)$ provides the feedback as a time-dependent boundary condition for the mantle thermal evolution. If the mantle model is also of the volume-averaged type, these can be solved backward in time as well as forward. However, if the mantle model is 2D or 3D, only forward time solution is possible.

Figure 59 show the results of a thermal evolution calculation of the isolated core type. The model assumes 11 TW of heat flow at the cmb and no radioactive heat sources. Backward time integration "melts" the inner core, which disappears around 0.85 Ga. For greater ages, the core cools by sensible heat loss alone, at a rate of nearly -125 K/Ga (a fast cooling rate because of the large core heat loss assumed). This model suggests very high core temperatures (a sort of thermal catastrophe) and large core superheats in the deep past.

Figure 60 shows isolated evolution models of the core in which the dissipation is fixed. Increasing Ohmic dissipation shortens the predicted inner core age approximately as $\tau_i \sim \Phi^{-1/2}$. In the same way as with the constant core heat loss models, the core temperature rises rapidly with increasing age prior to inner core nucleation in these fixed dissipation models. Note that a very old (>4 Ga) inner core requires an efficient geodynamo, dissipating 0.1 TW or less.

Figure 61 shows the results of a thermal evolution model by Xie and Tackley (2004) with core-mantle coupling. Mantle dynamics is represented in this model using 2D infinite Prandtl number thermal subsolidus convection in a cylindrical shell. The core is represented by a volume averaged thermal evolution model as described previously. Coupling is through continuity of the average heat flow and the temperature at the cmb. The core is assumed to have an initially large superheat, much of which is maintained over the 4.5 Ga of the simulation. As the mantle radioactivity diminishes, internal heat production decreases and the Earth surface heat flow decreases, although with a non-monotonic trend. Heat flow at the cmb also decreases slowly with time, and has small fluctuations on the same time scales as the fluctuations of the mantle convection. The calculated entropy production in the core shows similar mantle-driven fluctuations, which would be balanced by Ohmic dissipation, according to our dynamo energy considerations.

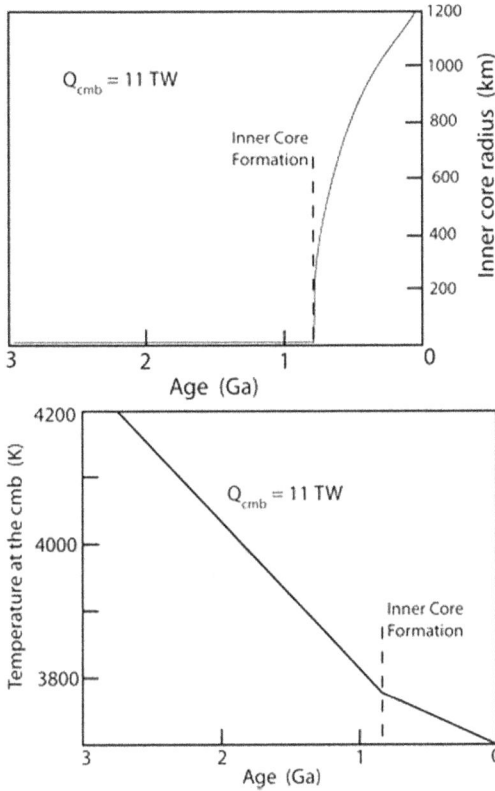

Fig. 59. Isolated core evolution model. Upper plot: Inner core radius growth. Lower plot: cmb temperature over time.

Note that, for a time early in the calculation, the entropy production in the core falls below the threshold for dynamo maintenance. According to this model, the geodynamo would fail during that time.

5.13. Summary of results

1. Present-day core heat flow may be high, in the range $Q_{cmb} \sim 9\text{--}12$ TW.

2. According to dynamo energy considerations, the geodynamo can theoretically operate with lower heat flow, for example
   i. $\Phi = 0.1$ TW implies $Q_{cmb} \sim 2\text{--}4$ TW
   ii. $\Phi = 0.5$ TW implies $Q_{cmb} \sim 4\text{--}6$ TW

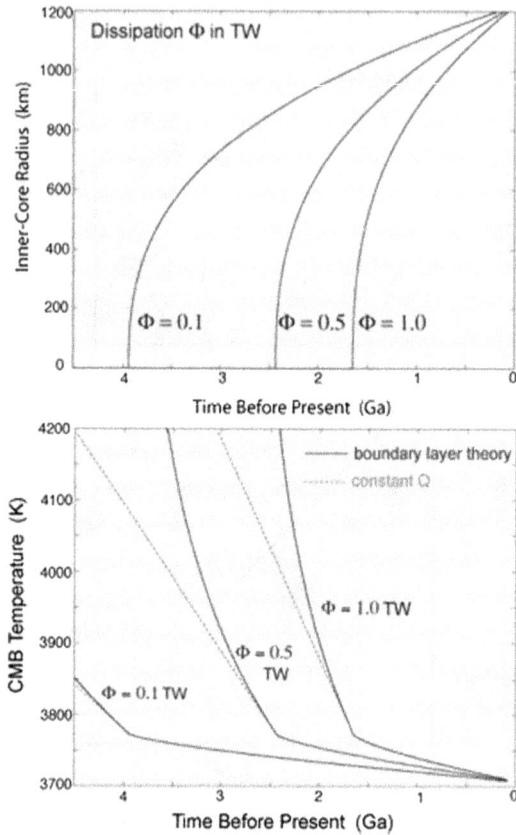

Fig. 60. Core thermal evolution models with fixed Ohmic dissipation, after Buffett (2002). First: inner core radius versus age for different dissipation values. Second: core-mantle temperatures calculated for different dissipations, assuming constant Q (constant core-mantle heat flow) and also a boundary layer model of core-mantle core heat flow).

3. Core dissipation of $\Phi > 0.5$ TW requires additional energy sources in the deep past, prior to nucleation of the inner core. Otherwise the rapid loss of sensible heat would imply unacceptably high core temperatures.

4. Potassium concentrations of a few 100 ppm (the range of concentrations inferred by some geochemical and cosmochemical models) do not strongly affect these energy calculations and do not enormously increase inner core age. Larger potassium concentrations (>400 ppm) do have an effect on core evolution, especially in softening the problematic early thermal catastrophe.

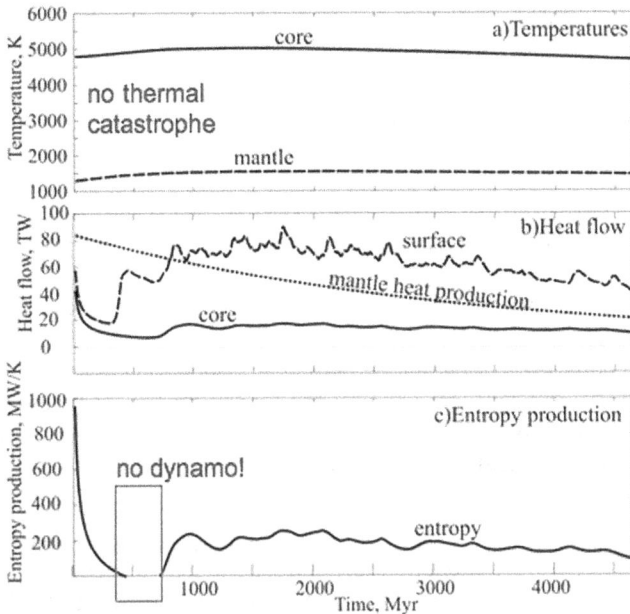

Fig. 61. Coupled core-mantle evolution model of Xie & Tackley (2004).

References

Buffett, B.A., Bloxham, J. 2002. Energetics of numerical geodynamo models, Geophysical Journal International **149**(1), 211–224. doi:10.1046/j.1365-246X.2002.01644.x

Buffett, B.A. 2002. Estimates of heat flow in the deep mantle based on the power requirements of the geodynamo, Geophys. Res. Lett. **29**(12), 1566–1570.

Buffett, B.A. 2000. Earth's core and the geodynamo, Science **288**, 2007–2012.

Buffett, B.A. 2003. The thermal state of the Earth's core, Science **299**, 1675–1676.

Christensen, U.R. and Tilgner, A. 2004. Power requirement of the geodynamo from ohmic losses in numerical and laboratory dynamos, Nature **439**, 169–171.

Gubbins, D., Alfe, D., Masters, G., Price, G.D. and Gillan, M.J. 2003. Can the Earth's dynamo run on heat alone? Geophys. J. Intl. **155**, 609–22.

Jones, C.A. 2007. Thermal and compositional convection in the outer core, in: *Treatise on Geophysics*, vol. 8 ch. 5, Elsevier B.V., pp. 131–176.

Labrosse, S., Poirier, J.P. and LeMouel, J.L. 2001. The age of the inner core, Earth Planet. Sci. Lett. **190**, 111–123.

Labrosse, S. 2003. Thermal and magnetic evolution of the Earth's core, Phys. Earth Planet. Inter. **140**, 127–143.

Loper, D. 1978. Some thermal consequences of the gravitationally powered dynamo,. J. Geophys. Res. **83**, 5961–5970.

Nimmo, F. 2007. Energetics of the core, in: *Treatise on Geophysics*, vol. 8, ch. 2, Elsevier B.V.

Nimmo, F., Price, G.D., Brodholt, J. and Gubbins, D. 2004. The influence of potassium on core and geodynamo evolution, Geophys. J. Int. **156**, 363–376.

Roberts, P.H., Jones, C.A. and Calderwood, A.R. 2003. Energy fluxes and ohmic dissipation in the earth's core, in: *Earth's Core and Lower Mantle*, C. Jones, A. Soward and K. Zhang, eds., Taylor and Francis.

Nimmo, F. 2007. Energetics of the core, in: *Treatise on Geophysics*, vol. 8, ch. 2, Elsevier Press.

Verhoogen, J. 1980. *Energetics of the Earth*, National Academy Press, Washington, DC, 139 pp.

Xie, S.X. and Tackley, P.J. 2004. Evolution of U-Pb and Sm-Nd systems in numerical models of mantle convection and plate tectonics, J. Geophys. Res. **109**, B11204.

6. Planetary dynamo scaling laws

Dynamo scaling laws are an old quest. More than six decades ago, Elsasser (1946; Fig. 62, left) proposed that planetary magnetic fields equilibrate at an intensity that corresponds to a balance between Coriolis and Lorentz forces, which is now known as the constant Elsasser number assumption. If we assert that dynamo equilibration occurs at an Elsasser number $\Lambda = \sigma B^2/\rho\Omega \sim 1$, then the equilibrium internal field strength in a planetary dynamo is

$$B_{rms} \simeq \sqrt{\rho\Omega/\sigma} \qquad\qquad (6.1)$$

As previously noted, the equilibrium field strength is independent of the energy source with this assumption. Shortly thereafter, Blackett (1947; Fig. 62, right) argued against the idea that planetary magnetic fields result from fluid dynamo action. He proposed an alternative theory, in which magnetic fields are intrinsic properties of rotating planets; in essence he proposed a modification of Maxwell's equations. His theory predicted that the dipole moment (denoted by M here) should be proportional to the planetary core angular momentum, L ie. $M \propto L$.

It is interesting to test these scaling laws by comparing their predictions against known and estimated planetary dipole moments. The planetary magnetic dipole moment predicted by each scaling can be expressed as products of the physical parameters (core radius, rotation rate, conductivity, and density) to fractional

Fig. 62. Left: Walter Elsasser. Right: Patrick Blackett.

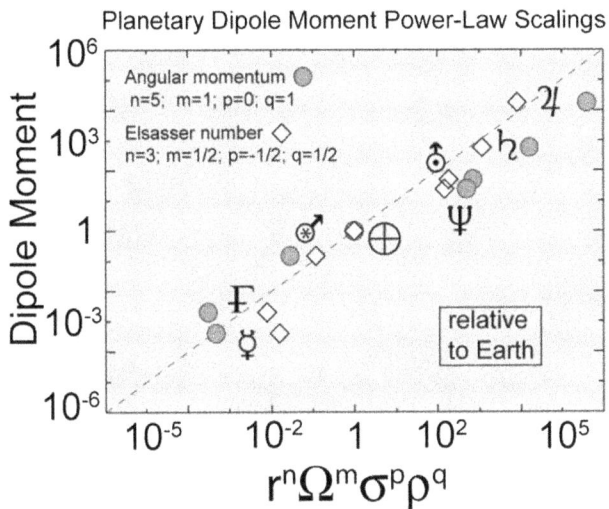

Fig. 63. Dipole moment scaling laws for both the constant Elsasser number assumption and the angular moment law, compared with dipole moments of Mercury, Earth, Ancient Mars, Jupiter, Saturn, Uranus, Neptune, Ancient Moon, and Ganymede. Dipole moments of the extinct dynamos are estimated from their crustal magnetization. Values are plotted relative to the Earth. On the x-axis, r, Ω, σ, and ρ are core radius, rotation rate, electrical conductivity, and density, respectively.

Table 9

Planetary core parameters (some estimated). Rotation rate Ω, core radius r_c, electrical conductivity σ, core viscosity ν, dipole moment M, and heat flow Q

Planet	Ω (s^{-1})	r_c (km)	ρ_c $(kg\,m^{-3})$	σ (Sm^{-1})	ν $(m^2\,s^{-1})$	M $(Z\,Am^2)$	Q (mW/m^2)
Mercury	1.24×10^{-6}	1800	10×10^3	6×10^5	1×10^{-6}	0.03	40
Earth	7.29×10^{-5}	3485	11×10^3	6×10^5	1×10^{-6}	78	45
Jupiter	1.76×10^{-4}	50000	1.8×10^3	1×10^4	1×10^{-6}	1.4×10^6	4×10^4
Saturn	1.64×10^{-4}	30000	1.8×10^3	1×10^4	1×10^{-6}	4.5×10^4	3×10^4
Uranus	1.01×10^{-4}	17000	2×10^3	1×10^4	1×10^{-6}	3.9×10^3	4×10^2
Neptune	1.09×10^{-4}	16000	2×10^3	1×10^4	1×10^{-6}	1.8×10^3	4×10^2

powers. Figure 63 compares the predicted and observed (or inferred) dipole moments for the two scaling laws, using the parameter values in Table 9. The planetary dipole moments have been normalized by the Earth's dipole moment. The insert in Fig. 63 gives the exponents used in each scaling, and the dashed line indicates a perfect correlation. The first thing to notice is that both of these scaling laws appear to be broadly reasonable when plotted this way. However, much of the apparent correlation stems from the fact that the variables on both axes depend strongly on planetary core radius.

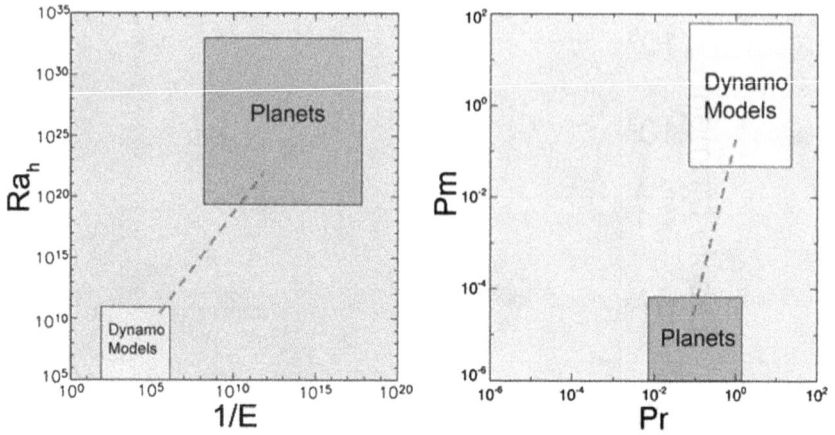

Fig. 64. The parameter regimes of the planets and dynamo modeling are orders of magnitude apart.

Why do we need dynamo scaling laws? First, we need them for theoretical reasons. It is important to determine what are the unifying physics that underly the variety of planetary magnetic fields. Second, we need them for numerical modeling of planetary dynamos. Figure 64 clearly shows why: numerical dynamos are far from the planets in 3 of the governing parameters, Ra, E, and Pm.

There are many scaling possible relationships for planetary dynamos that can be derived from dimensional and similarity arguments, because dynamo theory contains so many physical parameters (Busse, 1976; Russell, 1979; Stevenson, 1983; Sano, 1993; Cain et al., 1995, Starchenko and Jones, 2002). Obviously, the most useful of such relationships are the ones that can be tested using theoretical and numerical models and conform to planetary magnetic field measurements.

Recall that there are 4 independent nondimensional control parameters (plus at least one geometrical parameter; here we chose the radius ratio $r* = r_i/r_c$) and we are interested in at least 2 or 3 output parameters. The control parameters include the Rayleigh Ra (6.2), Ekman E (6.3), magnetic Prandtl Pm (6.4), and Prandtl Pr (6.5) numbers, and the geometric factor $r*$ (6.6). Here we choose a Rayleigh number defined in terms of the superadiabatic heat flow q':

$$Ra = \frac{\alpha g_0 d^4 q'}{k \kappa \nu} \tag{6.2}$$

$$E = \frac{\nu}{\Omega d^2} \tag{6.3}$$

$$Pm = \frac{\nu}{\lambda} \tag{6.4}$$

$$Pr = \frac{\nu}{\kappa} \tag{6.5}$$

$$r* = r_i / r_c \tag{6.6}$$

The output parameters include the magnetic Reynolds number Rm (6.7) for the dynamo fluid velocity and the Elsasser number Λ (6.8) defined using the rms internal field. For reference, we include the definition of the dipole moment M (6.9), which can be represented in terms of the rms of the cmb dipole field:

$$Rm = \frac{ud}{\lambda} \tag{6.7}$$

$$\Lambda = \frac{\sigma B_{rms}^2}{\rho \Omega} \tag{6.8}$$

$$M = \frac{4\pi r_c^3 B_{rms}^{dipole,cmb}}{\sqrt{2}\mu_0} \tag{6.9}$$

6.1. Power law scaling

6.1.1. Similarity relationships

Derivation of power law scaling relationships can be made using dimensional arguments alone, without reference to the dynamo equations, although scaling the individual terms in the dynamo equations adds additional constraints. Let us assume that, for a given planetary dynamo, we can write the Elsasser and magnetic Reynolds numbers as functions of products of the control parameters raised to powers, i.e.,

$$\Lambda = F_1(Ra^\alpha E^\beta Pr^\gamma Pm^\delta) \tag{6.10}$$

and

$$Rm = F_2(Ra^\alpha E^\beta Pr^\gamma Pm^\delta) \tag{6.11}$$

If the functions F_1 and F_2 are themselves power laws of their respective arguments, then we can write them in terms of the physical parameters as

$$(F_1, F_2) \propto \nu^{(\beta-\alpha+\gamma+\delta)} \kappa^{(-2\alpha-\gamma)} \lambda^{(-\delta)} q'^{(\alpha)} \Omega^{(-\beta)} d^{(4\alpha-2\beta)} \tag{6.12}$$

In order to specify F_1 or F_2, we need a total of 4 similarity constraints. Physical intuition, observational experience, or judicious scaling of the terms in the dynamo equations suggest what these constraints might be. Here we derive the consequences of three frequently-invoked sets of constraints, one for the convective velocity, the other two for the magnetic field intensity

Case 1: Assume that the fluid velocity is independent of all 3 diffusivities and the shell thickness. Setting the exponents of these factors to zero yields

$$\alpha = 1/2, \quad \beta = 1/2, \quad \gamma = -1, \quad \delta = 1 \tag{6.13}$$

Using $q'/\rho H_T = F$, where F is the buoyancy flux, back substitution into (6.11) yields a standard velocity scaling for rotating convection,

$$u \propto \left(\frac{F}{\Omega}\right)^{1/2} \tag{6.14}$$

Case 2: Magnetic field intensity is independent of the small diffusivities (thermal and viscous), shell thickness, and the rotation rate. Setting the exponents of these factors to zero yields

$$\alpha = 1/2, \quad \beta = 1, \quad \gamma = -1, \quad \delta = 1/2 \tag{6.15}$$

which, when back substituted into (6.10) yields

$$B \propto \left(\frac{\mu_0 \rho q'}{\sigma H_T}\right)^{1/4} \tag{6.16}$$

Case 3: Magnetic field intensity is independent of all diffusivities and the rotation rate. These constraints give

$$\alpha = 1/6, \quad \beta = 1, \quad \gamma = 0, \quad \delta = 1/6 \tag{6.17}$$

so that

$$B \propto (\rho \mu_0)^{1/2} \left(\frac{q'd}{\rho H_T}\right)^{1/3} \tag{6.18}$$

Note that both of these laws predict weak dependence on the convection (through q'), and both differ significantly from either the constant Elsasser number or angular momentum laws.

6.1.2. Scaling the dynamo equations

By balancing sets of terms in the dynamo equations, additional types of scaling for the fluid velocities and the magnetic field strengths can be derived. Two examples of velocity scaling are called MAC (Magnetic, Archemedies, Coriolis) and Columnar Convection scaling (Jones, 2007). Unexpectedly perhaps, neither of these involves explicit dependence on the magnetic field strength.

MAC velocity scaling assumes a balance between Coriolis and buoyancy forces in the Navier–Stokes equation. For thermal buoyancy, this means

$$2\Omega u \sim \alpha g T' \tag{6.19}$$

where T' is the temperature perturbation. The assumed heat equation balance is convective, i.e.

$$q' \sim \rho C_p u T' \tag{6.20}$$

Combining these gives

$$u_{MAC} \sim \left(\frac{q'}{\rho H_T \Omega}\right)^{1/2} \sim \left(\frac{F}{\Omega}\right)^{1/2} \tag{6.21}$$

as in Case 1 above. For thermal convection in the core, $u_{MAC} \sim 1 \times 10^{-3}\ \mathrm{m\,s^{-1}}$.

Quasi-geostrophic (columnar-style) convection assumes an elongated columnar velocity structure, with the length scale of the flow parallel to the rotation axis being very much larger than the transverse length scales. To find the velocity and length scales of this style of flow, we form a three-way balance between inertia, Coriolis, and buoyancy terms in the axial vorticity equation, which gives

$$\frac{u^2}{\delta^2} \sim \frac{\alpha g T'}{\delta} \sim \frac{\Omega u}{d} \tag{6.22}$$

The heat balance is again convective

$$q' \sim \rho C_p u T' \tag{6.23}$$

The above two relationships provide 3 independent balance constraints, from which we can find the velocity

$$u \sim \left(\frac{d}{\Omega}\right)^{1/5} \left(\frac{q'}{\rho H_T}\right)^{2/5} \tag{6.24}$$

the transverse length scale of the flow

$$\delta \sim \left(\frac{d}{\Omega}\right)^{3/5} \left(\frac{q'}{\rho H_T}\right)^{1/5} \tag{6.25}$$

and also the temperature or buoyancy perturbation (not shown). Using Earth's core parameters, these results predict $u_{columnar} \sim 10^{-2}$ m/s and $\delta \sim 10$ km.

6.2. Scaling with numerical dynamo models

Scaling laws can also be obtained empirically, from the results of numerical dynamo models or from the results of laboratory experiments. Although we are free to use any complete set of the control parameters, including the basic set defined above, it has been found that certain control parameter definitions lead to better results than others. For example, recent efforts to scale numerical dynamo models (Christensen and Aubert, 2006; Olson and Christensen, 2006) adopt the following set. The first control parameter is a flux-based Rayleigh number Ra_Q

$$Ra_Q = \frac{r_c F}{r_i d^2 \Omega^3} \tag{6.26}$$

where, again F is the buoyancy flux. The flux-based definition allows both thermal and compositional convection dynamos to be scaled. For thermal convection, in terms of the convective heat flow q' we have

$$F = \frac{\alpha g q'}{\rho C_p} \tag{6.27}$$

The second control parameter is the magnetic Ekman number

$$E_\lambda = \frac{\lambda}{\Omega d^2} = \frac{E}{Pm} \tag{6.28}$$

The other two control parameters are Pr and Pm. The preferred output parameters are the Rossby number for the fluid velocity

$$Ro = \frac{u}{\Omega d} = E_\lambda Rm \tag{6.29}$$

and a local (small scale) Rossby number based on the characteristic wavenumber or spherical harmonic degree of the flow l_u:

$$Ro_l = \frac{l_u}{\pi} Ro \tag{6.30}$$

For the magnetic field, we use the Lorentz number

$$Lo = \frac{B_{rms}}{\sqrt{\rho \mu_0} \Omega d} = \sqrt{\Lambda E_\lambda} \tag{6.31}$$

The Lorentz number can be defined in terms of the rms internal field, the rms field at the cmb, or the rms of the dipole field at the cmb.

6.3. Scaling results

There are now enough different numerical dynamos in the literature to construct some preliminary scaling laws in terms of these parameters. Figure 65 shows the data coverage in terms of Ra, E, Pr and Pm from a compilation by Olson and Christensen (2006). One insert key indicates the choice of boundary conditions and the presence or absence of internal heat generation.

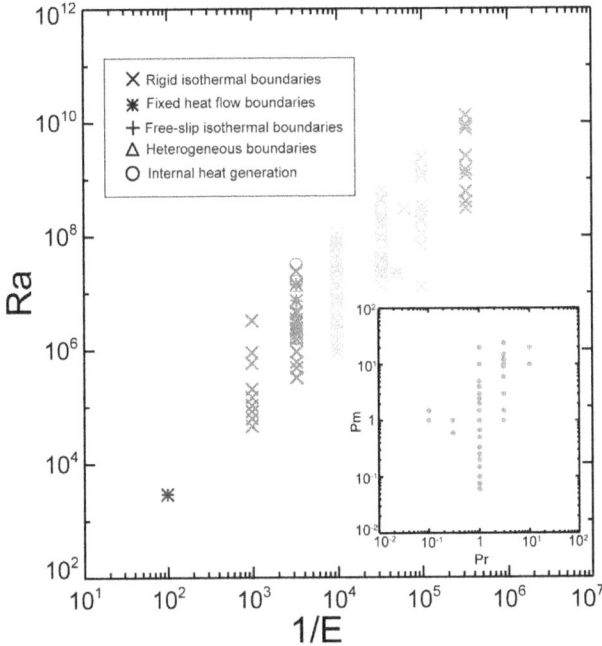

Fig. 65. The data coverage of a compilation of dynamo simulations.

The first question we can address is: how good is the constant Elsasser number assumption? Here we use the dipole field on the cmb to define Λ and scale this parameter against Ra_Q. We can see from Fig. 66 that this scaling assumption is not very good. There is at least two orders of magnitude variation in this parameter, corresponding to one order of magnitude in the dimensionless dipole field strength. In addition, there does not seem to be a simple relationship between Λ and Ra_Q.

A second question is the scaling for the fluid velocity. The top panel of Fig. 67 shows Ro versus Ra_Q for the dynamo models from Fig. 65. Here there is good evidence of power-law scaling. The Rossby number of the rms velocity increases as $Ra_Q^{2/5}$ approximately. The coefficient in this relationship for dynamos driven by basal heating (crosses, etc) is $\beta = 0.85$, about twice the value of the coefficient for internally-heated dynamos (circles). In dimensional terms, this scaling is equivalent to equation (6.24).

We now return to the problem of the scaling for the magnetic field. The bottom panel of Fig. 67 shows Lo_{dip}, the Lorentz number for the dipole field at the cmb versus Ra_Q for all dynamo models with dipole-dominated fields from

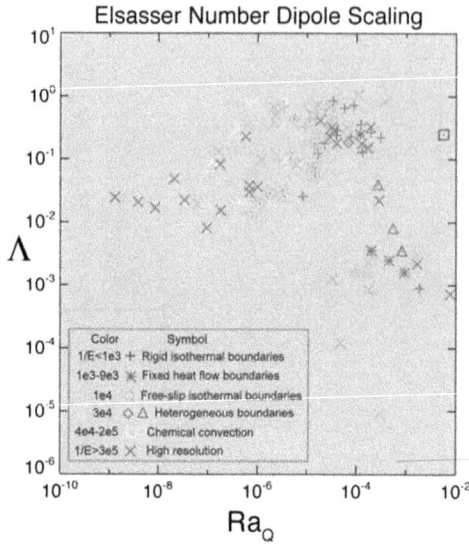

Fig. 66. Test of the constant Elsasser number dipole scaling.

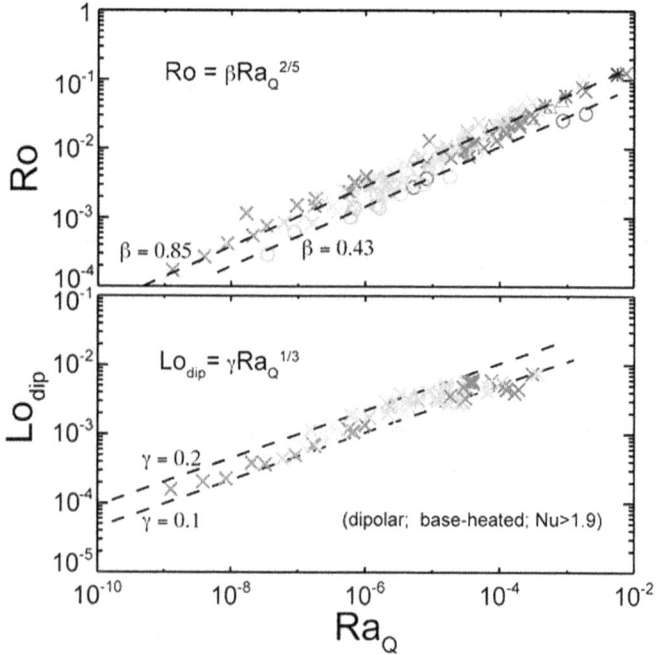

Fig. 67. Upper: Rossby number velocity scaling; Lower: Lorentz number dipole field scaling, versus buoyancy flux Rayleigh number.

Table 10

Planetary scaling results

Planet	E_λ	Ra_Q	Rm	M_{calc} (Z Am^2)	M_{obs}
Venus	8.9×10^{-7}	4×10^{-6}	5650	0	0
Mercury	5.2×10^{-6}	3×10^{-7}	430	3.1×10^{-2}	3.1×10^{-2}
Earth	3.6×10^{-9}	3×10^{-13}	400	73	78
Young Mars	1.4×10^{-8}	2×10^{-12}	350	16	12?
Jupiter	4.5×10^{-13}	1×10^{-16}	8000	$1.4 \times 10^{+6}$	$1.4 \times 10^{+6}$
Saturn	3.2×10^{-12}	1×10^{-16}	2350	$9.5 \times 10^{+4}$	$4.5 \times 10^{+4}$
Uranus	1.0×10^{-11}	8×10^{-17}	1500	$5.5 \times 10^{+3}$	$3.9 \times 10^{+3}$
Neptune	9.6×10^{-12}	8×10^{-17}	1450	$2.0 \times 10^{+3}$	$1.9 \times 10^{+3}$
Young Moon	3.5×10^{-6}	1×10^{-8}	100	7.0×10^{-2}	8×10^{-2}?
Ganymede	6.5×10^{-8}	8×10^{-13}	40	0	0.16

Fig. 65. Here there is evidence of power-law scaling over a limited region, with $Lo_{dip} \sim Ra_Q^{1/3}$ in that region. In terms of dimensional variables, for the dipole field at the cmb, this is equivalent to equation (6.18). But in this case, the scaling is not asymptotic, since the power-law trend shown in the lower panel of Fig. 67 terminates near $Ra_Q \simeq 10^{-3}$, where the dynamos transition from dipole-dominant to a multipolar structure.

Applying the $Lo_{dip} \sim Ra_Q^{1/3}$ law to the planets using the measured and estimated properties from the Planetary Core Table 9 gives the results are shown in the Planetary Scaling Table 10. In several cases, the predicted dipole moments are in fair agreement with the observed moments. Table 10 also gives convective velocities for planetary cores obtained using the $Ro \sim Ra_Q^{2/5}$ scaling law. It shows that the magnetic Reynolds numbers are expected to be supercritical for dynamo action in all the magnetic planets (as expected) although the predicted velocity is close to critical in the smaller objects such as Ganymede and the ancient Lunar dynamo.

References

Planetary Magnetic Field Scaling

Blackett, P.M.S. 1947. The magnetic field of massive rotating bodies, Nature **159**, 658–666.

Busse, F.H. 1976. Generation of planetary magnetism by convection, Phys. Earth Planet. Inter. **12**, 350–358.

Cain, J.C., Beaumont, P., Holter, W., Wang, Z. and Nevanlinna, H. 1995. The magnetic Bode fallacy, J. Geophys. Res. **100**, 9439–9454.

Christensen, U.R. and Tilgner, A. 2004. Power requirement of the geodynamo from ohmic losses in numerical and laboratory dynamos, Nature **439**, 169–171.

Christensen, U.R. and Aubert, J. 2006. Scaling properties of convection-driven dynamos in rotating spherical shells and application to planetary magnetic fields, Geophys J. Int. **166**, 97–114.

Elsasser, W.M. 1946. Induction effects in terrestrial magnetism, Phys. Rev. **70**, 202–212.

Olson, P. and Christensen, U.R. 2006. Dipole moment scaling for convection-driven planetary dynamos, Earth Planet. Sci. Lett. **250**, 561–571.

Russell, C.T. 1979. Scaling law test and two predictions of planetary magnetic moments, Nature **281**, 552–553.

Sano, Y. 1993. The magnetic fields of the planets: a new scaling law of the dipole moments of the planetary magnetism, J. Geomag. Geoelectr. **45**, 65–77.

Starchenko, S.V. and Jones, C.A. 2002. Typical velocities and magnetic field strengths in planetary interiors, Icarus, **157**, 426–435.

Stevenson, D.J. 1983. Planetary magnetic fields, Rep. Prog. Phys. **46**, 555–620.

7. Gravitational and tomographic dynamo examples

7.1. Introduction

In this part we examine the behavior of simple gravitational and mantle-controlled dynamos. The term "simple" means, in this context, numerical dynamo models that lie within the most accessible part of parameter space (as defined by our dimensionless numbers), with all the unessential complexity removed from them. There are some compelling reasons to study relatively simple numerical dynamo models for planetary applications. First, such models are usually smaller, cheaper, and faster to compute, and therefore allow more extensive exploration of the parameter space. Second, longer simulation times are possible with simpler models. This is crucial for the geodynamo, where the paleomagnetic record offers important observational constraints on the long-term behavior of the geodynamo, but it also may be important for dynamos on other planets, in situations where evolution is significant. Finally, as we have seen, even the largest dynamo models need to be extrapolated in order to reach realistic planetary conditions, so the challenges of applying large and small dynamo models to the planets are nearly the same. In spite of these challenges, numerical dynamos are in wide use for simulating the geodynamo (see Kono and Roberts, 2002; Glatzmaier and Roberts, 2002; Christensen and Wicht, 2007; Takahashi et al., 2005).

7.2. Goals for numerical models of the geodynamo

It is useful to summarize what it is we are trying to simulate and understand about the geodynamo (Dormy et al., 2000). First, we want to understand the structure of the core field, as exemplified by the present-day radial field intensity on the cmb, as shown in Fig. 12 (the Oersted 2000 satellite field). This figure shows that the core field is dipole dominant, but the nondipole field is nevertheless important. The present-day dipole moment is about 78 Z Am^2 and on the cmb the dipolarity, the ratio of the rms dipole to the rms field intensity, is about 0.6. The present-day dipole tilt is 10.8^o, but like the dipole moment, the tilt is variable and averages to zero on a time scale of several thousand years. The non-dipole terms also tend to average toward zero, so the long-term field approximates GAD (Geocentric Axial Dipole) configuration. Actually, the GAD is an approximation, since there is paleomagnetic evidence for a small but persistent axial quadrupole, and also evidence for persistent non-axial fields, which we consider later, using a tomographic dynamo model.

Second, we want to understand the origin of variability in the geodynamo, specifically, the time variations of the geomagnetic and paleomagnetic fields. The core field changes over a broad spectrum of time scales, ranging from a few years (exemplified by geomagnetic jerks) to a few hundred million years. Evidence for this variation includes the secular variation of the non-dipole field, for example, the development and evolution of reversed magnetic flux spots on the cmb in the last 150 years (Fig. 15), which are related to the historical decrease in the dipole moment (recall, this decrease is more than 10 times free decay). Decade-scale oscillations of the field, possibly related to torsional oscillations in the core, are another example of geodynamo variability that dynamo models seek to explain.

On paleomagnetic time scales, the dipole exhibits large amplitude, broad-band intensity variations and occasional large directional variations, the excursions and reversals. Figure 17 shows a record of this lower frequency variability over the past 2 Ma, derived from the magnetization of deep sea sediments. The 100 Myr modulation of the frequency of reversals and the occurrence of long, constant polarity "superchrons" represents an even longer time scale of variability. Finally, the effects on the geodynamo of the secular evolution of the core, including the nucleation and growth of the inner core on Gyr time scales, needs to be understood.

7.3. Gravitational dynamo model equations

The conservation of momentum, magnetic induction, mass and magnetic field continuity, and buoyancy transport equations for chemical convection and magnetic field generation in an electrically conducting Boussinesq fluid in a rotating, self-gravitating spherical shell can be written in dimensionless form as

$$E\left(\frac{\partial \mathbf{u}}{\partial t} + \mathbf{u} \cdot \nabla \mathbf{u} - \nabla^2 \mathbf{u}\right) + 2\hat{\mathbf{z}} \times \mathbf{u} + \nabla P$$
$$= E Pr^{-1} Ra \frac{r}{r_o} \chi \hat{\mathbf{r}} + E Pm^{-1} (\nabla \times \mathbf{B}) \times \mathbf{B}, \tag{7.1}$$

$$\frac{\partial \mathbf{B}}{\partial t} = \nabla \times (\mathbf{u} \times \mathbf{B}) + Pm^{-1} \nabla^2 \mathbf{B}, \tag{7.2}$$

$$\nabla \cdot (\mathbf{u}, \mathbf{B}) = 0, \tag{7.3}$$

$$\frac{\partial \chi}{\partial t} + \mathbf{u} \cdot \nabla \chi = Pr^{-1} \nabla^2 \chi - 1 \tag{7.4}$$

Here $\mathbf{B}$, $\mathbf{u}$, P and χ are the dimensionless magnetic induction, fluid velocity, pressure perturbation, and light element concentration variable, t is time, $\mathbf{r}$ is the radius vector, and E, Pr, Pm, and as in previous lectures, Ra are the Ekman, Prandtl, magnetic Prandtl, and Rayleigh numbers, respectively. The conservation of momentum (7.1) is the Boussinesq form of the Navier–Stokes equation in a spherical (r, θ, ϕ) coordinate system rotating at angular velocity $\Omega \hat{\mathbf{z}}$ in which gravity increases linearly with radius. It includes the inertial and Coriolis accelerations, plus the pressure, viscous, buoyancy, and Lorentz forces. Equations (7.1)–(7.4) have been nondimensionalized using the shell thickness $d = r_o - r_i$ as the length scale (r_o and r_i are the inner and outer radii) and the viscous diffusion time d^2/ν as the time scale (ν is the kinematic viscosity). The fluid velocity is scaled by ν/d and the magnetic field by $(\rho_o \Omega/\sigma)^{1/2}$, where ρ_o is outer core mean density and σ is electrical conductivity. As before, we have written the light element concentration as the sum of a slowly varying average χ_o and a perturbation χ. Even though the light element variable χ is naturally without dimension (it is a concentration) we still need to scale it properly. In the above equations, D represents the light element diffusivity and χ is scaled by $d^2 \dot{\chi}_o/\nu$, then $\dot{\chi}_o = -1$ is the light element "sink" term in (7.4). With this scaling, the chemical Rayleigh number has the same definition as in previous chapters:

$$Ra = \frac{\beta g_o d^5 \dot{\chi}_o}{D \nu^2} \tag{7.5}$$

Figure 68 is a qualitative depiction of the various regimes of gravitational dynamos, as defined by the dipole moment, from numerical solutions of the above equations. Here we are fixing E, Pr, Pm and varying only the Rayleigh number. With increasing Ra, the primary regimes are:
• No convection, i.e. conductive, subcritical for both convection and dynamo action.
• Convective by no dynamo, i.e., supercritical for convection but subcritical for dynamo action.
• Balanced dynamos, i.e., dynamos with constant, or nearly constant total magnetic and kinetic energy. These dynamos usually have fixed flow and magnetic field patterns that drift either prograde (eastward) or retrograde (westward) with respect to the rotating coordinate system.
• Nonreversing dynamos with chaotic time behavior. These dynamos exhibit broad-band fluctuations, but generally the fluctuations are not large enough to precipitate polarity reversals. These dynamos are usually dipole-dominant.

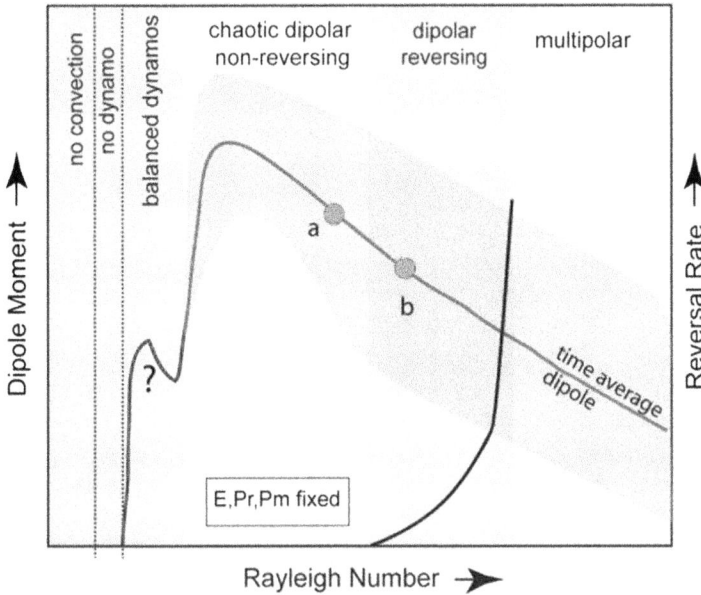

Fig. 68. Depiction of gravitational dynamo regimes, in terms of dipole moment versus Rayleigh number. Solid line and shading indicate time average and time variability of the dipole moment, respectively. Two cases are examined: (a) non-reversing, lower Ra, (b) reversing, higher Ra.

- Reversing dipolar dynamos. These are generally the most interesting and the most readily applicable to the geodynamo. The dark solid curve indicates qualitatively the frequency of reversals in this regime.
- Multipolar dynamos, in which the dipole field reverses, but is generally smaller than higher field multipoles.

For some types of convective forcing, the reversing dipolar regime is either missing or is restricted to a very narrow range of Ra-values. However, for gravitational-type dynamos, it is often a distinct regime. In general, the frequency of reversals increases with Ra. This can be understood in terms of the change in the ratio of the dipole fluctuations to the dipole mean value, s_{dipole}. In the figure, the shaded region denotes the dipole standard deviation, the central curve its time average. As Ra increases, s_{dipole} is first very small, but it strongly increases in the chaotic regime. At high Ra, there is finite probability of prolonged times when the dipole field is very weak compared to other field components. Since polarity reversals occur when the dipole field is weak for a finite period of time, they are far more probable at high Ra.

7.4. Behavior of chaotic gravitational dynamos

In this section, results of long-duration numerical solutions to (7.1-35) by Olson (2007) are discussed. The numerical dynamo model (MAG; source code available at *www.geodynamics.org*, originally developed by G. Glatzmaier) has been previously benchmarked. Boundary conditions for the composition variable are fixed light element concentration at the inner core boundary, and zero light element flux at the core-mantle boundary. Mechanical and electrical conditions on both spherical boundaries are no-slip and electrically insulating. These calculations use 25 radial grid intervals and harmonic truncation $n_{max} = 32$. The calculations were initialized with random buoyancy perturbations and an axial dipole magnetic field. The dynamos were run until their statistical fluctuations became stationary. The next several figures summarize the behavior of gravitational dynamos with $E = 6.5 \times 10^{-3}$, $Pr = 1$ and $Pm = 20$ for two different Rayleigh numbers Ra. The relatively large E-value permits low spatial resolution, hence long-time simulations, while the large Pm-value ensures a large magnetic Reynolds number and chaotic time behavior.

The first set of Figs. 69 show a non-reversing gravitational dynamo at $Ra = 1.7 \times 10^4$; the second set 70 show a reversing case at $Ra = 2.3 \times 10^4$. The radial field on the model core-mantle boundary is shown by color contours with the tangent cylinder of the inner core shown by the solid circle (continents have been added for visual reference only; their longitude has no significance in these calculations). Equatorial sections showing light element composition, axial vorticity and axial magnetic field are also shown. The time series show the behavior of the dipole moment and the tilt of the dipole over about 35 dipole decay times (each about 20 kyr, in the core). Vertical lines indicate the time of the images.

Both cases produce chaotic dynamos, with variable, tilted dipoles. In both cases the dipole field comes primarily from two strong, high latitude flux bundles in each hemisphere, as in the present-day geomagnetic field. In the equatorial sections the magnetic flux is concentrated in the two strong negative axial vortices. These vortices drift to the west, in the retrograde sense. However, in the higher Ra reversing case the two equatorial flux concentrations have different polarity. The polarity reversals in this case occur when the relative strength of these structures change.

7.5. Tomographic dynamo models

We have argued previously that the strong heterogeneity in the D''-layer of the lower mantle makes it likely that the heat flow on the cmb is spatially variable. Dynamical interpretation of lower mantle heterogeneity usually ascribes high seismic shear velocities V_s to regions of mantle downwellings, hence high heat

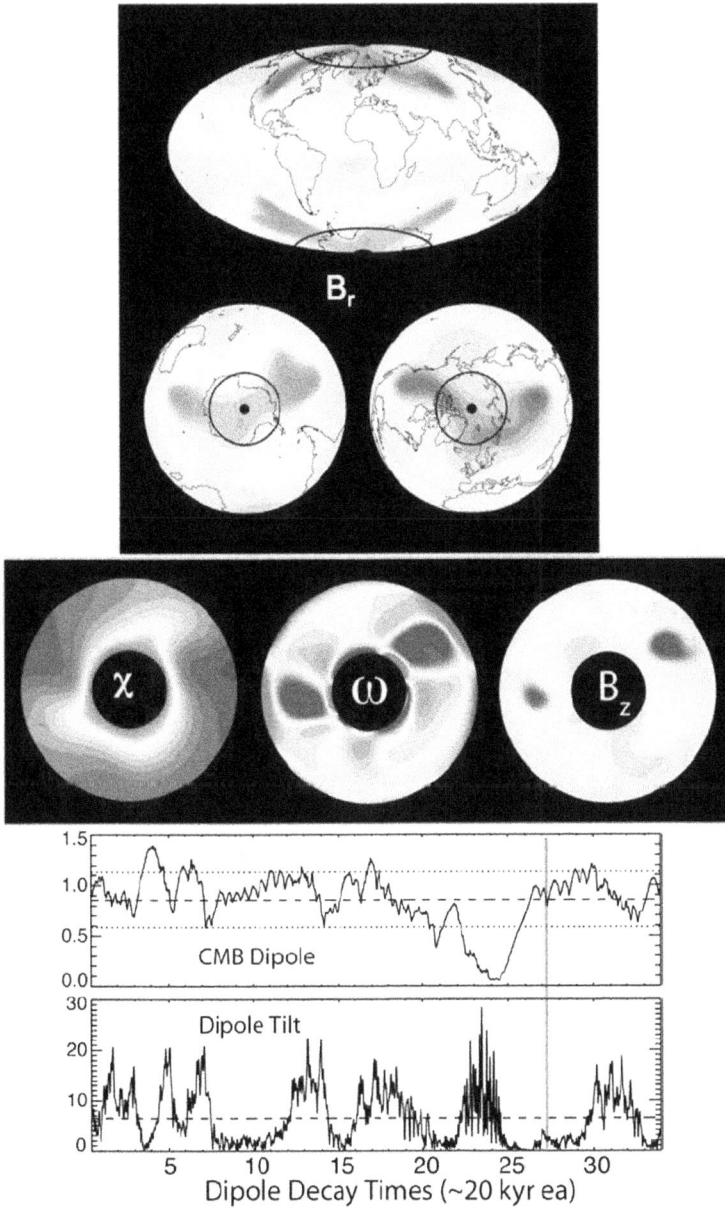

Fig. 69. Non-reversing gravitational dynamo with $Ra = 1.7 \times 10^4$.

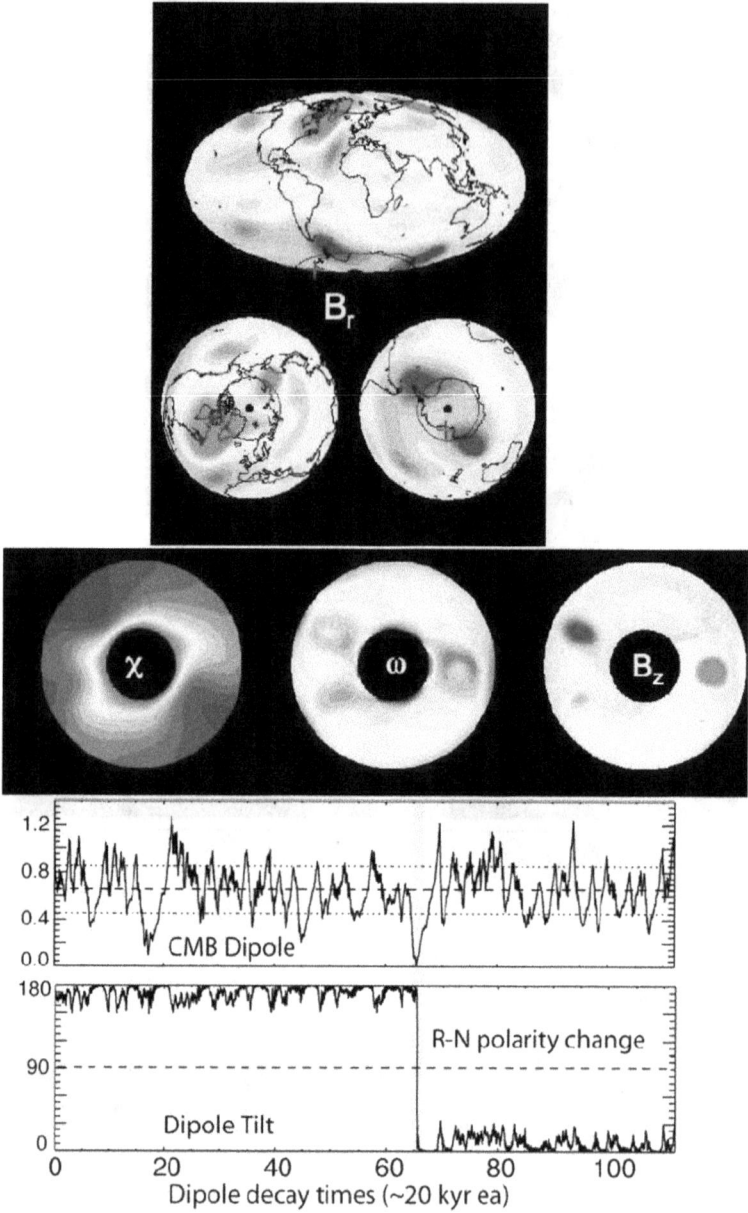

Fig. 70. Reversing gravitational dynamo with $Ra = 2.3 \times 10^4$.

flow on the cmb. Conversely, low shear velocities in D″ are usually interpreted as mantle upwellings, with smaller than average cmb heat flow. The same dynamics can produce topographic coupling (Kuang and Chao, 2001), although thermal interaction with D″ heterogeneity is usually cited as the most important mantle control on the geodynamo (Bloxham, 2000; Buffett, 2007). To test this concept, we construct a "tomographic" thermal boundary condition for a dynamo model by assuming the following linear relationship

$$\frac{\delta q_{cmb}}{q_{cmb}} \propto \frac{\delta V_s}{V_s} \tag{7.6}$$

The constant of proportionality in this relationship is not well constrained, so dynamo modelers usually explore a range of values for this parameter. Positive values of the proportionality factor imply thermal coupling as described above, whereas a negative value of this parameter is more consistent with a chemical interpretation of D″ heterogeneity.

Figure 71 shows the result of this boundary condition on numerical dynamos driven by thermal convection (from Olson and Christensen, 2002). The upper 3 shows the long-time average structure of a dynamo with a uniform heat flow cmb condition. The lower 3 shows the long-time average structure of a dynamo with the tomographic boundary condition interpreted as thermal coupling, i.e., a positive coefficient in (7.6) The images show the radial velocity U_r, the streamfunction ψ of the horizontal velocity (both just beneath the cmb), and the radial magnetic field B_r on the cmb. The pattern of cmb heat flow driving the tomographic case is taken from the Masters et al. (2000) lower mantle tomography.

The homogeneous cmb heat flow condition does not remove the azimuthal degeneracy of the time-average dynamo state, and so the dynamo structure approaches axisymmetry in its time average state, although the dynamo is not axisymmetric at any instant in time. The time average radial velocity is essentially zero everywhere except on the tangent cylinder, where it is downward, and the flow is upward inside the tangent cylinder. The streamfunction of the horizontal flow is also essentially zero outside the tangent cylinder, but is strongly retrograde inside.

The tomographic dynamo differs in having a large-scale horizontal time average flow, essentially thermal winds driven by the mantle heat flow heterogeneity. In addition, the time average radial field reflects this flow pattern, with field concentrations where large-scale downwellings occur. The tomographic boundary condition has been shown to have other substantial effects on dynamos. In reversing dynamos, it produces a bias that makes the dipole axis appear to favor paths where the boundary heterogeneity produces large-scale fluid downwellings (Kutzner and Christensen, 2004; Glatzmaier and Coe, 2007).

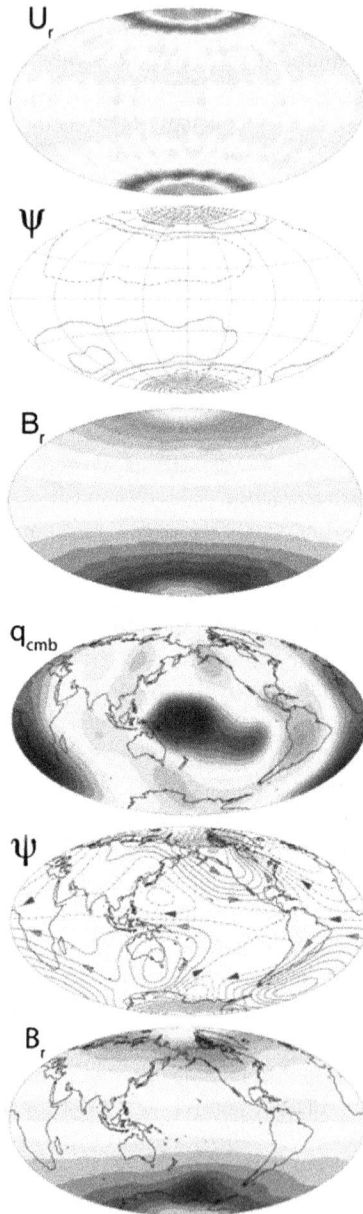

Fig. 71. Long time averages of cmb radial velocity U_r, streamfunction ψ, radial magnetic field B_r, and heat flow q_{cmb}. Upper 3 panels: uniform cmb heat flow case; Lower 3 panels: tomographic cmb heat flow case.

Acknowledgments

I thank Hagay Amit, Peter Driscoll, and Renaud Deguen for their help in preparing and reviewing these course notes.

References

Bloxham, J. 2000. Sensitivity of the geomagnetic axial dipole to core-mantle thermal interactions, Nature **405**, 63–65.

Buffett, B.A. 2007. Core-Mantle interactions, in: *Treatise on Geophysics*, vol. 8, ch. 12, Elsevier B.V., pp. 345–358.

Dormy, E., Valet, J.P. and Courtillot, V. 2000. Numerical models of the geodynamo and observational constraints, Geochem. Geophys. Geosys. **1**.

Christensen, U.R. and Wicht, J. 2007. Numerical dynamo simulations, in: *Treatise on Geophysics*, vol. 8, ch. 8, Elsevier Press, pp. 245–282.

Glatzmaier, G.A. and Coe, R.S. 2007. Magnetic polarity reversals in the core, in: *Treatise on Geophysics*, vol. 8, ch. 9, Elsevier Press, pp. 283–298.

Glatzmaier, G. and Roberts, P. 2002. Simulating the geodynamo, Contemp. Phys. **38**, 269–288.

Kono, M. and Roberts, P. 2002. Recent geodynamo simulations and observations of the geomagnetic field, Rev. Geophys. **40**, 1013.

Kuang, W. and Chao, B. 2001. Topographic core-mantle coupling in numerical dynamo models, Geophys. Res. Lett. **28**, 1871–1874.

Kutzner, C. and Christensen, U.R. 2004. Simulated geomagnetic reversals and preferred virtual geomagnetic pole paths, Geophys. J. Int. **157**, 1105–1118.

Masters, G., Laske, G., Bolton, H., Dziewonski, A. 2000. The relative behavior of shear velocity, bulk sound speed, and compressional velocity in the mantle: Implications for chemical and thermal structure, in: Karato, S., Forte, A.M., Liebermann, R.C., Masters, G., Stixrude, L., eds., *Earth's Deep Interior*, AGU monograph 117, AGU, Washington D.C.

Olson, P. 2007. Gravitational dynamos and the low frequency geomagnetic secular variation, Proc. Nat. Acad. Sci., 10.1073/pnas.0709081104.

Olson, P. and Christensen, U.R. 2002. The time-averaged magnetic field in numerical dynamos with non-uniform boundary heat flow, Geophys. Jour. Intl. **151**, 809–823.

Takahashi, F., Matsushima, M., Honkura, Y. 2005. Simulations of a quasi-Taylor state geomagnetic field including polarity reversals on the Earth simulator, Science **309**, 459–461.

Course 4

ASTROPHYSICAL DYNAMOS

Anvar Shukurov[1] and Dmitry Sokoloff[2]

[1] *School of Mathematics and Statistics, Newcastle University, Newcastle upon Tyne, NE1 7RU, UK*
[2] *Department of Physics, Moscow State University, Moscow, Moscow, 119992, Russia*

Ph. Cardin and L.F. Cugliandolo, eds.
Les Houches, Session LXXXVIII, 2007
Dynamos
© *2008 Published by Elsevier B.V.*

Contents

1. Introduction

The dynamo theory is relatively young: the ability of a flow of conducting fluid to maintain magnetic field was conjectured 90 years ago [23], but the first example of self-sustained dynamo is only 50 years old [17]. Both physical and mathematical aspects of the theory are often complicated and may seem unnatural to an excessively sceptical observer. However, the need for dynamo action to maintain magnetic fields of the Earth and the Sun is so evident (in particular because the global magnetic fields of these objects exhibit time variation that is inconsistent with any other viable option) that the paradigm of dynamo theory has been very widely accepted for planets and stars. The situation is different with other astrophysical objects—accretion discs, galaxies and galaxy clusters. Firstly, magnetic fields in these remote objects are more difficult to detect and explore. Secondly, the size of the parent objects is almost invariably so large (with the exception of accretion discs in stellar objects) that we only have information about the spatial structures of astrophysical magnetic fields and, in vast majority of cases, any time variation can only be hypothesized. Therefore, it is not surprising that, until recently, astrophysical dynamos appeared to be exotic creatures in the world of astrophysics, and the idea of primordial magnetic fields was preferred by many, either explicitly or implicitly. The situation is now changing: more researchers would be prepared to accept that most astrophysical objects host a dynamo as dynamo theory becomes more detailed and capable to provide testable predictions.

Many recent developments in dynamo theory arise from extensive numerical simulations whose complexity approaches, in many respects, that of the laboratory experiment. Therefore, numerical experiments need to be interpreted with the same care and caution as laboratory experiments or astronomical observations. We discuss in Section 7 an approach to quantifying the morphology of (random) structures based on Minkowski functionals. Furthermore, it would be difficult to understand the outcome of complicated models, experiments or observations without a range of simplified analytical models. The recent explosive development of numerical approaches to astrophysical dynamos, based on the growth in the computing power available, has resulted in a reduced interest in simplified analytical models. We believe that this hampers proper interpretation of the numerical results, now often presented in the form of aesthetically appealing images (perhaps arranged into a time sequence called a movie) accompanied by a subjective qualitative description. We present, in Section 6, a simple but surprisingly accurate analytical approximate solution of the mean-field dynamo equations for thin discs and spherical shells; such solutions can be useful in both interpretations of numerical experiments and in various applications where a simple analytical structure of magnetic field is needed, rather than a three-dimensional data cube.

This text preserves the flavour of lecture notes; in particular, we do not attempt to provide extensive references. The depth of the presentation varies. For example, the reviews of Sections 2–3 only touch upon the observations of astrophysical magnetic fields and the hydrodynamic modelling of stars and galaxies; we only represent facts required to construct solutions of dynamo equations discussed in the second half of the text. We discuss in some detail dynamo models for the Sun, spiral galaxies and galaxy clusters. A review of dynamo action in accretion discs can be found in Refs [9,41], and dynamos in elliptic galaxies are reviewed in Ref. [41].

We shall be using CGS units in this text, with the unit of magnetic flux density of $1\,\mathrm{G} = 10^{-4}$ T; in application to galactic magnetic fields, a smaller unit $1\,\mu\mathrm{G} = 10^{-6}\,\mathrm{G} = 0.1$ nT is often convenient. In the context of stellar physics, the Solar radius $R_\odot = 7 \times 10^{10}$ cm is a convenient unit length, whereas $1\,\mathrm{pc} = 3 \times 10^{18}$ cm ≈ 3.26 light years is a suitable length scale in the case of galaxies, with $1\,\mathrm{kpc} = 10^3$ pc. One parsec is the distance from which the Earth orbit around the Sun has the angular diameter (*par*allax) of one *sec*ond of arc.

2. Observations of astrophysical magnetic fields

2.1. Zeeman splitting

Measuring the splitting of spectral lines in magnetic field is historically the first method of observation of cosmic magnetic fields. Only twelve years after the discovery of Zeeman, Hale [16] has succeeded in using it for measuring the magnetic field of sunspots. Comparison of intensities of a spectral line wings, produced by the Zeeman effect, has allowed Babcock [1] to detect it in the emission of distant peculiar magnetic stars. Ten years later Bolton and Wild [8] proposed to use the Zeeman splitting of the $\lambda 21$ cm neutral hydrogen absorption line to measure magnetic fields in the interstellar medium. Such measurements were achieved a further ten years later [53,54].

In the absence of external fields, atomic energy levels do not depend on the direction of the total angular momentum (orbital L plus spin S) of electrons. In other words, the energy levels are degenerate with respect to the momentum direction. In the magnetic field B, that distinguishes a certain direction, an atom acquires the additional energy $-\mu(\mathbf{L} + 2\mathbf{S}) \cdot \mathbf{B}$ which depends on the orientation of the angular momentum with respect to the magnetic field (here, $\mu = e\hbar/2m_{\mathrm{e}}c = 9.3 \times 10^{-21}$ erg G^{-1} is the Bohr magneton). The energy levels split into $2j + 1$ equidistant levels, where j is the quantum number of the total angular momentum $\mathbf{J} = \mathbf{L} + \mathbf{S}$. The energy levels are given by [21]

$$E_H = E_0 \pm \mu g M B, \quad M = 0, 1, \ldots, j.$$

The factor g is called the Lande factor,

$$g = 1 + \frac{j(j+1) + s(s+1) - l(l+1)}{2j(j+1)},$$

with l and s the quantum numbers of the orbital and spin momenta. In particular, the Lande factor appears because the mechanical and magnetic momenta are related differently for the electron's orbital motion (M_l, l) and spin (M_s, s): $M_l = -(e/2m_{\mathrm{e}})l$ and $M_s = -(e/m_{\mathrm{e}})s$, where m_{e} is the electron mass.

The quantum selection rules only allow transitions between the levels for which M changes by $\Delta M = 0, \pm 1$. If the Lande factor is the same for the upper and lower levels, the spectral line of the basic frequency v_0 is split into a triplet (v_π, v_σ) (the *normal* Zeeman effect):

$$v_\pi = v_0, \qquad v_\sigma = v_0 \pm g \frac{e}{4\pi m_{\mathrm{e}} c} B = v_0 \pm 1.4 \, g \left(\frac{B}{10^{-6}\,\mathrm{G}}\right) \mathrm{Hz},$$

where c is the speed of light. In a general case, when the upper and lower levels have different Lande factors the number of components may be larger (the *anomalous* Zeeman effect). The component separation is proportional to the difference in gM between the energy levels involved, $\Delta(gM)$, but remains proportional to magnetic field strength.

The main obstacle in the observations of the Zeeman splitting is the thermal broadening of the spectral lines which can exceed the separation of the multiplet components. For the interstellar $\lambda 21$ cm line of neutral hydrogen, the Zeeman splitting in the field 10^{-5} G is about 30 Hz, while the line half-width due to the thermal Doppler broadening is $\Delta v = v_0 v_T/c \approx 10^4$ Hz for $T = 100$ K (where v_T is the thermal velocity). Thus, what is often observed in practice is the broadening of spectral lines by the Zeeman effect rather than their splitting. It is therefore important that the components of the spectral lines split by the Zeeman effect are polarized, which helps with their detection because the wings of a spectral line broadened by the Zeeman effect have different polarizations.

In the Solar atmosphere, the Zeeman splitting is observable where magnetic field strength exceeds about 1500 G; magnetic field in sunspots can reach 3000 G. For weaker magnetic fields, polarimetric observations of the Zeeman broadening are feasible. Detailed discussion of the Zeeman effect and its applications in solar physics can be found in the book of Stix [47]. In the interstellar space the normal Zeeman effect is observed in the $\lambda 21$ cm neutral hydrogen absorption line, and the anomalous Zeeman effect, in the $\lambda 18$ cm OH molecule line. In dense, cold star-forming regions, with gas number density in excess of $n = 10^5–10^6$ cm^{-3}, where magnetic field strength exceeds 1 mG, the Zeeman splitting can be detected in the CO and CN molecular radio lines. At lower densities, the Zeeman broadening of the spectral lines of neutral hydrogen and the hydroxyl OH is observable, but the required gas densities and magnetic fields are still rather high, $n > 10$ cm^{-3}, $B > 1\,\mu$G. These values should be compared with the typical density $n \simeq 0.1$ cm^{-3} in the diffuse warm interstellar medium and $n \simeq 10^{-3}$ cm^{-3} in the hot interstellar gas. Altogether, the Zeeman effect provides the most important method in the observational studies of the Solar magnetic fields and plays prominent role in the observations of magnetic fields of other stars. In the interstellar space, however, this method is useful only when applied to relatively dense regions with rather strong magnetic fields. Although interstellar gas clouds are the site of many important processes (including star formation), they occupy a negligible fraction of the total volume of the interstellar space.

2.2. Synchrotron emission and Faraday rotation

Estimates of magnetic field strength in the diffuse interstellar medium of the Milky Way and other galaxies are most efficiently obtained from the intensity and Faraday rotation of synchrotron emission. The total I and polarized P synchrotron intensities and the Faraday rotation measure RM are weighted integrals of magnetic field over the path length L from the source to the observer, so they provide various average measures of magnetic field in the emitting or magneto-active volume:

$$I = K \int_L n_{cr} B_\perp^2 \, ds, \quad P = K \int_L n_{cr} \bar{B}_\perp^2 \, ds, \quad RM = K_1 \int_L n_e B_\parallel \, ds, \quad (2.1)$$

where n_{cr} and n_e are the number densities of relativistic and thermal electrons, $B = \bar{B} + b$ is the total magnetic field comprising regular $\bar{B}$ and random b parts, with $\bar{B} = \langle B \rangle$, $\langle b \rangle = 0$ and $\langle B^2 \rangle = \langle B \rangle^2 + \langle b^2 \rangle$, where angular brackets denote averaging, subscripts $\perp$ and $\parallel$ refer to magnetic field components perpendicular and parallel to the line of sight, and K and $K_1 = e^3/(2\pi m_e^2 c^4) = 0.81$ rad m^{-2} cm^3 μG^{-1} pc^{-1} are certain dimensional constants (an explicit expression for K is omitted here; it can be found, e.g., in Ref. [27]). The degree of polarization p is related to the degree of regularity of the magnetic field. In the simplest case of $n_{cr} = $ const, an expression often used is

$$p \equiv \frac{P}{I} \approx p_0 \frac{\bar{B}_\perp^2}{\langle B_\perp^2 \rangle} = p_0 \frac{\bar{B}_\perp^2}{\bar{B}_\perp^2 + \frac{2}{3}\langle b^2 \rangle}, \qquad (2.2)$$

where the random field b has been assumed to be isotropic in the last equality, and $p_0 \approx 0.75$ weakly depends on the spectral index of the emission. This widely used relation is only approximate. In particular, it does not allow for any anisotropy of the random magnetic field, for the dependence of n_{cr} on B, and for depolarization effects; some generalizations are discussed in Ref. [44].

The orientation of the apparent large-scale magnetic field in the sky plane is given by the observed B-vector of the polarized synchrotron emission. As polarized radio emission propagates through magnetized plasma, its polarization plane rotates because of what is known as the Faraday effect (i.e., magnetized plasma is birefringent). The rotation angle is given by $\Delta \psi = \mathrm{RM} \, \lambda^2$, where λ is the emission wavelength. The Faraday rotation measure can be obtained from measurements of the differences in the polarization angles ψ between several wavelengths. The special importance of the Faraday rotation measure, RM, is that this observable is sensitive to the direction of B (the sign of $\bar{B}_\parallel$) and this allows one to determine not only the orientation of $\bar{B}$ but also its direction. Thus, analysis of Faraday rotation measures can reveal the three-dimensional structure of the magnetic vector field.

Since n_{cr} is difficult to measure, it is often assumed that magnetic field and cosmic rays are in pressure equilibrium or energy equipartition; this allows to express n_{cr} in terms of B. The physical basis of this assumption is the fact that cosmic rays (charged particles of relativistic energies) are confined by magnetic fields. The cosmic ray number density n_{cr} in the Milky Way can be determined independently from the γ-ray emission produced when cosmic ray particles interact with the interstellar gas [48]. Then magnetic field strength can be obtained without assuming equipartition; the results are generally consistent with the equipartition values.

In the Milky Way, the dispersion measures of pulsars, $\mathrm{DM} = \int_L n_e \, ds$ provide information about the mean thermal electron density, but the accuracy is limited by our uncertain knowledge of distances to pulsars. Estimates of the strength of the regular magnetic field in the Milky Way are often obtained from the Faraday rotation measures of pulsars simply as

$$\bar{B}_\parallel = \frac{\mathrm{RM}}{K_1 \, \mathrm{DM}}. \qquad (2.3)$$

This estimate is meaningful if magnetic field and thermal electron density are statistically uncorrelated. If the fluctuations in magnetic field and thermal electron density are correlated with each other, they will contribute positively to RM and Eq. (2.3) will yield overestimated $\bar{B}_{\parallel}$. In the case of anticorrelated fluctuations, their contribution is negative and Eq. (2.3) is an underestimate. Physically reasonable assumptions about the statistical relation between magnetic field strength and electron density can lead to Eq. (2.3) being in error by a factor of 2–3 [5].

Magnetic fields in the Solar corona can also be measured via their rotation of the polarization plane of the radio emission of background extragalactic radio sources [46]. At the wavelength of $\lambda 21$ cm, a magnetic field of 0.03 G in the corona (thermal electron density of 1.5×10^4 cm^{-3}, path length of ten solar radii) produces RM $\simeq 10$ rad m^{-2}, with the corresponding rotation angle of the polarization plane of 25°.

2.3. Results of observations

2.3.1. The Sun and stars

The magnetic fields that are most readily observable in the Sun are those in the sunspots, where magnetic fields of a strength exceeding 1500 G makes the Zeeman spectral mutliplets observable. Weaker fields are detectable through the Zeeman broadening. Despite strong local magnetic fluctuations associated with the granulation, a weaker overall magnetic field of the Sun can be measured; it is dominated by a dipolar component of a strength 1 G near the poles. The overall magnetic field is oscillatory with a period of about 22 years, and is described as a dynamo wave propagating from latitudes $\pm(30°–35°)$ towards the Solar equator down to the latitude of about $\pm(5°–10°)$ in each hemisphere; an additional, weaker branch of the dynamo wave propagates polewards from the mid-latitudes. The strong magnetic field of the sunspots is believed to be a surface manifestation of the strong toroidal magnetic field produced in the Solar interior. Unlike the dipolar poloidal magnetic field, the overall toroidal magnetic field does not penetrate outside the Solar surface (except in the sunspots), in agreement with the vacuum boundary conditions often employed in modelling the large-scale magnetic field of the Sun. The overall symmetry of the Solar magnetic field is approximately dipolar, $\bar{B}_r(\theta) = -\bar{B}_r(-\theta)$, $\bar{B}_\theta(\theta) = \bar{B}_\theta(-\theta)$, $\bar{B}_\phi(\theta) = -\bar{B}_\phi(-\theta)$ in terms of spherical coordinates with θ the latitude ($\theta = 0$ at the equator and $\theta = \pm 90°$ at the poles). Although weak, deviations from the perfect equatorial antisymmetry and axial symmetry are noticeable; these are described as a quadrupolar component of the magnetic field and 'active longitudes'.

Magnetic fields in other stars can be detected using various proxies. For example, spectral lines of ionized oxygen and calcium, O VI and Ca II are produced in the magnetically heated plasmas in the chromosphere. The emission flux in these lines is known to be proportional to the square root of the magnetic field strength.

Observations in these lines reveal stellar activity cycles in late-type stars G0–K7, which have an outer convection zone. Magnetic activity cycles result in a cyclic variation of the area covered by starspots, and hence to cyclic photometric variations; the techniques of Doppler imaging allow the production of maps of the stellar surface showing large starspots [29].

2.3.2. Spiral galaxies

The observable quantities (2.1) have provided extensive data on magnetic field strengths in both the Milky Way and external galaxies. The average total field strengths in nearby spiral galaxies, obtained from total synchrotron intensity I, ranges from $B \approx 4 \, \mu$G in the galaxy M31 to about 15 μG in M51, with a mean of $B = 9 \, \mu$G for the sample of 74 galaxies [2]. The typical degree of polarization of synchrotron emission from galaxies at short radio wavelengths is $p = 10$–20%, so Eq. (2.2) gives $\bar{B}/B = 0.4$–0.5; these are always lower limits due to the limited resolution of the observations, and $\bar{B}/B = 0.6$–0.7 is a more plausible estimate. The total equipartition magnetic field in the Solar neighbourhood is estimated as $B = 6 \pm 2 \, \mu$G from the synchrotron intensity of the diffuse Galactic radio background. Combined with $\bar{B}/B = 0.65$, this yields a strength of the local regular field of $\bar{B} = 4 \pm 1 \, \mu$G. Hence, the typical strength of the local Galactic random magnetic fields, $b = (B^2 - \bar{B}^2)^{1/2} = 5 \pm 2 \, \mu$G, exceeds that of the regular field by a factor $b/\bar{B} = 1.3 \pm 0.6$. RM data yield similar values for this ratio.

Meanwhile, the values of $\bar{B}$ in the Milky Way obtained from Faraday rotation measures seem to be systematically lower than the above values. RM of pulsars and extragalactic radio sources yield $\bar{B} = 1$–2 μG in the solar vicinity, a value about twice smaller than that inferred from the synchrotron intensity and polarization. The discrepancy can be explained, at least in part, if the methods described above sample different volumes. The depth probed by the total synchrotron emission and Faraday rotation measures of pulsars and extragalactic radio sources is of the order of a few kpc. Polarized emission, however, may emerge from more nearby regions because emission from remote regions is *depolarized* by various propagation effects [44]. However, a more fundamental reason for the discrepancy can be a partial correlation between fluctuations in magnetic field and thermal electron density. Such a correlation can arise from statistical pressure balance in the interstellar medium: if the total pressure is constant on average, regions with larger gas density (and hence larger gas pressure) usually have weaker magnetic field (and hence lower magnetic pressure), and vice versa. The term $\langle b_\parallel n_e \rangle$ then differs from zero and contributes to the observed RM leading to underestimated $\bar{B}$ [5]. In a similar manner, correlation between B and the cosmic ray number density n_{cr} biases the estimates of magnetic field from synchrotron intensity and polarization [44]. Altogether, $\bar{B} = 4 \, \mu$G and $b = 5 \, \mu$G

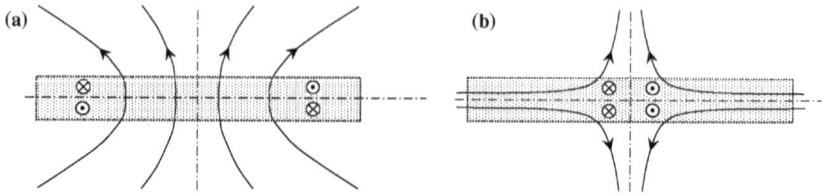

Fig. 1. A schematic representation of the magnetic lines of the meridional magnetic field (solid) of (a) dipolar and (b) quadrupolar symmetry in a thin disc (shaded). The direction of the azimuthal magnetic field on both sides of the slab's midplane is also shown: $\odot$, out of the page and $\otimes$, into the page. The symmetry remains unchanged if all the field directions are simultaneously reversed.

seem to be acceptable estimates of magnetic field strengths for a region within several kiloparsecs from the Sun.

Unlike the Solar magnetic field, which has a dipolar parity, galactic magnetic fields appear to be *quadrupolar* [13]; the difference is illustrated in Fig. 1. This general fact that the global magnetic fields of spherical objects (including starts and planets) are likely to be predominantly dipolar, whereas those of flat objects (spiral galaxies) are quadrupolar can be elegantly explained by the dynamo theory (see Sect. 6.1.2).

To summarize, magnetic fields of spiral galaxies have the following typical parameters. At scales much larger than the turbulent scale of about 100 pc, the strength of the global magnetic field is $\bar{B} \simeq 3\text{--}7\,\mu G$. With the total field strength of $B \simeq 5\text{--}12\,\mu G$, the ratio of energy densities in the random and regular magnetic fields is $\langle b^2 \rangle / \bar{B}^2 \simeq 3$. The global magnetic field is likely to have a global quadrupolar parity, but this has been verified observationally only for the Milky Way. The global magnetic pattern has the form of a spiral trailing with respect to the galactic rotation (similarly to the galactic spiral arms), with the pitch angle $p_B = \arctan \bar{B}_r / \bar{B}_\phi = -(10°\text{--}30°)$. Galactic magnetic fields exhibit a variety of complicated spatial structures (e.g., magnetic arms, field reversals between the disc and the halo, etc.). There is a widespread misconception that the strength of the regular magnetic field near the Sun, $\bar{B}_\odot \approx 2\,\mu G$, is representative of all spiral galaxies. In fact, the Sun is close to a reversal of the large-scale magnetic field; magnetic fields at smaller galactocentric radii are significantly stronger than that in the immediate vicinity of the Sun.

2.3.3. Galaxy clusters

Galaxy clusters are the largest gravitationally bound systems in the universe, having masses of order $10^{14}\text{--}10^{15}\,M_\odot$. Observations of clusters in X-rays reveal that they have an atmosphere of hot gas with temperatures $T \simeq 10^7\text{--}10^8$ K, extending over scales of order $1\,\mathrm{Mpc} = 10^6\,\mathrm{pc}$. Succinct reviews of the observational data on cluster magnetic fields can be found in Refs [10, 15]. The central parts of a relatively small fraction of galaxy clusters emit radio synchrotron emission (i.e., possess radio halos) which directly indicates the presence of magnetic fields and relativistic electrons in their intergalactic medium. Cluster magnetic fields can also be probed using Faraday rotation studies of both cluster radio galaxies and background radio sources seen through the cluster. Clear contribution of the intracluster medium to the Faraday rotation has been detected in many clusters, so that it seems plausible that magnetic fields (unlike relativistic electrons) are common in clusters of galaxies.

Typical number density and temperature of the interstellar gas of such rich galaxy clusters as Coma are $n = 10^{-3}\,\mathrm{cm}^{-3}$ and $T = 10^6\,\mathrm{K}$. The radius of the synchrotron halo in Coma is $L \simeq 500\,\mathrm{kpc}$. Under the assumption of energy equipartition between the cosmic rays and magnetic fields, magnetic field strength is of order $2\,\mu\mathrm{G}$. Over the path length $L = 500\,\mathrm{kpc}$, such a field would produce $\mathrm{RM} \simeq 10^3\,\mathrm{rad\,m}^{-2}$. However, the observed RM is ten times smaller being of order $100\,\mathrm{rad\,m}^{-2}$ [11]. The difference is explained by the fact that magnetic field is random. To justify this, we consider the autocorrelation function of the Faraday rotation measure in a random magnetic field. For this purpose, we introduce coordinates (x, y, z) with the z-axis directed towards the observer, and those in the plane of the sky, $X = (X, Y)$. We assume the magnetic field to be an isotropic, homogeneous, random field with zero mean value. Then its equal-time, two-point correlation tensor has the form $\langle B_i(x, t) B_j(y, t) \rangle = M_{ij}(r, t)$, where

$$M_{ij} = \left(\delta_{ij} - \frac{r_i r_j}{r^2}\right) M_{\mathrm{N}}(r, t) + \frac{r_i r_j}{r^2} M_{\mathrm{L}}(r, t).$$

Here $r = |x - y|$, $r_i = x_i - y_i$; $M_{\mathrm{L}}(r, t)$ and $M_{\mathrm{N}}(r, t)$ are known as the longitudinal and transverse correlation functions of the magnetic field, respectively ([26]; Section 34 of [22]). Since $\nabla \cdot B = 0$,

$$M_{\mathrm{N}} = \frac{1}{2r} \frac{\partial}{\partial r} \left(r^2 M_{\mathrm{L}} \right).$$

We further assume for simplicity that the electron density is constant. This is consistent with the fact that random gas motions in galaxy clusters are quite subsonic. The correlation function of RM is then

$$
\begin{aligned}
C(R) &= \langle \mathrm{RM}(X_1) \mathrm{RM}(X_2) \rangle \\
&= K_1^2 n_{\mathrm{e}}^2 \int_0^L \int_0^L B_z(X_1, z_1) B_z(X_2, z_2)\, dz_1\, dz_2
\end{aligned}
$$

$$= K_1^2 n_e^2 L \int_{-L}^{L} M_{zz}(R, \zeta) d\zeta$$

$$= K_1^2 n_e^2 L \int_{-L}^{L} \left(M_N \frac{R^2}{R^2 + \zeta^2} + M_L \frac{\zeta^2}{R^2 + \zeta^2} \right) d\zeta$$

$$= K_1^2 n_e^2 L \int_{-L}^{L} \left(M_L + \frac{R^2}{2r} \frac{dM_L}{dr} \right) d\zeta. \tag{2.4}$$

Here we have assumed that L is much larger than the correlation length of the magnetic field, $\zeta = z_1 - z_2$, $R = |X_1 - X_2|$ and $r^2 = R^2 + \zeta^2$.

For the sake of illustration, consider the longitudinal correlation function of the form

$$M_L = \tfrac{1}{3} b^2 \exp \left(-\frac{r^2}{2l_B^2} \right),$$

which corresponds to the one-dimensional magnetic spectrum of the form $M_k \propto k^4 \exp(-k^2 l_B^2/2)$ [26]; here $b^2 = \langle B^2 \rangle$. We note that M_k attains maximum at a wavenumber $k_m = 2/l_B$ (or a scale $2\pi/k_m = \pi l_B$), whereas the longitudinal correlation scale is given by $l_L = [M_L(0)]^{-1} \int_0^\infty M_L(r)\,dr = l_B \sqrt{\pi/2}$.

Straightforward calculation then yields

$$C(R) = \frac{\sqrt{2\pi}}{3} K_1^2 n_e^2 b^2 L l_B \left(1 - \frac{R^2}{2l_B^2} \right) \exp \left(-\frac{R^2}{2l_B^2} \right). \tag{2.5}$$

The root-mean-square value of RM can be obtained from Eq. (2.4) or (2.5) at $R = 0$:

$$\sigma_{RM}^2 = K_1^2 n_e^2 L \int_{-L}^{L} M_L(R, \zeta)|_{R=0}\, d\zeta = \frac{\sqrt{2\pi}}{3} K_1^2 n_e^2 b^2 L l_B. \tag{2.6}$$

Thus, the standard deviation of RM grows with the square root of the path length L, $\sigma_{RM} \propto L^{1/2}$. This happens because the polarization angle ψ of the radio emission propagating through the random magnetic field experiences random walk because of the Faraday rotation, and hence the amount of rotation accumulated is proportional to $N^{1/2}$, where $N \simeq L/l_B$ is the number of correlation cells on the path length. Since RM $\propto \Delta\psi$, where $\Delta\psi$ is the difference in ψ between two wavelengths, the resulting standard deviation of RM is also proportional to $N^{1/2}$. If the value of RM produced in a single correlation cell is $RM_0 \simeq K_1 n_e b l_B$, we obtain $\sigma_{RM} \simeq RM_0 N^{1/2}$, which agrees with Eq. (2.6).

Using $b = 2\,\mu G$, $n_e = 10^{-3}\,cm^{-3}$, $L = 500\,kpc$ and $\sigma_{RM} = 100\,rad\,m^{-2}$ in Eq. (2.6), we obtain the magnetic correlation length as $l_B \simeq 10\,kpc$. Thus, the correlation length of magnetic fields in the intracluster gas of galaxy clusters is much smaller than the size of a cluster (but is rather comparable to the size of a galaxy). An estimate of the field strength in galaxy clusters obtained from Faraday rotation measurements is [11]

$$b \simeq 5(l_B/10\,\text{kpc})^{-1/2}\,\mu\text{G}.$$

In conclusion, there is considerable evidence that galaxy clusters are magnetized with the field root-mean-square strength ranging from a few μG to several tens of μG in the central parts of some clusters, and with coherence scales of order 10 kpc. These fields, if not maintained by some mechanism, will evolve as decaying MHD turbulence, and perhaps decay on the appropriate Alfvén time scale of about 10^8 yr, much shorter than the age of the cluster. Even though the scale of the magnetic field is comparable to the size of a galaxy, these magnetic fields cannot result from stripping of the interstellar gas together with its magnetic field: the strength of any magnetic field stripped from a galaxy decreases by a factor of order ten as the gas expands from the interstellar densities of order $0.1\,\text{cm}^{-3}$ to the intergalactic densities of about $10^{-3}\,\text{cm}^{-3}$: $B \propto n^{2/3}$ if magnetic field is frozen into the gas and the expansion is spherically symmetric. Thus, even under optimistic assumptions the stripping could account for at most 0.1 of the observed intergalactic magnetic field strength. Magnetic fields in galaxy clusters need dynamo action to be produced [32, 50].

3. Astrophysical flows

As discussed elsewhere in this volume, the generation of a magnetic field at a scale comparable to the size of the parent object is a rather subtle process: since the regular magnetic field is not mirror symmetric, its generation is a symmetry-breaking process. (If the magnetic field is of a small scale, the system remains mirror-symmetric on the average, and no systematic deviations from the mirror symmetry are required to maintain a small-scale magnetic field—see Sect. 7.) To appreciate the significance of mirror symmetry, look at the face of a clock through a mirror. The numbers on the dial's reflection look differently from the original. However, the sense of rotation of the hands is the same in the clock and in its mirror image. Under a mirror reflection of the Cartesian reference frame, $(x, y, z) \rightarrow (x, y, -z)$, the velocity components transform similarly, $(v_x, v_y, v_z) \rightarrow (v_x, v_y, -v_z)$, and so the linear velocity v is a true vector. But the angular velocity or vorticity $\omega = \nabla \times v$ change differently under the reflection, $(\omega_x, \omega_y, \omega_z) \rightarrow (-\omega_x, -\omega_y, \omega_z)$, and so these vector fields are *not* mirror-symmetric. [Write out the vorticity components in terms of the partial derivatives of v_x, v_y, v_z to obtain the above symmetry relations from those for r and v.] The angular velocity is a *pseudo*-vector. Magnetic field is a pseudo-vector too: consider the reflections of a linear electric current j and the associated magnetic field B, with $j = \nabla \times B$. Similarly, helicity of motion $v \cdot \omega$ is a pseudo-scalar as it changes sign upon mirror reflection (indeed, the reflection of a right-handed screw is a left-handed screw).

The mirror asymmetry of the magnetic field has far-reaching physical consequences: a system that is perfectly mirror symmetric (i.e., lacks any pseudo-vectorial or pseudo-scalar properties) cannot generate magnetic field at its own scale. A pseudo-vectorial property ubiquitous in astrophysical systems is rotation, resulting in the intrinsic connection of regular magnetic fields and rotation.

Another feature of electrically conducting flows important for magnetic fields is the randomness or Lagrangian chaos: the trajectories of elementary volumes in a random or chaotic flows diverge exponentially. In a fluid of high electric conductivity, magnetic field is (almost) frozen into the flow, and the divergence of the trajectories can lead to the exponential stretching and, therefore, exponential amplification of magnetic field embedded into the flow. A type of randomness widespread in nature is turbulence; hence, the importance of turbulent dynamos.

In this section we briefly discuss the properties of plasma motions in the Sun, spiral galaxies and galaxy clusters important for the generation of magnetic fields. Our focus will be on differential rotation, the α parameter of small-scale random motions (a measure of their deviation from mirror symmetry) and turbulent magnetic diffusivity.

3.1. Solar convection zone

The Sun's magnetic field is maintained by convective motions in its part known as the convection zone which extends from a radius of $0.7R_\odot$ to about $0.95R_\odot$ (almost the Solar surface). The angular velocity of rotation in the convection zone has been determined using methods of helioseismology. The angular velocity slightly increases with the radius within about $30°$ of the equator and decreases closer to the poles. With the mean angular velocity of $\Omega \simeq 4.6 \times 10^{-6}\,\mathrm{s}^{-1}$, the magnitude of the differential rotation across the convection zone is $\Delta\Omega \simeq 0.1\Omega$. A thin region between the convection zone and the radiative zone, known as the tachocline, is a site of especially strong differential rotation; this makes this region especially important for the solar dynamo [18].

The scale and velocity of the convective motions associated with the granulation are $l \simeq 10^3\,\mathrm{km}$ and $v \simeq 1\,\mathrm{km\,s^{-1}}$, respectively. With the granulation time scale $\tau \simeq 500\,\mathrm{s} \approx 10\,\mathrm{min}$, the Rossby number of these motions is $\mathrm{Ro} = (2\Omega\tau)^{-1} \simeq 300$. Convective structures of a larger scale, mesogranules, have $l \simeq 3.5 \times 10^4\,\mathrm{km}$, $v \simeq 0.5\,\mathrm{km\,s^{-1}}$, $\tau \simeq 5 \times 10^4\,\mathrm{s}$ and $\mathrm{Ro} \simeq 3$. For the deep convection zone, the relevant parameters are: the pressure scale height $h \simeq 5 \times 10^9$ cm, the gas density $\rho \simeq 0.2\,\mathrm{g\,cm^{-3}}$, and the convection velocity $v \simeq 20\,\mathrm{m\,s^{-1}}$ and scale $l \simeq 2 \times 10^9$ cm. Magnetic field strength corresponding to energy equipartition with the kinetic energy of the convection is $B = (4\pi \rho v^2)^{1/2} \simeq 3000\,\mathrm{G}$, and the Rossby number is Ro $\simeq 0.6$. The relatively small value of the Rossby number indicates that convective motions deep in the convection zone are significantly modified by rotation; in particular, they acquire significant helicity.

3.2. Spiral galaxies

3.2.1. Turbulence and multi-phase structure
The interstellar medium (ISM) is much more inhomogeneous and active than stellar and planetary interiors. The reason for that is ongoing star formation: massive young stars evolve rapidly (in about 10^6 yr) and then explode as supernova stars (SN) releasing large amounts of energy ($E_{SN} \simeq 10^{51}$ erg per event). These explosions control the structure of the ISM.

SN remnants are filled with hot, overpressured gas that starts by expanding supersonically; at this stage the gas surrounding the blast wave is not perturbed. When pressure inside a SN remnant reduces to values comparable to that in the surrounding gas, the remnant disintegrates and merges with the ISM—at this stage the expanding SN remnant drives motions in the surrounding gas, and its energy is partially converted into the kinetic energy of the ISM. Since SN occur at (almost) random times and positions, the result is a random force that drives random motions in the ISM, which eventually become turbulent. The size of an SN remnant when it has reached pressure balance, determines the energy-range turbulent scale,

$$l \simeq 0.05\text{–}0.1 \text{ kpc}.$$

A fraction $f = 0.07$ of the SN energy is converted into the ISM's kinetic energy. With the SN frequency of $\nu_{SN} \sim (30\,\mathrm{yr})^{-1}$ in the Milky Way (i.e., one SN per 30 yr), the kinetic energy supply rate per unit mass is $\dot{e}_{SN} = f \nu_{SN} E_{SN} M_{gas}^{-1} \sim 10^{-2}\,\mathrm{erg\,g^{-1}\,s^{-1}}$, where $M_{gas} = 4 \times 10^9\,M_\odot$ is the total mass of gas in the galaxy. This energy supply can drive turbulent motions at a speed v such that $2v^3/l = \dot{e}_{SN}$ (where the factor 2 allows for equal contributions of kinetic and magnetic turbulent energies), which yields

$$v \simeq 10\text{–}30\,\mathrm{km\,s^{-1}},$$

a value similar to the speed of sound at a temperature $T = 10^4$ K or higher. The corresponding turbulent diffusivity follows as

$$\eta_t \simeq \tfrac{1}{3} l v \simeq (0.5\text{–}3) \times 10^{26}\,\mathrm{cm^2\,s^{-1}}. \tag{3.1}$$

Supernovae are the main source of turbulence in the ISM. Stellar winds is another significant source, contributing about 25% of the total energy supply.

The time interval between supernova shocks passing through a given point is about [24]

$$\tau = (0.5\text{–}5) \times 10^6 \text{ yr}.$$

After this period of time, the velocity field at a given position completely renovates to become independent of its previous form. Therefore, this time can be identified with the correlation time of interstellar turbulence. The renovation time is 2–20 times shorter than the 'eddy turnover' time $l/v \simeq 10^7$ yr. This means that the short-correlated (or δ-correlated) approximation, so often employed in turbulence and dynamo theory, can be quite accurate in application to the ISM—this is a unique feature of the interstellar turbulence. Note that the standard estimate (3.1) is valid if the correlation time is l/v. If the renovation time was used instead, the result would be $\eta_t \simeq l^2/\tau \simeq 10^{27} \text{ cm}^2\text{ s}^{-1}$, a value an order of magnitude larger than the standard estimate.

Another important result of supernova activity is a large amount of gas heated to a temperature $T = 10^6$ K. The gas is so tenuous that the collision rate of the gas particles is low, and so its radiative cooling time is very long and exceeds τ: the hot bubbles produced by supernovae can merge before they cool. The result is a network of hot tunnels that form the hot component of the ISM. Altogether, the interstellar gas is found in several distinct states, known as 'phases' (this usage may be misleading as most of them are not proper thermodynamic phases) whose parameters are presented in Table 1. Some of the parameters (especially the volume filling factors) are not known confidently, so estimates of Table 1 should be approached with healthy caution. The warm diffuse gas can be considered as a background against which the ISM dynamics evolves; this is the primary phase that occupies a connected (percolating) region in the disc, whereas the hot gas

Table 1

The multi-phase ISM. The origin and parameters of the most important phases of interstellar gas: n, the mid-plane number density in hydrogen atoms per cm^3; T, the temperature in K; c_s, the speed of sound in km s^{-1}; h, the scale height in kpc; and f_V, the volume filling factor in the disc of the Milky Way, in percent

Phase	Origin	n	T	c_s	h	f_V
Warm		0.1	10^4	10	0.5	60–80
Hot	Supernovae	10^{-3}	10^6	100	3	20–40
Hydrogen clouds	Compression	20	10^2	1	0.1	2
Molecular clouds	Self-gravity, thermal instability	10^3	10	0.3	0.075	0.1

may or may not fill a connected region. The warm gas is ionized by the stellar ultraviolet radiation and cosmic rays; its degree of ionization is about 30% at the Galactic midplane. The hot gas is so hot that it is fully ionized by gas particle collisions.

The locations of SN stars are not entirely random: 70% of them cluster in regions of intense star formation (known as OB associations as they contain large numbers of young, bright stars of spectral classes O and B) where gas density is larger than on average in the galaxy. Collective energy input from a few tens (typically, 50) of SN within a region about 0.5–1 kpc in size produces a superbubble that can break through the galactic disc. This removes the hot gas into the galactic halo and significantly reduces its filling factor in the disc (from about 70% to 10–20%). This also gives rise to a systematic outflow of the hot gas to large heights where the gas eventually cools, condenses and returns to the disc after about 10^9 yr in the form of cold, dense clouds of neutral hydrogen. This convection-type flow is known as the galactic fountain, and it can plausibly support a mean-field dynamo of its own [45]. The local vertical velocity of the hot gas at the base of the fountain flow is 100–200 km s^{-1}. Thus, galactic discs are open systems that exchange matter and magnetic fields with the galactic halos (cf. [20]). This exchange can be important for the magnetic helicity balance and galactic dynamo action [42].

3.2.2. Galactic rotation

Spiral galaxies have conspicuous flat components because they rotate rapidly enough. The Sun moves in the Milky Way at a velocity of about $V_\odot = r_\odot \Omega_\odot = 220$ km s^{-1}, to complete one orbit of a radius $r_\odot \approx 8.5$ kpc in $2\pi/\Omega_\odot - 2.4 \times 10^8$ yr. These values are representative for spiral galaxies in general. The Rossby number is estimated as

$$\text{Ro} = \frac{v}{l\Omega_\odot} \sim 4.$$

Ro $= 1$ at a scale 0.4 kpc in the warm gas, which is similar to the scale height of the gas layer. This implies that rotation significantly affects turbulent gas motions, making them helical on average, so that they are capable of producing large-scale magnetic fields via the α-effect of the mean-field dynamo theory. A convenient estimate of the α-effect can be obtained from Krause's formula,

$$\alpha_0 \simeq \frac{l^2\Omega}{h} \approx 0.5 \text{ km s}^{-1}, \tag{3.2}$$

where Ω is the angular velocity, and the numerical estimate refers to the Solar neighbourhood of the Milky Way. Thus, $\alpha_0 \simeq 0.05v$ near the Sun and increases in the inner Galaxy together with Ω. This estimate of α_0 will be used to calculate the dynamo number and, hence, to assess the efficiency of dynamo action in the Galaxy.

The spatial distribution of galactic rotation is known for thousands galaxies [43] from systematic Doppler shifts of various spectral lines emitted by stars and gas. In this respect, galaxies are much better explored than any star or planet (including the Sun and the Earth) where reliable data on the angular velocity in the interior are much less detailed and reliable or even unavailable. The radial profile of the galactic rotational velocity is called the rotation curve. Rotation curves of most galaxies are flat beyond a certain distance from the axis, so $\Omega \propto r^{-1}$ is a good approximation for $r \gtrsim 5 \, \mathrm{kpc}$.

3.3. Galaxy clusters

Clusters of galaxies do not exhibit any rotation. Correspondingly, magnetic field in the intracluster gas is random, without any mean component. Theoretical models strongly suggest that the intracluster gas is turbulent [12, 50]. The turbulence is mainly driven by the recent or ongoing merger events where large clumps of matter merge to form the cluster. The scale and velocity of the turbulent motions are estimated as $l \simeq 250\text{--}150 \, \mathrm{kpc}$ and $v \simeq 300\text{--}150 \, \mathrm{km \, s^{-1}}$; the latter is useful to compare with the speed of sound (or the thermal velocity) in the gas, $c_s \approx 1000 \, \mathrm{km \, s^{-1}}$. Since the turbulent Mach number is as small as 0.3 or even less, the compressibility effects are relatively weak and the turbulent fluctuations in the gas density can be neglected. The intracluster plasma is so tenuous that the mean free path is of order $10 \, \mathrm{kpc}$. Nevertheless, magnetohydrodynamic description remains meaningful because, in a magnetized plasma, the role of the mean free path is played by the Larmor radius which is very small even in magnetic fields much weaker than those observed in galaxy clusters. However, the corresponding effective pressure becomes anisotropic, and this can lead to interesting (and largely unexplored) effects [35].

4. The necessity of dynamo action

The necessity of dynamo action in the Earth and the Sun is practically obvious, in part because of the time variation of the magnetic fields of these objects: the geomagnetic magnetic field is known to change its polarity at irregular time intervals, whereas the solar magnetic field drives the 11-year activity cycle and changes its polarity every 22 years. Even without any other arguments in favour of planetary and stellar dynamos, the time variation would be sufficient to treat seriously applications of dynamo theory to planets and stars.

The situation is different with galaxies where the time scales involved are by far too long to be useful for this purpose and the only clues to the origin of galactic magnetic fields come from their spatial structures. Nevertheless, there are several lines of evidence that consistently indicate that the large-scale galactic magnetic fields need to be maintained by ongoing dynamo action [41] (see, however, [20]).

It is sometimes claimed that magnetic field does not need any support if the electric resistivity of the medium is small enough, i.e., the magnetic Reynolds number is large enough. In the case of the interstellar gas, the magnetic diffusivity of a fully ionized gas, $\eta = 10^7 (T/10^4 \text{ K})^{-3/2} \text{ cm}^2 \text{ s}^{-1}$, is so small that the magnetic Reynolds number at the scale equal to the scale height of the warm gas $h = 500 \text{ pc}$ is as large as $R_{\mathrm{m}} \simeq 10^{20}$, and the decay time of the large-scale magnetic field would seem to follow as 10^{27} yr. However, this estimate is hardly useful because the ISM is turbulent, and the corresponding decay time of the large-scale magnetic field is only $h^2/\eta_{\mathrm{t}} \simeq 5 \times 10^8$ yr. More generally, since magnetic energy in any three-dimensional turbulent flow rapidly cascades towards small scales where it dissipates, any three-dimensional, turbulent, magnetized system needs some form of dynamo action to maintain its magnetic field in a steady state.

Another argument in favour of dynamo action is related to the fact that the large-scale magnetic fields of spiral galaxies are only mildly wrapped up by the differential rotation, with the pitch angle $p_B = \arctan \bar{B}_r / \bar{B}_\phi = -(10°\text{–}30°)$, where the negative sign means that the magnetic spiral is trailing with respect to the galactic rotation. Near the Sun, the Milky Way galaxy gas made about $N = 30$ (differential) rotations during its lifetime. If the galactic large-scale magnetic field was primordial and its spiral shape was produced by the differential rotation, its pitch angle would be of the order of $p_B \simeq -1/N \simeq -2°$. This suggests that the large-scale magnetic field observed in the Milky Way and in spiral galaxies in general cannot be just a primordial magnetic field twisted by differential rotation.

If an external, quasi-uniform magnetic field is to have quadrupolar symmetry with the respect to the disc's midplane, as appropriate to spiral galaxies, it has to be in the plane of the galaxy. Then an initially (quasi-)uniform magnetic field would be twisted into a nonaxisymmetric configuration with the azimuthal wave number $m = 1$ (a bisymmetric structure). Meanwhile, dynamo models in a thin disc consistently favour axially symmetric magnetic structures, $m = 0$. Early observations of galactic magnetic fields seemed to indicate that the global magnetic structures are predominantly bisymmetric, in contradiction with the galactic dynamo theory. However, the improved quality of observations and their interpretation since the 1990's have unexpectedly revealed that magnetic structures in most (if not all) spiral galaxies can be described as variously distorted axisymmetric magnetic fields [4]: what seemed to be a weakness of the dynamo theory turned out to be its strength!

The recently discovered magnetic arms in the spiral galaxy NGC 6946 (see [3]), where the large-scale magnetic field (unlike the total field) is stronger between the gaseous spiral arms, i.e., where the gas density is lower, directly indicates that the regular magnetic field is not frozen into the ISM and therefore must be maintained against turbulent diffusion.

The complicated magnetic structure in the spiral galaxy M51, where the large-scale magnetic fields in the disc and the halo are almost oppositely directed [7] also requires an explanation more complicated than just a quasi-uniform primordial magnetic field twisted by the galactic differential rotation.

5. Dynamo parameters

Using the parameters of the solar convection zone presented in Sect. 3.1, one obtains $\alpha \simeq l^2 \Omega / h \simeq 2 \times 10^3$ cm s^{-1} (which is close to the convection velocity), and $\eta_t \simeq \frac{1}{3} l v \simeq 10^{12}$ cm^2 s^{-1}. This yields the following crude estimates of the dimensionless numbers that controls the mean-field dynamo action:

$$ R_\alpha = \frac{\alpha H}{\eta_t}, \qquad R_\omega = \frac{\Delta \Omega H^2}{\eta_t}, \qquad D = R_\alpha R_\omega \simeq 4000, $$

where $H = 0.3 R_\odot$ is the thickness of the convection zone. Here R_α and R_ω are the turbulent magnetic Reynolds numbers that characterize the intensities of helical small-scale motions and differential rotation, respectively. Their product, the dynamo number D, quantifies the efficiency of the mean-field dynamo action in systems with strong differential rotation, $|R_\omega| \gg |R_\alpha|$. Here we assume that the solar dynamo acts in the bulk of the convection zone. Dynamo models that explicitly include the tachocline are reviewed in Ref. [51].

For spiral galaxies, assuming a flat rotation curve, $V_0 = r\Omega = $ const, we similarly obtain the following estimates:

$$ R_\omega \simeq -3 \frac{V_0}{v} \frac{h^2}{l R_0} \simeq -15, \qquad R_\alpha \simeq 3 \frac{V_0}{v} \frac{l}{R_0} \simeq 0.5, $$

where $V_0 = 200$ km s^{-1} is the typical rotational velocity, $v = 10$ km s^{-1}, $l = 0.1$ kpc, $h = 0.5$ kpc and $R_0 = 10$ kpc. Similarly,

$$ D = R_\alpha R_\omega \simeq 10 \frac{h^2}{v^2} r\Omega \frac{d\Omega}{dr} \approx -10 \left(\frac{h}{R_0} \right)^2 \left(\frac{V_0}{v} \right)^2 \simeq -10. \qquad (5.1) $$

Note that the dynamo number is independent of the turbulent scale and only depends on parameters reasonably well known from observations.

In the case of galaxies, it is useful to define the *local* dynamo parameters R_α and R_ω, as functions of the galactocentric radius r. These are obtained when the r-dependent, local values of the parameters are used instead of the characteristic ones, for example α can be replaced by $l^2(r)\Omega(r)/h(r)$. Then the local dynamo number is given by

$$D_L = \frac{\alpha(r)G(r)h^3(r)}{\eta_t^2(r)} \simeq -10\left(\frac{\Omega h}{v}\right)^2, \tag{5.2}$$

where $G = r d\Omega(r)/dr$ ($G = -\Omega$ for $\Omega \propto r^{-1}$). The local dynamo number rapidly grows towards the galactic centre (roughly as r^{-1}) mainly due to the increase in Ω. The estimate (5.1) is based on parameter values typical of the Solar vicinity of the Milky Way. In other galaxies and in other parts of our Galaxy, this can be a poor measure of the dynamo activity; unfortunately, this is often forgotten and the single estimate (5.1) is used to represent the whole diverse world of galactic dynamos.

6. Perturbation solutions for mean-field dynamos

6.1. Disc dynamos

In this section we shall develop an approximate solution of the mean-field dynamo equation

$$\frac{\partial \bar{B}}{\partial t} = \nabla \times (\bar{V} \times \bar{B}) + \nabla \times (\alpha \bar{B}) - \nabla \times \eta_t \nabla \times \bar{B}. \tag{6.1}$$

It is convenient to introduce cylindrical polar coordinates (r, ϕ, z) with the z-axis parallel to the angular velocity vector Ω. In a thin disc, all spatial derivatives can be neglected in comparison with those in z, $\partial/\partial z \gg \partial/\partial r$, $\partial/r\partial\phi$. Then the local dynamo equations, written in the $\alpha\omega$-approximation, reduce to the following dimensionless form (for $\eta_t = \text{const}$):

$$\frac{\partial \bar{B}_r}{\partial t} = -R_\alpha \frac{\partial}{\partial z}(\alpha \bar{B}_\phi) + \frac{\partial^2 \bar{B}_r}{\partial z^2}, \tag{6.2}$$

$$\frac{\partial \bar{B}_\phi}{\partial t} = R_\omega \bar{B}_r + \frac{\partial^2 \bar{B}_\phi}{\partial z^2}, \tag{6.3}$$

$$\frac{\partial \bar{B}_z}{\partial t} = \frac{\partial^2 \bar{B}_\phi}{\partial z^2}, \tag{6.4}$$

where the units of z and t are h and h^2/η_t, respectively (see [4, 41] for details). Equation for $\bar{B}_z$ splits from the system and Eqs. (6.2) and (6.3) can be solved separately. The vertical magnetic field in a thin disc is supported through the radial and azimuthal components via their radial derivatives which are neglected in Eqs (6.2)–(6.4). These equations are supplemented with the vacuum boundary conditions

$$\bar{B}_\phi = 0, \qquad \bar{B}_r \approx 0, \qquad \frac{\partial \bar{B}_z}{\partial z} = 0 \quad \text{at } z = \pm 1, \tag{6.5}$$

and the symmetry conditions at the disc midplane $z = 0$:

$$\frac{\partial \bar{B}_r}{\partial z} = \frac{\partial \bar{B}_\phi}{\partial z} = \bar{B}_z = 0 \quad \text{at } z = 0 \quad \text{(quadrupolar)}, \tag{6.6}$$

and

$$\bar{B}_r = \bar{B}_\phi = \frac{\partial \bar{B}_z}{\partial z} = 0 \quad \text{at } z = 0 \quad \text{(dipolar)}. \tag{6.7}$$

At the kinematic stage of the dynamo, when the velocity field remains unaffected by the growing magnetic field, we have

$$\bar{B} = \mathcal{B} \exp(\gamma t).$$

We further rescale the radial magnetic field $\bar{B}_r \to R_\alpha \bar{B}_r$ to obtain the following boundary value problem involving the dynamo number $D = R_\alpha R_\omega$:

$$
\begin{aligned}
\gamma \mathcal{B}_r &= -\frac{\partial}{\partial z}(\alpha \mathcal{B}_\phi) + \frac{\partial^2 \mathcal{B}_r}{\partial z^2}, \\
\gamma \mathcal{B}_\phi &= D\mathcal{B}_r + \frac{\partial^2 \mathcal{B}_\phi}{\partial z^2}, \\
\mathcal{B}_r(1) &= \mathcal{B}_\phi(1) = 0,
\end{aligned}
\tag{6.8}
$$

together with (6.6) and (6.7). Thus, we have formulated a one-dimensional boundary value problem with the eigenvalue γ and vectorial eigenfunction $\mathcal{B}(z)$.

6.1.1. Free decay modes

Equations (6.8) can easily be solved in the absence of sources, $\alpha = D = 0$. The resulting solutions, known as *free decay modes*, are doubly degenerate since two distinct eigenfunctions $\mathcal{B}_n$ and $\mathcal{B}'_n$ correspond to each eigenvalue. The pairs of odd modes are given by

$$\mathcal{B}_n^{(d)} = \begin{pmatrix} \sqrt{2}\sin(\pi n z) \\ 0 \end{pmatrix}, \qquad \mathcal{B}_n^{(d)'} = \begin{pmatrix} 0 \\ \sqrt{2}\sin(\pi n z) \end{pmatrix},$$

$$\gamma_n^{(d)} = -\pi^2 n^2, \quad n = 1, 2, \ldots,$$

whereas the free decay modes of even parity are

$$\mathcal{B}_n^{(q)} = \begin{pmatrix} \sqrt{2}\cos\left[\pi\left(n+\tfrac{1}{2}\right)z\right] \\ 0 \end{pmatrix}, \qquad \mathcal{B}_n^{(q)'} = \begin{pmatrix} 0 \\ \sqrt{2}\cos\left[\pi\left(n+\tfrac{1}{2}\right)z\right] \end{pmatrix},$$

$$\gamma_n^{(q)} = -\pi^2\left(n+\tfrac{1}{2}\right)^2, \quad n = 0, 1, 2, \ldots.$$

The eigenfunctions $\mathcal{B}_n = (\mathcal{B}_{rn}, \mathcal{B}_{\phi n})$ have been normalized to have $\int_0^1 \mathcal{B}_n^2\, dz = 1$ for both dipolar and quadrupolar cases. The free-decay eigenfunctions form an orthonormal set of basis functions which are used below to develop a perturbation solution for $\alpha, D \neq 0$.

The dipolar mode with $n = 0$ is trivial as the horizontal magnetic field of the corresponding eigenfunction is identically zero, $(\bar{B}_r^{(d)}, \bar{B}_\phi^{(d)}) = 0$ even for $\alpha \neq 0$ and $D \neq 0$. The trivial dipolar solution consists of a uniform vertical magnetic field $B_z = $ const which is not affected by magnetic diffusion and therefore neither grows nor decays in this approximation.

The lowest quadrupolar mode decays four times as weakly as the lowest non-trivial dipolar one. This fact is closely associated with the property of the lowest quadrupolar mode to be generated preferentially (at a larger growth rate for a given dynamo number) than the dipolar ones. Therefore, large-scale magnetic fields of even parity dominate in spiral galaxies. The preference of even, quadrupolar modes is a specific feature of the disc geometry; in spherical bodies, such as the Sun and the Earth, the dipolar mode is preferred, in agreement with observations.

6.1.2. The perturbation expansion

For $|D| \ll 1$, terms containing α and D on the right-hand sides of equations (6.8) can be treated as a small perturbation, and an approximate solution can be obtained by perturbing the free-decay modes obtained for $\alpha = D = 0$. To isolate the perturbation operator, we introduce a new variable $\tilde{B}_\phi = |D|^{-1/2}B_\phi$, so that B_r and $\tilde{B}_\phi$ are of the same order of magnitude in D. Preserving the original notation for the renormalized azimuthal field component, we rewrite the dynamo equations in the matrix-operator form

$$\gamma \mathcal{B} = \left(\widehat{W} + |D|^{1/2}\widehat{V}\right)\mathcal{B}, \tag{6.9}$$

where

$$\widehat{W} = \begin{pmatrix} \dfrac{d^2}{dz^2} & 0 \\ 0 & \dfrac{d^2}{dz^2} \end{pmatrix}, \qquad \widehat{V}\mathcal{B} = \begin{pmatrix} 0 & -\dfrac{d}{dz}(\alpha B_r) \\ B_\phi\,\mathrm{sign}\,D & 0 \end{pmatrix},$$

are the unperturbed (free-decay) and perturbation operators, respectively. The perturbed solution is represented as a superposition of free decay modes. Since each free-decay eigenvalue is doubly degenerate, the perturbation $\epsilon \widehat{V}$ first removes the degeneracy, giving an $O(\epsilon)$ correction to the eigenvalue but and $O(1)$ correction to the eigenfunction (Sect. 33 of [21]). Thus, to the first order, the perturbed leading eigenfunction and eigenvalue have the form

$$\mathcal{B} \approx C_0 \mathcal{B}_0 + C_0' \mathcal{B}_0', \quad \gamma \approx \gamma_0 + \epsilon \gamma_1, \tag{6.10}$$

where C_0, C_0', γ_0 and γ_1 are constants of order unity in ϵ, and the lowest (quadrupolar) free-decay modes $\mathcal{B}_0$ and $\mathcal{B}_0'$ are given in Section 6.1.1 [we have dropped the superscript (q) to simplify the notation]. To calculate the expansion coefficients, these forms are substituted into Eq. (6.9); to the zeroth order in ϵ, this yields $\gamma_0 = \lambda_0$. Terms of order ϵ are then isolated, their dot product is taken with $\mathcal{B}_0$ and then with $\mathcal{B}_0'$, and the results are integrated over z from 0 to 1. This brings us to a system of two homogeneous algebraic equations for C_0 and C_0':

$$(\gamma_1 - V_{00})C_0 - V_{00'}C_0' = 0, \qquad -V_{0'0}C_0 + (\gamma_1 - V_{0'0'})C_0' = 0,$$

where $V_{nm} = \int_0^1 \mathcal{B}_n \cdot \widehat{V} \mathcal{B}_m \, dz$ (and likewise for $V_{n'm}$, but with $\mathcal{B}_n$ replaced by $\mathcal{B}_n'$, etc.) are the matrix elements (note that $V_{nn} = V_{n'n'} = 0$). The solvability condition of this system (vanishing of the determinant) yields γ_1:

$$\gamma_1 = \pm\sqrt{V_{0'0}V_{00'}}, \quad \text{and} \quad C_0' = \pm C_0\sqrt{\frac{V_{0'0}}{V_{00'}}}. \tag{6.11}$$

Since we are interested in solutions that decay slower as $|D|$ increases, and then grow when $|D|$ is large enough, we select the upper sign in these relations to have $\gamma_1 > 0$.

A similar solution can be obtained for the dipolar mode, $\mathcal{B} \approx C_1 \mathcal{B}_1^{(d)} + C_1' \mathcal{B}_1^{(d)'}$.

Having calculated the matrix elements for $\alpha = z$, we obtain

$$\gamma^{(d)} \approx -\pi^2 + \sqrt{-\tfrac{1}{2}D}, \tag{6.12}$$

$$\gamma^{(q)} \approx -\tfrac{1}{4}\pi^2 + \sqrt{-\tfrac{1}{2}D} \tag{6.13}$$

for solutions of dipolar and quadrupolar symmetry, respectively, where we have chosen the sign in front of the square root corresponding to solutions growing for $D < 0$.

The solutions are non-oscillatory, Im $\gamma = 0$, and they grow if $|D| > |D_c|$ (note that D, $D_c < 0$ in galactic discs), where

$$D_c^{(d)} \approx -2\pi^4 \approx -195, \qquad D_c^{(q)} \approx -\tfrac{1}{8}\pi^4 \approx -12,$$

for the dipolar and quadrupolar modes, respectively. The preference of the quadrupolar modes in a thin disc is now obvious: $|D_c^{(q)}| \ll |D_c^{(d)}|$.

Strictly speaking, this approximate solution should not be extended to estimate D_c because $|D_c|$ is not small. However, such bold extensions of asymptotic solutions often yield useful results. In particular, the above estimates are rather close to those obtained from numerical solutions of the dynamo equations (see below). The reason for that is that the dependence of the growth rate on D has the same form for both $|D| \ll 1$ and $|D| \gg 1$, namely $\gamma = \text{const} + |D|^{1/2}$, where the constant can be neglected for $|D| \gg 1$ [31]. Thus the dependence of γ on D is reasonably approximated by our perturbation solution even for those values of $|D|$ were it is not formally applicable.

Of course, the critical dynamo number depends on the form of $\alpha(z)$. A perturbation solution similar to that given above, but now for $\alpha = \sin \pi z$, gives $D_{cr}^{(q)} \approx -\tfrac{1}{4}\pi^3 \approx -8$. Numerical solutions show that the critical dynamo number for the lowest quadrupolar mode lies between approximately -4 and -12 for various forms of $\alpha(z)$. The rather low generation threshold $D_{cr}^{(q)} \approx -4$ is obtained if the α-effect is concentrated at halfway between the symmetry plane and the surface of the slab, $\alpha = \delta(z - \tfrac{1}{2}) - \delta(z + \tfrac{1}{2})$. Smooth distributions of $\alpha(z)$ give higher generation thresholds. For $\alpha = \sin \pi z$, the critical dynamo number obtained numerically is very close to the above approximate value, $D_{cr}^{(q)} \approx -8$, while $D_{cr}^{(q)} \approx -11$ for $\alpha = z$ (again in good agreement with the approximate solution). If $\alpha(z)$ is piecewise constant, $\alpha = \theta(z) - \theta(-z)$, $D_{cr}^{(q)} \approx -6$.

Given the above results for the eigenvalues, the coefficients of the expansion in free-decay modes are related by

$$C_0' = -\sqrt{2}C_0, \quad \text{for } \alpha = z,$$

for both the quadrupolar and dipolar modes. Restoring the original scaling of the field components ($\mathcal{B}_r \rightarrow R_\alpha \mathcal{B}_r$ and $\mathcal{B}_\phi \rightarrow |D|^{1/2}\mathcal{B}_\phi$), the lowest-order eigenfunctions are obtained as

$$\begin{pmatrix} \mathcal{B}_r \\ \mathcal{B}_\phi \end{pmatrix} \approx C_0\sqrt{2} \begin{pmatrix} R_\alpha \\ -\sqrt{2}|D|^{1/2} \end{pmatrix} \times \begin{cases} \sin \pi z & \text{(odd modes)}, \\ \cos \tfrac{1}{2}\pi z & \text{(even modes)}, \end{cases} \quad (6.14)$$

for $\alpha = z$, where C_0 remains an arbitrary constant. This results in the following estimate of the pitch angle of magnetic lines in the growing (kinematic) solution, the same for both dipolar and quadrupolar solutions:

$$p_B = \arctan \frac{\bar{B}_r}{\bar{B}_\phi} \approx -\arctan \frac{1}{\sqrt{2}} \sqrt{\frac{R_\alpha}{|R_\omega|}}. \tag{6.15}$$

For $R_\alpha = 1$ and $R_\omega = -20$, this yields $p_B \approx -10°$ in a good agreement with the pitch angles observed in spiral galaxies. For $\alpha = \sin \pi z$, a similar estimate differs insignificantly from the above (prefactor $\sqrt{\pi}/2$ instead of $1/\sqrt{2}$ in the estimate of p_B).

The accuracy of the perturbation solutions developed here is quite satisfactory even for $D \simeq D_c$, so it is worth considering the next approximation in ϵ. In particular, we show in Sect. 6.3 that the radial component, $\bar{B}_r$, of any growing eigenfunction must change sign near the disc surface (given the vacuum boundary conditions). Since this detail of the eigenfunction appears to be essential for the dynamo action, it is useful to develop a solution that captures this feature. To provide more examples, we present the second-order results for a different choice of the α-coefficient, $\alpha = \sin \pi z$. The second-order quadrupolar solution has the form

$$\mathcal{B} \approx \tilde{\mathcal{B}}_0 + \epsilon \sum_{n=1}^{\infty} (C_n \mathcal{B}_n + C_n' \mathcal{B}_n'),$$

$$\gamma \approx \gamma_0 + \epsilon \gamma_1 + \epsilon^2 \gamma_2,$$

which is useful to compare with Eq. (6.10). Here $\tilde{\mathcal{B}}_0 = C_0 \mathcal{B}_0 + C_0' \mathcal{B}_0'$ is the properly normalized first-order eigenfunction, $\int_0^1 \tilde{\mathcal{B}}_0^2 \, dz = 1$:

$$\tilde{\mathcal{B}}_0 = \sqrt{\frac{2}{1 + 4/\pi}} \begin{pmatrix} 1 \\ -2/\sqrt{\pi} \end{pmatrix} \cos \frac{\pi z}{2} \quad \text{for } \alpha = \sin \pi z.$$

As before, these expansions are substituted into the dynamo equations, terms of order ϵ^2 are isolated, dot products with $\mathcal{B}_k$ and $\mathcal{B}_k'$ (with $k \neq 0$) are evaluated and then integrated over z from 0 to 1. This leads to algebraic equations for C_n and C_n', which yield

$$C_n = \frac{V_{n\tilde{0}}}{\lambda_0 - \lambda_n}, \quad C_n' = \frac{V_{n'\tilde{0}}}{\lambda_0 - \lambda_n},$$

and, from the solvability condition,

$$\gamma_2 = \sum_{n=1}^{\infty} \frac{V_{n\tilde{0}} V_{\tilde{0}n} + V_{n'\tilde{0}} V_{\tilde{0}n'}}{\lambda_0 - \lambda_n},$$

where $V_{n'\tilde{0}}$ denotes the matrix element involving $\mathcal{B}'_n$ and $\tilde{\mathcal{B}}'_0$ and similarly for the other matrix elements. Direct calculation yields

$$V_{n\tilde{0}} = \frac{1}{2}\sqrt{\frac{\pi}{1+4/\pi}} \times \begin{cases} 1, & n = 0, \\ 3, & n = 1, \\ 0, & n \neq 0, 1, \end{cases}$$

$$V_{n'\tilde{0}} = -\frac{1}{\sqrt{1+4/\pi}} \times \begin{cases} 1, & n = 0, \\ 0, & n \neq 0, \end{cases}$$

$$V_{\tilde{0}n} = \frac{2}{\sqrt{\pi+4}} \times \begin{cases} 1, & n = 0, \\ 0, & n \neq 0. \end{cases}$$

Thus it can be shown that $\gamma_2 = 0$ for any form of $\alpha(z)$, whereas, for $\alpha = \sin \pi z$,

$$C_n = C'_n = 0 \text{ for } n \neq 1, \quad C_1 = \frac{3}{4\pi^{3/2}\sqrt{1+4/\pi}}.$$

For $D < 0$ and $\alpha = \sin \pi z$, to the second order in $\epsilon = |D|^{1/2}$, the corresponding quadrupolar solution written in terms of the physical variables follows as

$$\mathcal{B}_r = R_\alpha C_0 \left(\cos\frac{\pi z}{2} + \frac{3}{4\pi^{3/2}}\sqrt{-D}\cos\frac{3\pi z}{2} \right) + O(D), \tag{6.16}$$

$$\mathcal{B}_\phi = -2C_0\sqrt{-\frac{D}{\pi}}\cos\frac{3\pi z}{2} + O(D), \tag{6.17}$$

$$\gamma = -\frac{\pi^2}{4} + \frac{1}{2}\sqrt{-\pi D} + O(|D|^{3/2}). \tag{6.18}$$

This solution is remarkably accurate even for D close to the critical value; in particular, it yields $D_c \approx -7.8$, as compared with the numerically obtained value of -8, and the eigenfunction shown in Fig. 2 for $|D| \lesssim 20$ is practically indistinguishable from the numerical solution.

6.2. Spherical shell dynamos

Perturbation solutions similar to those developed in Sect. 6.1 can be obtained for other dynamo systems. Here we illustrate the techniques using Parker's model of the mean-field dynamo in a thin spherical shell, where the dominant modes are oscillatory, and the unperturbed state is not degenerate. In terms of the (scaled) azimuthal components of the vector potential, $\bar{A}_\phi = \mathcal{A}\exp\gamma t$, and magnetic

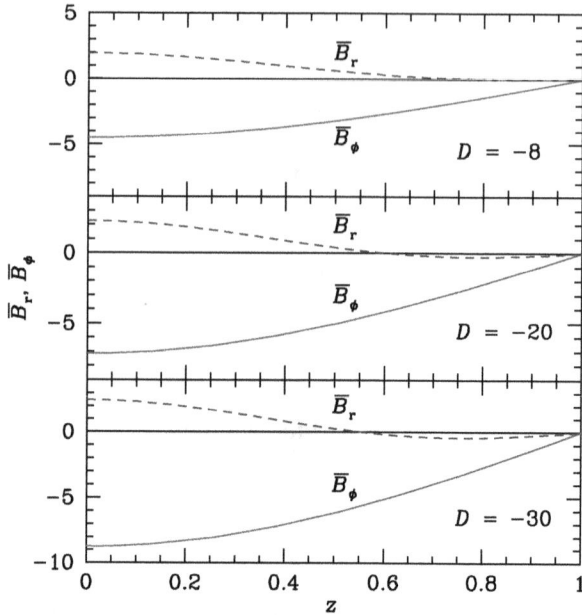

Fig. 2. The approximate eigenfunctions $\bar{B}_r$ (dashed) and $\bar{B}_\phi$ (solid) for $\alpha = \sin \pi z$ from Eqs (6.16)–(6.17) for $D = -8, -20, -30$ (with $D_c \approx -8$), $C_0 = 1$ and $R_\alpha = 1$.

field, $\bar{B}_\phi = \mathcal{B} \exp \gamma t$, the $\alpha\omega$-dynamo equations in a thin shell can be written as [28]

$$\gamma \mathcal{A} = \alpha(\theta)\mathcal{B} + \frac{\partial^2 \mathcal{A}}{\partial \theta^2}, \tag{6.19}$$

$$\gamma \mathcal{B} = -D \cos\theta \frac{\partial \mathcal{A}}{\partial \theta} + \frac{\partial^2 \mathcal{B}}{\partial \theta^2}, \tag{6.20}$$

where θ is the latitude, D (< 0) is the dynamo number, the angular velocity is assumed to depend on the spherical radius alone, and we consider axially symmetric solutions. The simplest model assumes that $\alpha = \sin\theta$, and we shall be adopting this form in what follows.

The boundary conditions at the equator,

$$\mathcal{B}(0) = 0, \quad \frac{\partial \mathcal{A}}{\partial \theta}(0) = 0,$$

isolate dipolar modes as appropriate for the Sun. Obviously,

$$\mathcal{B}(\pi/2) = 0,$$

but the boundary condition for $\mathcal{A}$ at the pole is more difficult to establish. We are interested here in modelling the main branch of the dynamo wave that propagates from mid-latitudes towards the equator. It is launched away from the pole, at about $\theta = 30°$, and there is another branch that propagates from the mid-latitudes to the pole. Thus, the solution we are interested in does not reach the pole, and the boundary condition for it could be posed at some intermediate latitude not known in advance. To simplify the model, we still pose it at $\theta = \pi/2$ but select it as to obtain a desired migratory wave solution. Our experimentation with different forms of this boundary condition suggests that the perturbation solution is oscillatory for

$$\left(\frac{\partial \mathcal{A}}{\partial \theta} + i\xi \mathcal{A}\right)\Bigg|_{\theta=\pi/2} = 0,$$

where ξ is an arbitrary real constant.

As in Sect. 6.1, we renormalize $\mathcal{B} = \mathcal{B}'\sqrt{|D|}$ to recast Eqs (6.19) and (6.20) in the form (we drop prime at $\mathcal{B}'$)

$$\gamma \begin{pmatrix} \mathcal{A} \\ \mathcal{B} \end{pmatrix} = \widehat{W} \begin{pmatrix} \mathcal{A} \\ \mathcal{B} \end{pmatrix} + \epsilon \widehat{V} \begin{pmatrix} \mathcal{A} \\ \mathcal{B} \end{pmatrix},$$

where

$$\epsilon = |D|^{1/2}, \quad \widehat{W} = \begin{pmatrix} \dfrac{\partial^2}{\partial \theta^2} & 0 \\ 0 & \dfrac{\partial^2}{\partial \theta^2} \end{pmatrix}, \quad \widehat{V} = \begin{pmatrix} 0 & \alpha \\ \cos\theta \dfrac{\partial}{\partial \theta} & 0 \end{pmatrix}.$$

For $\epsilon = 0$, Eqs. (6.19) and (6.20) decouple and the free-decay modes can easily be found. One of them is given by

$$\begin{pmatrix} \mathcal{A}_m \\ \mathcal{B}_m \end{pmatrix} = \begin{pmatrix} 0 \\ 2\pi^{-1/2}\sin 2m\theta \end{pmatrix}, \quad \lambda'_m = -4m^2, \quad m = 0, 1, \ldots,$$

and the other is

$$\begin{pmatrix} \mathcal{A}_n \\ \mathcal{B}_n \end{pmatrix} = C \begin{pmatrix} \cos(\sqrt{-\lambda_n}\theta) \\ 0 \end{pmatrix}, \quad \sqrt{-\lambda_n}\tan\left(\pi\sqrt{-\lambda_n}/2\right) = i\xi,$$

where C is the normalization constant. The transcendental equation for λ_n can be solved in approximate manner for $|\xi| \gg 1$ with the *ansatz* $\sqrt{-\lambda_n} = 1 - x$ with $|x| \ll 1$ to yield $\sqrt{-\lambda_n} \approx 1 + 2i/(\pi\xi) + 4n$, or

$$\lambda_n \approx -(1+4n)^2 - \frac{4i}{\pi\xi}(1+4n), \quad n = 0, 1, \ldots.$$

The normalization $\int_0^{\pi/2} |\mathcal{A}_n|^2 \, d\theta = 1$ then yields $C \approx 2/\sqrt{\pi}$.

To lowest order, the perturbed solution that contains both azimuthal and meridional components has the form

$$\begin{pmatrix} A \\ B \end{pmatrix} = C_0 \begin{pmatrix} A_0 \\ 0 \end{pmatrix} + C_1 \begin{pmatrix} 0 \\ B_1 \end{pmatrix},$$

which yields, for the leading eigenfunction,

$$\gamma \approx \lambda_0 - \epsilon^2 \frac{V_{01} V_{10}}{(\lambda_0 - \lambda_1')^2},$$

where

$$V_{01} \approx \tfrac{1}{2}, \qquad V_{10} \approx -\frac{2}{\pi} \left(1 + \frac{2i}{\pi \xi} \right).$$

Thus,

$$\gamma \approx -1 - \frac{4i}{\pi \xi} - \frac{1}{9\pi} D,$$

so that the critical value of the dynamo number, corresponding to $\mathrm{Re}\,\gamma = 0$, is $D_c \approx -9\pi$. The cycle frequency of this solution, $\omega = 4/(\pi \xi)$, is controlled by the magnitude of ξ. In terms of dimensional variables, to obtain the cycle period $T = 22\,\mathrm{yr}$, we need $\xi = 2T\eta_t/(\pi H)^2 \approx 0.3$, where $\eta_t = 10^{12}\,\mathrm{cm}^2\,\mathrm{s}^{-1}$ is the turbulent magnetic diffusivity in the Solar convection zone and $H = 0.2R_\odot$ is the thickness of the convection zone. We note that the value of ξ required to fit the period of the dynamo cycle is not much larger than one as assumed when deriving the above solution.

6.3. Diffusion in mean-field dynamos

Integrating the thin-disc dynamo equations (6.2) and (6.3), written in the dimensional form, over the interval $0 < z < h$ for a smooth function $\alpha(z)$ gives:

$$\frac{\partial}{\partial t} \int_0^h \bar{B}_r \, dz = \eta_t \frac{\partial \bar{B}_r}{\partial z}(h), \tag{6.21}$$

$$\frac{\partial}{\partial t} \int_0^h \bar{B}_\phi \, dz = G \int_0^h \bar{B}_r \, dz + \eta_t \frac{\partial \bar{B}_\phi}{\partial z}(h), \tag{6.22}$$

where $G = r\,d\Omega/dr$ and we have used the quadrupolar symmetry conditions (6.6). It is notable that α does not enter the integrated equations because $\alpha(0) = 0$ and $\bar{B}_\phi(1) = 0$.

The integral form of the dynamo equations (6.21) and (6.22) highlights the role of magnetic diffusivity in the dynamo mechanism. It would seem at first glance that magnetic diffusion can be neglected for the growing solutions. However, setting $\eta_t = 0$ in (6.21) (more precisely, supposing that both the turbulent and Ohmic diffusivities vanish), results in $\int_0^h \bar{B}_r \, dz = \text{const}$, and then (6.22) shows that $\int_0^h \bar{B}_\phi \, dz$ can grow only linearly in time. In other words, the solution cannot grow exponentially if $\eta_t = 0$ (and thus the dynamo action is impossible). In the more general case where η_t varies with z, the dynamo action requires that $\eta_t(h) \neq 0$.

Consider a function $\bar{B}_\phi(z)$ that has no zeros and, say, $\bar{B}_\phi > 0$ for $0 < z < h$. It is expected that such a function corresponds to the lowest excited mode. Since $\bar{B}_\phi(h) = 0$, this implies that

$$\frac{\partial \bar{B}_\phi}{\partial z}(h) < 0.$$

Then Eqs (6.21) and (6.22) imply, for $G < 0$, that any growing even solution must satisfy the following inequalities:

$$\int_0^h \bar{B}_r \, dz < 0, \qquad \eta_t \frac{\partial \bar{B}_r}{\partial z}(h) < 0. \tag{6.23}$$

Hence, the radial component of a growing magnetic field must change its sign near the disc surface.

The second inequality of (6.23) shows that the dynamo action requires non-vanishing flux of the radial magnetic field across the disc surface. The role of diffusion can also be seen directly from the dynamo equations. For definiteness, assume again that $G < 0$, $B_\phi > 0$, and, near the symmetry plane, $B_r < 0$. The α-effect generates, via the term $-\partial(\alpha \bar{B}_\phi)/\partial z$, a positive radial field, i.e., a radial field opposite to that of the growing solution near the symmetry plane. The positive $\bar{B}_r$ near the surface produces, through differential rotation $-|G|\bar{B}_r$, a negative contribution to $\partial \bar{B}_\phi/\partial t$, which can be compensated only by the viscous term, $\eta_t \partial^2 \bar{B}_\phi/\partial z^2$. In order to provide such a compensation, the latter term must be positive near the disc surface, i.e., the field must be transported outwards. It also becomes clear that, in addition to conditions (6.23), $\bar{B}_\phi(z)$ must have an inflection point at somewhat smaller z than the zero of $\bar{B}_r$.

On the other hand, the diffusivity should not be excessively large, otherwise the field would rapidly decay within the main part of the disc and would be carried out toward the disc surface.

The discussion above referred to the growing solutions. In the stationary case ($\partial/\partial t = 0$), we have

$$\eta_t \frac{\partial \bar{B}_r}{\partial z} = 0, \qquad \eta_t \frac{\partial^2 \bar{B}_\phi}{\partial z^2} = 0 \quad \text{at } z = h,$$

i.e., both horizontal (parallel to the disc midplane) field components have fixed (and opposite) signs within the disc.

The forms of $\bar{B}_r(z)$ and $\bar{B}_\phi$ shown in Fig. 2 for a large, moderate and critical value of $|D|$ illustrate these properties of the growing and marginal solutions of the dynamo equations.

7. Turbulent magnetic fields in galaxies and galaxy clusters

7.1. The fluctuation dynamo

The evolution of a magnetic field embedded into a flow of conducting fluid is controlled by the magnetic Reynolds number,

$$R_{\mathrm{m}} = \frac{v_0 l_0}{\eta},$$

where v_0 and l_0 are the representative velocity and scale in the flow and η is the turbulent magnetic diffusivity. In turbulent flows without any mean velocity, it is convenient to choose l_0 and v_0 as the integral scale and velocity at that scale (these quantities were denoted l and v above; in this section, we label them with subscript zero). The larger is R_{m}, the better is the magnetic flux freezing approximation, i.e., the better magnetic lines follow the fluid particles.

The generation of a random magnetic field by a random flow, called the *fluctuation dynamo*, is a result of a random stretching of magnetic field by the local velocity shear (see reviews in [9, 38, 56]).[1] This type of dynamo does not require any mean flow, rotation or helicity, but only needs a random flow. Magnetic field produced by the fluctuation dynamo is purely random, i.e., its mean value vanishes. The root-mean-square magnetic field (or, equivalently, mean magnetic energy density) can grow under a fairly weak condition $R_{\mathrm{m}} > R_{\mathrm{mc}} \simeq 30\text{--}100$ (where the variation within the range depends on the form of the velocity correlation function).

A turbulent flow consists of a broad spectrum of motions, with v_l the velocity at a scale l. The e-folding time for the magnetic field is roughly equal to the 'eddy turnover time' l/v_l. In the Kolmogorov turbulence, where $v_l \propto l^{1/3}$, the e-folding time is shorter at smaller scales, $l/v_l \propto l^{2/3}$, and so smaller eddies amplify the field faster. As a result, most of the magnetic energy produced by the fluctuation dynamo at its *kinematic* stage (i.e., the stage of exponential growth) is at small scales comparable to the magnetic dissipation scale of order $R_{\mathrm{m}}^{-1/2}$. At the kinematic stage, the root-mean-square magnetic field grows as $b \propto \exp \sigma t$

---

[1] This type of dynamo is also called the *small-scale dynamo,* with reference to the fact that the scale of the magnetic field does not exceed l_0.

with $\sigma \simeq \frac{2}{3}(v_0/l_0)\ln(R_m/R_{mc})$ for $R_m \gg 1$ [30]; numerical results show that this form is quite accurate for R_m of the order of or even smaller than R_{mc} [55].

Since $\eta \ll \nu$ in rarefied astrophysical plasmas, where ν is the kinematic viscosity (see, e.g., [9]), we have $R_m \gg \text{Re}$, where $\text{Re} = v_0 l_0/\nu$ is the kinematic Reynolds number. Therefore, $R_m > R_{mc}$ if Re is large enough. Turbulent systems necessarily have large Re, and random motions in galaxies and galaxy clusters will be a fluctuation dynamo for any Reynolds number which is large enough to make them turbulent.

The fluctuation dynamo is sensitive to the value of the magnetic Prandtl number, $\text{Pr}_m = \nu/\eta = R_m/\text{Re}$ [36]. For $\text{Pr}_m > 1$ (intergalactic and interstellar gas) magnetic spectrum extends to smaller scales than the kinetic energy spectrum, whereas for $\text{Pr}_m < 1$ the Ohmic dissipation scale is larger than the viscous scale. The dynamo action for $\text{Pr}_m \geq 1$ has been demonstrated convincingly with various analytical and numerical models. For $\text{Pr}_m < 1$, the dynamo action is also possible but requires larger values of R_m than for $\text{Pr}_m > 1$ [19]. The situation at very small values of Pr_m remains unclear, but asymptotic results obtained for a δ-correlated velocity field suggest that $R_{mc} \to 400$ for $\text{Pr}_m \to 0$ [30].

In the kinematic regime, the fluctuation dynamo produces intermittent magnetic field: the size of the magnetic structures is, in at least one dimension, as small as the resistive scale

$$l_\eta = l_0 R_m^{-1/2} \qquad (7.1)$$

in a single-scale flow [56]. We note that magnetic field at the small Ohmic diffusion scale l_η is produced by the shear of the flow at a larger scale l_0. In a turbulent flow, where a broad spectrum of motions is present, flow at each scale l would produce magnetic structures at scales down to the corresponding Ohmic scale. In the kinematic regime this would correspond to a set of eigenfunctions, each with a distinct growth rate v_l/l. The fastest growing eigenfunction is due to stretching by the smallest eddies with scale l such that $R_m(l) > R_{mc}$, where $R_m(l) = R_m(l/l_0)^{3/4}$ for the Kolmogorov spectrum. These are the viscous scale eddies, with $l = l_\nu = l_0\text{Re}^{-3/4}$, provided $R_m/\text{Re} > R_{mc}$. However in the nonlinear regime, when the fastest growing mode saturates, magnetic modes of larger scales still can grow. Since most of the kinetic energy is contained at the scale l_0, the dominant magnetic scale could still be determined by dynamo action due to eddies of scale l_0 and, especially, by the subtle details of the dynamo saturation. An estimate of the scale of magnetic structures similar to (7.1) but now with allowance for a broad flow spectrum can be obtained from the balance of the stretching and dissipation terms in the induction equation. With l_B the scale of magnetic field, this balance yields $|(\boldsymbol{B} \cdot \nabla)\boldsymbol{v}| \simeq |\eta\nabla^2\boldsymbol{B}|$, or $l_B v(l_B) \simeq \eta$ provided $l_B > l_\nu$ (this inequality may hold also for $\text{Pr}_m > 1$ in the nonlinear state). In a

flow with kinetic energy spectrum

$$E(k) \propto k^{-s} \tag{7.2}$$

(with $s = 5/3$ corresponding to the Kolmogorov spectrum), we have $v^2(l) = k^{-1}E(k)$, so that $v(l) = v_0(l/l_0)^{(s-1)/2}$. This leads to

$$l_B \simeq l_0 R_m^{-2/(s+1)}. \tag{7.3}$$

Nonlinear effects can modify the resulting magnetic structures, although it is as yet not clear in what way [9, 38]). A simple model of Subramanian [49] suggests that the smallest scale of the magnetic structures will be renormalized in the saturated state to become

$$l_B \simeq l_0 R_{mc}^{-2/(s+1)}, \tag{7.4}$$

instead of the resistive scale l_η.

7.2. Shapefinders

Magnetic field produced by a (kinematic) fluctuation dynamo [55] is illustrated in Fig. 3. The velocity field used in this model has a well defined and controllable power-law range (7.2) and allows us to test the theoretical predictions, such as (7.1), (7.3) and (7.4). The structures generated by the fluctuation dynamo are

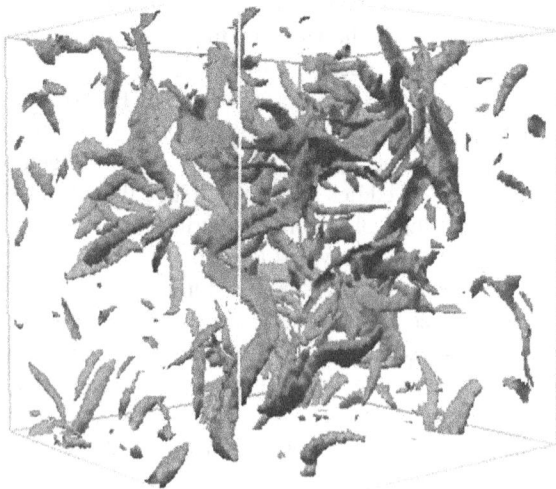

Fig. 3. Isosurfaces $B^2 = 3\langle B^2 \rangle$ from a kinematic fluctuation dynamo model [55].

evidently elongated; they were variously described as magnetic filaments, sheets or ribbons from the visual inspection of magnetic isosurfaces and application of heuristic morphology indicators. A mathematically justifiable approach to the morphology of random magnetic structures based on the *Minkowski functionals* was employed in Ref. [55]. This tool has previously been applied to galaxy distribution and cosmological structure formation [25, 33, 40].

As an example, consider statistical properties of the isosurfaces $B^2 = \text{const}$ similar to those shown in Fig. 3. The topological and geometrical properties of structures in three dimensions can be fully quantified using the four Minkowski functionals [25]:

$$V_0 = \iiint dV, \qquad\qquad V_1 = \frac{1}{6} \iint dS,$$

$$V_2 = \frac{1}{6\pi} \iint (\kappa_1 + \kappa_2)\, dS, \qquad V_3 = \frac{1}{4\pi} \iint \kappa_1 \kappa_2\, dS, \qquad (7.5)$$

where integration is over the volume and surface of the structures, respectively, and κ_1 and κ_2 are the principal curvatures of the surface. V_0 is the total volume enclosed by the structures, V_1 is their surface area, V_2 is the integral mean curvature of their surfaces, and V_3 is the integral Gaussian curvature (related to the Euler characteristic). A simple method to compute the Minkowski functionals for structures given on a grid is based on the intersection formula of Crofton [39]:

$$V_0 = n_3, \qquad\qquad V_1 = \frac{2(n_2 - 3n_3)}{9N},$$

$$V_2 = \frac{2(n_1 - 2n_2 + 3n_3)}{9N^2}, \qquad V_3 = \frac{n_0 - n_1 + n_2 - n_3}{N^3},$$

where n_0 is the number of grid vertices within the structures, n_1 is the number of complete edges, n_2 is the number of complete grid cell faces, n_3 is the number of complete grid cubes within the structures, and N is the total number of grid points in the domain.

The Minkowski functionals can be used to calculate the typical thickness, width and length of the structures, as $T = V_0/2V_1$, $W = 2V_1/\pi V_2$, and $L = 3V_2/4V_3$, respectively. Then, useful dimensionless measures of 'planarity' P, and 'filamentarity' F can be defined as [33]

$$P = \frac{W - T}{W + T}, \qquad F = \frac{L - W}{L + W}.$$

In idealized cases and for convex surfaces, values of P and F lie between zero and unity. For example, an infinitely thin pancake has $(P, F) = (1, 0)$, a perfect filament has $(P, F) = (0, 1)$, whereas $(P, F) = (0, 0)$ for a sphere. We note that the unit cube has $T = 3/4$, $W = 2/\pi$, $L = 1/2$, thus we do not always have $T < W < L$. Deviations from this ordering are relatively rare for random fields studied here, yet to avoid confusion we introduce the notation $l_1 = \min(T, W, L)$, $l_2 = \text{med}(T, W, L)$ and $l_3 = \max(T, W, L)$.

As discussed in Ref. [55], the filamentarity of magnetic structures produced by the fluctuation dynamo, F, increases faster than P with R_m, so the structures are better described as filaments, especially at the larger values of R_m— see Eq. (7.6) below. Remarkably, velocity field structures of that model are not filamentary since the velocity field at small scales is nearly isotropic. Correspondingly, the isosurfaces of v^2 have negligible planarity and filamentarity. The isosurface of vorticity $\mathbf{\Omega}$ with $\Omega^2 = 4\langle\Omega^2\rangle$ has $(P, F) = (0.18, 0.11)$; similarly the isosurface of the total strain ($S^2 = S_{ij}S_{ij}$) with $S^2 = 4.5\langle S^2\rangle$ has $(P, F) = (0.11, 0.16)$. Thus, the morphology of the magnetic field is controlled by the nature of the dynamo action rather than by immediate features of the velocity field. The isosurfaces of the electric current density $\mathbf{J} = \nabla \times \mathbf{B}$ are ribbon-like, with $(P, F) = (0.57, 0.82)$ at a level $J^2 = 4\langle J^2\rangle$ for $R_m \approx 1500$.

Using the Minkowski functionals, we can also reliably measure the characteristic length scales of magnetic structures and explore their scalings with R_m and s. In the left panel of Fig. 4, we show l_1 versus R_m at two instants in time for a flow with $s = 5/3$. Whilst the behavior for $R_m \ll R_{mc}$ shows variations in time, we observe for $R_m \gtrsim 200$ a time-independent scaling of the thickness of magnetic structures: $l_1 \propto R_m^{-3/4}$ in agreement with Eq. (7.3).

The right panel of Fig. 4 shows variation with R_m of the characteristic width l_2 and length l_3 of the magnetic structures. For $R_m \gtrsim 200$ we observe another

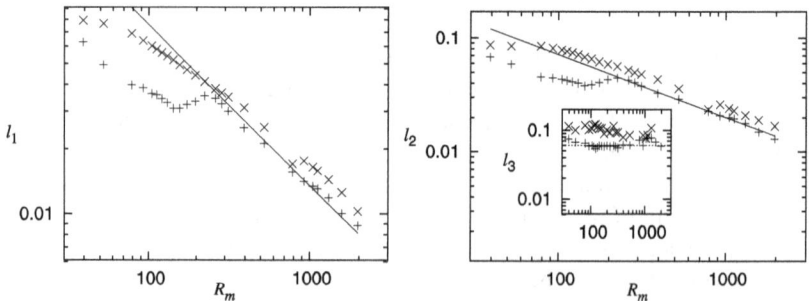

Fig. 4. Average values of the thickness l_1 (left panel), width l_2 (main frame of the right panel) and length l_3 (inset) of the magnetic isosurfaces versus R_m at two instants in time, from the fluctuation dynamo model of Ref. [55]. Data marked '×' ('+') are results attained when the smallest eddies have made 25 (50) revolutions. The solid straight line has a slope of $-3/4$ in the left panel and -0.55 in the right one. The unit length is the size of the computational box. The data were obtained by averaging over eight isosurface families $B^2 = q\langle B^2\rangle$ with $q = 1.5, 2, 2.5, \ldots, 5$.

time-independent scaling, $l_2 \propto R_{\mathrm{m}}^{-0.55}$. This distinct behavior of the width of magnetic structures has not been obtained in earlier analytical or numerical studies of the fluctuation dynamo. The simultaneous decrease of l_2 and l_1, coupled with the approximately R_{m}-independent behavior of l_3 (inset in the right panel of Fig. 4) supports the notion that the magnetic structures become filamentary as R_{m} increases. Indeed, using $l_1 \simeq 2.4 R_{\mathrm{m}}^{-0.75}$, $l_2 \simeq 0.9 R_{\mathrm{m}}^{-0.55}$ and $l_3 \simeq 0.05 R_{\mathrm{m}}^{0}$ in the definitions of P and F, we obtain, for $s = 5/3$,

$$P \simeq 1 - 2\left[1 + \tfrac{3}{8} R_{\mathrm{m}}^{0.2}\right]^{-1}, \qquad F \sim 1 - 2\left[1 + \tfrac{1}{18} R_{\mathrm{m}}^{0.55}\right]^{-1}, \qquad (7.6)$$

so that $F > P$ for $R_{\mathrm{m}} \gtrsim 200$.

To investigate how the scaling laws identified via the Minkowski functionals compare with those inferred from other measures, we calculated the inverse 'integral scale' of the magnetic field, $2\pi/l_I = \int k M(k)\, dk / \int M(k)\, dk$, where $M(k)$ is the magnetic spectral density defined similarly to $E(k)$. We found that l_I follows a scaling of $R_{\mathrm{m}}^{-0.42}$, which understandably differs from the scalings of l_1, l_2 and l_3. The scale l_I is a poor measure of the dimension of anisotropic magnetic structures such as filaments. We note that the above scaling of l_I is maintained for *all* subcritical and supercritical values of R_{m}, unlike the results illustrated in Fig. 4 which display well-defined, time-independent scalings only for $R_{\mathrm{m}} \gtrsim 200$. The scaling (7.3) emerges for $s = 5/3, 2, 3$ [55]. On the contrary, $l_2 \sim R_{\mathrm{m}}^{-0.55}$ independently of s. Asymptotic solutions [30] suggest that the small-scale dynamo (with $\mathrm{Pr_m} < 1$) is only possible for $s > 3/2$. The results of Ref. [55] show that, for high effective $\mathrm{Pr_m}$, the dynamo action is possible for $s = 1$ as well, although a scaling different from Eq. (7.3) is exhibited. It appears that the nature of the

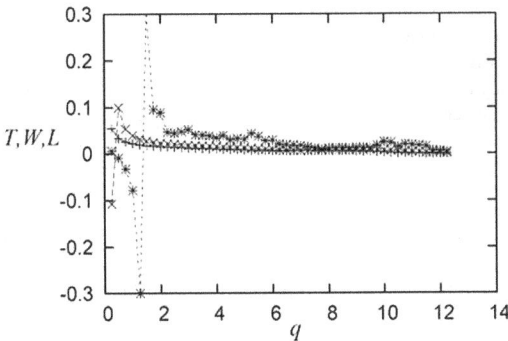

Fig. 5. T, W and L (lines marked with +, × and *, respectively) as a function of the isosurface level, $B = q\langle B^2 \rangle^{1/2}$ for a dynamo-generated field ($R_{\mathrm{m}} = 1300$). Calculations for $q < 1$ generate anomalous results, whereas the range $1.5 < q < 5$ appears to be acceptable for generating the averages as we discuss in the text.

asymptotic solution, rather than the possibility of a dynamo, is different at $s = 1$ from that at $s > 3/2$.

7.3. Turbulence in galaxy clusters: three evolutionary stages

7.3.1. The epoch of major mergers

Theories of hierarchical structure formation suggest that clusters of galaxies have been assembled relatively recently. N-body simulations indicate that the clusters form at the intersection of dark matter filaments in the large-scale structure, and result from both major mergers of objects of comparable mass (of order $10^{15} M_\odot$) and the accretion of smaller clumps onto massive protoclusters. It is likely that intense random vortical flows, if not turbulence, are produced in the merger events. Their plausible properties are summarized in Table 2 (see Ref. [50] for details). The structure of magnetic field at this stage is illustrated in the left-hand panel of Fig. 6. What is shown is the statistically steady state of magnetic field produced by dynamo action in turbulent flow with the Reynolds number about 400 and the magnetic Prandtl number equal to unity. Similar magnetic structure plausibly occur in the turbulent wakes of subclusters and galaxies as well.

It is not quite clear whether random flows driven during major merger events and at later stages of evolution will develop into turbulence. The nature of the flow depends on the value the Reynolds number which is difficult to estimate reliably for the collisionless, magnetized plasma of the intracluster space where plasma instabilities can be responsible for anomalous viscosity and resistivity [37]. The problem is further complicated by the possibility of dynamo action, since the magnetic field can affect both viscosity and magnetic diffusivity. This

Table 2

Summary of turbulence and magnetic field parameters at various stages of cluster evolution: duration of the stage (the last two stages represent steady states), the root-mean-square velocity v_0 and scale l_0 of turbulence and eddy turnover time $t_0 = l_0/v_0$ (for the decaying turbulence, values for the middle of the decay stage are given, 2 Gyr after its start), the equipartition magnetic field $B_{eq} = (4\pi\rho v_0^2)^{1/2}$ with ρ the gas density (i.e., maximum field strength within a turbulent cell), thickness of magnetic filaments and sheets l_B for the statistically steady state of the dynamo, the root-mean-square magnetic field within a turbulent cell B_{rms}, and finally the standard deviation of the Faraday rotation measure σ_{RM} (calculated for the volume filling turbulence along path length of 750 kpc through the central parts of a cluster in the first two lines, and assuming one transverse wake along the line of sight in the last two lines). A subcluster mass of $3 \times 10^{13} M_\odot$ has been assumed

Evolution stage	Length (Gyr)	v_0 (km/s)	l_0 (kpc)	t_0 (Gyr)	B_{eq} (μG)	l_B (kpc)	B_{rms} (μG)	σ_{RM} (rad/m^2)
Major mergers	4	300	150	0.5	4	25	1.8	200
Decaying turbulence	5	130	260	2.0	2	44	0.8	120
Subcluster wakes		260	200	0.8	4	34	1.6	110
Galactic wakes		300	8	0.03	4	1.4	1.6	5

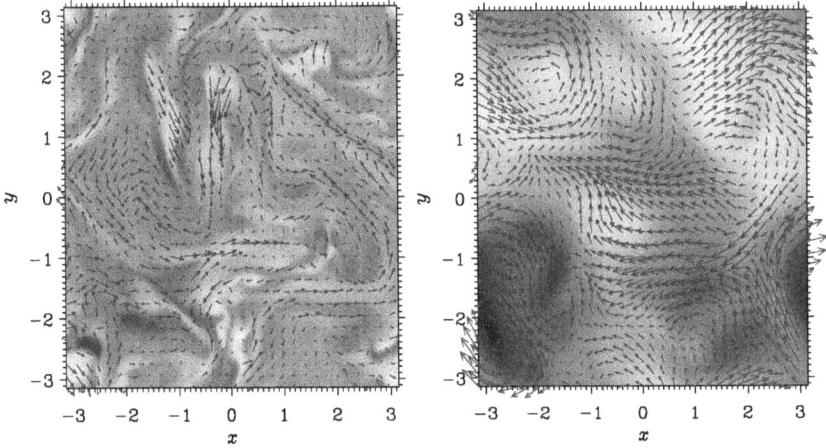

Fig. 6. Snapshots of magnetic field in a cross-section through the middle of the computational do-main in a numerical simulation of turbulence driven by an imposed random force and its dynamo action [50]. The left-hand panel shows a statistically steady state at a time $t/t_{0i} = 0.30$ whereas the right-hand panel illustrates magnetic field structures in turbulence at a late stage of decay, $t/t_{0i} \approx 60$. Here t_{0i} is the eddy turnover time before the start of the turbulence decay (given in Table 2). The dimensionless energy injection scale in these simulations is about 4 (with the domain size of 2π), so each frame contains a few turbulent cells. The strongest magnetic field within the frame is close to the equipartition value with respect to the turbulent energy. The magnitude of the field component perpendicular to the plane of the figure is shown color coded (in shades of grey) with black corre-sponding to field pointing into the figure plane, and lighter shades, to field pointing out of the plane. The field in the plane of the figure is shown with vectors whose length is proportional to the field strength.

may lead to the growth of the magnetic diffusivity and reduction of viscosity as the magnetic field is being amplified by the dynamo, so that the magnetic Prandtl number tends to unity.

7.3.2. Decaying turbulence

Random flows produced by major mergers decay after the end of the merger event. Unlike a laminar flow that decays exponentially in time due to viscosity, turbulent kinetic energy decays slower, as a power law [14, 22]. The reason for this is that kinetic energy mainly decays at small scales, to where it is constantly supplied by the turbulent cascade. As a result, the energy decay rate depends nonlinearly on the energy itself, which makes the decay a power law in time. Our simulations confirm that the power-law decay occurs even for the Reynolds number as small as Re $\approx$ 100 [50]. At this stage of evolution, the turbulent scale l_0 grows with time, whereas turbulent energy density E reduces, together with the turbulent speed v_0, typically as

$$E \simeq \tfrac{1}{2} v_0^2 \propto (t/t_{0i})^{-6/5}, \quad l_0 \propto (t/t_{0i})^{2/5} \quad \text{for } t/t_{0i} \gg 1,$$

where subscript 'i' refers to the start of the evolution, t_{0i} is a certain dynamical time scale, which can be identified with the initial turnover time of the energy-containing eddies, $t_{0i} = l_{0i}/v_{0i}$, subscript '0' refers to the energy-range (correlation) scale of the motion. The structure of magnetic field in the decaying flow is shown in the right-hand panel of Fig. 6, and parameters of the flow and magnetic field are shown in the second line of Table 2.

7.3.3. Turbulent wakes of subclusters and galaxies

At the final stage of the evolution, when the cluster enters a steady state, turbulence is maintained only in the wakes of galaxies and smaller mass clumps that continue to accrete onto the cluster. The wakes become weaker as the gas within the clumps or galaxies is stripped by the ram pressure of intracluster gas. The radius of a wake at its head is close to the radius within which gas of the mass clump or galaxy remains intact. We estimate the stripping radius as $R_0 \simeq 100 \, \mathrm{kpc}$ for clumps of a mass $10^{13} M_\odot$ (which fall into a cluster every 3 Gyr) and $R_0 = 3\text{–}5 \, \mathrm{kpc}$ for massive elliptical galaxies. If the flow within the wake becomes turbulent (so that it can be described in terms of Prandtl's theory of turbulent wakes [22]), the wake length X is controlled by the magnitude of the Reynolds number via

$$X/R_0 \simeq (\mathrm{Re_i}/\mathrm{Re_c})^3,$$

where $\mathrm{Re_c} \approx 400$ [52] is the marginal Reynolds number with respect to the onset of turbulence. This value of $\mathrm{Re_c}$ was obtained for a flow around a solid sphere; $\mathrm{Re_c}$ for gas spheres is not known. The strong dependence of the wake parameters on the Reynolds number makes the estimates somewhat uncertain. On the other hand, it implies that galactic wakes can be very sensitive to the detailed parameters of the galactic motion and intergalactic gas, so that clusters with very similar parameters can have vastly different wake structures.

The area covering and volume filling factors, f_S and f_V, respectively, of $N = 5$ wakes, produced by $10^{13} M_\odot$ subclusters, within the virial radius $r \approx 3 \, \mathrm{Mpc}$, are estimated as

$$f_S \simeq 0.15 \, \frac{N}{5} \left(\frac{R_0}{100 \, \mathrm{kpc}} \right)^6 \left(\frac{\mathrm{Re_c}}{400} \right)^{-4} \left(\frac{\tilde{\lambda}}{1 \, \mathrm{kpc}} \right)^{-4},$$

$$f_V \simeq 0.02 \, \frac{N}{5} \left(\frac{R_0}{100 \, \mathrm{kpc}} \right)^8 \left(\frac{\mathrm{Re_c}}{400} \right)^{-5} \left(\frac{\tilde{\lambda}}{1 \mathrm{kpc}} \right)^{-5},$$

where $N \approx 5$ is consistent with models of hierarchical structure formation, and $\tilde{\lambda}$ is an effective mean free path in the intracluster gas (its introduction is an attempt to allow for our insufficient understanding of viscosity mechanisms). The covering and filling factors strongly depend on $\tilde{\lambda}$ and $\mathrm{Re_c}$. Furthermore, both f_S and f_V depend on high powers of another poorly known parameter, the stripping radius R_0. Hence, as noted above, properties of the subcluster wakes can be rather different in apparently similar clusters. In addition, numerical simulations of turbulent wakes should be treated with caution as otherwise reasonable approximations, numerical resolution, and numerical viscosities can strongly affect the results. On the other hand, it is plausible that $f_S = O(1)$ but $f_V \ll 1$, so that a typical line of sight through the cluster intersects at least one turbulent region (where our estimate of the r.m.s. turbulent speed is 200–300 km s^{-1}) despite the fact that turbulence occurs only in a small fraction of the cluster volume. A possible signature of such spatially intermittent turbulence could be a specific shape of spectral lines, with a narrow core, produced in quiescent regions, accompanied by nonthermally broadened wings.

The area covering factor of galactic wakes within the gas core radius, 180 kpc, is unity if

$$X/R_0 \simeq 30\text{–}15, \quad X \simeq 100\text{–}70\,\mathrm{kpc}, \tag{7.7}$$

and the volume filling factor of such wakes is $f_V \simeq 0.07$. The length of galactic wakes required to cover the projected cluster area, given by Eq. (7.7), does not seem to be unrealistic. For example, a wake has been observed behind a massive elliptical galaxy (mass of order $2 \times 10^{12} M_\odot$) moving through the intracluster gas at a speed about $v_c \simeq 1000$ km s^{-1} [34]. The length of the detectable wake is about $X \simeq 130$ kpc (assuming that it lies in the sky plane), and its mean radius is 40 kpc (obtained from the quoted volume of about 2×10^6 kpc^3). The projected area of the wake is about 10^4 kpc^2, as compared to 10^3 kpc^2 for the wake parameters derived above. This wake has been detected only because it is exceptionally strong, and it is not implausible that weaker but more numerous galactic wakes can cover the projected area of the central parts of galaxy clusters.

We conclude that subcluster wakes are likely to be turbulent, but galactic wakes can be laminar if the viscosity of the intracluster gas is as large as Spitzer's value. Given the uncertainty of the physical nature (and hence, estimates) of the viscosity of the magnetized intracluster plasma, we suggest that turbulent galactic wakes remain a viable possibility. Both types of wake have low volume filling factor but can have an area covering factor of order unity. Parameters of turbulence and magnetic fields produced within the wakes are given in the last two lines of Table 2.

7.4. Magnetic fields in the intracluster gas

Parameters of magnetic fields produced by the fluctuation dynamo at various stages of the cluster evolution are presented in Table 2 [50]. This model implies that the correlation scale of random motions in the intracluster gas, l_0, is larger than that assumed earlier. With $l_0 \simeq 150\,\text{kpc}$ (Table 2), only 5 turbulent cells occur along a path length of $L = 750\,\text{kpc}$. The resulting degree of polarization of radio emission from clusters with synchrotron halos can be estimated as $p \simeq \frac{1}{2}p_0/n^{1/2} \simeq 0.2$, where $p_0 \approx 0.7$, $n \simeq L/l_0$ is the number of magnetic structures along the line of sight (assuming that one magnetic sheet with well-ordered magnetic field occurs in each turbulent cell and that the linear resolution is better than l_0), and a factor $1/2$ allows, in a very approximate manner, for the volume-filling magnetic field outside the magnetic sheet which only produces unpolarized emission. Depolarization by Faraday dispersion and beam depolarization can reduce the degree of polarization to a fraction of percent at long wavelengths. However, polarization observations at wavelengths 3–6 cm (where Faraday depolarization is sufficiently weak) can reveal magnetic structures produced by the dynamo action if the angular resolution is high enough.

7.5. Interstellar turbulent magnetic fields

Using parameters typical of the warm phase of the ISM, theory of Sect. 7.1 predicts that the small-scale dynamo would produce magnetic flux ropes of the length (or the curvature radius) of about $l_0 = 50$–$100\,\text{pc}$ and thickness 3–5 pc from Eq. (7.4) for $R_{\text{mc}} = 50$ and $s = 5/3$. The volume filling factor of the ropes is $f \simeq l_0 l_B^2/l_0^3 \simeq R_{\text{mc}}^{-3/2} \simeq 3\%$ assuming that there is one flux rope per turbulent cell, and $3n\%$ if there are n ropes. The field strength within the ropes, if at equipartition with the turbulent energy, has to be of order $1.5\,\mu\text{G}$ in the warm phase ($n = 0.1\,\text{cm}^{-3}$, $v_0 = 10\,\text{km s}^{-1}$) and $0.5\,\mu\text{G}$ in the hot gas ($n = 10^{-3}\,\text{cm}^{-3}$, $v_0 = 40\,\text{km s}^{-1}$). Note that some heuristic models of the small-scale dynamo admit solutions with magnetic field strength within the ropes being significantly above the equipartition level, e.g., because the field configuration locally approaches a force-free one, $|(\nabla \times \boldsymbol{B}) \times \boldsymbol{B}| \ll B^2/l_B$ [6].

The small-scale dynamo is not the only mechanism producing random magnetic fields (e.g., §4.1 in Ref. [4], and references therein). Any mean-field dynamo action producing magnetic fields at scales exceeding the turbulent scale also generates small-scale magnetic fields. Similarly to the mean magnetic field, this component of the turbulent field presumably has a filling factor close to unity in the warm gas and its strength is expected to be close to equipartition with the turbulent energy at all scales. This component of the turbulent magnetic field may be confined to the warm gas, the site of the mean-field dynamo action, so magnetic field in the hot phase may have a better pronounced ropy structure.

The overall structure of the interstellar turbulent magnetic field in the warm gas can be envisaged as a quasi-uniform fluctuating background with one percent of the volume occupied by flux ropes (filaments) of a length 50–100 pc containing a well-ordered magnetic field. This basic distribution would be further complicated by compressibility, shock waves, MHD instabilities (such as Parker instability), the fine structure at subviscous scales, etc.

The site of the mean-field dynamo action is plausibly the warm phase rather than the other phases of the ISM. The warm gas has a large filling factor (so it can occupy a percolating global region), it is, on average, in a state of hydrostatic equilibrium, so it is an ideal site for both the small-scale and mean-field dynamo action. Molecular clouds and dense clouds of neutral hydrogen have too small a filling factor to be of global importance. The time scale of the small-scale dynamo in the hot phase is $l_0/v_0 \simeq 10^6$ yr for $v_0 = 40 \, \text{km s}^{-1}$ and $l_0 = 0.04 \, \text{kpc}$ (the width of the hot, 'chimneys' extended vertically in the disc). This can be shorter than the advection time due to the vertical streaming of the hot gas in the galactic fountain flow, $h/U_z \simeq 10^7$ yr with $h = 1 \, \text{kpc}$ and $U_z = 100 \, \text{km s}^{-1}$. Therefore, the small-scale dynamo action should be possible in the hot gas. However, the growth time of the mean magnetic field must be significantly longer than l_0/v_0, reaching a few hundred Myr. Thus, the hot gas can hardly contribute significantly to the mean-field dynamo action in the disc and can drive the dynamo only in the halo [45]. The main rôle of the fountain flow in the disc dynamo is to enhance magnetic connection between the disc and the halo.

8. Conclusions

A remarkable property of systems with high electric conductivity (or large magnetic Reynolds number R_m) is that the decay time of magnetic field due to Ohmic resistivity can be very long. Since astrophysical plasmas usually have extremely large values of R_m, the Ohmic decay time often exceeds the age of the Universe. Does this make dynamos unnecessary? We believe that the answer to this question is negative because astrophysical plasmas are most often turbulent. A fundamental property of turbulence is the energy cascade to small scales (in three dimensions). If magnetic field is weak and $R_m \gg 1$, the turbulent motions will inevitably tangle magnetic fields and the magnetic energy will be transferred from the energy-range scale l to small scales, where Ohmic dissipation is rapid. (In systems with large magnetic Prandtl number, this will be the viscous dissipation scale, where the magnetic Reynolds number can still be large; however, this scale is by far smaller than that of observable astrophysical magnetic fields. If the initial magnetic field has a scale much larger than l, it will be reduced to l at the relatively short turbulent diffusion time.) The time scale of magnetic field decay is then controlled by the cascade time, i.e., eddy turnover time. If, on the other hand, magnetic field is strong enough, any externally maintained turbulence is not needed as the Lorentz force will induce motions which will become turbulent if the magnetic field is non-homogeneous enough. The magnetically-induced turbulence will then drain its parent magnetic field by dissipating its energy in few eddy turnover times as above. Altogether, any three-dimensional, turbulent, magnetized system must host a dynamo (unless its magnetic field is maintained by external electric currents or decays). Indeed, turbulent flows can drive the large- and small-scale dynamos, but magnetic fields produced by them are controlled by the dynamo mechanism rather than by the initial magnetic field. In this sense, the properties of the initial magnetic field in a turbulent system are unimportant as long as it can provide a suitable seed for the dynamo. In other words, initial conditions are forgotten in a dynamo system (as in any other unstable system) unless the initial magnetic field is strong enough to make the dynamo nonlinear from the very beginning.

Acknowledgements

AS is grateful to Barbara Ferreira, Waleed Mouhali, Adolfo Ribeiro and Christophe Gissinger for their help in removing errors from the calculations presented here. Barbara Ferreira has carefully read the text and made helpful suggestions. Useful discussions with A. Brandenburg and K. Subramanian are gratefully acknowledged. AS thanks the organizers of the Summer School for the privilege to present the lectures, and for financial support. This work was partially supported by the Royal Society and the Leverhulme Trust grant F/00125/N.

References

[1] H.W. Babcock, Zeeman effect in stellar spectra, Astrophys. J. **105**, 105–191 (1947).

[2] R. Beck, Magnetic fields in normal galaxies, Phil. Trans. Roy. Soc. London A **358**, 777–796 (2000).

[3] R. Beck, Magnetism in the spiral galaxy NGC 6946: magnetic arms, depolarization rings, dynamo modes, and helical fields, Astron. Astrophys. **470**, 539–556 (2007).

[4] R. Beck, A. Brandenburg, D. Moss, A. Shukurov and D. Sokoloff, Galactic magnetism: recent developments and perspectives, Ann. Rev. Astron. Astrophys. **34**, 155–206 (1996).

[5] R. Beck, A. Shukurov, D. Sokoloff and R. Wielebinski, Systematic bias in interstellar magnetic field estimates, Astron. Astrophys. **411**, 99–107 (2003).

[6] M.P. Belyanin, D.D. Sokoloff and A.M. Shukurov, Asymptotic steady-state solutions to the nonlinear hydromagnetic dynamo equations, Russ. J. Math. Phys. **2**, 149–174 (1994).

[7] E.M. Berkhuijsen, C. Horellou, M. Krause, N. Neininger, A.D. Poezd, A. Shukurov and D.D. Sokoloff, Magnetic fields in the disk and halo of M51, Astron. Astrophys. **318**, 700–720 (1997).

[8] J.G. Bolton and J.P. Wild, On the possibility of measuring interstellar magnetic fields by 21-cm Zeeman splitting, Astrophys. J. **125**, 296–297 (1957).

[9] A. Brandenburg and K. Subramanian, Astrophysical magnetic fields and nonlinear dynamo theory, Phys. Rep. **417**, 1–209 (2005).

[10] C.L. Carilli and G.B. Taylor, Cluster magnetic fields, Ann. Rev. Astron. Astrophys. **40**, 319–348 (2002).

[11] T.E. Clarke, P.P. Kronberg, H. Böhringer, A new radio-X-ray probe of galaxy cluster magnetic fields, Astrophys. J. Lett. **547**, L111–L114 (2001).

[12] T.A. Enßlin and C. Vogt, Magnetic turbulence in cool cores of galaxy clusters, Astron. Astrophys. **453**, 447–458 (2006).

[13] P. Frick, R. Stepanov, A. Shukurov and D. Sokoloff, Structures in the rotation measure sky, Mon. Not. Roy. Astron. Soc. **325**, 649–664 (2001).

[14] U. Frisch, *Turbulence. The Legacy of A.N. Kolmogorov*, Cambridge Univ. Press, Cambridge, 1995.

[15] F. Govoni and L. Feretti, Magnetic fields in clusters of galaxies, Int. J. Mod. Phys. **13**, 1549–1594 (2004).

[16] G.E. Hale, On the probable existence of a magnetic field in sun-spots, Astrophys. J. **28**, 315–343 (1908).

[17] A. Herzenberg, Geomagnetic dynamos, Proc. Roy. Soc. Lond. **250A**, 543–583 (1958).

[18] D.W. Hughes, R. Rosner and N.O. Weiss, eds., *The Solar Tachocline*, Cambridge Univ. Press, Cambridge, 2007.

[19] A.B. Iskakov, A.A. Schekochihin, S.C. Cowley, J.C. McWilliams and M.R.E. Proctor, Numerical demonstration of fluctuation dynamo at low magnetic Prandtl numbers, Phys. Rev. Lett. **98**, 208501 (2007).

[20] R.M. Kulsrud and E.G. Zweibel, On the origin of astrophysical magnetic fields, Rep. Prog. Phys. **71**, 046901 (2008).

[21] L.D. Landau and E.M. Lifshitz, *Quantum Mechanics. Non-Relativistic Theory*, Pergamon Press, Oxford, 1974.

[22] L.D. Landau and E.M. Lifshitz, *Fluid Mechanics*, Pergamon Press, Oxford, 1975.

[23] J. Larmor, How could a rotating body such as the Sun become a magnet? Rep. 87th Meeting Brit. Assoc. Adv. Sci., Bournemouth, 1919 Sept. 9–13, John Murray, London, pp. 159–160.

[24] C.F. McKee and J.P. Ostriker, A theory of the interstellar medium—Three components regulated by supernova explosions in an inhomogeneous substrate, Astrophys. J. **218**, 148–169 (1977).

[25] K.R. Mecke, T. Buchert and H. Wagner, Robust morphological measures for large-scale structure in the Universe, Astron. Astrophys. **288**, 697–704 (1994).

[26] A.S. Monin and A.M. Yaglom, *Statistical Fluid Mechanics*, Vol. 2, Dover Publ., 2007.

[27] A.G. Pacholczyk, *Radio Astrophysics. Nonthermal Processes in Galactic and Extragalactic Sources*, Freeman, San Francisco, 1970.

[28] E.N. Parker, Hydromagnetic dynamo models, Astrophys. J. **122**, 293–314 (1955).

[29] N. Piskunov and O. Kochukhov, Doppler imaging of stellar magnetic fields. I. Techniques, Astron. Astrophys. **381**, 736–756 (2002).

[30] I. Rogachevskii and N. Kleeorin, Intermittency and anomalous scaling for magnetic fluctuations, Phys. Rev. E **56**, 417–426 (1997).

[31] A.A. Ruzmaikin, A.M. Shukurov and D.D. Sokoloff, *Magnetic Fields of Galaxies*, Kluwer Academic Publ., Dordrecht, 1988.

[32] A. Ruzmaikin, D. Sokoloff and A. Shukurov, The dynamo origin of magnetic fields in galaxy clusters, Mon. Not. Roy. Astron. Soc. **241**, 1–14 (1989).

[33] V. Sahni, B.S. Sathyaprakash and S.F. Shandarin, Shapefinders: a new shape diagnostic for large-scale structure, Astrophys. J. **495**, L5–L8 (1998).

[34] I. Sakelliou, D.M. Acreman, M.J. Hardcastle, M.R. Merrifield, T.J. Ponman and I.R. Stevens, The cool wake around 4C 34.16 as seen by XMM-Newton, Mon. Not. Roy. Astron. Soc. **360**, 1069–1076 (2005).

[35] A.A. Schekochihin and S.C. Cowley, Turbulence, magnetic fields, and plasma physics in clusters of galaxies, Phys. Plasmas **13**, 056501–056501-8 (2006).

[36] A.A. Schekochihin, N.E.L. Haugen, A. Brandenburg, S.C. Cowley, J.L. Maron and J.C. McWilliams, Onset of small-scale turbulent dynamo at low magnetic Prandtl numbers, Astrophys. J. Lett. **625**, L115–L118 (2005).

[37] A.A. Schekochihin, S.C. Cowley, R.M. Kulsrud, G.W. Hammett and P. Sharma, Plasma instabilities and magnetic field growth in clusters of galaxies, Astrophys. J. **629**, 139–142 (2005).

[38] A.A. Schekochihin, A.B. Iskakov, S.C. Cowley, J.C. McWilliams, M.R.E. Proctor and T.A. Yousef, Fluctuation dynamo and turbulent induction at low magnetic Prandtl numbers, New J. Phys. **9**, 300 (2007).

[39] J. Schmalzing and T. Buchert, Beyond genus statistics: a unifying approach to the morphology of cosmic structure, Astrophys. J. **482**, L1–L4 (1997).

[40] J. Schmalzing, T. Buchert, A.L. Melott, V. Sahni, B.S. Sathyaprakash and S.F. Shandarin, Disentangling the cosmic web. I. Morphology of isodensity contours, Astrophys. J. **526**, 568–578 (1999).

[41] A. Shukurov, Introduction to galactic dynamos, in: *Mathematical Aspects of Natural Dynamos*, E. Dormy and A.M. Soward, eds., Chapman and Hall/CRC, London, 2007, pp. 313–359.

[42] A. Shukurov, D. Sokoloff, K. Subramanian and A. Brandenburg, Galactic dynamo and helicity losses through fountain flow, Astron. Astrophys. **448**, L33–L36 (2006).

[43] Y. Sofue and V. Rubin, Rotation curves of spiral galaxies, Ann. Rev. Astron. Astrophys. **39**, 137–174 (2001).

[44] D.D. Sokoloff, A.A. Bykov, A. Shukurov, E.M. Berkhuijsen, R. Beck and A.D. Poezd, Depolarisation and Faraday effects in galaxies and other extended radio sources, Mon. Not. Roy. Astron. Soc. **299**, 189–206 (1998) [Erratum: op. cit. **303**, 207–208 (1999)].

[45] D.D. Sokoloff and A. Shukurov, Regular magnetic fields in coronae of spiral galaxies, Nature **347**, 51–53 (1990).

[46] S.R. Spangler, A technique for measuring electrical currents in the Solar corona, Astrophys. J. **670**, 841–848 (2007).

[47] M. Stix, *The Sun*, Springer, Berlin, 2002.

[48] A.W. Strong, I.V. Moskalenko and V.S. Ptuskin, Cosmic-ray propagation and interactions in the Galaxy, Ann. Rev. Nucl. Particle Sys. **57**, 285–327 (2007).

[49] K. Subramanian, Unified treatment of small- and large-scale dynamos in helical turbulence, Phys. Rev. Lett. **83**, 2957–2960 (1999).

[50] K. Subramanian, A. Shukurov and N.E. L. Haugen, Evolving turbulence and magnetic fields in galaxy clusters, Mon. Not. Roy. Astron. Soc. **366**, 1437–1454 (2006).

[51] S. Tobias and N. Weiss, Stellar dynamos, in: *Mathematical Aspects of Natural Dynamos*, E. Dormy and A.M. Soward, eds., Chapman and Hall/CRC, London, 2007, pp. 281–311.

[52] A.G. Tomboulides and S.A. Orszag, Numerical investigation of transitional and weak turbulent flow past a sphere, J. Fluid Mech. **416**, 45–73 (2000).

[53] G.L. Verschuur, Positive determination of an interstellar magnetic field by measurement of the Zeeman splitting of the 21-cm hydrogen line, Phys. Rev. Lett. **21**, 775–778 (1968).

[54] G.L. Verschuur, Observations of the Galactic magnetic field, Fund. Cosm. Phys. **5**, 113–191 (1979).

[55] S.L. Wilkin, C.F. Barenghi and A Shukurov, Magnetic structures produced by the fluctuation dynamo. Phys. Rev. Lett. **99**, 134501 (2007).

[56] Ya.B. Zeldovich, A.A. Ruzmaikin and D.D. Sokoloff, *The Almighty Chance*, World Sci., Singapore, 1990.

Course 5

TURBULENCE AND DYNAMO

B. Dubrulle

DRECAM/SPEC/CEA Saclay, and CNRS (URA2464), F-91190 Gif sur Yvette Cedex, France

Ph. Cardin and L.F. Cugliandolo, eds.
Les Houches, Session LXXXVIII, 2007
Dynamos

301

Contents

1. Introduction: turbulence AND dynamo?

Turbulence and dynamo have been both studied for a long time, and would each deserve separately a whole school of their own- when you hit the word "turbulence" (resp. "dynamo") onto Google, you find 23 100 000 (resp. 1 940 000, excluding football team . . . .) entries. However, the connection between turbulence and dynamo has been more scarcely investigated −300 000 entries on Google for both "turbulence" and "dynamo"-, while it has recently regain a boost of interest, due to recent experimental advances with liquid sodium dynamos (Riga, Karlsruhe, VKS). In these experiments, the Reynolds number routinely reaches a few 10^6, still far away from astrophysical or geophysical Reynolds numbers, but way into what one usually calls a "fully turbulent regime". The purpose of this course is therefore not to present an exhausting review about turbulence, nor about dynamo but rather a panorama of some issues connecting turbulence to dynamo, that have been raised by the recent results of the Riga, Karlsruhe and VKS experiments. Most of the results or graphs presented in this lecture have been obtained during PhD thesis of Jean-Philippe Laval, Louis Marié, Nicolas Leprovost, Florent Ravelet and Romain Monchaux, within the Groupe Instabilité and Turbulence at CEA Saclay and the VKS collaboration. I refer the reader interested in more details to the corresponding thesis manuscripts [1–5].

2. Turbulence

2.1. Generalities

A classical joke states that before Einstein died, he prepared two questions to be asked to St Peter at the entrance of paradise. These questions, sorted by increasing order of importance were: i) how to quantify gravity? ii) what is the solution to turbulence? Unfortunately, neither Einstein, nor St Peter came back to provide us the answer to these fundamental questions, and we are left with the ever-anguishing question: What is turbulence, then, exactly, then? Apart from "Forty-two", I have personally never encountered a relevant answer in all my career. I shall therefore carefully avoid to give any detailed definition of "turbulence", and, rather, focus on its properties and consequences. A first striking

B. Dubrulle

Fig. 1. One of the two water von Karman experiments used in Saclay.

feature of turbulence is its ubiquity: you can meet it in the morning, while you are mixing sugar in your cup of coffee with a spoon, at lunch, when you are washing dishes in your sink, in the evening when your are emptying your bath, around your car, your plane or your bicycle on your way to this summer school, above your heads in the planetary surface layer, during your hikes inside mountains torrents, in the ocean, in the sun, in the stars, in the galaxies, inside the human body, in industrial devices, ... In the laboratory, a very efficient way to obtain turbulence is to stir a bucket of water with rotating impellers. This is called the von Karman flow and is illustrated on Fig. 1. In this device, when you rotate the impellers at a few dizains of Hertz, you obtain a turbulence equivalent to what is reached inside the "soufflerie de Modane", a huge wind-tunnel 14 meters long and 8 meters wide.

On your computer, you may also obtain turbulence by integrating the so-called Navier–Stokes equations:

$$\partial_t \mathbf{v} + \mathbf{v} \cdot \nabla \mathbf{v} = -\frac{1}{\rho_0} \nabla P + \nu \nabla^2 \mathbf{v} + f, \tag{2.1}$$

where ρ_0 is the (constant) density, $\mathbf{v}$ is the velocity, P the pressure, ν the viscosity, and f a well-chosen forcing. For example, if you choose f as a stochastic noise, you'll get a turbulence with no average velocity. If you choose f proportional to the Taylor–Green vortex $\mathbf{v}^{TG} = (\sin x \cos y \cos z, -\cos x \sin y \cos z, 0)$, you get a Taylor–Green flow, in which the mean flow resembles the von Karman mean flow (Fig. 2).

Like in the real von Karman flow, however, you will have to struggle a little bit until you reach the turbulent regime. How much you need to struggle

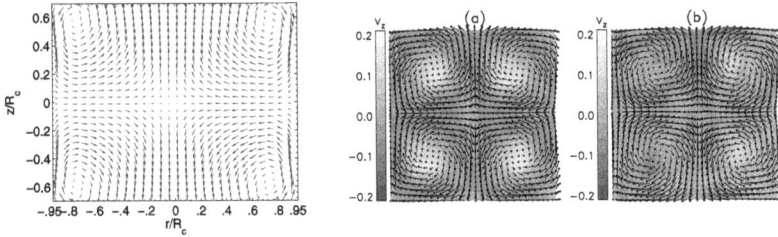

Fig. 2. Left: Time-averaged von Karman flow at $Re = 10^6$. The plot is a meridian cut taken at the middle of the cylinder. Right: Two time averages of the Taylor–Green flow at $Re = 100$. a) with no energy in the mode $k = 1$; b) with energy in the mode $k = 1$. The plots are horizontal cuts, taken at $z = 0$.

depends on the famous Reynolds number: $Re = VL/\nu$, where V and L are characteristic velocities and scale. In the von Karman flow (VK flow), you may take $L = R$, the cylinder radius, and $V = 2\pi RF$ where F is the impeller rotating frequency;[1] in the Taylor–Green flow (TG flow), you may take $V = \sqrt{2E/3}$, the r.m.s. velocity based on the total kinetic energy $E = \int E(k)dk$ and $L = (3\pi/4) \int E(k)/kdk/E$, the integral scale of the turbulent flow. Turbulence then takes place when Re is larger than 10^4 for the von Karman flow, and $Re > 80$ for the Taylor–Green flow. In the laboratory, this is easily reached when you rotate your impeller faster than a few hundredth of a Hertz; on your computer, however, this means that you have to use a powerful machine with a lot of memory, since you need to go to resolutions higher than 128^3. Why it is so can be understood by looking at your turbulent flow.

2.2. Spectra and number of degrees of freedom

2.2.1. Observations

Consider an instantaneous picture of a turbulent flow inside a von Karman apparatus (Fig. 3). One sees roughly a messy thing, with some zones with large velocities, other with smaller velocities, and a lot of spatial disorder. If you stick a probe inside this flow, so as to record a temporal signal of the velocity, you also see temporal disorder, with variations over different time scales (Fig. 4). Things clarify a little bit when you take the energy spectrum of this measurements, whether spatial, or temporal (Fig. 5).

Let us first consider the spatial energy spectrum. One sees clearly a power-law for wavenumbers larger than the injection wavenumber k_0. The slope of the power-law is approximately $k^{-5/3}$, corresponding to a now-famous "Kolmogorov

---

[1] In the VKS2 experiment, L is actually taken as the radius of the inner cylinder and $V = 0.6 \cdot 2\pi RF$.

B. Dubrulle

Profil instantané sans anneau (ech : 1/2)

Profil instantané avec anneau (ech : 1/2)

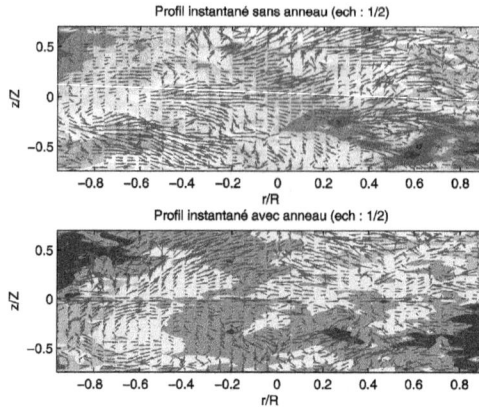

Fig. 3. Instantaneous velocity field in a von Karman experiment with counter-rotating impellers. Upper panel: without annulus. Lower panel: with a thin annulus inserted in the middle of the shear layer, at $z = 0$, see Fig. 2.

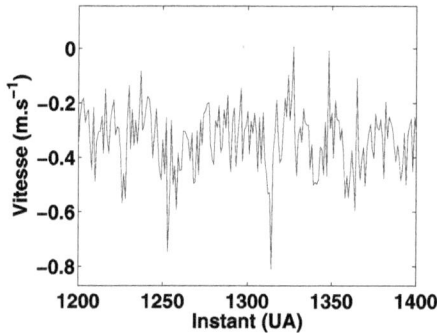

Fig. 4. Time record of velocity in a von Karman experiment, for a probe located near the midplane, near the cylinder boundary.

spectrum". The measurements performed in the present case did not reach a high enough resolution to see how far this spectrum extends. One usually considers that the power law extends through the so-called "inertial range" up to the Taylor scale λ, and then bends in an exponential fashion over the "dissipative range" until the Kolmogorov wavenumber k_η where energy is dissipated (see Fig. 6). This power-law spectrum reveals the existence of a wide range of scales in a turbulent flow, from the injection scale $1/k_0$ to the dissipative scale $1/k_\eta$. The resolution needed to capture all the flow dynamics is therefore $(k_\eta/k_0)^3$ that grows roughly like $Re^{9/4}$. This explains the needs for large spatial resolution when simulating a turbulent flow.

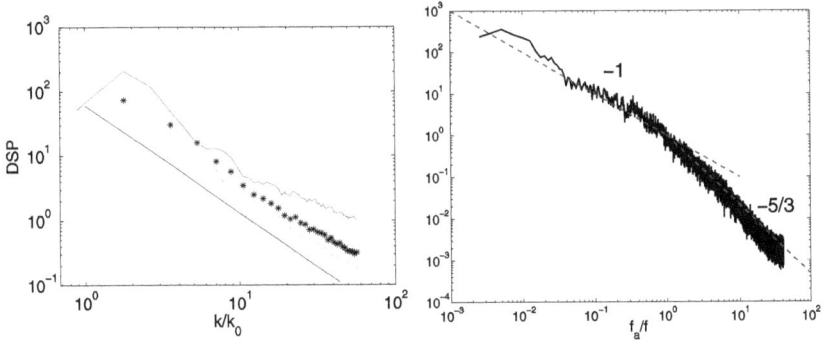

Fig. 5. Spatial (left) and temporal (right) energy spectrum in a von Karman experiment. The continuous straight line on the left is $k^{-5/3}$. The continuous line, stars and dotted on the left are spectra made over boxes of respectively 50%, 70% and 100% of the experimental domain.

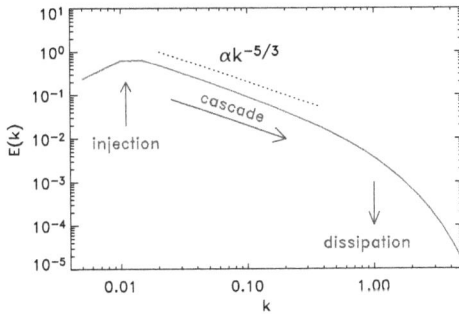

Fig. 6. Schematic view of a Kolmogorov spectrum.

Another difficulty stems from the temporal dynamics. Consider indeed the temporal energy spectrum. For frequencies f_a larger than the injection frequency f, one observes a steep power-like decrease of the energy, with an exponent close $-5/3$. This is usually explained via the spatial spectrum, and by saying that everything is advected at the mean velocity $\overline{V}$ so that $f_a \sim \overline{V}k$ (Taylor's hypothesis). In any case, there is, like in the spatial case, a whole hierarchy of time scales extending from the injection time scale to a dissipative time scale. However, there is also a "slow-time dynamics", apparent from the part of the spectrum for frequencies smaller than the injection frequencies. This inverse cascade of energy extends down to $0.01f$, via an f_a^{-1} spectrum. It traces very slow motions within the flow, probably large coherent vortices in the mixing layer in between the two propellers. From a numerical point of view, this slow

B. Dubrulle

dynamics represents another challenge, for it means that very long integration time are needed to capture it. Numerical simulations of such a turbulent flow would then require both monstrous resolutions (of the order of 10^{15} grid points!), and monstrous integration times-to integrate for 10 times $100/f$ to get convergence of the slowliest structures, one would require 6×10^5 years of cpu with current computers.

2.3. Mean-flow

Among all the disorder occurring in a turbulent flows, one can however define more regular and ordered things by considering average quantities. If you indeed take 5000 different instantaneous images in the von Karman flow and average them, you get something like Fig. 7. There, you clearly see a well-defined poloidal circulation, made of four cells, with a pumping towards the impellers at the center, centrifugal ejection of fluids at the impellers and descent towards the mid-plane at the boundary of the cylinder. The toroidal flow is made of two counter-rotating zones. This global topology is only slightly affect by the presence of an annulus at the mid-plane. However, the convergence towards this mean flow is more deeply affected by the annulus. Indeed, when one averages over only 20 images (Fig. 8), one sees already the general topology of the mean flow in the case where the annulus is present, while there are still some sort of coherent structures at the mid-plane when the annulus is removed. These coherent structures are in fact the large scale vortices present in the shear layer, that are responsible for the slow time dynamics. The presence of the annulus stabilizes

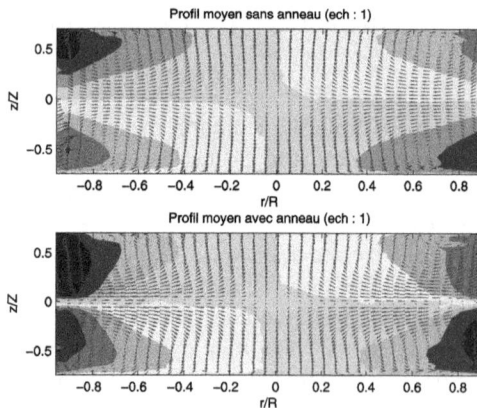

Fig. 7. Velocity field averaged over 5000 images in a von Karman experiment. Upper panel: without annulus. Lower panel: with an annulus.

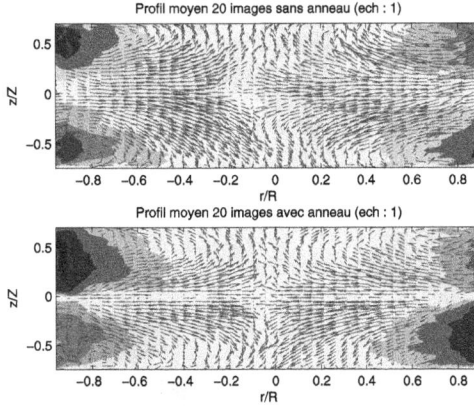

Fig. 8. Velocity field averaged over 20 images in a von Karman experiment. Upper panel: without annulus. Lower panel: with an annulus.

these structures, and allows a faster convergency towards the mean flow, a feature that will prove important in the search of a turbulent dynamo.

2.4. Fluctuations

In the case where there is no mean flow, fluctuations *are* the turbulence. In the case where there is a mean flow, fluctuations are the necessary component to understand the flow dynamics. A classical way to characterize their effect is though the turbulence intensity:

$$i(x) = \frac{\overline{(v - \overline{v})^2}}{\overline{v}^2}, \tag{2.2}$$

where $\overline{X}$ refers to time average of X. In a classical homogeneous, isotropic turbulence, this quantity is independent of the space, and fully characterizes the fluctuations. In the von Karman flow, however, the fluctuations are strongly inhomogeneous and anisotropic (Fig. 9) so that this turbulence intensity strongly varies over the experimental flows (Fig. 10). Moreover, in the vicinity of the central mixing layer, the average velocity is close to zero, and the turbulence intensity takes very large values.

In order to take into account the spatial inhomogeneity, it is therefore tempting to consider instead space-averaged quantities. An analog to the turbulent intensity is then:

$$\delta(t) = \frac{<v^2>}{<\overline{v}^2>}, \tag{2.3}$$

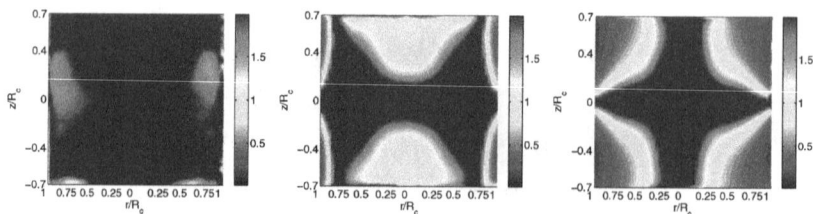

Fig. 9. Fluctuations of the three velocity components in the von Karman experiment with TM73 propellers in contrarotation (−). From left to right: v_r, v_θ, v_z.

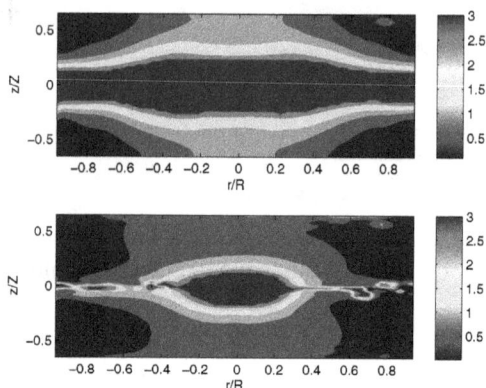

Fig. 10. Turbulence intensity in the von Karman experiment with TM73 propellers in contrarotation (−). The color code has been saturated whenever the turbulence intensity exceeds 300%. Upper panel: without annulus. Lower panel: with an annulus.

where now $< X >$ refers to spatial average of X. This is just the ratio of the total kinetic energy, to the kinetic energy of the mean-flow. This quantity still widely fluctuates in time (Fig. 11). However, centered, reduced histograms do not vary with Reynolds number and are nearly Gaussian (Fig. 11).

This means that only two scalar quantities are needed to characterize the distribution of $\delta(t)$: its average:

$$\delta = \frac{\overline{< v^2 >}}{< \overline{v}^2 >},$$

(2.4)

and its r.m.s.:

$$\delta_2 = \frac{\sqrt{\overline{< v^2 >^2} - \overline{< v^2 >}^2}}{< \overline{v}^2 >}.$$

(2.5)

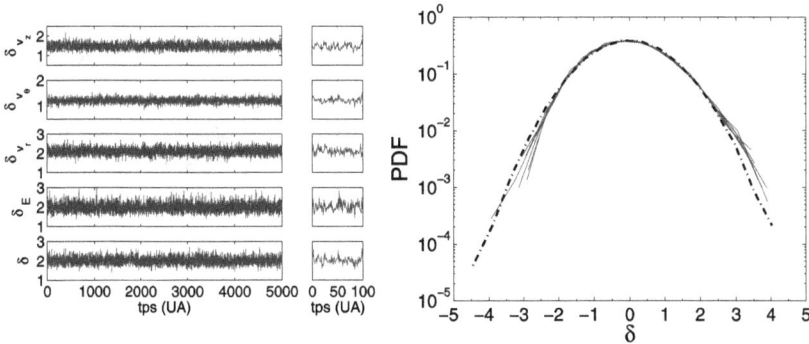

Fig. 11. Left: Time variation of δ (bottom) and analog quantities in a von Karman experiment without annulus with TM73 propellers rotating in the contra direction at 5 Hz. The small insert is a zoom. Right: PDF of the δ parameter at different Reynolds number ranging from $1 \cdot 10^5$ to $5 \cdot 10^5$. The PDF are centered and reduced. The dotted line is a Gaussian.

Note that in an homogeneous flow $\delta = i + 1$ so that it is a generalization of the turbulent intensity. For real flows like the von Karman flow, δ contains however an additional information regarding how close the instantaneous flow is to the mean flow. Indeed,

$$
\begin{aligned}
(\delta - 1) < \overline{v}^2 > \quad &= \quad \overline{< (v - \overline{v})^2 >} \\
&= \quad \overline{\frac{1}{N} \sum_{i,j} (v_{i,j} - \overline{v}_{i,j})^2},
\end{aligned}
\tag{2.6}
$$

where the sum runs over the N grid points of the measurements. The quantity under the overline is nothing but the square of the mean distance (using norm 2) between the instantaneous flow and the mean flow in the functional space. $\delta - 1$ is therefore a measure of how close on average the instantaneous velocity field is to its mean value. If δ is close to one, one therefore expects the instantaneous flow to strongly resemble the mean flow. If δ is much greater than 1, the instantaneous field will be more remote from the mean flow. For example, δ for the velocity field of Fig. 3 is 2 for the flow without an annulus, and 1.5 for the flow with an annulus, thereby quantifying the convergency effect we saw in previous sections. δ can also vary with the Reynolds number, since as the Reynolds number increases, fluctuations gets larger, and the instantaneous flow becomes more liable to variations from its mean. In fact, one can show that it is only the case for flows with a lot of large scale or slow time scale fluctuations. The other fluctuations are less energetics, and contribute only marginally to the value of δ. This can be illustrated using for example the Taylor–Green flow: when the Reynolds number is increased, the value of δ increases from 1 until a value of the order of

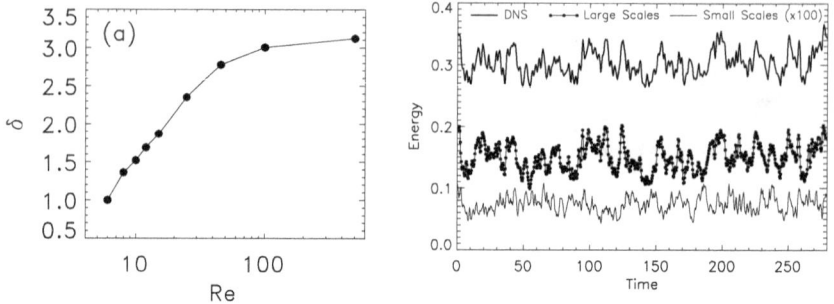

Fig. 12. Left: Noise intensity δ, as a function of the Reynolds number, measured in DNS and LES simulations of a Taylor–Green flow. Right: Instantaneous kinetic energy of a large scale (keep all $k \leq 4$) or small-scale (keep all $k > 8$) Navier–Stokes noise at $Re = 25$, compared with the kinetic energy of a full DNS at $Re = 25$. For better visibility, the energy of the small scale noise has been multiplied by 100.

3 for the largest Reynolds number simulated (Fig. 12). However, in the definition of δ, scales corresponding to only the first four modes contribute to sixty percent of the instantaneous kinetic energy, while all modes with waves number larger than 8 only contribute to 0.3 percent of the kinetic energy.

2.5. Transport properties

Other interesting things can be observed in a turbulent flow when you stick in some spot of ink. The ink is a passive scalar: it will not change the properties of the turbulent flow, but serves as a good indicator to trace the turbulent movements. Figure 13 shows the evolution of such an ink spot, in a turbulent flow with no average velocity. One sees that the ink spot develops rapidly finer and finer structures, changing the initial round spot into a very complicated and convoluted thing-some even think that it is a fractal-. If one wait for a little longer after this stage, one will eventually see nothing else than a uniform pattern of constant concentration throughout the flow: the turbulence is a very efficient way to mix things. This is mainly because, through the action of the velocities at all scales, it has enabled the formation of very small scales, that are then small enough for viscous diffusion to take place. This distinction is important: Turbulence in itself does not diffuse the spot. It just acts as a catalyst, for diffusion to be more efficient. A traditional way to quantify this effect is to consider that the turbulence has created a sort of "turbulent diffusivity", of the order of $D_t = LV$, where V and L are characteristic length and velocity of the turbulence. Note however that the real dissipation occurs at scales governed by the real diffusivity, and not by the turbulent diffusivity (that would give much larger scales), so that this approximation has to be taken with caution in numerical models.

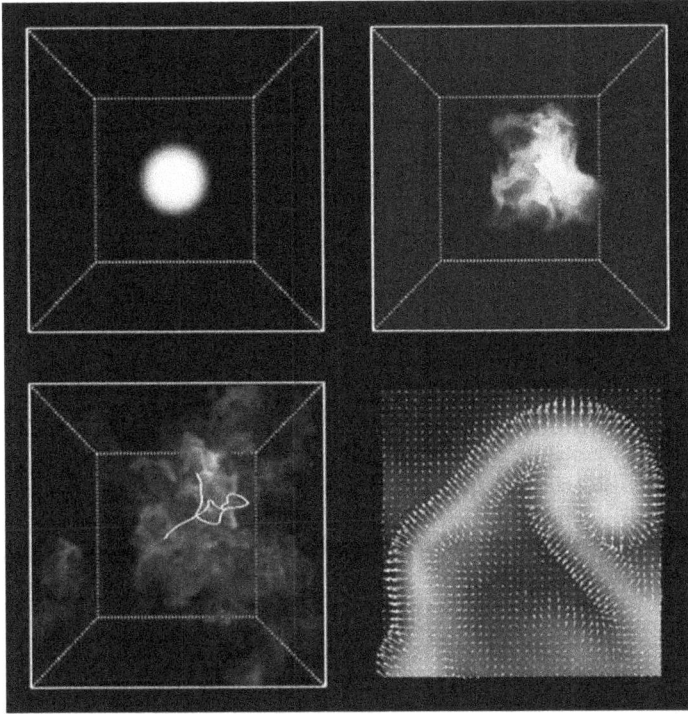

Fig. 13. Evolution of a spot of ink in a turbulent flow, starting from the left upper panel, to the upper right panel, and then the lower left panel. The lower right panel is an zoom inside the upper right panel. The white line is the trajectory of a passive particle. The arrows are instantaneous snapshots of the gradient of the scalar spot.

Besides passive scalar, turbulence also acts on transport properties of passive vector-like the gradient of scalar shown in Fig. 13, or active scalar-like temperature in convection or active vectors, like magnetic field or velocity itself. In the latter case, the effect can be measured by direct monitoring of the power consumption- or equivalently monitoring of the torque in a rotating experiment. Figure 14 shows the evolution of the non-dimensional torque $K_p = T/\rho_0 R^5 (2\pi F)^2$ in the von Karman experiment, for two different forcing mechanism through impellers fitted with blades presenting their concave ($-$, anticontrarotation) or convex ($+$, contrarotation) side to the flow during impeller rotation. These two forcing generate different type of turbulence. For example, the factor δ is of the order 2.7 for the anticontrarotation, and 2.2 for the contrarotation. Yet, the two turbulence give qualitatively the same behavior: a steep decay, $K_p \sim 1/Re$ at low Reynolds number, followed by a transition to a more or

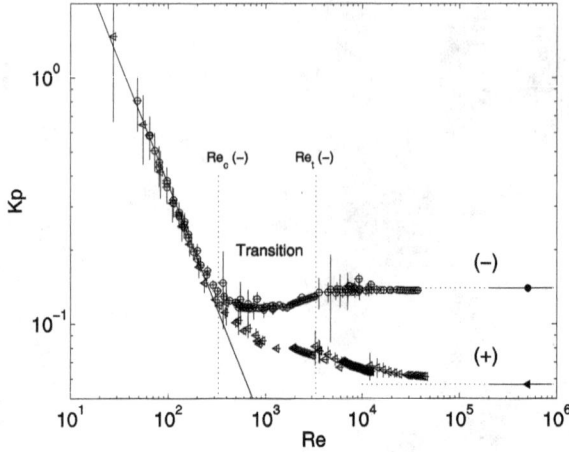

Fig. 14. Non-dimensional torque K_p as a function of Reynolds number in a von Karman experiment with TM60 propellers. circle: anticontrarotation $(-)$; triangle: contrarotation $(+)$. The line is a fit $K_p = 36.9/Re$.

less constant value at large Reynolds number. This behavior can be understood qualitatively by the introduction of the "turbulent viscosity": when the flow is laminar, the torque behaves like: $T \sim < \nu \nabla v >$ so that $K_p \sim 1/Re$; when it is turbulent, the torque behaves now like: $T \sim \nu_t < \nabla v >$. With $\nu_t \sim R^2 F$, this gives $K_p \sim cte$.

The different levels of saturation for K_p can also be understood by considering a slightly more refined formula for the torque, that is actually the sum of a viscous contribution, and a contribution due to Reynolds stress:

$$\frac{T}{\rho_0 R^3} \sim < \nu \nabla \bar{v} + \overline{v'^2} >, \tag{2.7}$$

so that $K_p \sim A/Re + C(\delta - 1)$. Taking into account that δ varies with the Reynolds number (see Fig. 12) and reaches different saturation level in contra and anticontrarotation, this gives Fig. 15 for the two different sense of rotation. It is qualitatively very similar to the real measurements.

This means that the torque measurements and δ are probably strongly correlated, a property that can be interesting in situation when δ is not directly accessible, like in von Karman filled with sodium.

2.6. Theories of turbulence

There have been many attempts to try and build THE theory of turbulence and the reader can find a large variety of books describing one or several of them [6–9].

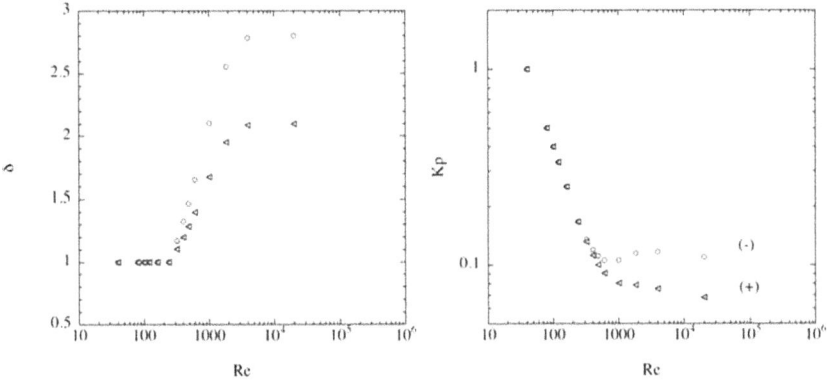

Fig. 15. Left: Synthetic δ for two rotating directions; Right: corresponding K_p built using the phenomenological formula $K_p = 40/Re + 0.06(\delta - 1)$. Same symbols as in Fig. 14.

Here, I describe two approaches, whose only merit is that they have later been used in connection with the turbulent dynamo problem.

2.6.1. *Dimensional analysis*
Dimensional analysis is not actually a theory per se, but it can be seen as a formal analysis, yielding interesting, and sometimes even useful results. For example, some of the features of the spectra observed above can be understood by dimensional analysis. The key point of such an approach is to first identify the relevant dimensional quantities needed to describe our turbulent flow. In the case of the von Karman flow, for example, the main global characteristic of the experiment are: the cylinder radius R and height H, the power delivered by its motors P, the frequencies at which the impellers are rotating, f_0, the fluid that is considered, or equivalently, its viscosity ν and density ρ_0. With these numbers, one can build three non dimensional numbers: the aspect ratio $\Gamma = H/R$, the Reynolds number $Re = R^2 f_0/\nu$, and the efficiency $E_{ff} = P/\rho_0 R^5 f_0^3$.

When considering *spatial spectrum*, one introduces two additional dimensional quantities: the wavenumber k and the velocity component at this wavenumber $v(k)$, resulting in two additional non-dimensional quantities: $kv(k)/Rf_0$ and kR. Through the famous Buckingham theorem, one can now state that there is a functional relation linking these 5 non-dimensional quantities, such as:

$$\frac{v(k)}{Rf_0} = F(\Gamma, Re, E_{ff}, kR). \tag{2.8}$$

For a range of wavenumbers in the inertial range, the fluid does not know how it will be dissipated (it doesn't know about ν), nor how it has been stirred (it doesn't know about f_0 either). This imposes:

$$
\begin{aligned}
(v(k))^3 &= \frac{P}{\rho_0 R^2} G(\Gamma, kR), \\
&= \epsilon k^{-1}(kR)G(\Gamma, kR),
\end{aligned}
\tag{2.9}
$$

where we have introduced the energy dissipation rate $\epsilon = P/\rho_0 R^3$. Since we are in the inertial range $kR \gg 1$. Calling $(C_K(\Gamma))^{3/2} = \lim_{kR \to \infty} kRG(\Gamma, kR)$, we therefore get:

$$
v(k) = C_K^{1/2}(\Gamma)\epsilon^{1/3}k^{-1/3},
\tag{2.10}
$$

resulting in $E(k) = v^2(k)/k \sim C_K \epsilon^{2/3} k^{-5/3}$, the Kolmogorov spectrum.

When considering *temporal spectrum*, one introduces now the two dimensional quantities: the frequency f and the velocity component at this wavenumber $v(f)$, resulting in two additional non-dimensional quantities: $v(f)/Rf_0$ and f/f_0. The Buckingham theorem now gives:

$$
\frac{v(f)}{Rf_0} = F(\Gamma, Re, E_{ff}, f/f_0).
\tag{2.11}
$$

If we impose, like the above, that the fluid does depend neither on v, nor f_0, we then get the analog of (2.9)

$$
\begin{aligned}
(v(f))^3 &= \frac{P}{\rho_0 R^2} G(\Gamma), \\
&= \epsilon RG(\Gamma),
\end{aligned}
\tag{2.12}
$$

so that $v(f)$ still depends on R on or H. This relation may explain the f^{-1} spectrum observed at small f in the von Karman spectrum, since relation (2.12) gives $v(f) \sim \epsilon^{1/3} R^{1/3}$, resulting in $E(f) \sim (\epsilon R)^{2/3} f^{-1}$. However, in the inertial range, one would like eventually to see R disappear along with f_0 from the relation. A way to do that is to consider now:

$$
\begin{aligned}
\left(\frac{v(f)}{Rf_0}\right)^2 &= \frac{P}{\rho_0 R^5 f_0^3} \frac{f_0}{f} G(\Gamma), \\
v^2(f) &= \epsilon f^{-1} G(\Gamma),
\end{aligned}
\tag{2.13}
$$

resulting in $E(f) \sim \epsilon f^{-2}$, only slightly steeper than a Kolmogorov spectrum inferred from the spatial spectrum with Taylor hypothesis (that is strictly valid only in presence of a well-defined mean flow). To distinguish between the two spectra, it is therefore more convenient to vary the power (hence ϵ) and see whether $E(f)$ varies like $\epsilon^{2/3}$ ($f^{-5/3}$ Kolmogorov spectrum) or like ϵ (f^{-2} spectrum).

2.6.2. *Stochastic description of turbulence*

Consider a turbulent flow, with velocity field $v_i(x, t)$, and introduce an (arbitrary) filtering procedure so as to separate it into a large-scale field $U_i = \overline{v_i}$ and a small-scale component $u_i = v_i - U_i$. Such small-scale motion varies over time scale t, while large scale vary over time scale T. In any reasonable turbulent flow, the ratio of the typical time scale of the two components varies like a power of the scale ratio, as $t/T \sim (l/L)(U/u) \sim (l/L)^{2/3}$. Therefore, small scales vary much more rapidly than large scale. From the point of view of the largest scales, the small scales may then be regarded as a noise. Hence the idea to simply replace them by an a priori chosen noise, with well-defined properties. One classical way is though a generalized Langevin equation:

$$\dot{u}_i = A_{ij}u_j + \xi_i, \tag{2.14}$$

where A is a generalized friction operator, and ξ is a noise. In the sequel, we explore various models characterized by different value of A and ξ.

2.6.2.1. *Obukhov model*

The simplest model one can imagine is to take $A = 0$ and ξ as a Gaussian white noise, isotropic and homogeneous in space, with short time correlation:

$$< \xi_i(x, t)\xi_j(x', t) >= 2\Delta\delta_{ij}\delta(t - t'). \tag{2.15}$$

This model has been first introduced in 1959 by Obukhov. It leads to a number of interesting properties.

2.6.2.1.1. *Richardson law and Kolmogorov spectrum*

Consider for example a cloud of passive scalar particles, embedded in such a flow. After a time t, this cloud of particles will have evolved into a situation where its velocity distribution obeys a Gaussian statistics, with variance scaling like square root of time: $\delta u = \sqrt{<u^2> - <u>^2} \sim t^{1/2}$. In parallel, the cloud of particles experienced a spread by a factor $r \sim \sqrt{<x^2> - <x>^2} \sim t^{3/2}$. This last law is nothing but the famous Richardson law, an empirical law describing the dispersion of passive tracers in the atmosphere. Moreover, we may combine the two simple relation to obtain that $\delta u \sim r^{1/3}$, implying a velocity spectrum $E(k) \sim k^{-5/3}$, i.e. the Kolmogorov spectrum. We see that with virtually no effort, Obukhov model reproduces the two more robust experimental results obtained so far in turbulence!

2.6.2.1.2. *Limitations*

Richardson law and Kolmogorov spectra are representative of velocities which do not differ from the mean by a large amount. The actual range of validity of the Obukhov model arises when considering higher

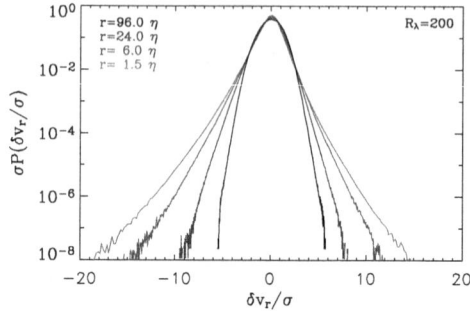

Fig. 16. Centered and reduced PDF of velocity increments $\delta v(r) = v(x + r) - v(x)$ for different values of r/η (η is the Kolmogorov scale) from a numerical simulation of isotropic, homogeneous turbulence at $Re = 200$. If the Obukhov model were to be correct, all PDF should collapse onto a Gaussian.

moments, involving rarer, but more violent events. Since velocities in this model are Gaussian, their moments obey a simple scaling relation: $< u^{2n} > \sim < u^2 >^n$, at variance with observations in real turbulent flows (Fig. 16). This simple hierarchy law disappears as soon as one allows for spatial or temporal correlation, as recently proved in the Kraichnan model of turbulence.

2.6.2.2. Kraichnan model   The Obukhov model is frictionless in essence. The Kraichnan model can be viewed as the opposite limit, with a very large friction $A_{ij} = -\gamma \delta_{ij}$, $\gamma \gg 1$, and a noise with spatial correlation

$$< \xi_i(x, t)\xi_j(x', t) > = 2\Delta_{ij}(x, x')\delta(t - t'). \tag{2.16}$$

Due to the large friction, the inertial term in the Langevin equation becomes negligible and the velocity adiabatically adjusts to the noise as: $u_i \sim \gamma \xi_i$. The Kraichnan model is thus made of small-scale delta-correlated Gaussian white noise, with spatial correlation.

2.6.2.2.1. Intermittency and conservation laws   Contrarily to Obukhov model, Kraichnan model leads to intermittency for the high order moments. The physical reasons have been recently reviewed in [10]. They are rooted in the spatial correlation, which induce a memory effect onto Lagrangian trajectories, and lead to the apparition of conservation laws within sets of Lagrangian particles. Since the moment of order $2n$ is associated with conservation laws of sets of $2n$ particles, and since conservation laws of sets of particles of different sizes are not simply related, this induces a breaking of the hierarchical structure of the moments.

2.6.2.2.2. Turbulent transport Another less well known property of Kraich-nan model concerns turbulent transport. Suppose we focus on the evolution of the vorticity in such a model. In classical turbulence, the vorticity obeys the equation

$$\partial_t \Omega_i = -v_k \partial_k \Omega_i + \Omega_k \partial_k v_i + \nu \partial_k \partial_k \Omega_i, \tag{2.17}$$

where ν is the molecular viscosity, and v is the sum of the large scale component U and the (small-scale) noise. Because of the presence of noise, eq. (2.17) admits stochastic solution, whose dynamic can be fully specified by the probability distribution function. Ignoring the viscosity and using standard techniques [44], one can derive the evolution equation for $P(\Omega, x, t)$, the probability of having the field Ω at point x and time t:

$$
\begin{aligned}
\partial_t P \quad = \quad & -U_k \partial_k P - (\partial_k U_i) \partial_{\Omega_i} [\Omega_k P] + \partial_k [\beta_{kl} \partial_l P] \\
& + 2 \partial_{\Omega_i} [\Omega_k \alpha_{lik} \partial_l P] \\
& + \mu_{ijkl} \partial_{\Omega_i} [\Omega_j \partial_{\Omega_k} (\Omega_l P)].
\end{aligned}
\tag{2.18}
$$

For simplicity, we assumed homogeneity of the fluctuations and we introduced the following turbulent tensors:

$$
\begin{aligned}
\beta_{kl} \quad &= \quad \langle u_k u_l \rangle, \\
\alpha_{ijk} \quad &= \quad \langle u_i \partial_k u_j \rangle, \\
\mu_{ijkl} \quad &= \quad \langle \partial_j u_i \partial_l u_k \rangle.
\end{aligned}
\tag{2.19}
$$

Due to incompressibility, the following relations hold: $\alpha_{kii} = \mu_{iikl} = \mu_{ijkk} = 0$.

To illuminate the signification of this complicated equation, let us consider the first moment of eq. (2.18), obtained by multiplication with Ω_i and integration:

$$
\begin{aligned}
\partial_t \langle \Omega_i \rangle \quad = \quad & -U_k \partial_k \langle \Omega^i \rangle + (\partial_k U_i) \langle \Omega_k \rangle - 2\alpha_{kil} \partial_k \langle \Omega_l \rangle \\
& + \beta_{kl} \partial_k \partial_l \langle \Omega_i \rangle.
\end{aligned}
\tag{2.20}
$$

In addition to the standard vorticity advection and stretching by the large scale, one recognize two additional effect: one proportional to α, resulting in large-scale vorticity generation through the AKA instability [11]; one proportional to β, akin to a turbulent viscosity. Within the Kraichnan model, one therefore naturally recovers the well-known formulation of turbulent transport, without resorting to scale separation [12]. In this very simple model, where the viscosity has been ignored, one can show that the tensor β is always positive: the turbulent viscosity always enhances turbulent transport. In actual viscid flows, the turbulent viscosity tensor is actually fourth order, and can be negative [12].

2.6.2.2.3. Limitation This digression about turbulent transport shows that the way we prescribe velocity correlation in Kraichnan model somehow determines the turbulent transport properties of the flow. It is a kind of adjustable parameter. In that respect, it would be nice to devise a model devoid of this freedom of choice, by ensuring for example that the turbulent transport somehow adjusts itself to the way energy is injected and dissipated, as in real turbulence. In the sequel, we present a model where the noise is dynamically computed at each time scale, thereby removing the arbitrariness of the Langevin model.

2.6.2.3. Stochastic RDT model

2.6.2.3.1. Description This method is based on the observation that small scales are mostly slaved to large scale via linear processes akin to rapid distortion. This observation is substantiated by various numerical simulations and is linked with the prominence of non-local interactions at small scale [13]. Specifically, let us decompose our small-scale velocity field into wave packets, via a localized Fourier transform:

$$\hat{u}_i(x, k) = \int h(x - x')e^{ik(x-x')}u_i(x')dx',$$

where h is a filtering function, which rapidly decays at infinity. Using incompressibility and non-locality of interaction, one can derive the following equation of motion for the wave-packet [13, 14]:

$$\begin{aligned} \dot{x}_i &= U_i, \\ \dot{k}_i &= -k_j\partial_j U_i, \\ \dot{\hat{u}}_i &= -\nu_t k^2 \hat{u}_i + \hat{u}_j\partial_j\left(2\frac{k_i k_m}{k^2}U_m - U_m\right) + \hat{\xi}_i. \end{aligned} \tag{2.21}$$

Here ν_t is a turbulent viscosity describing the local interactions between small-scales, and ξ is a forcing stemming from the energy cascade. Its expression only involves large-scale non-linearities (in fact the aliasing) via $\xi_i = \partial_j(U_j U_i - \overline{U_i U_j})$. By eq. (2.21) the wave-packet is transported by the large-scale flow, its local wavenumber is distorted by the large-scale velocity gradients, and its amplitude is modified through the action of local and non-local interactions. The different processes at work are summarized in Fig. 17.

The equation describing its amplitude evolution is a generalized Langevin equation, with friction generated by turbulent viscosity and with both multiplicative and additive noise stemming from interaction with large scale. Because these two noises are of same origin, they are correlated. One can show that this correlation is responsible for a skewness in the probability distribution of the small

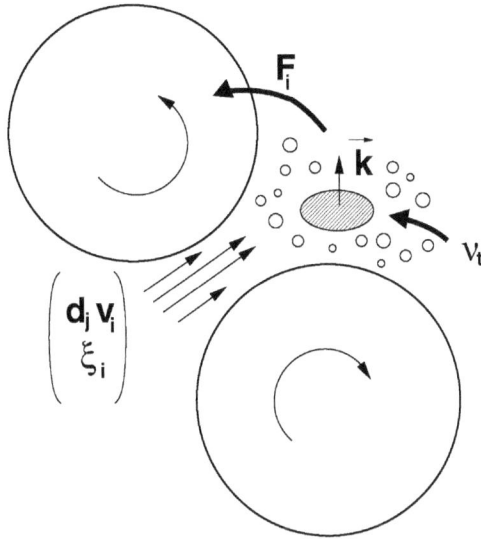

Fig. 17. Schematic representation of RDT processes.

scale [13]. Note also that in some sense, our Langevin model can be viewed as a generalization of Rapid Distortion Theory equation, with inclusion of turbulent viscosity and stochastic forcing. No wonder, interesting analytical properties are available in precisely the same case where Rapid Distortion Theory is the most useful, namely rotating, or stratified shear flows [15, 16].

2.6.2.3.2. Analogy with MHD Another interest of the stochastic model can be seen by rephrasing the model in the physical space, and considering the vorticity $\omega = \nabla \times u$ instead of the velocity. The RDT equation then writes:

$$\partial_t \omega = \nabla \times (\mathbf{U} \times \omega) + \nu_t \Delta \omega + \nabla \times \xi. \tag{2.22}$$

Moreover, one must not forget that the small scales react back onto the large scale through the Reynolds stress tensor. This may be summarized as:

$$\partial_t \mathbf{U} + \mathbf{U} \cdot \nabla \mathbf{U} = -\nabla P + \nu \nabla^2 \mathbf{U} + \mathbf{u} \times \omega + \mathbf{f}. \tag{2.23}$$

This set of coupled equation looks conspicuously alike the set of MHD equations to be considered in the next section, providing one identifies ω with the magnetic field and the Reynolds stress $\mathbf{u} \times \omega$ with the Laplace force $\mathbf{j} \times \mathbf{B}$. A noticeable difference is that $u = curl^{-1}\omega$ instead of $j = \nabla \times B$ in the MHD world. This difference is responsible for different conservation laws, and provides different laws of saturations (see below). However, everything that will be said in the linear stage of dynamo will be also true for the RDT model. Note that such an analogy between vorticity and MHD has been known for a long time, but in situation where U and u were equal, so that ω and U were not independent as they are here. The analogy is therefore much more profound in the RDT case, than in the general case where $U = u$. This is another argument to say that turbulence and dynamo ARE strongly linked.

2.6.2.3.3. Conservation laws   In the unforced $\xi = f = 0$ and inviscid case $\nu = \nu_t = 0$, the set of equation (2.22)–(2.23) may be shown to conserve several global quantities, namely the total energy E, the large and small scale helicity, H and h, and some cross helicities H_c and H_c:

$$H = \int \mathbf{U} \cdot \Omega \, d\mathbf{x}, \qquad\qquad (2.24)$$

$$h = \int \mathbf{u} \cdot \omega \, d\mathbf{x},$$

$$H_c = \int (\mathbf{U} \cdot \omega + \mathbf{u} \cdot \Omega) \, d\mathbf{x},$$

$$E = \frac{1}{2} \int (\mathbf{U}^2 + \mathbf{u}^2) \, d\mathbf{x}.$$

2.7. Influence of rotation

2.7.1. Generalities

Most natural objects (planets, stars, galaxies) are rotating. It is therefore interesting to investigate the influence of rotation onto turbulence. In a rotating flow, a new dimensional parameter becomes relevant: the rotation rate Ω. A new non-dimensional parameter is therefore required to characterize the corresponding turbulent flow. One may choose for example, the Rossby number $Ro = V/L\Omega$, or its inverse, the rotation number $\theta = L\Omega/U$ or the Ekman number $E = \nu/\Omega L^2 = Ro/Re$. The first two numbers are generally favored by astrophysicists since it does not require the knowledge of the viscosity to compute it while the last number is generally preferred by geophysicists. In the sequel, I shall follow the astrophysicists mood, mainly because I prefer to work with numbers of order unity.

The Rossby number is nothing but the ratio of a characteristic time scale of rotation over a characteristic time scale of turbulence. Therefore, when $Ro > 1$, the rotation time scale is longer than the turbulence time scale, and one may think that the turbulence does not really feel the rotation. On the other hand, when $Ro < 1$, turbulent eddies have no time to overturn in a rotation period, and one may think that rotation will become important. How and why depends of course on the geometry: since the Coriolis force only acts in a plane perpendicular to the rotation axis, this induces a natural anisotropy in the problem. In the sequel, we shall not review the vast literature on the subject, but focus on features observed in the rotating von Karman experiment.

2.7.2. *Rotating von Karman flow*

There are actually two ways to obtain an experimental rotating von Karman flow: one is to rotate the outer cylinder at some rotation frequency F_{rot}, keeping the two propellers in exact counter-rotation at frequency F; a second one is to rotate one propeller at a frequency F_1 and the second propeller at the frequency F_2. During his thesis, L. Marié showed that this second operating procedure was equivalent to the first one (within dissipation at the cylinders boundaries, that are negligible at large Reynolds number), providing one identifies F to $(F_1 + F_2)/2$ and F_{rot} to $(F_1 - F_2)/2$. The second procedure is much simpler than the first one, but requires a very fine control of the rotation frequencies to reach Rossby number less than unity, that is regime where rotation is dominant. In the VKS experiment, only the second procedure has been used so far, allowing the rotation number $\theta = (F_1 - F_2)/(F_1 + F_2)$ to vary between -1 and 1-corresponding to regime of turbulence predominant over rotation. We have shown on Fig. 18 the operating regime of the VKS experiments, to be compared with other natural objects. One sees that VKS operates in rotation number equivalent to that of sun-like-stars, but that another operating set up should be used to reach the parameters regime of the earth.

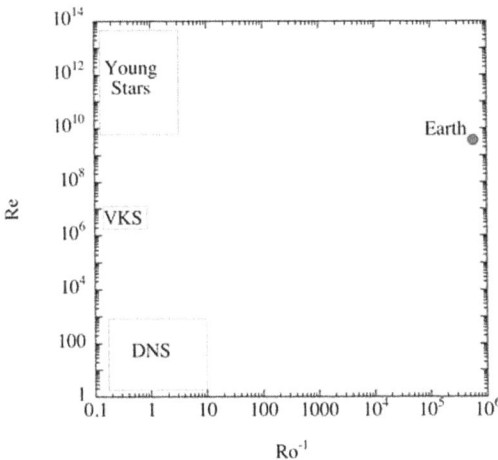

Fig. 18. Typical range of Reynolds number and Rossby number for young stars, the earth, VKS2 experiment and DNS simulations.

2.7.3. Mean-flow modifications

A first noticeable influence of rotation is a modification of the mean-flow: one sees on Fig. 19 that both the toroidal part and the poloidal part of the flow are increased by the rotation, albeit with some anisotropy. At $Ro = 1/6$, the maximum of the toroidal (resp. poloidal) component has been increased by a factor 3 (resp. 5), so that rotation tends to increase the ratio poloidal/toroidal. Another

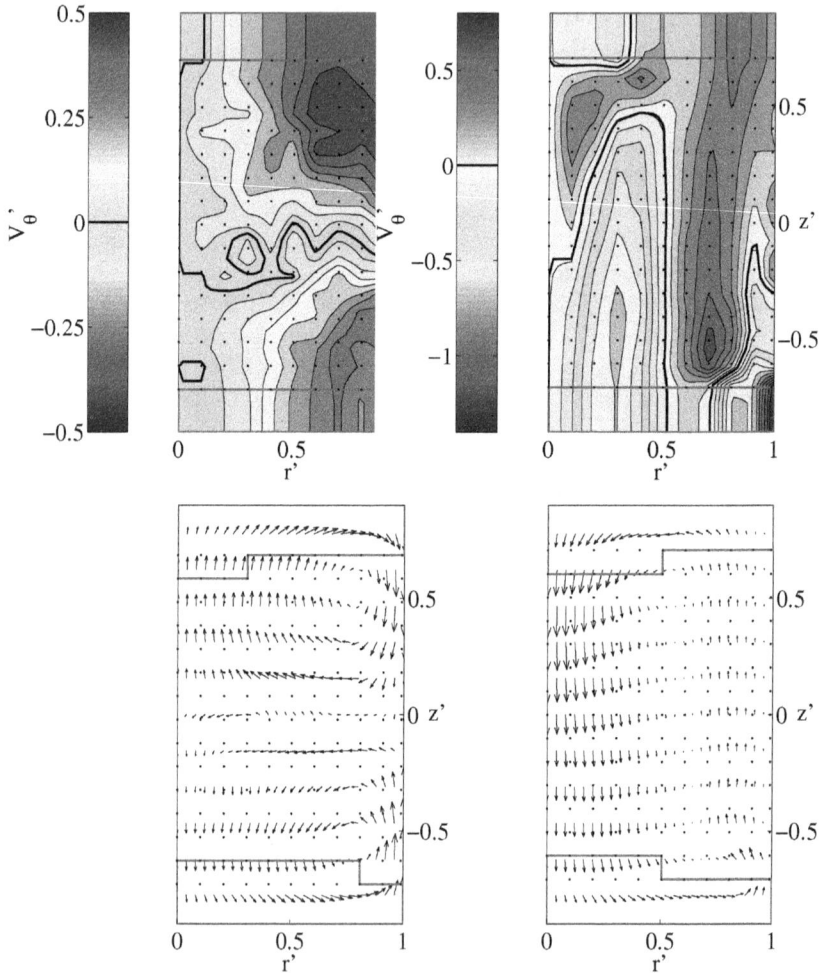

Fig. 19. Comparison of mean velocity field between a non-rotating von Karman flow (left) and a rotating von Karman flow at $Ro = 1/6$ (right). Upper panel: azimuthal field. Lower panel; poloidal field.

striking feature is the bi-dimensionalization of the flow: velocities tends to become independent of the coordinate along the vertical axis. This is in agreement with the Taylor-Proudman theorem.

2.7.4. Fluctuations modifications

With regards to fluctuations, the situation looks unclear. From a visual point of view (Fig. 20), rotation does not seems to decrease fluctuations in any direction.

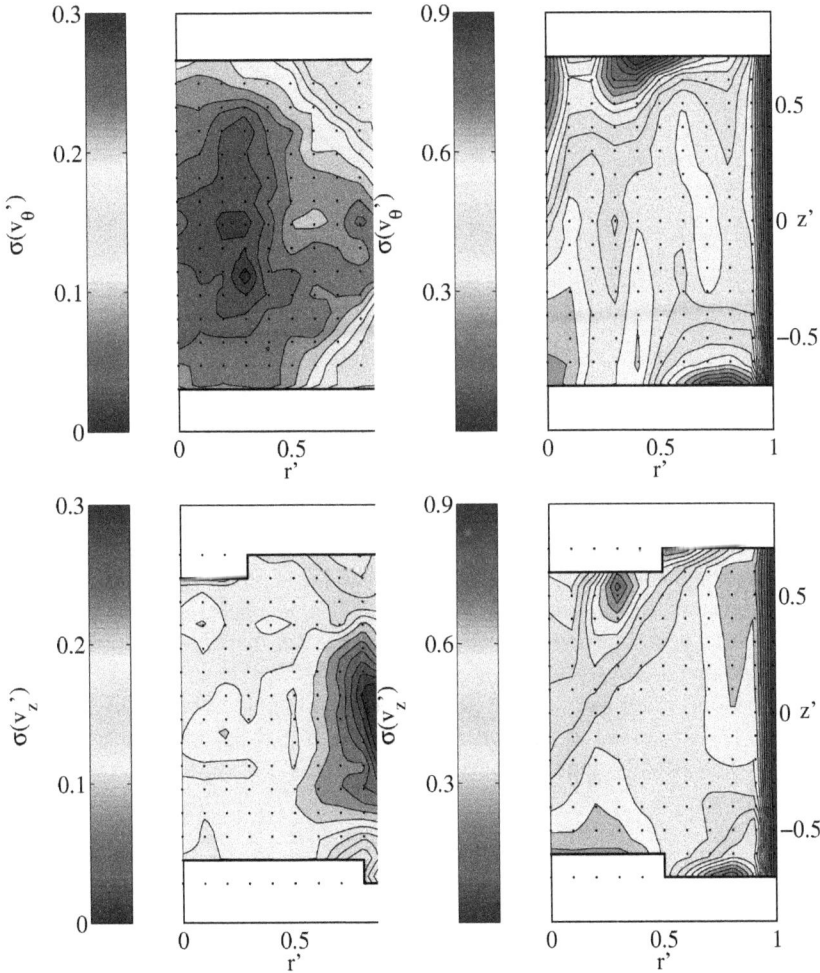

Fig. 20. Comparison of fluctuations between a non-rotating von Karman flow (left) and a rotating von Karman flow at $Ro = 1/6$ (right). Upper panel: azimuthal fluctuations. Lower panel; vertical fluctuations.

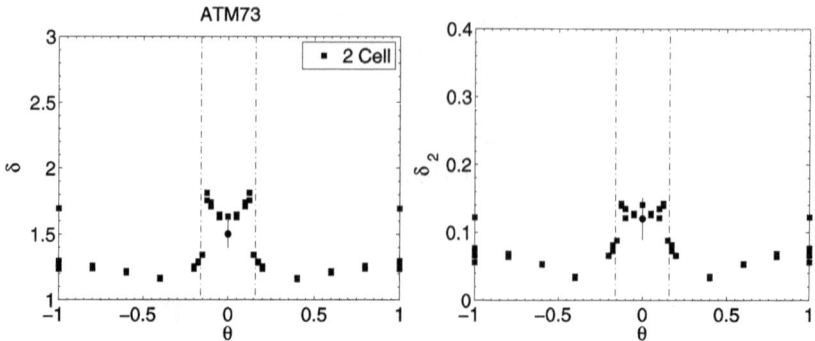

Fig. 21. Variation of the parameter δ (left) and δ_2 (right) with the rotation number θ in a von Karman flow with an annulus and TM73 propellers, in contrarotation. The value of δ has been obtained by statistics over 400 images. The dot is the value of δ and δ_2 at $\theta = 0$ measured over 5000 images. The factor 1.13 in between the two measurements is an indication that the present δ is not fully converged. In the sequel, we correct its value by the factor 1.13 in all applications.

However, the parameter δ is strongly affected by rotation (Fig. 21), decreasing by over 30 percent with respect to the non-rotating case.

A possible explanation to conciliate these two results is that rotation preferentially kills large scale fluctuations, since they have the smallest Rossby number. This results in a large reduction of δ, while visual inspection of velocity profile, that is more sensitive to small scales, does not show any change.

2.7.5. Transport modification

The result on the transport may look even more surprising (Fig. 22): in the presence of rotation, the non-dimensional torque K_p is increased with respect to the non-rotating case. In previous sections, we have seen that K_p is proportional to $\delta - 1$, and that δ is decreased by the rotation!

This shows that there is a missing ingredient to explain the torque increase, and that a serious look to angular momentum transport balance has to be taken at. This was actually done by L. Marié in his thesis. He showed that the actual balance writes:

$$\Gamma \approx \nu \int \rho_0 \partial_r < v_\theta > dS + \int \rho_0 < v_\theta v_z > dS + 2\Omega \int \rho_0 \partial_r < v_r > dS.$$

$$(2.25)$$

We recognize in the first term our viscous contribution, and in the second term the fluctuation contribution, already considered previously. The surprise comes from the third term, that represents angular momentum transport by the mean-flow. Now we understand where the surprise comes from: in the non-rotating

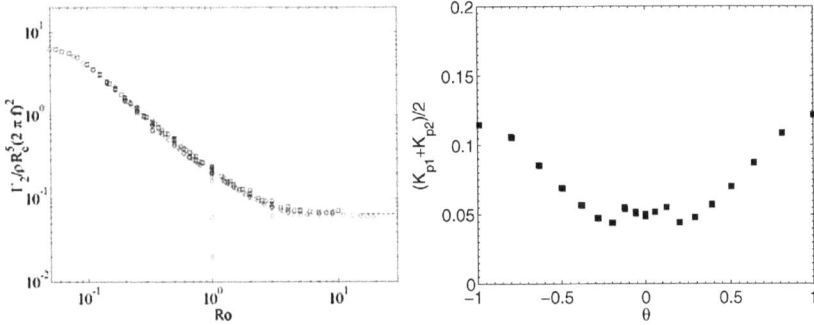

Fig. 22. Left: Non-dimensional torque at a function of the Rossby number $Ro = F/F_{rot}$ in a VK experiment: blue squares: in VKR, co-rotating propellers TM60 at frequency F, with overall rotation F_{rot}; red diamonds: in a VK experiment with TM60 propellers rotating at different speed F_1 and F_2, and with $F = (F_1 + F_2)/2$, $F_{rot} = (F_1 - F_2)/2$. Right: Non-dimensional torque as a function of $\theta = 1/Ro$ in a von Karman experiment with TM73 propellers in contrarotation at different speed F_1 and F_2.

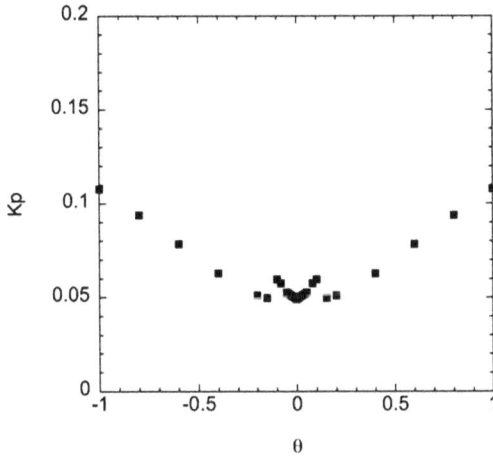

Fig. 23. Synthetic K_p as a function of δ from TM73 propellers in contra direction. K_p is built using the phenomenological formula $K_p = 0.07|\theta| + 0.03(\delta - 1)$ with $\delta(\theta)$ as given in Fig. 21.

case, the mean-flow is negligible in the mixing layer, and all the transport is ensured by the fluctuations. In the rotating case, the mean flow is increased with respect to the rotating case, and it becomes almost constant along the rotation axis, yielding a non-negligible contribution. The magnitude of this contribution can be phenomenologically estimated from the Coriolis force (the third term) as $U_r r\Omega/F^2 \propto |\theta|$, yielding a total K_p of:

$$K_p(\theta) = A/Re + C(\delta(\theta) - 1) + D|\theta)|. \tag{2.26}$$

With the variation of δ given in Fig. 21, this gives Fig. 23 that reproduces rather well the actual curve.

3. Dynamo

3.1. Generalities

3.1.1. Equations
In the case where a magnetic field is present, the equations to be considered are now the coupled equations:

$$\partial_t \mathbf{v} + (\mathbf{v} \cdot \nabla)\mathbf{v} = -\frac{1}{\rho_0}\nabla P + \mathbf{j} \times \mathbf{B} + \nu\Delta\mathbf{v} + \mathbf{f},$$
$$\partial_t \mathbf{B} = \nabla \times (\mathbf{v} \times (\mathbf{B}) + \eta\Delta\mathbf{B}, \tag{3.1}$$

where $\mathbf{v}$ is the fluid velocity, P is the pressure, $\mathbf{B}$ is the Alfven velocity (or equivalently $\sqrt{\rho\mu_0}\,\mathbf{B}$ is the magnetic field), $j = \nabla \times \mathbf{B}$ is the magnetic current, ν and η are the viscosity and the magnetic diffusivity, and ρ_0 is the (constant) fluid density. Note the analogy with the RDT equations (2.23)–(2.22) we derived earlier. In this MHD case, two new dimensionless parameter become necessary. For example, $Rm = LV/\eta$ (or equivalently $Pm = Rm/Re$), and the interaction parameter $N = RmB^2/V^2$. The former is the ratio of the magnetic field stretching by the velocity field over the magnetic field diffusion; the latter characterizes the ratio of the Lorentz force over the velocity advective term, and is a measure of the "non-linearity" of the MHD system: when $N \ll 1$, the magnetic field does not react back to the velocity field, and the two equations decouple.

In TG numerical simulations, one can reach values of Rm up to 100 (Pm varying from 1 to 0.01). In VKS experiment, Rm cannot exceed 50, with $Pm = 2 \times 10^{-5}$.

3.1.2. Conservation laws
In the non-rotating, inviscid, force free-limit, the set of equations (3.1) conserve at least three global quantities, namely:

$$H_m = \int \mathbf{A} \cdot \mathbf{B}\,d\mathbf{x}, \tag{3.2}$$
$$H_c = \int \mathbf{v} \cdot \mathbf{B}\,d\mathbf{x},$$
$$E = \frac{1}{2}\int (\mathbf{v}^2 + \mathbf{B}^2)\,d\mathbf{x},$$

where H_m is the magnetic helicity, H_c is the cross-helicity and E is the total energy. Note that due to the Lorentz force, the kinetic helicity is not conserved anymore, unlike in the pure hydrodynamical case (see Section 2.6.2.3.3.).

3.2. Linear theory: instability

In this section, we focus on the "linear" (also called "kinematic") case, where $N \ll 1$, so that v can be considered as independent of B.

3.2.1. Laminar vs turbulent

Like previously, we decompose the velocity field into a mean (time averaged) $\overline{U}$ and a fluctuating part u'. Since the velocity field is independent of B, we can study the evolution of B from the induction equations, that reads:

$$\partial_t \mathbf{B} = \nabla \times \left(\overline{\mathbf{U}} \times \mathbf{B} \right) + \nabla \times \left(\mathbf{u}' \times \mathbf{B} \right) + \eta \Delta \mathbf{B}. \tag{3.3}$$

This is a linear equation. Since $\overline{U}$ is by construction time-independent, it admits exponentially growing or decaying solutions in the absence of the second term of the r.h.s., like in any classical instability problem. The natural non-dimensional parameter to quantify the importance of the fluctuating term is $\epsilon = \delta - 1$. Therefore, when $\epsilon \ll 1$, we have a "laminar" instability, with exponential growth or decay. The frontier in between the two behavior is the dynamo threshold, that will be close to the instability threshold computed only the mean flow.

For ϵ of order unity, the fluctuating term becomes important, and the equation now includes a time-dependent, stochastic like behavior. The instability is now akin to an instability in presence of a multiplicative noise, and requires special tools to be detailed later.

From the behavior of the parameter $\delta - 1$ detailed in previous section, we see that the dynamo is probably laminar for Taylor–Green flow at $Re < 20$, or for VKS with an annulus and rotation, while it is probably turbulent for a TG flow with $Re > 50$ and for VKS without an annulus.

3.2.2. Laminar dynamo

Laminar dynamo are countless. Some, like Ponomarenko or Robert's dynamo, can even be studied analytically. Here we focus on the TG and VKS laminar dynamos. In the case of the TG flow, the laminar dynamo is characterized by two windows of instability [35, 36] (Fig. 24): the dynamo takes place for $Rm_{c1} < Rm < Rm_{c2}$ and for $Rm > Rm_{c3}$. The three critical magnetic Reynolds number have been computed for mean flows measured at different Reynolds number $6 < Re < 100$ and were found to be roughly independent of Re. With the forcing adopted in [35], one finds: $Rm_{c1} \sim 6$, $Rm_{c2} \sim 13$ and $Rm_{c3} \sim 25$.

Because it was at the heart of VKS optimization, the laminar dynamo has been studied with different codes, and different boundary conditions or propeller shape and size. The lower threshold were obtained for TM73 propellers. In addition, it was found that the addition of a layer of sodium at rest produces a significant reduction of the dynamo threshold from $Rm_c \sim 180$ to $Rm_c \sim 40$ [19, 20],

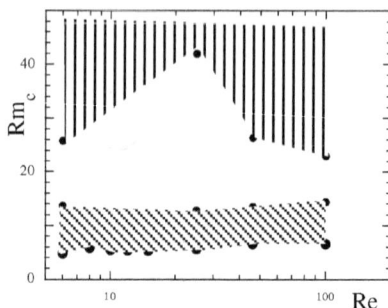

Fig. 24. Windows of kinematic dynamo action with a time-averaged TG flow, as a function of the Reynolds number Re. The dashed area corresponds to region of the parameter space where kinematic dynamo is observed, corresponding to positive values of the Lyapunov exponent.

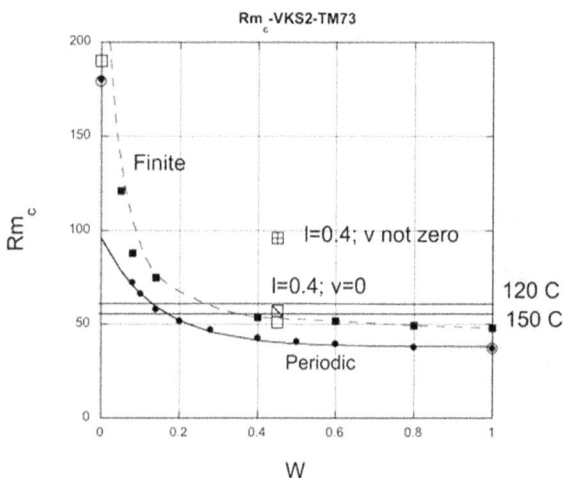

Fig. 25. Critical value of the magnetic Reynolds number as a function of the percentage of sodium at rest W from kinematic simulation with time-averaged von Karman flow with inox TM73 propellers rotating in the contra direction. Note that the kinematic simulation with iron propellers have not yet been done. Filled circle [20] and open circle (courtesy C. Nore): with periodic axial boundary conditions; Filled square: with finite axial boundary conditions [21]; open square (resp. square with cross): when taking into account the thin layer of fluid at rest (resp. stirred) behind the impellers [22]. The two solid line delimit the largest Rm that can be reached in the VKS2 experiment at the lowest (resp. largest) operating temperature 120 C (resp. 150 C).

and that the moving sodium behinds the propeller had a tendency to increase the dynamo threshold [21, 22]. This is summarized in Fig. 25.

Specifically, the various threshold found with the kinetic simulation based on the time-averaged velocity field with a layer of resting sodium of size $w = 0.4$ are:

• $Rm_c = 43 \pm 1$ for periodic axial boundary conditions in a homogeneous conducting domain [20];
• $Rm_c = 49 \pm 2$ for finite axial boundary conditions;
• $Rm_c = 57$ (resp. 95) when taking into account the thin layer of fluid at rest
(resp. stirred) behind the impellers [22,];
• $Rm_c = 46$ without the fluid behind the impellers for more realistic conditions:
finite axial boundary condition, 5 mm copper shell separating the flow and the
static conducting layer, copper container.
• $Rm_c = 55$ (resp. $Rm_c = 150$) for these conditions with the fluid behind the
impellers at rest (resp. stirred). These results are given by [21].
In the experiments, dynamo has been observed with iron propellers, with a threshold $Rm_c \sim 32$ in contra rotation. With inox propellers, no dynamo has been
observed in contrarotation. However, induction measurements with an external
applied field B_a can be used to estimate a dynamo threshold via the response B_i
as $B_a/B_i \sim \Lambda \sim Rm - Rm_c$. Linear fit to the induction curve B_a/B_i then gives
(see Fig. 26):
• $Rm_c = 127$ for TM73 inox propellers with no resting sodium and no annulus
(VKS2b campaign);
• $Rm_c = 67$ for TM73 inox propellers with $w = 0.4$ of resting sodium and no
annulus (VKS2a campaign);
• $Rm_c = 53$ for TM73 inox propellers with $w = 0.4$ of resting sodium and an
annulus (VKS2f campaign);
• $Rm_c = 32$ for TM73 iron propellers with $w = 0.4$ of resting sodium and an
annulus (VKS2g campaign).
The decrease of Rm_c seen between 2b and 2a suggests that indeed the resting
sodium is favorable to dynamo action. The difference between 2a and 2f threshold suggests that the turbulence (described by the parameter δ) has an impact on
the dynamo threshold. This is the subject of the next section.

3.2.3. Turbulent dynamo
3.2.3.1. Perturbation theories
We consider now a situation where fluctuation
are non-negligible. A first natural approach is to identify a small parameter ϵ in
the problem, and try and solve the full problem by perturbation theory. Specifically, one consider first the time-averaged of Eq. (3.3):

$$\partial_t \overline{\mathbf{B}} = \nabla \times \left(\overline{\mathbf{U}} \times \overline{\mathbf{B}} \right) + \nabla \times \left(\overline{\mathbf{u}' \times \mathbf{b}'} \right) + \eta \Delta \overline{\mathbf{B}}. \tag{3.4}$$

The main idea is to find the shape of $\overline{\mathbf{u}' \times \mathbf{b}'}$ as a function of $\overline{B}$ through the
perturbation expansion.

3.2.3.1.1. Scale separation
An historically successful approach is to con-
sider an ideal case where there is a scale separation between the typical scale l of

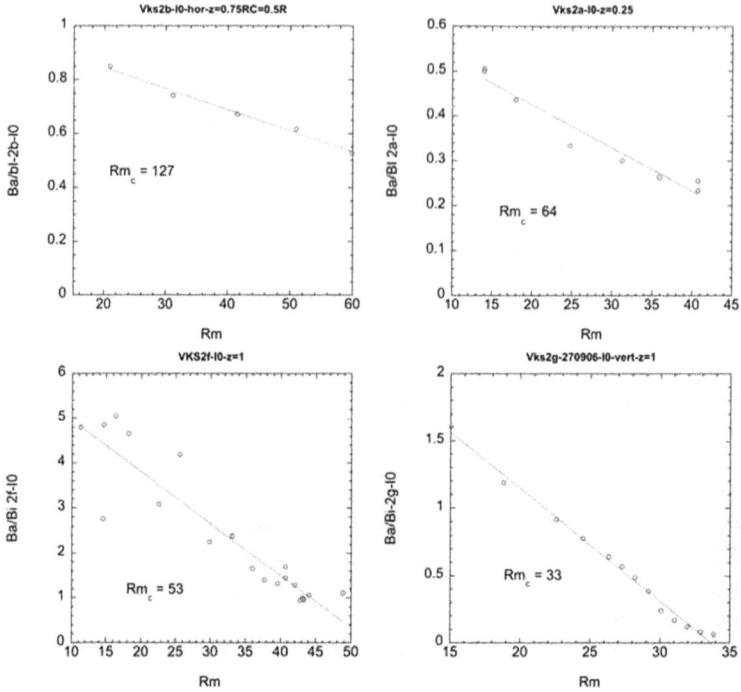

Fig. 26. Estimate of dynamo threshold from induction measurements in VKS2 through the quantity B_a/B_i in four different configuration: Upper left: 2b: no resting sodium, TM73 inox propellers, no annulus; Upper right: 2a: 40% of resting sodium, TM73 inox propellers, no annulus; Lower left: 2f: 40% of resting sodium, TM73 inox propellers, annulus; Lower right: 2g: 40% of resting sodium, TM73 iron propellers, annulus. The lines are linear fit, providing the value of Rm_c indicated on each plot.

(u', b') and the typical scale L of $\overline{B}$. The natural expansion parameter is therefore $\epsilon = l/L \ll 1$, or equivalently $\nabla \overline{B}$. Without any computations, one can then infer that

$$\epsilon_{ijk}\overline{u'_j \times b'_i} = \alpha_{ij}\overline{B_j} + \beta_{ijk}\nabla_j\overline{B_k} + O(\epsilon^2), \tag{3.5}$$

where α_{ij} and β_{ijk} are two tensors that depend on the velocity field and that can be computed through classical perturbation procedure applied to Eq. (3.3) (see e.g. [12]). When plugged back into (3.4), this expansion gives:

$$\partial_t\overline{B} = \nabla \times \left(\overline{U} \times \overline{B}\right) + \nabla_j\alpha_{ijk}\overline{B_k} + \nabla_j\nabla_k(\beta_{ijkl} + \eta\delta_{jk}\delta_{il})\overline{B_l}, \tag{3.6}$$

where $\alpha_{ijk} = \epsilon_{ijm}\alpha_{mk}$ and $\beta_{ijkl} = \epsilon_{ikm}\beta_{mjl}$. α is the famous alpha coefficient, while β is a turbulent diffusivity tensor, that need not be definite positive. This equation is the analog of Eq. (2.21) derived for the mean vorticity from the Kraichnan model. In the absence of mean flow, it usually leads to a large scale instability via the alpha effect.

3.2.3.1.2. Weak fluctuation We can also consider that the fluctuations are weak, so that $\epsilon = u'^2/U^2 = \delta - 1$. This case has been considered by Fauve and Petrelis [31] for the Roberts flow. The details of the perturbation expansion are rather technical and not very interesting. However, the results of the expansion regarding the dynamo threshold displacement with respect to the fully laminar case $\Delta Rm_c = Rm_c(\delta) - Rm_c(1)$ show that this displacement is linear in $\delta - 1$.

3.2.3.2. Stochastic computations

3.2.3.2.1. Philosophy Non-perturbation computations can be performed both analytically and numerically by replacing the true velocity fluctuations by some well chosen noise. We have seen in previous section that most global properties of the turbulence could be reproduced by a shortly correlated noise. Of course, real turbulence is characterized by temporal and spatial correlation that cannot be captured by such a simple noise. One can however hope that first order effects can be captured by our simple model. The reader can judge by himself from the final comparison. In any case, the advantage of these stochastic computations is twofold: first, they allow for non-perturbation analytical and numerical computations; second, their numerical cost is equivalent to the cost of a kinematic simulation. Simulation of 64^3 can then prove sufficient to explore a range of fluctuations equivalent to $Re = 10^7$, (i.e. that would require 10^{15} grid points).

3.2.3.2.2. Generalities We therefore now consider that u' is a white noise, characterized by:

$$\langle u_i'(x,t)u_j'(x',t')\rangle = 2G_{ij}(x,x')\delta(t-t'),\tag{3.7}$$

where the brackets denote ensemble average, over the realizations of the noise. Equation (3.3) then takes the shape of a stochastic partial differential equation for B, with multiplicative noise. The problems associated with this type of noise can be understood by looking at a simple unidimensional model:

$$\dot{x} = \mu x + \xi x,\tag{3.8}$$

where $\xi(t)\xi(t') = 2D\delta(t-t')$. In the absence of noise, x is exponentially increasing (unstable) as soon as $\mu > 0$. In the presence of noise, we can take different moments of the equation and get the following hierarchy:

$$
\begin{aligned}
<\dot{x}> &= (D+\mu)\\
<\dot{x}^2> &= 2(2D+\mu)
\end{aligned}
\tag{3.9}
$$

so that the $< x >$ (resp. $< x^2 >$) is unstable for $\mu > -D$ (resp. $\mu > -2D$). Therefore, its seems that the instability threshold depends on the moment we consider! One can in fact prove that this pathology is due to the absence of non-linear terms, and that in fact the correct threshold that would prevail with non-linear term is captured by considering the Lyapunov:

$$\Lambda = \partial_t < \ln x > . \tag{3.10}$$

Due to the convexity of the log, $\Lambda \leq \partial_t \ln < x >$, so that Λ is always smaller than the growth rate. The system is unstable as soon as $\Lambda > 0$.

3.2.3.2.3. Analytical computations   Analytical computation of the stochastic model have been done by Leprovost and Dubrulle [44]. In order to make the computations tractable, two approximation were made: i) a saturating term was added to the induction equation as $-cB^2 B_i$ because of symmetry consideration. In some sense, this modification is akin to an amplitude equation, and the cubic shape for the non-linear term could be viewed as the only one allowed by the symmetries. Such a procedure is motivated by the observation that the precise form of the nonlinear term does not affect the threshold value. ii) The diffusivity was ignored.

Using standard techniques [26,29], one can then derive the evolution equation for $P(\mathbf{B}, x, t)$, the probability of having the field $\mathbf{B}$ at point x and time t:

$$\begin{aligned} \partial_t P &= -\bar{U}_k \partial_k P - (\partial_k \bar{U}_i)\partial_{B_i}[B_k P] + \partial_k[\beta_{kl}\partial_l P] \\ &\quad + c\partial_{B_i}[B^2 B_i P] + 2\partial_{B_i}[B_k \alpha_{lik}\partial_l P] \\ &\quad + \mu_{ijkl}\partial_{B_i}[B_j \partial_{B_k}(B_l P)], \end{aligned} \tag{3.11}$$

with the following turbulent tensors:

$$\beta_{kl} = \langle u'_k u'_l \rangle, \quad \alpha_{ijk} = \langle u'_i \partial_k u'_j \rangle \quad \text{and} \quad \mu_{ijkl} = \langle \partial_j u'_i \partial_l u'_k \rangle. \tag{3.12}$$

Due to incompressibility, the following relations hold: $\alpha_{kii} = \mu_{iikl} = \mu_{ijkk} = 0$.

The physical meaning of these tensors can be found by analogy with the "Mean-Field Dynamo theory" [17, 30]. Indeed, consider the equation for the evolution of the mean field, obtained by multiplication of equation (3.11) by B_i and integration:

$$\begin{aligned} \partial_t \langle B_i \rangle &= -\bar{U}_k \partial_k \langle B^i \rangle + (\partial_k \bar{U}_i)\langle B_k \rangle - 2\alpha_{kil}\partial_k \langle B_l \rangle \\ &\quad + \beta_{kl}\partial_k \partial_l \langle B_i \rangle - c\langle B^2 B_i \rangle. \end{aligned} \tag{3.13}$$

This equation resembled the classical Mean Field Equation of dynamo theory, with generalized (anisotropic) "α" and "β". The first effect leads to a large scale instability for the mean-field, while the second one is akin to a turbulent diffusivity. Note that the tensor μ does not appear at this level.

The Lyapunov exponent can be computed in a similar way from (3.11) by changing variable $B_i = Be_i$, then multiplying the resulting equation by $B^{d-1} \ln B$ and integrating with respect to B. This yields:

$$\Lambda \equiv \partial_t \langle \ln B \rangle = \langle \partial_k \bar{U}_i e_i e_k \rangle_\phi + \langle \mu_{ijkl} (\Delta_{ik} e_j e_l + \Delta_{kj} e_i e_l) \rangle_\phi, \qquad (3.14)$$

where we used $\Delta_{ij} = \partial_{n_i}(n_j) = \delta_{ij} - e_i e_j$ an "angular Dirac tensor", and the symbol $\langle \bullet \rangle_\phi$ denotes verages over the angular variables.

3.2.3.2.4. Condition for dynamo action The condition for dynamo action in this model is $\Lambda > 0$. In the limit of zero noise, the term proportional to μ is negligible and one recovers the classical criterion for instability in a laminar dynamo in the infinite Prandtl number limit. Indeed, in such a case, the magnetic field will mainly grow in the direction given by the largest eigenvalue λ_{max} of $S_{ij} = \partial_j \bar{U}_i$, so that

$$\Lambda \approx \langle \partial_k \bar{U}_i e_i e_k \rangle_\phi = \lambda_{max}. \qquad (3.15)$$

There will be dynamo only if $\lambda_{max} > 0$.

Consider now a situation where you increase the noise level. Two different influences on the sign of Λ then result: one through the factor proportional by μ. According to the sign of this factor, it can therefore favor or hinder the dynamo. In isotropic homogeneous turbulence, μ is positive, so that it is in general favorable to dynamo action. Moreover, being proportional to derivatives of the noise, this term gets larger as the typical scale of the noise is small.

However, there exists another less obvious -and adverse- influence of the noise: the disorientation effect. Indeed, noise changes the distribution of magnetic field orientation. In the absence of noise, the latter tends to be oriented towards the most unstable direction. However, noise constantly drives the magnetic field away from this favorable direction, sometimes even driving it towards a stable direction, where the magnetic field exponentially decreases. The net result is a decrease of the effective growth rate of the magnetic field. A phenomenological way to quantify this effect is through the parameters $\delta - 1$ and δ_2. Indeed, the largest these coefficient are, the further away the instantaneous velocity field is from the averaged field, and the largest the disorientation effect can be. This effect is more important when the noise it at largest scale, since in that case the disorientation effect is more pronounced-one can get farther from the mean flow.

From this discussion, one expects large scale noise to be adverse to dynamo action-through the disorientation mechanism, while small scale noise should be favorable to dynamo action-through the μ effect.

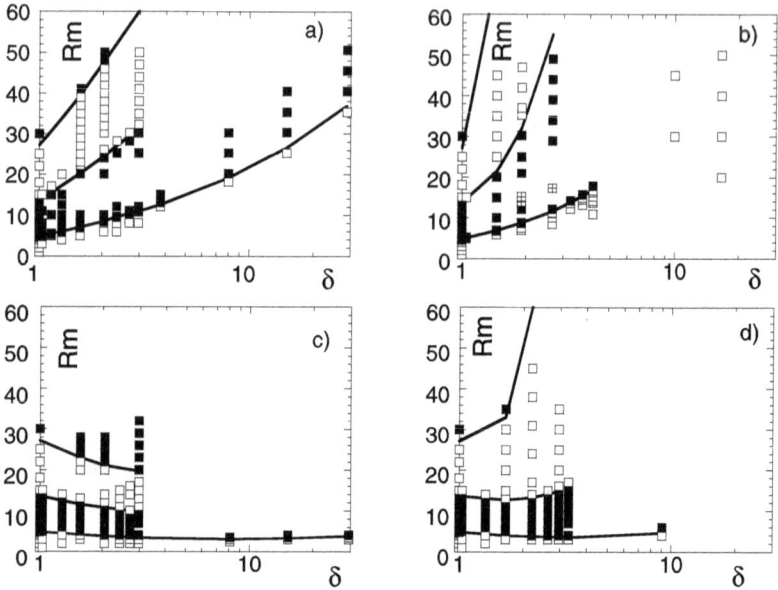

Fig. 27. Parameter space for noise at $Re = 6$ for different noise parameters: a) $\tau_c = 0$, $k_I = 1$; b) $\tau_c = 0.03$, $k_I = 1$; c) $\tau_c = 0$, $k_I = 16$; d) $\tau_c = 0.03$, $k_I = 16$. Open square: no-dynamo case; Square with cross: undecided state; Filled square: dynamo case. The full lines are zero-Lyapunov lines.

3.2.3.2.5. Numerical simulations   The previous analytical computation were tractable only in the limit $\eta \to 0$ $(Rm \to \infty)$. To investigate the more realistic case of finite diffusivity, one may resort to numerical simulations. This has been done by Laval et al. [35, 37] for the case of the Taylor–Green flow, without inclusion of the non-linear term in the induction equation. Two kinds of noise were tested: shortly correlated noise, like in the analytical case, and Markovian noise, with finite correlation time τ_c that can be varied from 0 to several eddy-turnover time. Two typical noise scale were also tested, one at the largest available scale of the system $k = 1$, and one of the order of the magnetic diffusive scale $k = 16$.

The results are summarized in Fig. 27 for time-correlated noises at large and small scale. In the case of the large scale noise, one sees that the two dynamo windows are lifted up by the noise, resulting in an increase of the dynamo threshold. It can also be shown that this effect becomes more pronounced as the correlation time of the noise increase until the mean eddy turn over time is reached. Above this, the effect does not change anymore. From the previous discussion, we can attribute this increase of the threshold to the disorientation effect. In contrast,

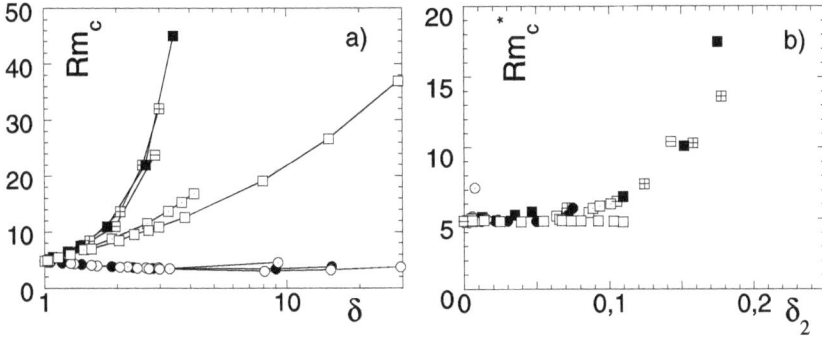

Fig. 28. Evolution of the dynamo threshold for KS simulations with $\bar{u}(Re = 6)$. a) Rm_c as a function of δ and b) Rm_c^* as a function of δ_2 for different noise parameters: $k = 1$: square $\tau_c = 0$; boxdot: $\tau_c = 0.1$ sec; boxminus: $\tau_c = 1$ sec; boxplus: $\tau_c = 8$ sec; black square $\tau_c = 50$ sec; $k = 16$: circle: $\tau_c = 0$; odot: $\tau_c = 0.1$ sec; bullet: $\tau_c = 50$ sec.

when the noise it at small scale, the dynamo threshold is -slightly- decreased with respect to the laminar case. This is probably a benefit of the μ effect.

The influence of the noise onto the first dynamo threshold can be summarized by plotting the critical magnetic Reynolds numbers as a function of the noise intensity (Fig. 28a). Large scale (resp. small-scale) noise tends to increase (resp. decrease) the dynamo threshold. For small noise intensities, the correction $Rm_c^{turb} - Rm_c^{MF}$ is linear in $\delta - 1$, in agreement with the perturbation theory [27]. Furthermore, one sees that for small scale noise, the decrease in the dynamo threshold is almost independent of the noise correlation time τ_c, while for the large scale noise, the increase is proportional to τ_c at small τ_c. At $\tau_c > 1$, all curves $Rm_c(\delta)$ collapse onto the same curve. We have further investigated this behavior to understand its origin. Increasing δ first increases of the flow "turbulent viscosity" $\overline{v_{rms}l_{int}}$ with respect to its mean flow value $V_{rms}L_{int}$. This effect can be corrected by considering $Rm_c^* = Rm_c V_{rms} L_{int}/\overline{v_{rms}l_{int}}$. Second, an increase of δ produces an increase of the fluctuations of kinetic energy, quantified by δ_2. This last effect is more pronounced at $k_I = 1$ than at $k_I = 16$. It is amplified through increasing noise correlation time. We thus re-analyzed our data by plotting Rm_c^* as a function of δ_2 (Fig. 28b). All results tend to collapse onto a single curve, independently of the noise injection scale and correlation time. This curve tends to a constant equal to Rm_c^{MF} at low δ_2. This means that the magnetic diffusivity needed to achieved dynamo action in the mean flow is not affected by spatial velocity fluctuations. This is achieved for small scale noise, or large scale noise with small correlation time scale. In contrast, the curve diverges for δ_2 of the order of 0.2, meaning that time-fluctuations of the kinetic energy superseding 20 percent of the total energy annihilate the dynamo.

B. Dubrulle

3.2.3.2.6. A new paradigm for turbulent dynamo   An obvious stationary solution of (3.11) is a Dirac function, representing a solution with vanishing magnetic field. Another stationary solution can be found for B such that $B_i = Be_i$
by setting $\partial_t P = 0$ in (3.11), with solution:

$$P(B) = \frac{1}{Z} B^{\Lambda/a-1} \exp\left[-\frac{c}{2a} B^2\right],$$ (3.16)

where Z is a normalization constant and $a = \langle \mu_{ijkl} e_i e_j e_k e_l \rangle_\phi$. This solution
can represent a meaningful probability density function-and therefore a dynamo
case- only if it can be normalized. Condition of integrability at infinity of (3.16)
requires a be positive. This illustrates the importance of the non-linear term
which is essential to ensure vanishing of the probability density at infinity. Condition of integrability near zero requires $\Lambda > 0$ be positive.

This is the dynamo condition identified before, that is obtained using the
mean field as control parameter. However, the shape of the PDF traces an interesting new paradigm for the turbulent dynamo (Fig. 29). Indeed, in the range
$0 < \Lambda < a$, the PDF is maximum at zero, meaning that the most probable value
for the magnetic field is zero. This is the signature of an "intermittent" dynamo,
with periods of large magnetic field followed by quiescent periods, in a way reminiscent to "on-off" intermittency. Above this second threshold, $\Lambda > a$, the PDF

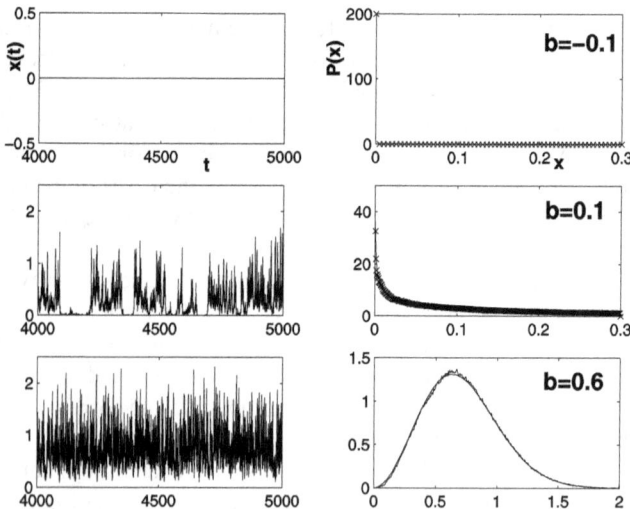

Fig. 29. Result of the surrogate 1D model $\partial_t x = [b + \xi(t)]x - \gamma x^3$: On the left side we show
time series for $a = 0.2$, $\gamma = 1$ and 3 different values of the parameter b. On the right side, the
corresponding PDF and the theoretical curve corresponding to equation (3.16), with $\Lambda = b$.

exhibits a non-zero value for its most probable value, meaning a more classical "turbulent stationary dynamo", with fluctuations of the magnetic field around a finite value. Note that the transition from one regime to another can be mediated by the value of $\delta - 1$: as this parameter is increased, the disorientation effect becomes more and more important, and Λ decreases. This remark is corroborated by recent stochastic computations of Aumaitre and Petrelis, who showed that the intermittent behavior could be switched off by changing the value of the noise spectrum at zero frequency, i.e. by removing large scale noise [38]. Note also that this new paradigm cannot be tested in the previous TG computations, since they did not include any non-linearities. A more serious question is also: Is this new paradigm an artifact of our synthetic turbulence, or is it something that one can actually see? To check this, one needs to resort to numerical simulations or experiments.

3.2.4. Numerical simulations

Direct numerical simulation of Eq. (3.1) for TG forcing have been made in [34–37]. The dynamo threshold Rm_c has been computed for different values of Re. The result is shown in Fig. 30. One sees that the dynamo threshold increases with Re until a value of the order $Re \sim 100$ where it seems to saturate. Dubrulle et al. [37] also performed computations at low Re but larger and larger Rm to try and detect a possible signature of the second laminar window of instability. At $Re = 6$, they detected a transition from an intermittent dynamo at $Rm = 25$, to a

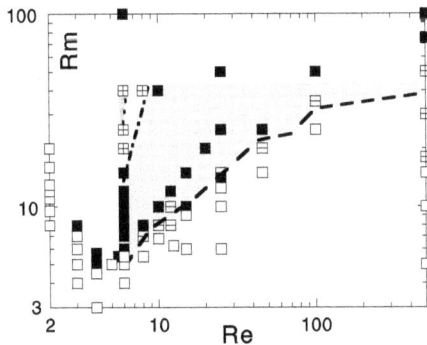

Fig. 30. Comparison between DNS and KS simulations with $Re = 6$ with $k_I = 1$, $\tau_c \geq 0.3$. Squares refer to DNS-MHD and LES-MHD simulations, and shaded areas to windows of dynamo action for kinematic-stochastic simulations at $Re = 6$ with $k_I = 1$, $\tau_c \geq 0.3$. Note the tiny dynamo window near $Re = 6$, $Rm = 40$. Open square: no-dynamo case; Square with cross: intermittent dynamo; Filled square : dynamo case; square with line: undecided state; $- Rm_c^{turb}$; $-- Rm_c^{MF}$; $- \cdot -\cdot$ end of the first dynamo window; $\cdots$ beginning of the second dynamo window.

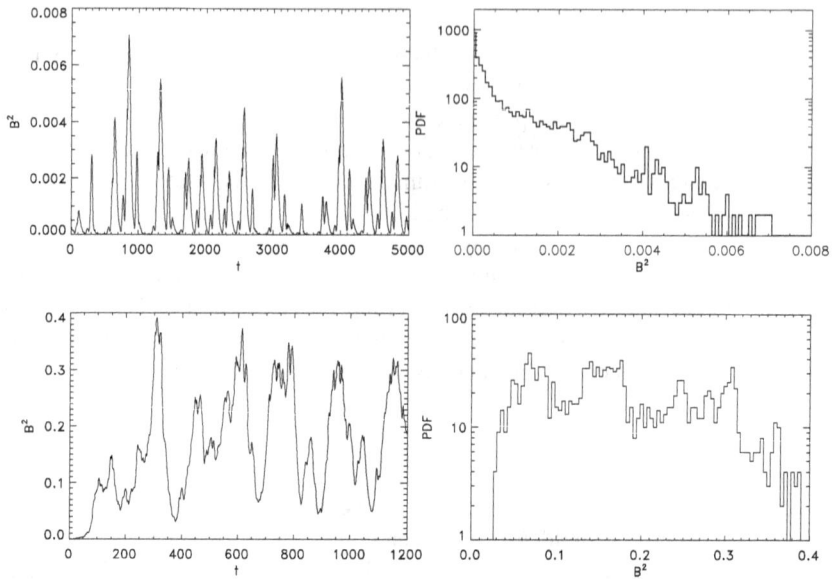

Fig. 31. Example of dynamos in TG flow. Left: Magnetic energy as a function of time. Right: PDF of the magnetic energy. Upper panel: Intermittent dynamo $Re = 6$, $Rm = 40$. Lower panel: Turbulent dynamo $Re = 25$, $Rm = 50$.

dynamo with a mean field at $Rm = 100$ (see Fig. 31). Moreover, one can see a remarkable correlation between the dynamo windows predicted by the stochastic numerical simulation and the direct numerical simulation. This is an indication that maybe the stochastic model does capture the main features of the turbulent dynamo transition.

3.2.5. Experiments

Up to now, three main configurations have been tested in VKS with the TM73 propellers and the layer of sodium at rest: with and without annulus, and with inox or iron propellers. In the inox case, no dynamo has been observed [40]. However, a critical magnetic Reynolds number could be estimated from induction measurements for configurations with positive and negative rotation. In the induction regime, the disorientation effect could be directly measured by following strong local magnetic field perturbations [39].

In the iron case, with an annulus in the midplane, different types of dynamo have been identified, in the rotating and non-rotating case [41,42] (see the lecture by J.F. Pinton for more details). Among them, intermittent dynamo have been observed around $\theta = 0.2$ (see Fig. 32).

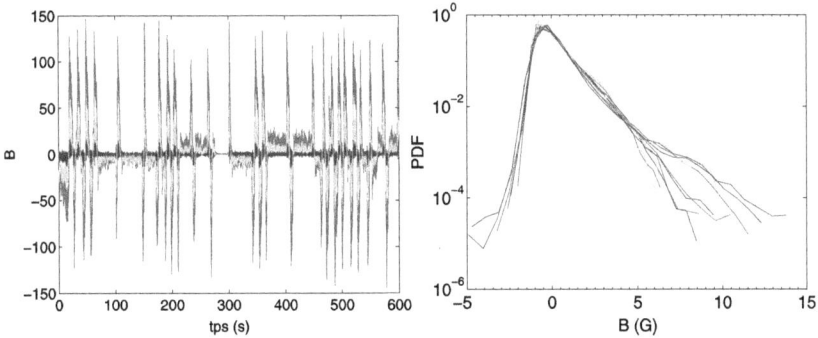

Fig. 32. Example of intermittent dynamo observed in the VKS2 with TM73 iron propellers at $\theta = 0.17$, $Rm = 32$. Left: Components of the magnetic field as a function of time (red: B_z, green B_θ; blue: B_r. Right: Corresponding PDF of the magnetic field components.

Note that the intermittent dynamo observed near $\theta = 0.2$ is characterized by the largest value of δ (see Fig. 21). In the previous section we argued that it was probably a good condition to observe, if any, the intermittent dynamo.

Regarding threshold for dynamo instability (transition towards stationary dynamo), it has been accurately measured so far in 3 cases: at $\theta = 0$, with impellers rotating in the $(+)$ or $(-)$ direction with respect to the pales curvature; at $\theta = -1$ with impellers rotating in the $(+)$ direction. Using the values of δ_2 measured in water, we can check whether the trend observed in TG numerical simulation (higher threshold for larger values of δ or δ_2, see Fig. 28) is also valid here. With the presently available data, the trend is indeed respected (Fig. 33, but more data is needed to confirm this point.

3.3. Non-linear dynamo: saturation and transport

We now consider situations where N is non-negligible anymore, so that B can react back on the velocity field, stopping the exponential growth and eventually reaching a saturated state. In the sequel, we focus on three questions regarding the magnitude of this saturated state: how is it reached? What are its characteristic (amplitude, PDF)? What are its consequences on transport properties? Generalization to magnetic field orientation along the lines provided here is possible, but much more involved.

From a general point of view, in any turbulent flow, the magnitude of the magnetic field near the transition obeys an equation like:

$$\partial_t B = (V/L) B F(Rm, Re, E_{ff}, \delta, \dots, N), \tag{3.17}$$

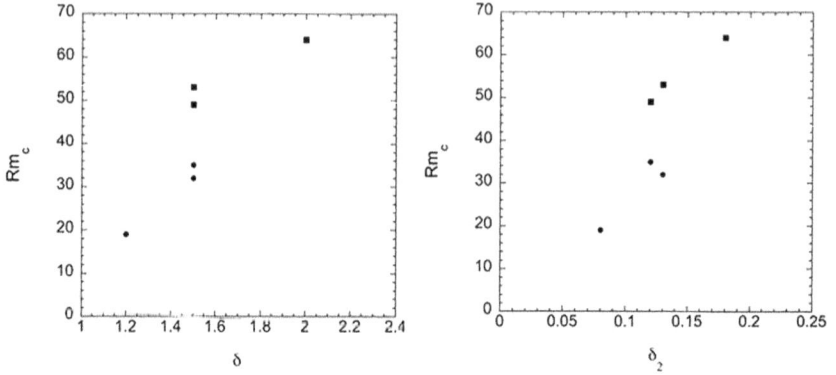

Fig. 33. Critical magnetic Reynolds number Rm_c as a function of δ (left) and δ_2 (right) in the VKS2 experiment for inox (square) and iron propellers (circle). The δ_2 have been estimated from the water model experiment. The Rm_c are measured in the case of iron propellers, and estimated using induction measurements for the inox propellers.

where the dots denotes possible other relevant non-dimensional parameters, such as Γ for VKS, or Ro in case of a rotating turbulence. The question is to find the shape of F.

3.3.1. Weakly non-linear: saturation
We first focus on the intermediate case where N is non-negligible, but not too large, so that one can consider expansion of (3.17) in N.

3.3.1.1. Transition:Super-critical vs sub-critical bifurcation   To first order in N the expansion of (3.17) gives:

$$\partial_t B = \Lambda B + a B^3, \tag{3.18}$$

where $\Lambda = (V/L)F(Rm, Re, E_{ff}, \delta, \ldots, 0)$ and $a = (Rm/LV)\partial_N F(Rm, Re, E_{ff}, \delta, \ldots, 0)$. If a is negative, $a = -|a|$, the behavior of solutions of (3.18) as a function of Λ is quite simple, characteristic of a super-critical bifurcation see Fig. 34: the magnetic field amplitude is 0 for $\Lambda < 0$, and takes the finite value $B = \sqrt{\Lambda/|a|} \sim (V/Rm^{1/2})G(Rm, Re, E_{ff}, \delta, \ldots)$ for $\Lambda > 0$. The behavior of such a saturated state with $Rm, Re, \ldots$ will be discussed below.

In the case where $a > 0$ we have a subcritical bifurcation (Fig. 34): there is a possibility of finite amplitude dynamo solution for $\Lambda < 0$ and we need to go to higher order in N to capture the saturation for $\Lambda > 0$:

$$\partial_t B = \Lambda B + |a|B^3 - |c|B^5, \tag{3.19}$$

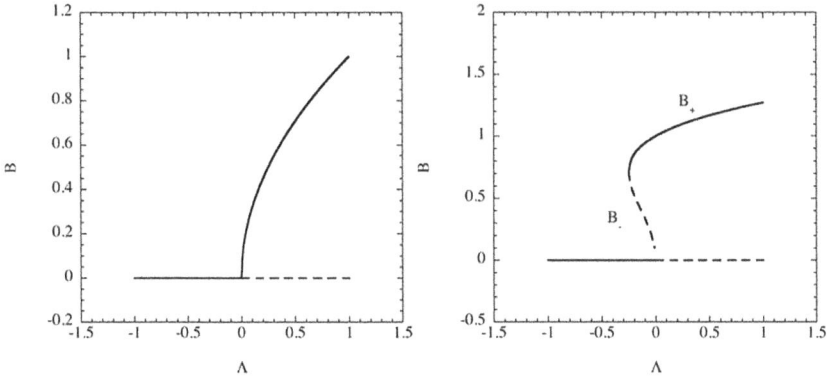

Fig. 34. Left: Supercritical bifurcation. Right: Subcritical bifurcation. The stable states are denoted with continuous line, unstable states with dotted line.

where $c = (Rm^2/2LV^3)\partial_N^2 F(Rm, Re, E_{ff}, \delta, \ldots, N)|_{N=0}$. When $\Lambda \ll -a^2/4|c|$, $B = 0$ is the only stable solution. As $-a^2/4|c| < \Lambda < 0$, there is apparition of two new solution: one unstable $B_-^2 = (1/2|c|)(|a| - \sqrt{a^2 + 4\Lambda|c|}$, bifurcating from 0 at $\Lambda = 0$, and one stable $B_+^2 = (1/2|c|)(|a| + \sqrt{a^2 + 4\Lambda|c|}$, bifurcating from a finite value. Finally, for $\Lambda > 0$, there is one stable solution B_+ and one unstable solution $B = 0$. Note that both B_- and B_+ are of the shape $(V/Rm^{1/2})G_\pm(Rm, Re, E_{ff}, \delta, \ldots)$.

In TG flow, subcritical dynamos have been observed at low Rm (see lectures of Y. Ponty). Supercritical dynamos seem to be observed only at higher Rm, at least at $Re = 6$ [37]. In VKS, the situation is not quite clear: at $\theta = 0$, both the curve in the anti-contra (+) and contra (−) rotation direction appear to look like supercritical dynamos (Fig. 35), but imperfections due to iron propellers (that bear a small but non-zero magnetization) could not be ruled out, see below.

3.3.1.2. Influence of an external magnetic field In any reasonable experimental situation, the earth magnetic field is present and so, it is interesting to wonder what would be the influence of an constant external magnetic field over the two generic bifurcations encountered above. It is therefore interesting to set $B \to B_0 + B$ in (3.18) and (3.19), where $B_0 \gg B$ is finite, and observe the changes.

In the supercritical case, we find to leading order:

$$\partial_t B = \Lambda B_0 - |a| B_0^3 + (\Lambda - 3|a| B_0^2) B - 3|a| B_0 B^2. \tag{3.20}$$

A first striking modification is that below the threshold, $B = 0$ is not solution anymore: we have magnetic induction due to the external field. For large enough

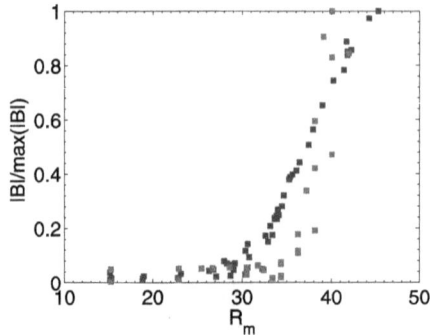

Fig. 35. Non-dimensional magnetic field as a function of Rm in VKS2 with iron TM73 propellers: blue: contra $(-)$ rotation direction; red: anticontra $(+)$ rotation direction.

Λ however, there is still possibility of exponentially growing solution. From the observation of (3.20), we see that the bifurcation is still supercritical, but with a larger threshold ($\Lambda_c = 3|a|B_0^2$). However, one finds at this order a saturated state $B = (\Lambda - \Lambda_c)/3B_0|a|$, that depends on B_0, so that it cannot describe the dynamo regime, where amplitudes do not depend on the applied field.

In the subcritical case, a similar expansion gives:

$$\partial_t B = \Lambda B_0 + |a|B_0^3 - |c|B_0^5 + \left(\Lambda + 3|a|B_0^2 - 5|c|B_0^4\right)B$$
$$+ \left(3|a|B_0 - 10|c|B_0^3\right)B^2. \tag{3.21}$$

Again, below the threshold, $B = 0$ is not solution anymore and the nature of the bifurcation depends on the intensity of B_0. The threshold is here increased if $5|c|B_0^4 - 3|a|B_0^2 > 0$ i.e $B_0 > B_c = \sqrt{3|a|/5|c|}$. However, in such a case $3|a|B_0 - 10|c|B_0^3 < 0$, and the bifurcation is transformed into a supercritical bifurcation. In the case $0 < B_0 < B_c$, the threshold is lowered, and the bifurcation is supercritical for $0 < B_0 < B_c/\sqrt{2}$ or subcritical for $B_c/\sqrt{2} < B_0 < B_c$.

This simple and oversimplified model leads us to suspect that an external magnetic field could produce changes in both the nature of the bifurcation, and in its threshold. This behavior was indeed obtained in the TG flow, where the application of a small B_0 over a subcritical dynamo was observed to decrease the threshold of instability and leads to a supercritical instability, while for larger B_0, the threshold is increased (Fig. 36). The real nature of this changes is for the moment unknown, and probably includes back reaction of the magnetic field onto the fluctuations (see Section 3.3.3), a feature that is not described by our simple toy model.

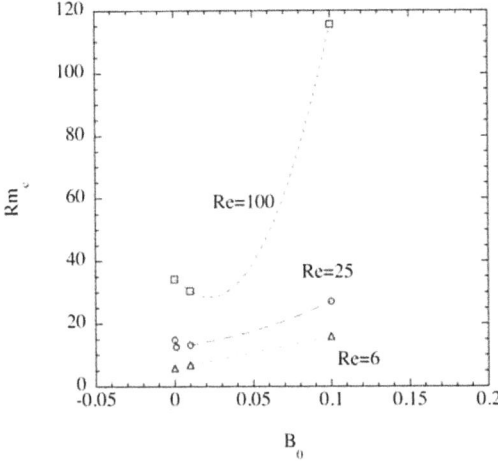

Fig. 36. Critical magnetic Reynolds number as a function of an external transverse applied field B_0 measured in TG DNS simulation at three different Reynolds number. The dots are smoothing of the data.

In a similar way, it has been observed in the VKS experiment that once the iron propellers are demagnetized, the threshold for dynamo action is increased and that the application of a non-zero external field could trigger bifurcation towards the dynamo.

3.3.2. Non-linear: saturation
3.3.2.1. Dimensional arguments and scalings Let us investigate a little bit further the scaling properties of the saturated field, with respect to the various parameters of the system. This has been done recently by Fauve and Petrelis [28,43] using a series of dimensional arguments that we adapt here.

In previous section, we have seen that the saturated field behaves like:

$$Rm B^2 = V^2 F(Rm, Re, E_{ff}, \delta, Ro, \ldots). \tag{3.22}$$

The final shape of F is certainly difficult to disentangle, but there are simple situations or hypothesis that help.

3.3.2.1.1. Near dynamo threshold Let us for example suppose that we are near the dynamo threshold: $Rm \approx Rm_c$. Since $B = 0$ at $Rm = Rm_c$, we can expand (3.22) as:

$$Rm B^2 = V^2 \frac{Rm - Rm_c}{Rm_c} G(Re, E_{ff}, \delta, Ro, \ldots), \tag{3.23}$$

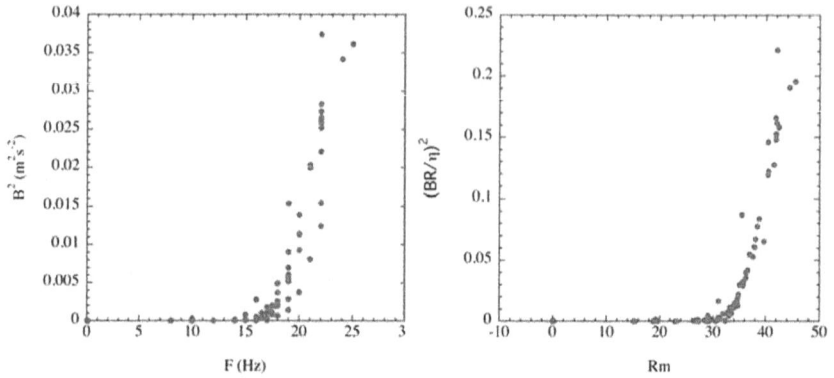

Fig. 37. Experimental check of the scaling relation in VKS2 experiment at $\theta = 0$ form iron TM73 propellers rotating in the contra direction. Left: square Alfven velocity vs propeller rotating frequency. Right: square Lindquist number vs magnetic Reynolds number.

or,

$$\frac{B^2 L^2}{\eta^2} = (Rm - Rm_c)G(Re, E_{ff}, \delta, Ro, \ldots). \tag{3.24}$$

This is the "turbulent scaling" proposed by Fauve and Petrelis. This scaling is interesting per se, because it means that if you are able to change the magnetic Reynolds number while keeping all other parameter constant, you should be able to rescale all your measurements by this simple law. Indeed, in VKS, this is feasible by a change in sodium temperature, that modifies η, (and also, but much less ν and ρ). Fig. 37 shows that indeed, by varying the sodium temperature near threshold, you obtain various magnetic field intensities, but as long as you rescale everything with η, all measurements fall onto a single line.

3.3.2.1.2. Inertial range dynamo If you now impose your dynamo to be created by "inertial range" turbulent where the viscosity does not matter, then, you find:

$$Rm\frac{B^2}{V^2} = G(Rm, E_{ff}, Ro, \ldots). \tag{3.25}$$

If far from the dynamo threshold, the turbulent diffusivity does not matter anymore either, one finds $B^2 \sim V^2 G(E_{ff})$. With $G \sim 1$, this is the scaling proposed by Bierman and Schuller that corresponds to equipartition between magnetic energy and kinetic energy.

3.3.2.1.3. Small scale dynamo If you now impose that your magnetic field does not see the largest scale, so that you ban V and L from your expression, you get:

$$B^2 = \sqrt{v\epsilon}G(Pm),\tag{3.26}$$

so that $B/V \sim Re^{-1/4}$ for $Pm \sim 1$. This is Batchelor's scaling.

3.3.2.1.4. Rotating dynamo In the case of a rotating dynamo ($Ro < \infty$), one may also expand G in eq. (3.25) as a function of $1/Ro$. One gets:

$$Rm\frac{B^2}{V^2} = G_0(Rm, E_{ff}, \ldots) + G_1(Rm, E_{ff}, \ldots)Ro^{-1}.\tag{3.27}$$

In the case of strong rotation $Ro^{-1} \gg 1$, this gives $B/V \sim (Rm\,Ro)^{-1/2}$. This scaling was used by Fauve and Petrelis to explain the results of Riga experiment.

3.3.2.2. Balance arguments Other scaling can be derived through physical balance in the dynamic equation. Consider first the non-rotating case. In the induction equation, the stationary state is obtained by balancing the induction term $\nabla \times (v \times B)$ with the dissipative term $\eta\Delta B$. Uncurling this balance gives $j = \nabla \times B \sim (1/\eta)v \times B$. In the Navier–Stokes equation, one must balance the Lorentz force $j \times B \sim (1/\eta)vB^2$ with either the non-linear term $v\nabla v$ or the dissipative term $v\Delta v$. In the former case, one finds:

$$B^2 \sim \eta\nabla v.\tag{3.28}$$

If the balance occurs at large scale (a situation typical of $Pm \ll 1$), $\nabla v \sim V/L$ so that $B/V \sim Rm^{-1/2}$ or also $N \sim 1$. This is in fact a rephrasing of the "turbulent large scale dynamo" of Fauve and Petrelis, with $Rm \gg Rm_c$. If the balance occurs at small scale, $\nabla v \sim (\epsilon/v)^{1/2}$ so that $B/V \sim \sqrt{v/\epsilon}/Pm$. This is the Batchelor's scaling for $Pm \sim 1$.

In the case where the balance is with the diffusive term, one finds:

$$B^2 \sim \eta v\Delta v/v.\tag{3.29}$$

The scaling is therefore $B/V \sim 1/\sqrt{Re\,Rm}$ if the balance occurs at large scale— this is the laminar dynamo of Fauve and Petrelis [28]—and $B/V \sim (\epsilon L/V^3)/Rm$ if the balance occurs at small scale.

In the case where rotation is present, a balance can now occur between the Lorentz force and the Coriolis force v, if the rotation is strong enough. In such a case, one finds:

$$B^2 \sim \eta\Omega,\tag{3.30}$$

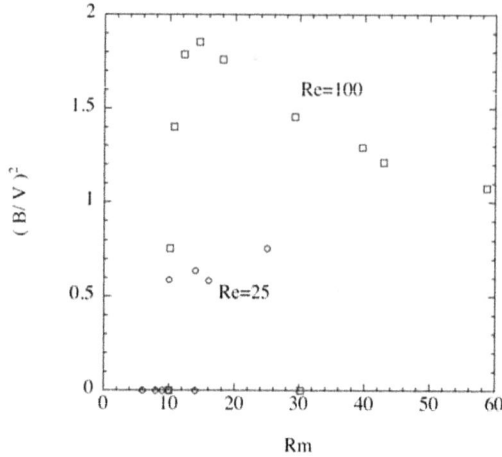

Fig. 38. Saturated field in TG numerical simulations, at $Re = 100$ (square) and $Re = 25$ (circle). The ratio of magnetic to kinetic energy is close to equipartition in both cases.

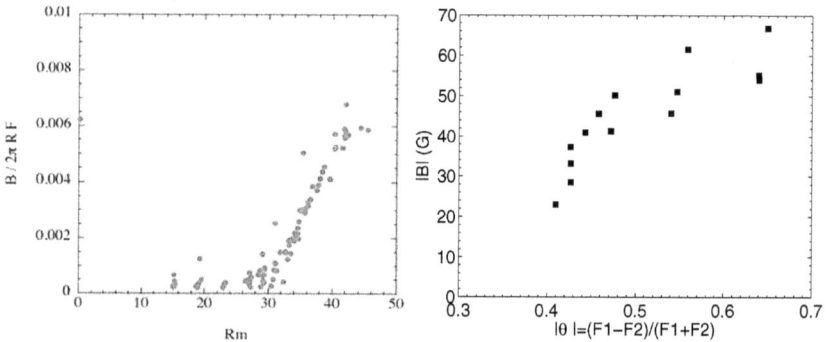

Fig. 39. Left: B/V in the von Karman experiment with iron TM73 propellers in contra direction for the stationary dynamo at $\theta = 0$. Right: Magnetic field intensity as a function of θ for $\theta > 0.4$ and $31 < Rm < 37$.

so that $B/V \sim (Rm\,Ro)^{-1/2}$, in agreement with the dimensional argument of previous section.

In TG flow, with $Pm \sim 1$, the fields at saturation are such that $B \sim V$ (Fig. 38). This is in favor of the "inertial range dynamo" scaling. In VKS, the ratio B/V is much smaller, of the order of 10^{-3} (Fig. 39). However, this corresponds to a value observed in one point of the experiment, and that may not be typical of the magnetic field within the experiment. It is therefore too early to conclude at this stage. However, in the rotating case, it was found [5] that for

$Ro^{-1} > 0.4$, $B \sim \sqrt{1/Ro - 0.4}$ (see Fig. 39), a scaling compatible with the strong rotation regime.

3.3.2.3. Ideal axisymmetric system   Exact non-linear analytical stationary solutions may be found in the axisymmetric case (relevant to the VKS experiment), and when forcing and dissipation are neglected [45]. This approximation is relevant in the bulk of large Reynolds number experimental flow, where the influence of scaling and dissipation is confined within walls or near the impellers.

3.3.2.3.1. Equations   In the axisymmetric case we consider, it is convenient to introduce the poloidal/toroidal decomposition for the fields $\mathbf{U}$ and $\mathbf{B}$:

$$\begin{aligned} \mathbf{U} &= \mathbf{U_p} + \mathbf{U_t} = \mathbf{U_p} + U\,\mathbf{e}_\theta, \\ \mathbf{B} &= \mathbf{B_p} + \mathbf{B_t} = \nabla \times (A\mathbf{e}_\theta) + B\,\mathbf{e}_\theta, \end{aligned} \tag{3.31}$$

where $\mathbf{A} = \mathbf{A_p} + A\,\mathbf{e}_\theta$ is the potential vector.

We also introduce alternate fields, built upon the poloidal and toroidal decomposition. They are: $\sigma_u = rU$, $\sigma_b = rA$, $\xi_u = \omega/r$ and $\xi_b = B/r$, where ω is the toroidal part of the vorticity field. In these variables, the ideal incompressible MHD equations (3.1) become, in the axisymmetric approximation, a set of four scalar equations:

$$\begin{aligned} \partial_t \sigma_b + \{\psi, \sigma_b\} &= 0, \\ \partial_t \xi_b + \{\psi, \xi_b\} &= \left\{\sigma_b, \frac{\sigma_u}{2y}\right\}, \\ \partial_t \sigma_u + \{\psi, \sigma_u\} &= \{\sigma_b, 2y\xi_b\}, \\ \partial_t \xi_u + \{\psi, \xi_u\} &= \partial_z\left(\frac{\sigma_u^2}{4y^2} - \xi_b^2\right) - \{\sigma_b, \Delta_*\sigma_b\}, \end{aligned} \tag{3.32}$$

where the fields are function of the axial coordinate z and the modified radial coordinate $y = r^2/2$ and ψ is a stream function: $\mathbf{U_p} = \nabla \times (\psi/r \;\; \mathbf{e}_\theta)$. We have introduced a Poisson Bracket: $\{f, g\} = \partial_y f \partial_z g - \partial_z f \partial_y g$. We also defined a pseudo Laplacian in the new coordinates:

$$\Delta_* = \frac{\partial^2}{\partial y^2} + \frac{1}{2y}\frac{\partial^2}{\partial z^2}. \tag{3.33}$$

3.3.2.3.2. Stationary solutions   Using properties of the Poisson bracket-namely $\{f, g\} = 0$ iif f is a function of g-, one may derive the shape for the non-linear stationary solutions. The first equations gives:

$$\sigma_b = F(\psi), \tag{3.34}$$

where F is an arbitrary function. When plugged into the second equation, this gives

$$\xi_b = \frac{F'(\psi)\sigma_u}{2y} + G(\psi), \tag{3.35}$$

where G is a second arbitrary function. The exact shape of ψ, σ_b, ... for given values of F and G can then be found by plugging these shapes into the resulting two equations.

The simplest solutions (presumably found at very large Reynolds number) are given for F linear $F(\psi) = \mu\psi$ and $G = 0$. In such case, one may check that $B = \mu U$ and $B_p = \mu U_p$, so that $\mathbf{U}$ and $\mathbf{B}$ are proportional.

3.3.2.3.3. PDF Using a formalism derived from statistical mechanics, one may also derive the behavior of fluctuations in this model. The fluctuations are Gaussian:

$$\rho = \frac{1}{Z} \exp\left\{ -\frac{\beta}{2}(\mathbf{u}'^2 + \mathbf{b}'^2) - \mu_c \mathbf{u}' \cdot \mathbf{b}' \right\} = \frac{1}{Z} \exp\left\{ \frac{1}{2} \sum_{i,j} x_i A_{ij} x_j \right\}, \tag{3.36}$$

where we defined a 6-dimensional vector: $x_i = (u'_1, u'_2, u'_3, b'_1, b'_2, b'_3)$ where β is the Lagrange parameter associated with energy conservation ($1/\beta$ plays the role of a "statistical temperature"), and the μ_c is the Lagrange parameter associated with cross helicity conservation. Using the Gaussian shape for the fluctuations, it possible to derive the mean properties of the fluctuations. We find that part of the energy is going into the fluctuations and that there is equipartition between the fluctuating parts of the magnetic energy and of the kinetic energy:

$$\langle u'^2 \rangle = \langle b'^2 \rangle = \frac{3\beta}{\beta^2 - \mu_c^2}. \tag{3.37}$$

One can also calculate the quantity of cross helicity going into the fluctuations:

$$\langle \vec{u}' \cdot \vec{b}' \rangle = -\frac{3\mu_c}{\beta^2 - \mu_c^2}. \tag{3.38}$$

The fractions of magnetic energy, cross helicity and kinetic energy going into the fluctuations are:

$$
\begin{aligned}
\frac{\langle b'^2 \rangle}{\int \bar{B}^2 \, d\mathbf{x}} &= \frac{\langle \mathbf{u}' \cdot \mathbf{b}' \rangle}{\int \bar{U} \cdot \bar{B} \, d\mathbf{x}} = \frac{3\beta}{\beta^2 - \mu_c^2} \mathcal{M}^{-1}, \\
\frac{\langle u'^2 \rangle}{\int \bar{U}^2 \, d\mathbf{x}} &= \frac{\beta^2}{\mu_c^2} \frac{3\beta}{\beta^2 - \mu_c^2} \mathcal{M}^{-1},
\end{aligned}
\tag{3.39}
$$

where $\mathcal{M} = \int \bar{B}^2 \, d\mathbf{x}$ is the magnetic energy of the coarse-grained field. The first equation shows that there is an equal fraction of magnetic energy and cross helicity which goes in the fluctuations and the positivity of the magnetic energy requires: $\beta^2 > \mu_c^2$ if $\beta > 0$ or $\beta^2 > \mu_c^2$ if $\beta < 0$. These two conditions are more interesting to interpret by noting that the ratio of the mean magnetic energy onto the mean kinetic energy in this case is simply:

$$\frac{E_m}{E_c} = \frac{\int \bar{B}^2 \, d\mathbf{x}}{\int \bar{U}^2 \, d\mathbf{x}} = \frac{\beta^2}{\mu_c^2}. \tag{3.40}$$

Therefore, positive temperature state correspond to super-equipartition ($E_m > E_c$) while sub-equipartition ($E_m < E_c$) correspond to negative temperature state. In 2D turbulence, these kind of states are associated with presence of coherent structure and condensation of vorticity into larger and larger scales. Maybe a similar behavior is present in MHD turbulence with accumulation of magnetic field at the largest scales due to inverse helicity cascade [32, 33].

From the behavior of fluctuations with β, we may also note that the no-dynamo laminar states ($E_m \to 0$ with negligible fluctuations) correspond to small β, i.e; to infinite temperature and the laminar dynamo state (finite E_m and $b' \to 0$) to $\beta \to \infty$ (zero temperature state). In contrast, the turbulent dynamo case is with $\beta \to \mu_c$, with energy equipartition and divergency of fluctuations.

3.3.2.3.4. Comparison with VKS data In the VKS experiment, there is no access to simultaneous measurements of velocity field and magnetic field, so that the comparison with the predicted stationary state is difficult. However, comparing the few available information regarding the neutral mode with velocity fields measured in water experiment, one see that the proportionality between the magnetic field and the velocity field is not to be excluded (see Fig. 40).

On the other hand, the PDF in VKS are indeed Gaussian (Fig. 41). However, no quantitative comparison can be made to check the equipartition at the fluctuation level. From the available one point measurements, we can nevertheless conclude that if the model is correct, and that the magnetic energy in the point we measured is representative of the total magnetic energy, then VKS dynamo are in a state of negative temperature since $E_m/E \ll 1$. Moreover, in that case, one can get an interesting interpretation of the fluctuations as:

$$\frac{1}{3} \frac{< b'^2 >}{\int \bar{B}^2 \, d\mathbf{x}} \approx \frac{-1}{\beta \int \bar{U}^2 \, d\mathbf{x}}, \tag{3.41}$$

providing therefore a simple measure of the statistical temperature in the VKS2 experiment. The result is provided in Fig. 41, for the contra and anti-contra rotation. One sees that in the anticontra rotation, the temperature before the transition

B. Dubrulle

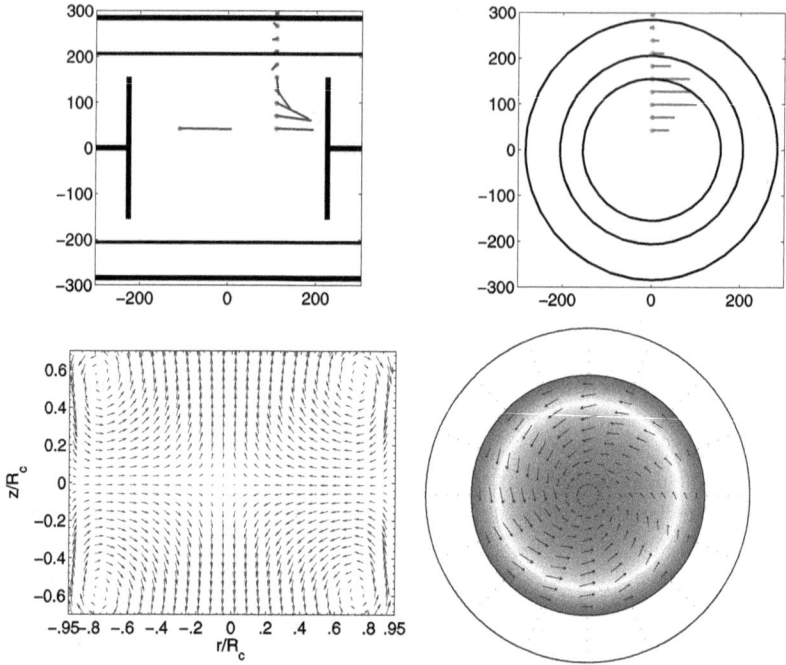

Fig. 40. Upper panel: A view of the saturated mode of the dynamo at $\theta = 0$ in VKS2 with iron TM73 rotating in the contra direction. Lower panel: A view of the velocity field in the water model experiment measured with SPIV. Left: Poloidal field. Right: Toroidal field. The three concentric circles are the propellers, the chemise and the cylinder radius.

is approximately equal to the temperature measured in the water model experiment, while in the contra case, the temperature before the transition is a little bit lower. This is a signature of the damping of the fluctuations by the mean magnetic field. At the transition, the temperature can become very large, both in the contra and anti-contra direction, like in a phase transition. This gives hope to be able to describe the dynamo instability as a phase transition.

3.3.3. Transport

Maybe an indirect measure of the fluctuation level could come from the modification of the transport properties induced by the magnetic field. Indeed, in the presence of a magnetic field, the torque now includes a magnetic contribution, that can roughly be estimated as proportional to $< \overline{b'^2} >$. Therefore, one can expect the non-dimensional K_p to scale in a way similar with the non-magnetic case $K_p \sim A/Re + C(\delta_M - 1)$ where now δ_M includes the contribution due to

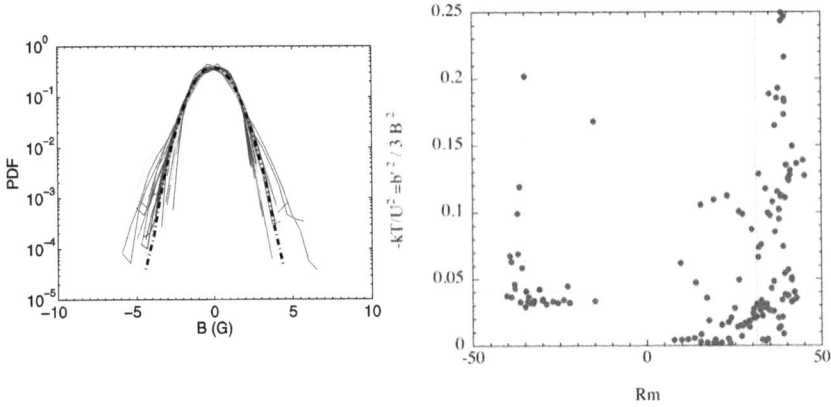

Fig. 41. Left:PDF of magnetic field components in VKS2 experiment with iron TM73 propellers at $\theta \approx 0$ in contrarotation. Right: Statistical temperature as a function of Rm in VKS2 with contra ($Rm > 0$) and anticontra rotation($Rm < 0$) iron TM73 propellers. The horizontal lines mark the value measured in a water model experiment. The vertical lines mark the dynamo threshold. Note the large temperature arising around that point.

magnetic fluctuations $\delta_M - 1 \sim < \overline{u'^2 + b'^2} > / < \overline{u}^2 >$. It is not easy to predict what will be the net result: on the one hand, strong magnetic field are known to reduce the level of fluctuation, thereby reducing $< \overline{u'^2} > / < \overline{u}^2 >$. On the other hand, since there will be a contribution due to magnetic fluctuation, they may compensate this loss. Fig. 42 shows the comparison of the dimensionless torque in the VK experiment in water (without magnetic field) and in sodium, at $\theta = 0$.

In fact, if we assume equipartition between the magnetic and kinetic fluctuation (cf previous section), we get $\delta_M - 1 \sim 2\delta$. So that, if we assume that neither A nor C are changed by the magnetic field, we get at large Reynolds number:

$$\frac{\delta(B) - 1}{\delta(0) - 1} = \frac{1}{2} \frac{K_p(B)}{K_p(0)}, \tag{3.42}$$

so that the comparison between the torque in a water experiment ($B = 0$) and the torque in a sodium experiment ($B \neq 0$) gives an estimate of the kinetic fluctuations. From Fig. 42, combined with the variation of B with Rm measured in one point, one may then get Fig. 43 showing the variation of δ with the magnetic field. One sees that this estimate leads to a lower level of fluctuations in the sodium, with respect to the water case (part of the energy is in the magnetic fluctuations), and that when one goes from an induction regime, to a dynamo regime, the level of fluctuation increases and then saturates. This is in strong contrast with the direct observation of the behavior of fluctuations made in a TG

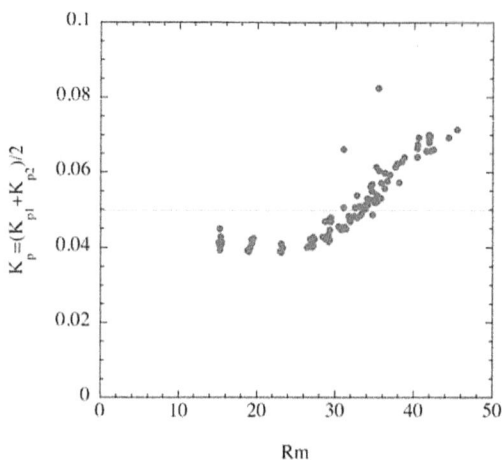

Fig. 42. Non-dimensional torque K_p as a function of Rm in VKS2 with TM73 iron contra propellers.

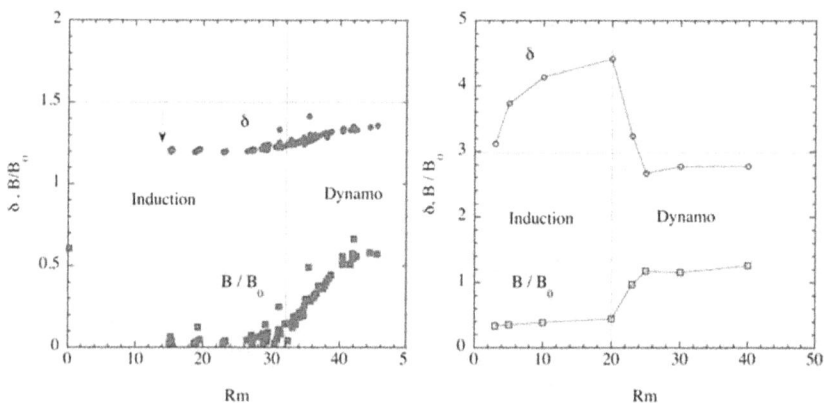

Fig. 43. δ and B/B_0 as a function of Rm in situation with an external applied field. Left: in VKS2 experiment with iron TM73 contra propellers. In that case, the applied field is the field of the experimental hall and δ has been estimated from the torque measurements. $B_0 = 100/(2\pi R F)$; Right: in a TG simulation at $Re = 100$ with applied field B_0.

simulation, where all quantities can be computed exactly (Fig. 43). This shows that our crude estimate must be taken with caution.

However, it may be useful to understand a yet-unexplained feature of some dynamos observed in VKS in the rotating case, for $\theta > 0.4$, where the power consumption of the dynamo is less than in the equivalent water experiment. This could be explained by noting that in (3.42), for $\delta(B) - 1 < 0.5(\delta(0) - 1)$, the non-dimensional torque in the sodium experiment becomes less that the non-dimensional torque in an equivalent water experiment: the growth of the field has been so such that all fluctuations have been smeared down, enabling a reduction of the torque. Clearly, all the questions will benefit from further improvements of the VKS experiment, with direct measurements of velocity fluctuations (quite a difficult measurement!!!).

Acknowledgements

The numerical data in the paper are obtained through a collaboration with J.-Ph. Laval, Y. Ponty, F. Daviaud, P. Blaineau and J-F. Pinton. The data from the von Karman water experiments have been obtained in Saclay in the group of F. Daviaud by A. Chiffaudel, F. Daviaud, L. Marié, F. Ravelet, R. Monchaux, P. Diribarne. The data from von Karman sodium experiment have been obtained in Cadarache by the VKS collaboration (CEA Saclay: S. Aumaitre, A. Chiffaudel, F. Daviaud, B. Dubrulle, C. Gasquet, L. Marie, R. Monchaux, V. Padilla and F. Ravelet; ENS Paris: M. Berhanu, S. Fauve, N. Mordant and F. Pétrélis; ENS Lyon: M. Bourgoin, Ph. Odier, J-F. Pinton, N. Plihon, R. Volk). The data interpretation is personal and does not engage the responsibility of any of these people.

References

[1]   J.-P. Laval, PhD Thesis, Paris VI, 1999.

[2]   L. Marié, PhD Thesis, Paris VI, 2002.

[3]   N. Leprovost, PhD Thesis, Paris VI, 2004.

[4]   F. Ravelet, PhD Thesis, Ecole Polytechnique, 2005.

[5]   R. Monchaux, PhD Thesis, Paris VI, 2007.

[6]   J.L. Lumley, *Stochastic Tools in Turbulence*, Academic Press, 1970.

[7]   A.S. Monin and A.M. Yaglom, *Statistical Fluid Mechanics*, MIT Press, 1973.

[8]   M. Lesieur, *Turbulence in Fluids*, Kluwer Academic Publishers, 1990.

[9]   U. Frisch, *Turbulence*, Cambridge University Press, 1995.

[10]   G. Falkovich, K. Gawedzki and M. Vergassola, Rev. Mod. Phys. **73**, 913 (2001).

[11]   U. Frisch, Z.-S. She and P.-L. Sulem, Physica D **28**, 382 (1987).

[12]   B. Dubrulle and U. Frisch, Phys. Rev. A **43**(10), 5355-5364 (1991).

[13]   J.-P. Laval, B. Dubrulle and S. Nazarenko, Phys. Fluids **13**, 1995-2012 (2001).

[14]   J.-P. Laval, B. Dubrulle and J.C. McWilliams, Phys. Fluids **15**(5), 1327-1339 (2003).

[15]   B. Dubrulle, Eur. Phys. J. B **28**(3), 361-367 (2002).

[16]   B. Dubrulle and F. Hersant, Eur. Phys. J. B **26**, 379-386 (2002).

[17]   F. Krause and K.-H. Rädler, *Mean Field MHD and Dynamo Theory*, Pergamon Press, 1980.

[18] L. Marié, J. Burgete, F. Daviaud and J. Leorat, Eur. Phys. J. B **33**, 469 (2003).

[19] R. Avalos-Zuniga, F. Plunian and A. Gailitis, PRE **68**, 066307 (2003).

[20] F. Ravelet, A. Chiffaudel, F. Daviaud and J. Leorat, Phys. Fluids **17**, 17104 (2005).

[21] R. Laguerre, C. Nore, J. Leorat and J.-L. Guermond, C. R. Mécanique **334**, 593 (2006).

[22] F. Stefani, M. Xu, G. Gerbeth, F. Ravelet, A. Chiffaudel, F. Daviaud and J. Leorat, Eur. J. Mech. B **25**, 894 (2006).

[23] D. Sweet et al., Phys. Rev. E **63**, 066211 (2001).

[24] E. Ott and J.C. Sommerer, Phys. Lett. A **188**, 39 (1994).

[25] A.P. Kazantsev, Sov. Phys. JETP **26**, 1031 (1968).

[26] S. Boldyrev, Astrophys. J. **562**, 1081 (2001).

[27] F. Pétrélis, PhD Thesis, Paris VI, 2002.

[28] S. Fauve and F. Petrélis, The dynamo effect, in: *Peyresq Lectures on Nonlinear Phenomena*, World Scientific, 2003.

[29] J. Zinn-Justin, *Quantum Field Theory and Critical Phenomena*, Oxford, 2002.

[30] H.K. Moffatt, *Magnetic Field Generation in Electrically Conducting Fluids*, Cambridge University Press, 1978.

[31] F. Pétrélis and S. Fauve, Eur. Phys. J. B **22**, 273 (2001).

[32] A. Pouquet, U. Frisch and J. Leorat, J. Fluid Mech. **77**, 321 (1976).

[33] E.J. Kim and B. Dubrulle, Physica D **165**, 213 (2002).

[34] Y. Ponty, P.D. Mininni, J.-F. Pinton, H. Politano and A. Pouquet, Phys. Rev. Lett. **94**, 164512 (2005).

[35] J.-P. Laval, P. Blaineau, N. Leprovost, B. Dubrulle and F. Daviaud, Phys. Rev. Let. **96**, 204503 (2006).

[36] Y. Ponty, P.D. Mininni, J.-F. Pinton, H. Politano and A. Pouquet, New J. Phys. **9**, 296 (2007).

[37] B. Dubrulle, P. Blaineau, O. Mafra Lopez, F. Daviaud, J-P. Laval and R. Dolganov, New J. Phys. **9**, 308 (2007).

[38] S. Aumaitre, F. Pétrélis and K. Mallick, Phys. Rev. Lett. **95**, 064101 (2005).

[39] R. Volk, F. Ravelet, R. Monchaux, M. Berhanu, A. Chiffaudel, F. Daviaud, S. Fauve, N. Mordant, Ph. Odier, F. Pétrélis and J.-F. Pinton, Phys. Rev. Lett. **97**, 074501 (2006).

[40] F. Ravelet, R. Volk, A. Chiffaudel, F. Daviaud, B. Dubrulle, R. Monchaux, M. Bourgoin, P. Odier, J.-F. Pinton, M. Berhanu, S. Fauve, N. Mordant and F. Pétrélis, Submitted to Phys. Fluids, arxiv/physics/07042565.

[41] M. Berhanu, R. Monchaux, S. Fauve, N. Mordant, F. Pétrélis, A. Chiffaudel, F. Daviaud, B. Dubrulle, L. Marié, F. Ravelet, M. Bourgoin, Ph. Odier, J.-F. Pinton and R. Volk, Eur. Phys. Rev. Lett. **77**, 59001 (2007).

[42] R. Monchaux, M. Berhanu, M. Bourgoin, Ph. Odier, M. Moulin, J.-F. Pinton, R. Volk, S. Fauve, N. Mordant, F. Pétrélis, A. Chiffaudel, F. Daviaud, B. Dubrulle, C. Gasquet, L. Marié and F. Ravelet, Phys. Rev. Lett. **98**, 044502 (2007).

[43] F. Pétrélis and S. Fauve, Europhys. Lett. **76**, 602 (2006).

[44] N. Leprovost and B. Dubrulle, Eur. Phys. J. B **44**, 395 (2005).

[45] N. Leprovost, B. Dubrulle and P.-H. Chavanis, Phys. Rev. E **73**, 046308 (2006).

Course 6

NUMERICAL MODELING OF LIQUID METAL DYNAMO EXPERIMENTS

Yannick Ponty

Laboratoire Cassiopée UMR 6202, Observatoire de la Côte d'Azur,
BP 4229, Nice Cedex 04, France

Ph. Cardin and L.F. Cugliandolo, eds.
Les Houches, Session LXXXVIII, 2007
Dynamos
© *2008 Published by Elsevier B.V.*

Contents

361

1. Preamble

Dynamo action, the self amplification of magnetic field due to the stretching of magnetic field lines by a fluctuating flow, is considered to be the main mechanism for the generation of magnetic fields in the universe [1]. In this respect many experimental groups have successfully reproduced dynamos in liquid sodium laboratory experiments [2–8]. The induction experiments [9–18] studying the response of an applied magnetic field inside a turbulent metal liquid also represent a challenging science. With or without dynamo instability the flow of a conducting fluid forms complex system, with a large degrees of freedom and a wide branches of non linear behaviors.

2. Introduction

For laboratory experiments, the numerical prediction of the dynamo threshold in realistic conditions is still out of reach. Nonetheless, the experiments [2] in Riga and Karlsruhe [5] found the onset to be remarkably close to the values predicted from numerical simulations based on the mean flow structure [19, 20], and this despite the fact that the corresponding flows are quite turbulent.

This has led several experimental groups seeking dynamo action in less constrained geometry, eventually leading to richer dynamical regimes [7, 8], to optimize the flow forcing using kinematic simulations based on mean flow measurements [21, 22]—the advantage being that mean flow profiles can be measured in the laboratory.

Actually, there are efforts in numerical methods to take account of real geometries and the shapes of the experimental apparatus, the effect of the fluid and the magnetic boundary conditions, the use of finite element, finite volume and finite difference mesh schemes [23–28]. However as we will shown in this lecture, it is possible to numerically study some aspects of experimental dynamo behavior without boundary conditions in a three-dimensional periodic space, and we shall seek the respective role of mean flow and the turbulence in this.

3. Numerical method

3.1. The periodic box numerical experiment

Incompressible turbulent flows have been studied intensively in a periodic space, which is a classical mathematical framework for analysis and theory [29, 30]. The energy cascade, the organisation of the energy transfer between the different scales are naturally handle in spectral space. The Navier–Stoke equation accumulates the maximum number of symmetries in a periodic space and this lack of real boundaries has been used for numerical simulations of isotropic and homogeneous turbulence [31]. The pseudo-spectral numerical method [33–36] is a global method and probably the most precise numerical method for a fix mesh size. For all these reasons, in this lecture we will concentrate on numerical simulations of incompressible conductor flow in a fully three-dimensional periodic box.

3.2. Fundamentals equations

Let us work with the incompressible magnetohydrodynamic equations (3.1)–(3.2)

$$\frac{\partial \mathbf{v}}{\partial t} + \mathbf{v} \cdot \nabla \mathbf{v} = -\nabla \mathcal{P} + \mathbf{j} \times \mathbf{B} + \nu \nabla^2 \mathbf{v} + \mathbf{F}, \qquad (3.1)$$

$$\frac{\partial \mathbf{B}}{\partial t} + \mathbf{v} \cdot \nabla \mathbf{B} = \mathbf{B} \cdot \nabla \mathbf{v} + \eta \nabla^2 \mathbf{B}, \qquad (3.2)$$

together with $\nabla \cdot \mathbf{v} = \nabla \cdot \mathbf{B} = 0$; a constant mass density $\rho = 1$ is assumed. Here, $\mathbf{v}$ stands for the velocity field, $\mathbf{B}$ the magnetic field (in units of Alfvén velocity), $\mathbf{j} = (\nabla \times \mathbf{B})/\mu_0$ the current density, ν the kinematic viscosity, η the magnetic diffusivity and $\mathcal{P}$ is the pressure. The forcing term $\mathbf{F}$ will be chosen between two different forcing,s respectively they are Taylor–Green vortex (TG) [37],

$$\mathbf{F}_{\mathrm{TG}}(k_0) = \begin{bmatrix} \sin(k_0 \, x) \cos(k_0 \, y) \cos(k_0 \, z) \\ -\cos(k_0 \, x) \sin(k_0 \, y) \cos(k_0 \, z) \\ 0 \end{bmatrix}, \qquad (3.3)$$

and the ABC flow [38, 39]

$$\mathbf{F}_{\mathrm{ABC}}(k_0) = \begin{bmatrix} A \sin(k_0 \, z) + C \cos(k_0 \, y) \\ B \sin(k_0 \, x) + A \cos(k_0 \, z) \\ C \sin(k_0 \, y) + B \cos(k_0 \, x) \end{bmatrix}, \qquad (3.4)$$

with $A = B = C = k_o = 1$.

There are several ways to implement a forcing term in the Navier–Stoke equation, but there are two superior, less artificial ways, which correspond to two possible experimental forcings. The first one is used in this lecture, it is called constant forcing or constant torque, where the forcing term in the Navier–Stoke equation is simply add at each time step to the rest of the momentum equation. The other one is the constant velocity, it is implemented generally in the spectral space, imposing constant values for selected wave vectors at each time step.

3.3. Non dimensional numbers

We solve the dimensional form of the MHD (3.1)–(3.2) equations, and compute the non dimensional numbers *a posteriori*, using the numerical output quantities. Working with the velocity in $m\,s^{-1}$, a 2π meter box and the viscosity in $m^2\,s^{-1}$, can try to simulate a real experiment however even if we can numerically reach the velocity values close to the experiment values (order one), Unfortunately, we are far away to handle the real viscosity values of water or liquid sodium ($10^{-6}\,m^2\,s^{-1}$). Actually, we can reach numerical values of the viscosity of 10^{-2} to $10^{-3}\,m^2\,s^{-1}$ which is just better than molasses ($1\,m^2\,s^{-1}$) or the honey ($10^{-1}\,m^2\,s^{-1}$).

Nevertheless, we can still define classical non dimensional numbers such as the Reynolds number $R_v = \frac{L\,V}{v}$ and $R_m = \frac{L\,V}{\eta}$ with a characteristic velocity V, length scale L and the viscosity v or the magnetic diffusivity η. We choose to use the root mean square velocity $V_{rms} = \sqrt{2E_v}$ based on the total kinetic energy E_v, or its average in time for fluctuated flow. We also choose the two following characteristic lengths scale: the integral length scale define by $L_{int} = \frac{1}{E_v} \sum \frac{E_v(k)}{k} dk$ which measures the largest eddy scale size, where $E_v(k)$ represent the uni-dimensional isotropic energy density, and the Taylor microscale, the scale where the viscous dissipation begins to affect the eddies, $L_\lambda = \sqrt{5E_v/\Omega_v}$ with $\Omega_v = \frac{1}{2}\int(\nabla \times \mathbf{v}(\mathbf{x},t))^2 dx^3$ is the enstrophy. In the fully turbulence regime, experimental results find the relation $R_\lambda \sim \sqrt{R_v}$.

Note two important non dimensional numbers for liquid metal dynamo process: Firstly, the magnetic Reynolds number $R_m = \frac{LV}{\eta}$ which represents the ratio of the eddy turn over time $\tau_{NL} = \frac{L_{int}}{V_{rms}}$ and the magnetic diffusion time $\tau_\eta = \frac{L_{int}^2}{\eta}$. This number is generally between 10 and 200 which implies that long hydrodynamic simulations are needed to achieved one or two magnetic diffusion times. Secondly, the magnetic Prandtl number $P_m = \frac{v}{\eta}$ which is the ratio of the magnetic diffusion time over the viscous time scale $\tau_v = \frac{L_{int}^2}{v}$. This number is very small (10^{-5}–10^{-6}) in liquid metal and this implies high Reynolds number regimes need to be reached to produce a reasonable magnetic Reynolds number above the dynamo onset threshold.

3.4. Numerical schemes

In the periodic box, we use the pseudo-spectral method initiated by the work of Orszag (1969, 1972) [32, 35] and Orszag and Patterson (1971) [36]. The success of this method is essential due to the high accuracy and the efficiency of the Fast Fourier Transform (FFT). The linear terms are treated in spectral space, the non linear terms in real space, and the FFT moves the fields between the two mathematical spaces. For the incompressible Navier–Stoke equation, the pressure is eliminated by applying the divergence free operator in Fourier space. This trick greatly simplifies the simulation of incompressible flow, but constrains us to recovering only periodic solutions. We use a "semi-exact" temporal scheme proposed by Basdevant (1982) [40] to implicitly treat the viscous or diffusion terms. This method has been used with success in homogeneous turbulence studies [31]. The equation diffusion system is written in spectral space as

$$\frac{dU_{\vec{k}}}{dt} = -\nu k^2 U_{\vec{k}} + G_{\vec{k}}(U) \tag{3.5}$$

where $U_{\vec{k}}$ is the velocity function in the spectral space for the wave vector $\vec{k}$. And the solution can be explicitly written with the exponential form as

$$\frac{d\, e^{\nu k^2 t} U_{\vec{k}}}{dt} = e^{\nu k^2 t} G_{\vec{k}}(U_{\vec{k}}(t), t) \tag{3.6}$$

Using the second order of Adams-Basford scheme, the temporal scheme can be written as

$$U_{\vec{k}}(t + \Delta t) = U_{\vec{k}}(t)e^{-\nu k^2 \Delta t} + e^{-\nu k^2 \Delta t}\Delta t\left[\frac{3}{2}G_{\vec{k}}(t) - \frac{1}{2}G_{\vec{k}}(t - \Delta t)e^{-\nu k^2 \Delta t}\right] \tag{3.7}$$

where Δt is time stepping. If we include the dealiasing there by removing high wave numbers using the 3/2 rule, we get a stable numerical scheme. These methods are described in detail in the chapter 7.2 of the Peyret's monography [34].

3.5. Turbulence and subgrid modeling

Nowadays the largest numerical simulation for the incompressible Navier–Stoke equation in periodic geometry reach 4096^3 grid points [41], with a such mesh the kinetic Reynolds number is bounded below $R_v \sim 65000$ or $R_\lambda \sim 1200$. This is still far from the minimum of $R_v \sim 10^6$ necessary to produce a dynamo in the low magnetic Prandtl number limit. One way around this difficulty is to resort to use Large Eddy Simulations (or LES) [42–45]. Such techniques are widely used in engineering contexts, as well as in atmospheric sciences and in some case in geophysics and astrophysics. Such subgrid model controls the energy transfer and obtains a larger inertial range, where the energy cascade has a constant dissipation rate. We have used one particular LES model [30,46,47] where the viscosity depends of the modulus of the wave-vectors $\vec{k}$. The turbulent viscosity becomes $v(k,t) = v_0 H(k)\sqrt{\frac{E(K_{\max},t)}{K_{\max}}}$ where $H(k)$ is called a "cups" function equal to unity at large scale and increases the dissipation a high wave number and v_0 depends on the forcing and the Kolmogorov constant. This eddy viscosity form is an asymptotic solution of a Navier–Stoke closure approximation. It is possible to take in to account all the closure terms, and solve the full integro-numerical system to obtain a dynamical and more precise LES model [48].

4. Magnetic induction

We consider the induction of a magnetic field in the flows of an electrically conducting fluid at low magnetic Prandtl number and large kinetic Reynolds number. The coupled magnetic and fluid equations (3.1)–(3.2) are solved using a mixed scheme, where the magnetic field fluctuations are fully resolved and the velocity fluctuations at small scale are modeled using a Large Eddy Simulation (LES) scheme as describe in (3.5). We study the response of a forced Taylor–Green flow (eq. (3.3)) to an externally applied field B_0 [49]. In a periodic box, the external magnetic field is implemented in spectral space by feeding energy to the $k_x = k_y = k_z = 0$ wave vector for the chosen magnetic field component.

Figure 1 shows the kinetic and magnetic spectra, at a given time of the simulation. The kinetic energy spectrum exhibits a $k^{-5/3}$ Kolmogorov scaling main-

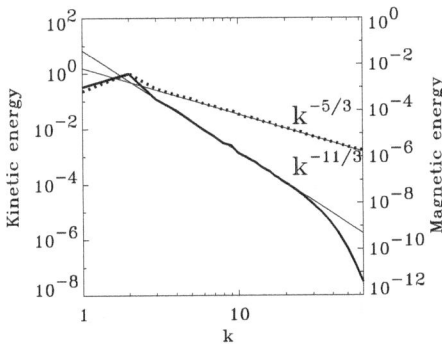

Fig. 1. Magnetic (solid line) and kinetic (dash line) energy spectra.

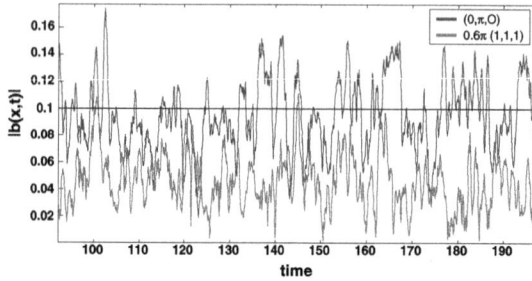

Fig. 2. Time traces of $|\mathbf{b}(\mathbf{x}, t)|$, for $\mathbf{B_0} = 0.1\,\hat{\mathbf{x}}$, at two fixed points.

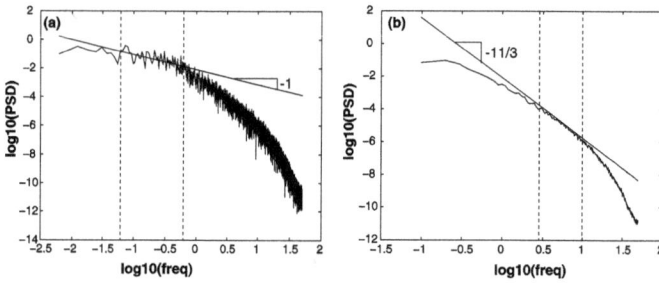

Fig. 3. Power spectral density of the magnetic field fluctuations of $b_x(\mathbf{x}, t)$ in time, recorded at space location. (a) PSD computed as averages over Fourier transforms calculated over long time intervals ($164t_0$) to emphasize the low frequency behavior; (b) PSD estimated from Fourier transforms over shorter time intervals ($10t_0$). The behavior is identical for the $b_y(\mathbf{x}, t)$ and $b_x(\mathbf{x}, t)$ field components.

tained by the LES scheme. The peak at low wave number, also visible on the magnetic field spectrum, is due the large scale TG forcing. The magnetic inertial range is well fitted by a $k^{-11/3}$ power law in agreement with the Kolmogorov phenomenology [1, 9], where the advection of the external magnetic field is balance by the magnetic diffusion $\mathbf{B_0}\nabla.\mathbf{v} \sim \eta\Delta\mathbf{b}$. Using this numerical "mixed scheme", we obtain low magnetic Prandtl numbers of $P_m \sim 10^{-3}$–10^{-4}, but the magnetic Reynolds number remains order one.

During the simulation, we record also the velocity and the magnetic fluctuation at fixed points, so as to be able to compare these experimental dates. In Fig. 2, we present a sample of the magnetic fluctuations. The amplitude of the magnetic fluctuations are easily above level of the external magnetic field. As shown in several experiment [9, 13, 17], the liquid metal turbulent flow represent a efficient magnetic amplifier.

In Fig. 3 we plot the power spectra of temporal fluctuations of the magnetic field component $b_x(\mathbf{x}, t)$ recorded one point fixed. The higher end of the time spectrum follows a $f^{-11/3}$ behavior (f is signal time frequency), as can be expected from the spatial spectrum using the Taylor hypothesis of "frozen" field lines advected by the mean flow [9]. For frequencies between roughly $1/t_0$ and $1/10t_0$, where the dynamical time t_0 is set to the magnetic diffusion time scale. The time spectrum develops a $1/f$ behavior (Fig. 3b), which is observed in experiment [13]. Note that the k^{-1} power law is not present on the spatial spectrum in Fig. 1.

5. Linear dynamo onset

5.1. *Static or turbulent kinematic dynamo*

The dynamo effect is an instability, initiated by a magnetic seed, where the geometric proprieties and the dynamics of the fluid amplify the magnetic field energy exponentially. We call this regime the linear phase, or kinematic dynamo. We must distinguish between the mathematical/static kinematic dynamo with the turbulent/dynamical kinematic dynamo. In the first case, a velocity defined mathematically or an time averaged velocity field from a fluid experiment or a numerical simulations are constant in time and inserted in the induction equation. In the second case the velocity field evolves in time it evolution coming from a numerical solution of the Navier–Stoke equation. However in the both cases the Lorentz force is not present, and so there is no back reaction on the flow.

There are historical examples of mathematical kinematic dynamos with the ABC flow, this being a helical Beltrami flow with chaotic Lagrangian trajectories [39]. The kinematic dynamo instability of the ABC flow, even with one of the amplitude coefficients set to zero ($2^{1/2}$ dimensional flow) [50, 51] has been study intensively [52–56], especially for fast dynamo investigation [57–61]. The magnetic field grows near the stagnation point of the flow, producing "cigar" shape structures aligned along the unstable manifold. For the ABC with all the coefficient equal to the unity ($A = B = C = k_0 = 1$), there are two windows of dynamo instability [52, 53], which disappear when the velocity field occurs on smaller scales ($k_0 > 2$) [54].

In the dynamic regime, where the velocity is fully resolved numerically at constant forcing F_{ABC} (3.4), and above a critical Reynolds number, the hydrodynamic system becomes unstable. After the first bifurcation, further increase of the kinematic Reynolds number, leads the system to jump to different attractors [62–65], until finally the fully turbulent regime is reached. Then the dynamo onset increase rapidly with the Reynolds number, until finally reaching a plateau [66].

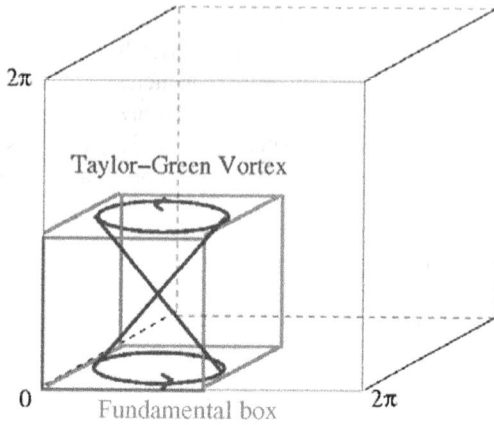

Fig. 4. Representation of fundamental Taylor–Green box in a fully periodic box.

5.2. Taylor–Green dynamo

The Taylor–Green flow included height fundamental box (for $k_0 = 1$) (see Fig. 4). In the fundamental box, the Taylor–Green vortex has similar hydrodynamic proprieties to the experimental Von Kàrmàn vortex [67]. It has been demonstrated that this forcing can produce a numerical dynamo [68, 69] at low Reynolds number and with a magnetic Prandtl number order one. Note the mathematical flow itself can not be a kinematic dynamo, it has one velocity component set to zero (eq. 3.3). The system needs the recirculation created by the inertial term, producing a three-dimensional velocity necessary for the dynamo instability. When the Reynolds number increases, the onset of the dynamo also increases until is reaching a plateau [70] (see Fig. 5).

We focus here on flows generated by a deterministic forcing at large scales for which a mean flow develops in addition to turbulent fluctuations which, observations show, develop over all spatial and temporal scales.

The spectra of the dynamical run and its average in time are shown in Fig. 5a, for the DNS calculation. While the dynamical flow has a typical turbulence spectrum, the time-averaged field is sharply peaked at the size of the TG cell. As the Reynolds number varies, the characteristics of the average flow are constant (Fig. 5b). An three-dimensional representation of a snapshot in time of the kinetic energy and the time average the kinetic energy along all the simulation are shown in Fig. 6.

Now, using the time averaged velocity for static kinematic dynamo studies, we observe the existence of two dynamo branches, Fig. 7, a behavior already noted for the ABC flow [50]. The first dynamo mode has larger scales than

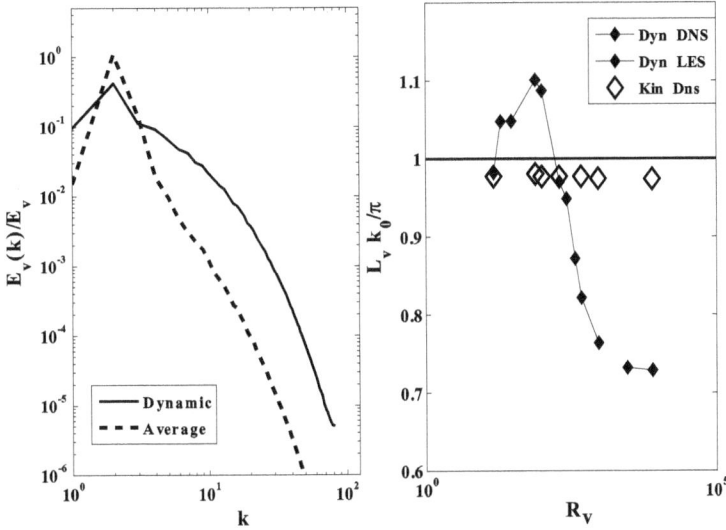

Fig. 5. (a) Kinetic energy spectra for TG1 ($\nu = 0.007$, $N = 256^3$); $E_{V,\mathrm{dyn}}(k, t = T)$ (solid line), $E_{V,\mathrm{kin}}(k)$ (dotted line); (b): integral length scales L_{dyn} and L_{kin}, normalized by the size of the unit TG cell, versus the flow Reynolds numbers.

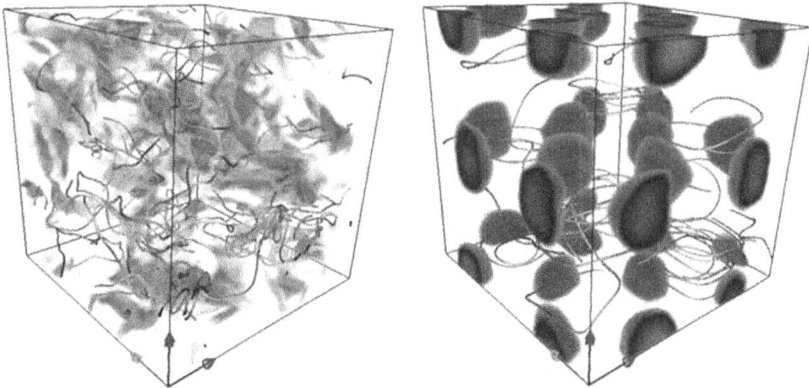

Fig. 6. (l.h.s) Snapshot of the velocity, shown in volume rendering of the kinetic energy, and some field line trajectories. the r.h.s displays the volume rendering of the kinetic energy for the time average velocity (same parameter as Fig. 4) (*imagery made with Vapor [77]*).

the fundamental box and then exist only by the collective effect of the height Taylor–Green vortex and the periodic boundaries. The second one is inside the fundamental box, and looks like the double bananas shape [72] found numeri-

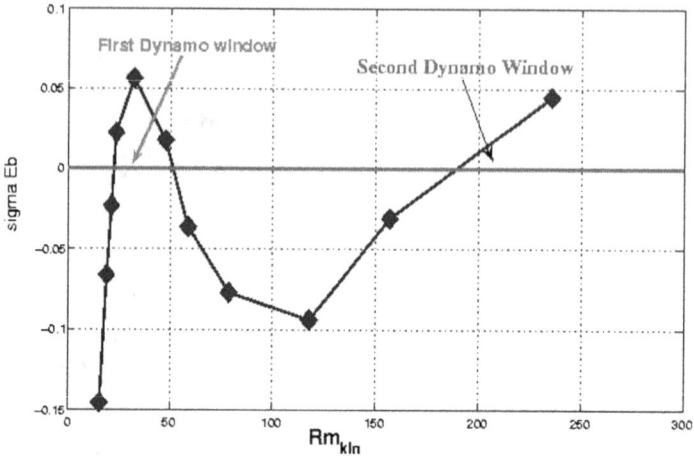

Fig. 7. Growth rates for the kinematic dynamo generated by mean flow versus the magnetic Reynolds number.

cally using the Von Kàrmàn time average velocity, recorded from water experiments [73,74].

We compare in Fig. 8 the evolution of all R_M^c the critical magnetic Reynolds numbers. At low R_V, the dynamo threshold for the dynamical problem lies within the low R_M dynamo window for the time-averaged flow. For R_V larger than 200, the dynamical dynamo threshold lies in the immediate vicinity of the upper dynamo branch (high R_M^c mode of the time-averaged flow). It is unclear from Fig. 8 whether the effect of the fluctuations in the dynamical runs is to increase the threshold of the first kinematic window, or decrease the threshold of the second one. In the work of Laval et al. [75], some artificial noise is added to the mean flow, and by this procedure the dynamo threshold increases with the level of noise. Unfortunately the realistic velocity fluctuation are not a simple noise, so at this stage we need further works to conclude.

In the dynamical runs, the magnetic energy spectrum peaks at scales smaller, than the hydrodynamic integral scale, it is suggested that turbulent fluctuations play a role on the dynamo effect. Looking only at the magnetic spectra the Taylor–Green dynamo, it looks like a fluctuation type dynamo. But in fact the Taylor–Green dynamo is easier to obtain than fluctuation type dynamo (Fig. 1b of [76]), and the localisation of growing magnetic energy inside the neutral plan of the Taylor–Green vortex, suggest that the mean flow plays a major role in the dynamo process [72]. We can only suggest at this stage, that certainly there is a double contribution to the dynamo, the mean flow mode and some additional fluctuating dynamo process.

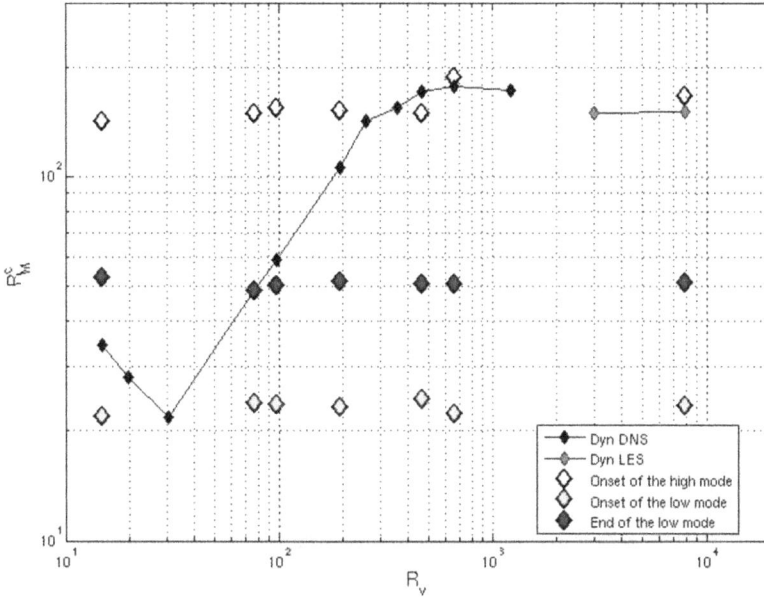

Fig. 8. Evolution of the critical magnetic Reynolds numbers for the kinematic runs from the time average velocity in diamond symbol and the dynamical runs in full line versus the Reynolds number R_V.

6. Non linear behavior

6.1. Subcritical dynamo

Previous works ([70–72, 75] or chap. 5.2) have explored the response of Taylor–Green forcing to a magnetic seed and computed the onset of the dynamo versus the Reynolds number. In the non linear regime when the magnetic field has reached sufficient amplitude, it can react back onto the velocity field, saturate the instability and reach a statistically stationary state, with approximate equipartition $E_M \sim E_V$ (Fig. 9, with the time less then 1000).

In Fig. 9, we have quenched the system at $t = 1000$, the magnetic diffusivity η is suddenly increased by a factor of 4, lowering R_M below R_M^c. After a short transient, both E_V and E_M decrease and reach a second statistically stationary state, with a non zero magnetic energy—a new dynamo state, for which equipartition is reached again. This behavior is evidence for global subcriticality [78]. The different levels of fluctuations in the two regimes suggest the possibility of different dynamo states, depending on the magnetic field or on history of the system. As subcritical bifurcations are also associated with hysteresis cycles, we have repeated the quenching procedure starting from the same dynamo state A, and obtain a clear hysteresis cycles shown in Fig. 10.

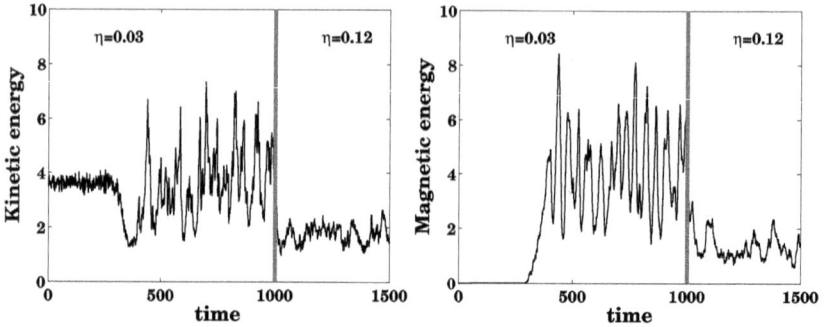

Fig. 9. After a dynamo is self-generated from infinitesimal perturbations, the induction equation is quenched at $t = 1000$ by a four-fold increase of the magnetic diffusivity. It corresponds to a sudden change from A to A_9—cf. Fig. 10.

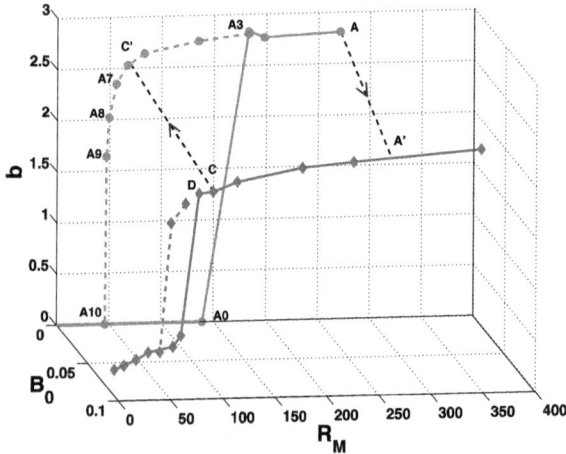

Fig. 10. Bifurcation curves and hysteresis cycles when an external magnetic field is applied (full diamond symbols) or without one (full circle symbols). In this case, the subcritical quenched states (see text) form the red line. Jumps between the two branches link A to A' and C to C'.

With an applied magnetic field the hysteresis cycle still exist but is smaller, and by changing suddenly the magnetic Reynolds number or the value of the applied magnetic field, it is possible to jump from branch to branch as Fig. 10.

A less deterministic behavior is observed when the system is operated in the vicinity of point D—shown along the blue curve in Fig. 10. We observe that the systems spontaneously switches between dynamo and non-dynamo periods, as shown in Fig. 11. This is reminiscent of the "on-off" bifurcation scenario. We present an extended example of this behavior in the next Section 6.2.

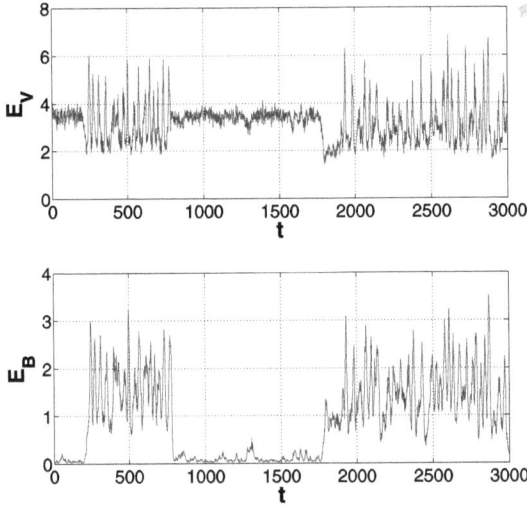

Fig. 11. Evolution in time of the kinetic (E_V) and magnetic energy (E_B) when the flow is occurs in the immediate vicinity of point D—see Fig. 10.

6.2. On-Off intermittency dynamo

On-Off intermittency has been observed in different physical experiments including electronic devices, electrohydrodynamic convection in nematics, gas discharge plasmas, and spin-wave instabilities [87]. In the MHD context, near the dynamo instability onset, the On-Off intermittency has been investigated by modeling of the Bullard dynamo [88], and experimental results have confirm a such behavior [18].

Using direct numerical simulation [84, 85], they were able to observe On-Off intermittency solution of the full MHD equations for the ABC dynamo ($F_{ABC}(k_0 = 1)$), (here we present an extended work of this particular case) [86].

Recent dynamo experiment results (VKS) [89] show some intermittent behavior, with features reminiscent of On-Off self-generation that motivated our study.

A simple and proven very useful way to model the behavior of the magnetic field during the on-off intermittency is using a stochastic differential equation (SDE-model) [82, 83, 90–98]:

$$\frac{\partial E_b}{\partial t} = (a + \xi)E_b - NL(E_b), \qquad (6.1)$$

where E_b is the magnetic energy, a is the long time averaged growth rate, ξ models the noise term typically assumed to be white (see however [97, 98]) and of amplitude D such that $\langle \xi(t)\xi(t') \rangle = 2D\delta(t - t')$. NL is a non-linear

Y. Ponty

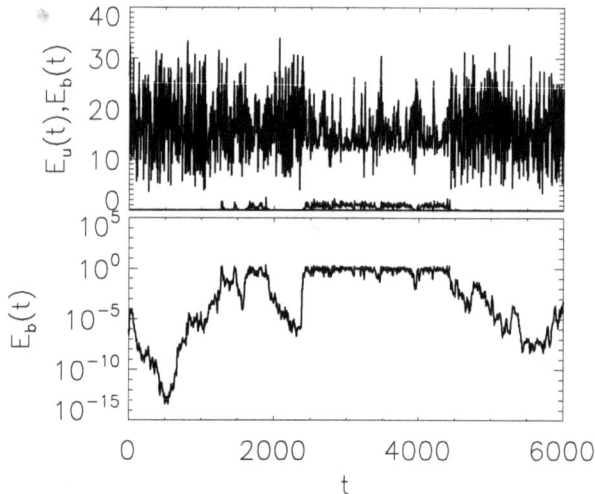

Fig. 12. A typical example of a burst. The top panel shows the evolution of the kinetic energy (top line) and magnetic energy (bottom line). The bottom panel shows the evolution of the magnetic energy in a log-linear plot. During the on phase of the dynamo the amplitude of the noise of the kinetic energy fluctuations is significantly reduced.

term that guaranties the saturation of the magnetic energy to finite values typically taken to be $NL(X) = X^3$ for investigations of supercritical bifurcations or $NL(X) = X^5 - X^3$ for investigations of subcritical bifurcations. In all these cases (independent of the non-linear saturation mechanism) the above SDE leads to stationery distribution function that for $0 < a < D$ has a singular behavior at $E_b = 0$: $P(E_b) \sim E_b^{a/D-1}$ indicating that the systems spends a lot of time in the neighborhood of $E_b = 0$ [93–96].

In the dynamical system eq. (6.1) studied, the noise amplitude or the noise proprieties do not depend on the amplitude of the magnetic energy.

However, in the MHD system, when the non-linear regime is reached, the Lorentz force has a clear effect on the flow by decreasing the small scale fluctuations, and decreases of the local Lyapunov exponent [79, 80]. In some cases, the flow is altered so strongly that the MHD dynamo system jumps into an other attractor, that cannot not no longer sustain the dynamo instability [81]. Although the exact mechanism of the saturation of the MHD dynamo is still an open question that might not have a universal answer, it is clear that both the large scales and the turbulent fluctuations are altered in the non-linear regime and need to be accounted in a model.

Figure 12 demonstrates this point, by showing the evolution of the kinetic and magnetic energy as the dynamo goes through On- and Off- phases. During

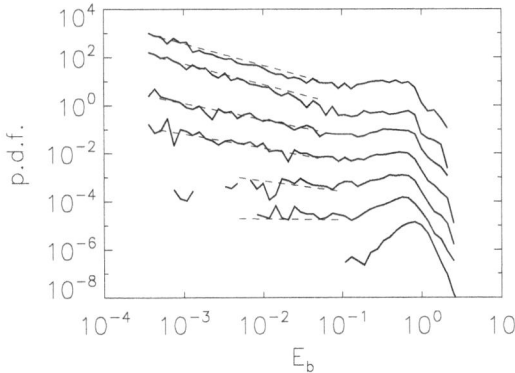

Fig. 13. The probability distribution functions of E_b, at seven different values of the magnetic Reynolds number The last case shows no on-off intermittency. The dashed lines shows the prediction of the SDE model.

the On phases, the mean value and the amplitude of the observed fluctuations of the kinetic energy are significantly reduced. As a result the On-phases last a lot longer than the SDE-model would predict. The effect of the long duration of the "On" times can also be seen in the pdfs of the magnetic energy. The probability distribution function (pdf) for the examined magnetic Reynolds number are shown in Fig. 13. For different values of the magnetic Reynolds number the pdf of the magnetic field is concentrated at large values $E_b \simeq 1$ producing a peak in the pdf curves. When the magnetic Reynolds number increases the On-Off power low disappears.

This peak can be attributed to the quenching of the hydrodynamic "noise" in the nonlinear stage. In principle the SDE model eq. (6.1) can be modified to include this effect: using an energy E_b depending of the amplitude of the noise. There many possible ways to model the quenching of the noise, however the nonlinear behavior might not be a universal behavior and we do not attempt to suggest a specific model.

7. Perspective

The periodic box remains an attractive tool to study MHD turbulence and the impact of velocity fluctuation on the dynamo onset. The absence of boundary condition could be modelled by new techniques introduced in pseudo-spectral codes such as the penalisation method [99] or pseudo-penalisation method [100]. The classical wishe in the numerical simulation is generally to increase the resolution to reach high Reynolds numbers. But there is a more challenging limit which is more relevant to dynamo experiment: to massively increase the run time of the simulations. Indeed, the dynamo experiments easily reach a thousand magnetic diffusion time ($\Delta T \sim 1000 \, \tau_\eta$). If we want to be able to compare data statistically between the experiments and the simulations, the simulation time needs to be extended ($\Delta T \sim 1000 \, R_m \tau_{NL}$) between ten thousand to hundred thousand of hydrodynamic eddy turn over time.

8. Acknowledgements

YP acknowledge the co-authors of our common articles [49, 70, 72, 78, 86] which represent the basis of this lecture (A. Pouquet, J-F Pinton, H. Politano, A. Alexakis, P. Mininni, J-P Laval, B. Dubrulle and F. Daviaud). All work presented here is supported by the French GDR Dynamo and the computer time was provided by IDRIS and the Mesocentre SIGAMM at Observatoire de la Côte d'Azur. YP thanks A. Miniussi for computing design assistance and M.S Rosin for reading and improving this manuscript. YP is grateful to the summer school organisers to give him the opportunity to have fruitful discussions with the lecturers and promising students, and spend great time with the family (Annick, Florian, Julien and Scooby (dog)).

References

[1] H.K. Moffatt, *Magnetic Field Generation in Electrically Conducting Fluids*, Cambridge University Press, Cambridge, 1978; F. Krause and K.-H. Radler, *Mean-Field Magnetohydrodynamics and Dynamo Theory*, Pergamon, Oxford, 1980; E.N. Parker, *Cosmical Magnetic Fields*, Clarendon, Oxford, 1979.

[2] A. Gailitis, O. Lielausis, S. Dement'ev, E. Platacis, A. Cifersons, G. Gerbeth, T. Gundrum, F. Stefani, M. Christen, H. Hänel and G. Will, Detection of a flow induced magnetic field eigenmode in the Riga dynamo facility, Phys. Rev. Lett. **84**, 4365 (2000).

[3] A. Gailitis, O. Lielausis, S. Dement'ev, A. Cifersons, G. Gerbeth, T. Gundrum, F. Stefani, M. Christen, H. Hänel and G. Will, Magnetic field saturation in the Riga dynamo experiment, Phys. Rev. Lett. **86**, 3024 (2001).

[4] A. Gailitis, O. Lielausis, E. Platacis, G. Gerbeth and F. Stefani, Riga dynamo experiment and its theoretical background, Physics of Plasmas **11**(5), 2838–2843 (2004).

[5] U. Müller and R. Stieglitz, Naturwissenschaften **87**, 381 (2000).

[6] R. Stieglitz and U. Müller, Experimental demonstration of a homogeneous two-scale dynamo, Phys. Fluids **13**, 561 (2001).

[7] R. Monchaux, M. Berhanu, M. Bourgoin, M. Moulin, Ph. Odier, J.-F. Pinton, R. Volk, S. Fauve, N. Mordant, F. Pétrélis, A. Chiffaudel, F. Daviaud, B. Dubrulle, C. Gasquet, L. Marié and F. Ravelet, Generation of a magnetic field by dynamo action in a turbulent flow of liquid sodium, Phys. Rev. Lett. **98**, 044502 (2007).

[8] M. Berhanu, R. Monchaux, S. Fauve, N. Mordant, F. Pétrélis, A. Chiffaudel, F. Daviaud, F. Ravelet, M. Bourgoin, Ph. Odier, J.-F. Pinton, R. Volk, B. Dubrulle and L. Marie, Magnetic field reversals in an experimental turbulent dynamo, Europhys. Lett. **77**, 59001 (2007).

[9] P. Odier, J.-F. Pinton and S. Fauve, Advection of a magnetic field by a turbulent swirling flow, Phys. Rev. E **58**, 7397–7401 (1998).

[10] N.L. Peffley, A.B. Cawthorne and D.P. Lathrop, Toward a self-generating magnetic dynamo, Phys. Rev. E **61**, 5287 (2000).

[11] N.L. Peffley, A.G. Goumilevski, A.B.C. and D.P. Lathrop, Characterization of experimental dynamos, Geoph. J. Int. **142**, 52–58 (2000).

[12] P. Frick, V. Noskov, S. Denisov, S. Khripchenko, D. Sokoloff, R. Stepanov and A. Sukhanovsky, Non-stationary screw flow in a toroidal channel: way to a laboratory dynamo experiment, Magnetohydrodynamics **38**(1/2), 143–162 (2002).

[13] M. Bourgoin, L. Marié, F. Pétrélis, C. Gasquet, A. Guiguon, J.-B. Luciani, M. Moulin, F. Namer, J. Burguete, F. Daviaud, A. Chiffaudel, S. Fauve, Ph. Odier and J.-F. Pinton, MHD measurements in the von Kàrmàn sodium experiment, Phys. Fluids **14**, 3046 (2002).

[14] M.D. Nornberg, E.J. Spence, R.D. Kendrick, C.M. Jacobson and C.B. Forest, Intermittent magnetic field excitation by a turbulent flow of liquid sodium, Phys. Rev. Lett. **97**, 044503 (2006).

[15] M.D. Nornberg, E.J. Spence, R.D. Kendrick and C.B. Forest, Measurements of the magnetic field induced by a turbulent flow of liquid metal, Phys. Plasmas **13**, 055901 (2006).

[16] R. Stepanov, R. Volk, S. Denisov, P. Frick, V. Noskov and J.-F. Pinton, Induction, helicity and alpha effect in a toroidal screw flow of liquid gallium, Phys. Rev. E **73**, 046310 (2006).

[17] R. Volk, P. Odier and J.-F. Pinton, Fluctuation of magnetic induction in von Kármán swirling flows, Phys. Fluids **18**, 085105 (2006).

[18] M. Bourgoin, R. Volk, N. Plihon, P. Augier, P. Odier and J.-F. Pinton, An experimental Bullard-von Kármán dynamo, New Journal of Physics **8**, 329 (2006).

[19] A. Gailitis et al., Project of a liquid Sodium MHD dynamo experiment, Magnetohydrodynamics **1**, 63 (1996).

[20] A. Tilgner, A kinematic dynamo with a small scale velocity field, Phys. Rev. A **226**, 75 (1997).

[21] L. Marié, J. Burguete, F. Daviaud and J. Léorat, Eur. J. Phys. B **33**, 469 (2003); F. Ravelet et al., Phys. Fluids **17**, 117104 (2005).

[22] R.A. Bayliss, C.B. Forest and P.W. Terry, Numerical simulations of current generation and dynamo excitation in a mechanically forced turbulent flow, Phys. Rev. Lett **75**, 026303–13 (2007).

[23] S. Kenjereś and K. Hanjalić, Numerical simulation of a turbulent magnetic dynamo, Phys. Rev. Lett **98**, 104501 (2007).

[24] S. Kenjereś and K. Hanjalić, Numerical insights into magnetic dynamo action in a turbulent regime, New J. Phys. **9**, 306 (2007).

[25] J.L. Guermond, J. Léorat and C. Nor,e A new Finite Element Method for magneto-dynamical problems: two-dimensional results, E. J. Mech. Fluids **22**, 555 (2003).

[26] J.L. Guermond, R. Laguerre, J. Léorat and C. Nore, An interior penalty Galerkin method for the MHD equations in heterogeneous domains, J. Comp. Phys. **221**, 349–369 (2007).

[27] A.B. Iskakov, S. Descombes and E. Dormy, An integro-differential formulation for magnetic induction in bounded domains: boundary element-finite volume method, J. Comp. Phys. **197**, 540–554 (2004).

[28] A.B. Iskakov and E. Dormy, On magnetic boundary conditions for non-spectral dynamo simulations, Geophys. Astrophys. Fluid Dyn. **99**, 481–492 (2006).

[29] U. Frisch, *Turbulence: The Legacy of A.N. Kolmogorov*, Cambridge University Press, 1996.

[30] M. Lesieur, *Turbulence in Fluids*, Kluwer, 1997.

[31] A. Vincent and M. Meneguzzi, The spatial structure and the statistical propierties of homege-neous turbulence, J. Fluid Mech. bf 225, 1–25 (1991).

[32] S.A. Orszag, Numerical methods for the simulation of turbulence, Phys. of Fluid **21**(Suppl. II), 250–257 (1969).

[33] C. Canuto, M.Y. Hussaini, A. Quarteroni and T.A. Zang, *Spectral methods in Fluid Dynamics*, Springer-Verlag, 1988.

[34] R. Peyret, *Spectral Method for Imcompressible Vicous Flow*, Applied Mathematical Sciences, vol. 148, Springer, 2002.

[35] S.A. Orszag, Comparison of pseudospectral and spectral approximations, Stud. Appl. Math. **51** (1972).

[36] S.A. Orszag and J.S. Patterson Jr., Numerical simulation of three-dimensional homogeneous isotropic turbulence, Phys. Rev. Lett. **28**, 76–79 (1972).

[37] M.E. Brachet, D.I. Meiron, S.A. Orszag, B.G. Nickel, R.H. Morf and U. Frisch, Small-scale structure of the Taylor–Green vortex, J. Mech. Fluids **130**, 411–452 (1983).

[38] V.I. Arnold, Comptes Rendus Acad. Sci. Paris **261**, 17 (1965).

[39] T. Dombre, U. Frisch, J.M. Greene, M. Henon, A. Mehr and A. Soward, Chaotic streamlines in the ABC flows, J. Fluid Mech. **167**, 353 (1986).

[40] C. Basdevant, Le modèle de simulation numérique de turbulence bidimensionnelle du L.M.D, Note interne LMD no 114, laboratoire de Météorologie Dynamique du CNRS, Paris, 1982.

[41] Y. Kaneda, T. Ishihara, M. Yokokawa, K. Itakura and A. Uno, Energy dissipation rate and energy spectrum in high resolution direct numerical simulations of turbulence in a periodic box, Phys. Fluids **15**, L21-L24 (2003).

[42] M. Lesieur and O. Métais, Ann. Rev. Fluid Mech. **28**, 45 (1996).

[43] C. Meneveau and J. Katz, Annu. Rev. Fluid Mech. **32**, 1 (2000).

[44] U. Piomelli, Prog. Aerosp. Eng. **35**, 335 (1999).

[45] P. Sagaut, *Large Eddy Simulation for Incompressible Flows*, 2nd ed., Springer-Verlag, Berlin, 2003.

[46] J.P.Chollet and M. Lesieur, Parameterisation for small scales of three dimensional isotropic turbulence using spectral closure, J. Atmos. Sci. **38**, 2747 (1981).

[47] O. Métais, Large-Eddy Simulations of Turbulence, in: *New trends in turbulence. Turbu-lence: nouveaux aspects Les Houches Session LXXIV 31 July–1 September 2000, Series: Les Houches—Ecole d'Ete de Physique Theorique*, Vol. 74, Lesieur, M., Yaglom, A., David, F., eds., Springer Jointly published with EDP Sciences, Les Ulis.

[48] J. Baerenzung, H. Politano, Y. Ponty and A. Pouquet, Spectral modeling of turbulent flows and the role of helicity, Phys. Rev. E under press (2008) (available arXiv:0707.0642).

[49] Y. Ponty, J. F Pinton and H. Politano, Simulation of induction at low magnetic Prandtl number, Phys. Re. Lett. **92** 14, 144503 (2004).

[50] D.J. Galloway and M.R.E. Proctor, Numerical calculations of fast dynamo for smooth velocity field with realistic diffusion, Nature **356**, 691–693 (1992).

[51] Y. Ponty, A. Pouquet and P.L. Sulem, Dynamos in weakly chaotic 2-dimensional flows, Geo-phys. Astrophys. Fluid Dyn. **79**, 239–257 (1995).

[52] V.I. Arnold and E.I. Korkina, The grow of magnetic field in a incompressible flow, Vestn. Mosk. Univ. Mat. Mekh. **3**, 43–46 (1983).

[53] D.J. Galloway and U. Frisch, Dynamo action in a family of flows with chaotic stream lines, Geophys. Astrophys. Fluid Dyn. **36**, 53–83 (1986).

[54] B. Galanti, P.L. Sulem and A. Pouquet, Linear and non-linear dynamos associated with the ABC flow, Geophys. Astrophys. Fluid Dyn. **66**, 183–208 (1992).

[55] V. Archontis, S.B.F. Dorch and A. Nordlund, Numerical simulations of kinematic dynamo action, Astron. Astrophys. **397**, 393–399 (2003).

[56] R. Teyssier, S. Fromang and E. Dormy, Kinematic dynamos using constrained transport with high order Godunov schemes and adaptive mesh refinement, J. Comp. Dyn. **218**, 44–67 (2006).

[57] S. Childress and A.D. Gilbert, *Stretch, Twist Fold: The Fast Dynamo*, Springer-Verlag, New York, 1995.

[58] H.K. Moffat and M.R. Proctor, Topological constraints associated with fast dynamo action, J. Fluid Mech. **154**, 493 (1985).

[59] B.J. Bayly and S. Childress, Geophys. Astrophys. Fluid Dyn. **44**, 211 (1988).

[60] J.M. Finn and E. Ott, Chaotic flows and fast magnetic dynamos, Phys. Fluids **31**, 2992 (1988).

[61] J.M. Finn and E. Ott, Chaotic flows and magnetic dynamos, Phys. Rev. Lett. **60**, 760 (1988).

[62] O.M. Podvigina and A. Pouquet, On the nonlinear stability of the $1 = 1 = 1$ ABC flow, Physica D **75**, 471–508 (1994).

[63] O.M. Podvigina, Spatially-periodic steady solutions to the three-dimensional Navier–Stokes equation with the ABC-force, Physica D **128**, 250–272 (1999).

[64] P. Ashwin and O. Podvigina, Hopf bifurcation with cubic symmetry and instability of ABC flow, Proc. R. Soc. London **459**, 1801–1827 (2003).

[65] O. Podgivina, P. Ashwin and D.J. Hawker, Modelling instability of ABC flow using a mode interaction between steady and Hopf bifurcations with rotational symmetries of the cube, Physica D **215**, 62–79 (2006).

[66] P.D. Mininni, Turbulent magnetic dynamo excitation at low magnetic Prandtl number, Physics of Plasmas **13**(5), 056502 (2006).

[67] S. Douady, Y. Couder and M.-E. Brachet, Direct observation of the intermittency of intense vorticity filaments in turbulence, Phys. Rev. Lett. **67**, 983 (1991).

[68] C. Nore, M. Brachet, H. Politano and A. Pouquet, Dynamo action in the Taylor–Green vortex near threshold, Phys. Plasmas **4**, 1 (1997).

[69] C. Nore, M.-E. Brachet, H. Politano and A. Pouquet, Dynamo action in a forced Taylor–Green vortex, in: *Dynamo and Dynamics, a Mathematical Challenge*, NATO Science Series II, vol. 26, P. Chossat, D. Armbruster and I. Oprea, eds., Kluwer Academic, Dordrecht, 2001. Proceedings of the NATO Advanced Research Workshop, Cargèse, France, 21–26 August 2000, pp. 51–58.

[70] Y. Ponty, P.D. Minnini, A. Pouquet, H. Politano, D.C. Montgomery and J.-F. Pinton, Numerical study of dynamo action at low magnetic Prandtl numbers, Phys. Rev. Lett. **94**, 164512 (2005).

[71] P.D. Mininni, Y. Ponty, D.C. Montgomery, J.-F. Pinton, H. Politano and A. Pouquet, Dynamo regimes with a non-helical forcing, The Astrophysical Journal **626**, 853–863 (2005).

[72] Y. Ponty, P.D. Minnini, J.-F. Pinton, H. Politano and A. Pouquet, Dynamo action at low magnetic Prandtl numbers: mean flow *versus* fully turbulent motions, New J. Phys. **9**, 296 (2007).

[73] F. Ravelet, A. Chiffaudel, F. Daviaud and J. Léorat, Toward an experimental von Kármàn dynamo: Numerical studies for an optimized design, Phys. Fluids **17**, 117104 (2005).

[74] F. Ravelet, A. Chiffaudel, F. Daviaud and L. Marié, Multistability and memory effect in a highly turbulent flow: experimental evidence for a global bifurcation, Phys. Rev. Letters **93**, 164501 (2004).

[75] J.-P. Laval, J.-P. Blaineau, N. Leprovost, B. Dubrulle and F. Daviaud, Influence of turbulence on the dynamo threshold, Phys. Rev. Let. **96**, 204503 (2006).

[76] A.A. Schekochihin, A.B. Iskakov, S.C. Cowley, J.C. McWilliams, M.R.E. Proctor and T.A. Yousef, Fluctuation dynamo and turbulent induction at low magnetic Prandtl numbers, New J. Phys. **9**, 300 (2007).

[77] Imagery using VAPOR code (www.vapor.ucar.edu) a product of the National Center for Atmospheric Research.

[78] Y. Ponty, J.-P. Laval, B. Dubrulle, F. Daviaud and J.-F. Pinton, Subcritical dynamo bifurcation in the Taylor–Green Flow, Phys. Rev. Lett. **99**, 224501 (2007).

[79] F. Cattaneo, D.W. Hughes and E.J. Kim, Suppression of chaos in a simpled nonlinear dynamo model, Phys. Rev. Lett. **76**, 2057–2060 (1996).

[80] E. Zienicke, H. Politano and A. Pouquet, Variable intensity of Lagrangian chaos in the nonlinear dynamo problem, Phys. Rev. Lett. **81**, 4640–4640 (1998).

[81] N.H. Brummell, F. Cattaneo and S.M. Tobias, Linear and nonlinear dynamo properties of time-dependent ABC flows, Fluid Dynamics Research **28**, 237–265 (2001).

[82] Y. Pomeau and P. Manneville, Intermittent transition to turbulence in dissipative dynamical systems, Commun. Math. Phys. **74**, 1889 (1980).

[83] N. Platt, E.A. Spiegel and C. Tresser, On-off intermittency: A mechanism for bursting, Phys. Rev. Lett. **70**(3), 279–282 (1993).

[84] D. Sweet, E. Ott, J.M. Finn, T.M. Antonsen, Jr. and D.P. Lathrop, Blowout bifurcations and the onset of magnetic activity in turbulent dynamos, Phys. Rev. E **63**, 066211 (2001).

[85] D. Sweet, E. Ott, T.M. Antonsen, Jr. and D.P. Lathrop, J.M. Finn, Blowout bifurcations and the onset of magnetic dynamo action, Physics of Plasmas **8**, 1944–1952 (2001).

[86] A. Alexakis and Y. Ponty, The Lorentz force effect on the On-Off dynamo intermittency, under press, Phys. Rev. E (available: http://fr.arxiv.org/abs/0710.0063 arXiv:0710.003).

[87] A.S. Pikovsky, Z. Phys. B **55**, 149 (1984); P.W. Hammer, N. Platt, S.M. Hammel, J.F. Heagy and B.D. Lee, Phys. Rev. Lett. **73**(8), 1095–1098 (1994). T. John, R. Stannarius and U. Behn, Phys. Rev. Lett. **83**(4), 749–752 (1999); D.L. Feng, C.X. Yu, J.L. Xie and W.X. Ding, Phys. Rev. E **58**(3), 3678–3685 (1998); F. Rodelsperger, A. Cenys and H. Benner, Phys. Rev. Lett. **75**(13), 2594–2597 (1995).

[88] N. Leprovost, B. Dubrulle and F. Plunian, Intermittency in the homopolar disc-dynamo, Magnetohydrodynamics **42**, 131–142 (2006).

[89] VKS Private communication, Les Houches, August 2007.

[90] H. Fujisaka and T. Yamada, Prog. Theor. Phys. **74**(4), 918–921 (1984); H. Fujisaka, H. Ishii, M. Inoue and T. Yamada, Prog. Theor. Phys. **76**(6), 1198–1209 (1986).

[91] E. Ott and J.C. Sommerer, Phys. Lett. A **188**, 39 (1994).

[92] L. Yu, E. Ott and Q. Chen, Phys. Rev. Lett. **65**, 2935 (1990).

[93] N. Platt, S.M. Hammel and J.F. Heagy, Phys. Rev. Lett. **72**, 3498 (1994).

[94] J.F. Heagy, N. Platt and S.M. Hammel, Phys. Rev. E **49**, 1140 (1994).

[95] S.C. Venkataramani, T.M. Antonsen, Jr., E. Ott and J.C. Sommerer, Phys. Lett. A **207**, 173 (1995).

[96] S.C. Venkataramani, T.M. Antonsen, Jr., E. Ott and J.C. Sommerer, Physica D **96**, 66 (1996).

[97] S. Aumaître, F. Pétrélis and K. Mallick, Low-frequency noise controls on-off intermittency of bifurcating systems, Phys. Rev. Lett. **95**, 064101 (2005).

[98] S. Aumaître, K. Mallick and F. Pétrélis, Effects of the low frequencies of Noise on On-off intermittency, Journal of Statistical Physics, **123**, 909–927 (2006).

[99] K. Schneider and M. Farge, Decaying two-dimensional turbulence in a circular container, Phys. Rev. Lett. **95**, 244502 (2005).

[100] R. Pasquetti, R. Bwemba and L. Cousin, A pseudo-penalization method for high Reynolds number unsteady flows, Applied Numerical Mathematics **33**, 207–216 (2007).

Course 7

TAYLOR'S CONSTRAINT AND TORSIONAL OSCILLATIONS

Mathieu Dumberry

Department of Physics, University of Alberta, Edmonton, Alberta, T6G 2G7, Canada

Ph. Cardin and L.F. Cugliandolo, eds.
Les Houches, Session LXXXVIII, 2007
Dynamos

Contents

1. Introduction

Taylor's constraint is a morphological condition on the Earth's magnetic field inside the core. It specifies that the field must be organized such that the Lorentz torque integrated over cylindrical surfaces aligned with the rotation axis must vanish. A magnetic field obeying such a Taylor state was once viewed as the way to find solutions of the geodynamo problem and Taylor's constraint is important partly for this historical reason. Another reason for its importance is that torsional oscillations, which are azimuthal oscillations of rigid cylindrical surfaces in the core, are an essential physical ingredient to a Taylor state. These oscillations are believed to be observed in the secular variation of the magnetic field at the Earth's surface and in the changes in the length of day. The presence of torsional oscillations is important not only because they suggest that the Earth's core is in a Taylor state, but also because they provide a window through which we can observe other aspects of core dynamics. In this article, the basic concepts behind Taylor's constraint and torsional oscillations are reviewed. Several other review articles have been written covering similar topics and where, in many cases, the theoretical discussions are pushed much further. Among these we list [1–4].

2. Taylor's constraint

The Navier–Stokes equation in the Earth's core, in it's simplest Boussinesq form, is

$$\rho \frac{\partial \mathbf{u}}{\partial t} + \rho\, \mathbf{u} \cdot \nabla \mathbf{u} + 2\rho\, \mathbf{\Omega} \times \mathbf{u} = -\nabla p + \mathbf{J} \times \mathbf{B} + F_T\, \hat{\mathbf{r}} + \rho \nu \nabla^2 \mathbf{u}, \qquad (2.1)$$

where t is time, p is pressure, $\mathbf{\Omega} = \Omega \hat{\mathbf{z}}$ is the Earth's frequency of rotation, ν is kinematic viscosity, and the vectors $\mathbf{u}$, $\mathbf{B}$ and $\mathbf{J}$ correspond respectively to velocity, magnetic field and current density. The unit vector $\hat{\mathbf{r}}$ is in the radial direction of a spherical coordinate reference frame (r, θ, ϕ) and the unit vector $\hat{\mathbf{z}}$ is in the along-axis direction of a cylindrical coordinate reference frame (s, ϕ, z). F_T represents the buoyancy force which is left unspecified but is assumed to be known.

In dimensionless form, (2.1) can be written as

$$R_o \left(\frac{\partial \mathbf{u}}{\partial t} + \mathbf{u} \cdot \nabla \mathbf{u} \right) + 2 \hat{\mathbf{z}} \times \mathbf{u} = -\nabla p + \mathbf{J} \times \mathbf{B} + F_T \, \hat{\mathbf{r}} + E \, \nabla^2 \mathbf{u}, \qquad (2.2)$$

where we have scaled length by the radius of the core r_c, time by r_c^2/η where η is magnetic diffusivity, and magnetic field by $\sqrt{2\Omega\rho\mu\eta}$ where μ is the magnetic permeability. With these scalings, typical core values of magnetic field of 2 mT and typical velocities of 3×10^{-4} m s^{-1} correspond respectively to non-dimensional values of 1 and 1000. The non-dimensional numbers $R_o = \eta/\Omega r_c^2$ and $E = \nu/\Omega r_c^2$ are the Rossby (or magnetic Ekman) and Ekman numbers. Using $r_c = 3.5 \times 10^6$ m, $\Omega = 7.27 \times 10^{-5}$ s^{-1} and typical molecular values for η and ν of respectively 1 and 10^{-6} m^2 s^{-1} [5], we obtain $R_o \approx 10^{-9}$ and $E \approx 10^{-15}$. This implies that to leading order, the (dimensional) Navier–Stokes equation can be reduced to

$$2\rho \, \mathbf{\Omega} \times \mathbf{u} = -\nabla p + \mathbf{J} \times \mathbf{B} + F_T \, \hat{\mathbf{r}}, \qquad (2.3)$$

which is the so-called magnetostrophic balance, representing the behavior of the fluid in the slow, long term, inviscid limit.

If we integrate the ϕ-component of the magnetostrophic balance on cylinders aligned with the rotation axis, as depicted in Fig. 1, we obtain

$$\int 2\rho (\mathbf{\Omega} \times \mathbf{u})_\phi \, d\Sigma = -\int (\nabla p)_\phi \, d\Sigma + \int (\mathbf{J} \times \mathbf{B})_\phi \, d\Sigma + \int (F_T \, \hat{\mathbf{r}})_\phi \, d\Sigma, \qquad (2.4)$$

where $d\Sigma = s \, d\phi \, dz$. The pressure-gradient integral term vanishes,

$$\int (\nabla p)_\phi \, d\Sigma = \int \left(\int_0^{2\pi} \frac{\partial p}{\partial \phi} d\phi \right) dz = 0. \qquad (2.5)$$

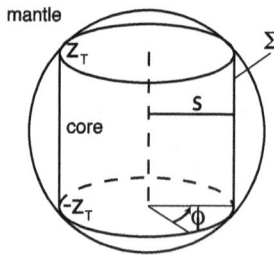

Fig. 1. Cylindrical surface of integration.

Since buoyancy is purely radial, it has no ϕ-component and

$$\int (F_T \, \hat{\mathbf{r}})_\phi \, d\Sigma = 0. \tag{2.6}$$

The Coriolis term integral can be written as

$$\int 2\rho (\mathbf{\Omega} \times \mathbf{u})_\phi \, d\Sigma = 2\rho\Omega \int u_s \, d\Sigma. \tag{2.7}$$

Since the normal (spherically radial) component of $\mathbf{u}$ over the two spherical caps associated with the cylinder surface Σ must be zero, we can extend the integral over the closed surface formed by the cylinder plus the two polar caps. Then, upon using Gauss' theorem,

$$2\rho\Omega \int u_s \, d\Sigma = 2\rho\Omega \int (\nabla \cdot \mathbf{u}) \, dV, \tag{2.8}$$

which must equal zero in the case of an incompressible fluid; the Coriolis integral also vanishes. Therefore, of the four integrals in (2.4), only the Lorentz term does not identically vanish and we must have

$$\int (\mathbf{J} \times \mathbf{B})_\phi \, d\Sigma = 0, \tag{2.9}$$

or

$$\int ((\nabla \times \mathbf{B}) \times \mathbf{B})_\phi \, d\Sigma = 0. \tag{2.10}$$

This result is known as Taylor's condition, or Taylor's constraint [6], and solutions that satisfy (2.9) are said to be in a Taylor state. This condition specifies that the magnetic torque on any cylindrical surface parallel to the rotation axis must vanish. (The integral of $\int (\mathbf{J} \times \mathbf{B})_\phi d\Sigma$ is often referred to as the Taylor torque.) It can also be interpreted as a morphological constraint on the magnetic field inside the core, which must be organized in such a way that (2.9) is satisfied. It is important to note that this constraint must be obeyed by having regions of positive and negative $(\mathbf{J} \times \mathbf{B})_\phi$ canceling one another over the whole of Σ, the only way to also have a dimensionless magnetic field strength of $\mathcal{O}(1)$ and a Lorentz force which is locally as large as the other terms in the magnetostrophic balance (2.3).

The theoretical ramifications of Taylor's constraint extend further. As Taylor showed, if a solution of $\mathbf{B}$ that satisfies (2.9) is found, then with the requirement that (2.9) remains satisfied at all times, the velocity $\mathbf{u}$ is completely determined in a diagnostic way by the magnetostrophic balance (2.3) and the induction equation. This allows solutions of the geodynamo problem to be found, though only in the magnetostrophic limit and without proper considerations for the evolution of the buoyancy force through the energy equation. Nevertheless, before the advent of numerical solutions of the geodynamo, Taylor's constraint thus represented the way towards a greater understanding of the geodynamo.

No solution has ever been found for which (2.9) is satisfied exactly and keeps being satisfied exactly at subsequent times. Although the theoretical construct of Taylor is correct, in practice it is extremely difficult to realize. The difficulty stems from the fact that if one finds a $\mathbf{B}$ that satisfies (2.9), though $\mathbf{u}$ is fully determined it tends to change $\mathbf{B}$ by advection and shear. Thus unless this process is exactly balanced by diffusion, (2.9) is no longer exactly satisfied at later times. However, many solutions satisfying (2.9) in an asymptotic sense have been found [7].

Solutions that approach (2.9) without satisfying it exactly are in any case a more realistic scenario. This is because we should not forget that (2.9) stems from the magnetostrophic balance where inertial and viscous terms have been neglected. These terms, though small, are expected to make a non-zero contribution to the torque balance on cylinders, and indeed are likely essential ingredients to maintain the solution in a Taylor state.

One possibility is to balance the Lorentz torque by viscous drag. The viscous forces in thin boundary layers at the core-mantle boundary (CMB) are much more important ($\propto E^{1/2}$) than the viscous forces between adjacent cylindrical surfaces inside the core ($\propto E$). The viscous drag from the differential azimuthal velocity between the fluid and the CMB produces a secondary poloidal flow through the whole core as a result of Ekman pumping and this results in a spin-up or spin-down of the whole cylinder of fluid. The torque balance (non-dimensional) in this case is given by

$$\int ((\nabla \times \mathbf{B}) \times \mathbf{B})_\phi \, d\Sigma = \frac{4\pi s E^{1/2}}{z_T^{1/2}} \mathcal{V}_\phi(s), \qquad (2.11)$$

where $\mathcal{V}_\phi(s)$ is the average or "rigid" azimuthal velocity of the cylinder surface. In the dynamo theory literature, $\mathcal{V}_\phi(s)$ is usually referred to as the geostrophic flow. The part of the Lorentz torque that does not cancel itself is thus balanced by a viscous torque from the friction between the geostrophic flow and the CMB. When the Lorentz torque in (2.11) largely cancels itself and the viscous torque is only equal to a small unbalanced part, then the system can be thought as being in a Taylor state. However, if no large cancellation in the Lorentz torque occurs, the geodynamo is said to be in an Ekman state. An equilibration of the Lorentz and viscous torque in an Ekman state requires a magnetic field amplitude of order $\mathcal{V}_\phi^{1/2} E^{1/4}$. Taking $E = 10^{-15}$ and $\mathcal{V}_\phi = 1000$, we get $B \sim 10^{-9/4} \sim 0.005$ or approximately 0.01 mT, which would be too small. However, if we instead adopt a turbulent value for viscosity and $E = 10^{-9}$, then $B \sim 10^{-3/4} \sim 0.17$

or approximately 0.5 mT. Hence, an Ekman balance may be appropriate for the core if turbulent values of viscosity are adopted.

If the balance to the Lorentz torque is achieved instead by the acceleration term, then we have

$$4\pi s z_T R_o \frac{\partial}{\partial t} \mathcal{V}_\phi(s, t) = \int ((\nabla \times \mathbf{B}) \times \mathbf{B})_\phi \, d\Sigma. \tag{2.12}$$

Although R_o is small, t can also be small compared to the magnetic diffusion time. Thus, a non-zero Lorentz torque generates a fast rigid acceleration of the cylindrical surface. The rigid acceleration has a feedback on the Lorentz torque: it shears B_s to induce a secondary field b_ϕ that reduces the total Lorentz torque. When b_ϕ is such that the Lorentz torque vanishes, $\partial \mathcal{V}_\phi / \partial t$ is likely non-zero. This means that the acceleration carries the cylindrical surface past the equilibrium position of zero Lorentz torque (Taylor state). This leads to an excess Lorentz torque in the reverse direction that builds up to oppose the motion. This balance between inertial acceleration and restoring magnetic force produces oscillations of rigid cylindrical surfaces about a Taylor state, as illustrated in Fig. 2. These oscillations in the geostrophic flow are the torsional oscillations which are the subject of the next section. The dynamo is then in a quasi-Taylor state, where (2.9) is satisfied except for the part providing the restoring force to the torsional oscillations. The damping of these oscillations by diffusion of the magnetic field or coupling with the mantle or inner core naturally brings the system back toward a Taylor state. This was the scenario that Taylor envisaged to maintain a Taylor state, a scenario that has been verified numerically [8, 9]. Torsional oscillations are therefore an essential ingredient of the geodynamo as they always allow the system to relax toward a state where (2.9) is satisfied everywhere in the core.

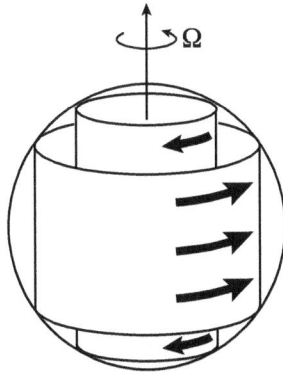

Fig. 2. Torsional oscillations: azimuthal oscillations of rigid cylindrical surfaces aligned with the rotation axis.

3. Torsional oscillations

Braginsky [10] was the first to exploit the theoretical concept of torsional oscillations in a quasi-Taylor state in an effort to explain geophysical observations. He
sought to explain the decade variations in the length of day (LOD) as exchanges
of angular momentum between the mantle and the core, where the angular momentum of the core is carried by torsional oscillations. This type of fluid motion,
he argued, would also be consistent with a part of the observed geomagnetic
secular variation. Braginsky established the wave equation for the torsional oscillations predicted by J.B. Taylor. The starting point is the momentum equation,
where we neglect viscous forces and buoyancy, and assume small $\mathbf{u}$,

$$\rho \frac{\partial \mathbf{u}}{\partial t} + 2\rho \, \mathbf{\Omega} \times \mathbf{u} = -\nabla p + \frac{1}{\mu} \mathbf{B} \cdot \nabla \mathbf{B}, \tag{3.1}$$

where the pressure p now includes a magnetic pressure term ($\mathbf{B} \cdot \mathbf{B}/2\mu$), and the
induction equation, where we neglect diffusion

$$\frac{\partial \mathbf{B}}{\partial t} = \nabla \times (\mathbf{u} \times \mathbf{B}). \tag{3.2}$$

Writing $\mathbf{B} = \mathbf{B}_p + \mathbf{B}_\phi + \mathbf{b}$, where $\mathbf{B}_p$ and $\mathbf{B}_\phi$ are the background "steady" poloidal
and azimuthal field and $\mathbf{b}$ a small perturbation, and looking for axisymmetric
solutions in $\mathbf{u}$, Braginsky showed that the leading order balance in the s-, z- and
ϕ-components of the momentum equation are, respectively,

$$2\Omega \, u_\phi = \frac{\partial p}{\partial s}, \tag{3.3}$$

$$0 = \frac{\partial p}{\partial z}, \tag{3.4}$$

$$\frac{\partial u_\phi}{\partial t} + 2\Omega \, u_s = \frac{1}{\rho \mu s} \nabla \cdot (s \mathbf{B}_p b_\phi). \tag{3.5}$$

Equation (3.4) indicates that p can only be a function of s. Thus, from (3.3), u_ϕ
is also only dependent on s. The leading order velocity is thus a rigid azimuthal
cylindrical flow. (u_s is smaller by a factor $\omega/2\Omega$, where ω is the frequency of
oscillation defined bellow.) It is noteworthy to point out that at this stage no
average in z has been taken and therefore the fact that u_ϕ is rigid to leading
order emerges as a natural consequence of the dynamics of the system. It is not
a surprise to recover such rigid motions because rapid rotation prevents large
variations in velocity along the rotation axis (The Proudman–Taylor theorem).

Proceeding with the development of the wave equation, integrating (3.5) over cylindrical surfaces eliminates the Coriolis term,

$$\int \frac{\partial u_\phi}{\partial t} \, d\Sigma = \frac{1}{\rho \mu s} \int \nabla \cdot (s \mathbf{B}_p b_\phi) \, d\Sigma. \tag{3.6}$$

The ϕ-component of the linearized induction equation is

$$\frac{\partial b_\phi}{\partial t} = s \mathbf{B}_p \cdot \nabla \frac{u_\phi}{s}, \tag{3.7}$$

and upon substituting (3.7) into $\partial(3.6)/\partial t$ we get

$$\frac{\partial^2 u_\phi}{\partial t^2} = \frac{1}{\rho \mu s^2 z_T} \frac{\partial}{\partial s} \left(s^3 z_T \{B_s^2\} \frac{\partial}{\partial s} \frac{u_\phi}{s} \right) - \frac{\partial f_\phi}{\partial t} \tag{3.8}$$

where f_ϕ is the force that the cylinder exerts on the boundary at the CMB from electromagnetic coupling with the mantle and

$$\{B_s^2\} = \frac{1}{4 \pi z_T} \int_{-z_T}^{z_T} \int_0^{2\pi} B_s^2 \, d\phi \, dz. \tag{3.9}$$

It is important to note that the B_s in $\{B_s^2\}$ comprises both the axisymmetric and non-axisymmetric parts. To simplify, let us assume that the mantle is a perfect insulator ($f_\phi = 0$), in which case we get an equation akin to elastic tension on a string

$$\frac{\partial^2 u_\phi}{\partial t^2} = \frac{1}{\rho \mu s^2 z_T} \frac{\partial}{\partial s} \left(T(s) \frac{\partial}{\partial s} \frac{u_\phi}{s} \right), \tag{3.10}$$

with $T(s) = s^3 z_T \{B_s^2\}$ representing the magnetic tension. The s-component of the magnetic field behaves as if it were elastic strings attached to the cylindrical surfaces, and provides a restoring force for the establishment of waves that propagate in the direction perpendicular to the rotation axis. These are the torsional oscillations. Note that the background B_ϕ field has no influence on these waves because it does not participate in the restoring force. Since the restoring force is purely magnetic, torsional oscillations are a type of Alfvén wave. The process for the restoring force of torsional oscillations is illustrated in Fig. 3.

To find the free modes of torsional oscillations, we can substitute in (3.10) solutions of the form

$$u_\phi = U_o e^{-i\omega t + iks}, \tag{3.11}$$

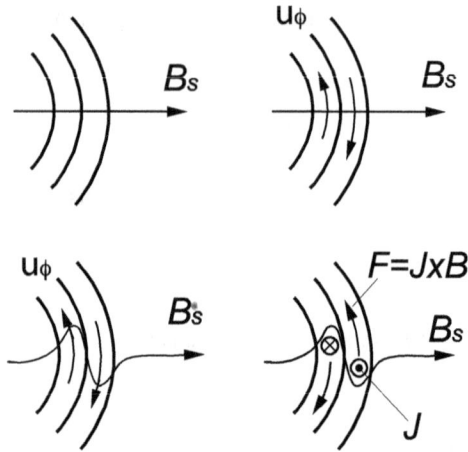

Fig. 3. The restoring force for torsional oscillations, viewed along the z-direction. A differential rotation of the cylinders (u_ϕ) shears the B_s field in the ϕ-direction. This creates a secondary field b_ϕ and an associated electrical current in the z-direction. By Lenz' law, the induced force opposes further relative motion between the cylindrical surfaces.

and we get

$$\omega^2 U_o \approx \frac{\{B_s^2\}}{\rho\mu} k^2 U_o. \tag{3.12}$$

The fundamental mode has wavenumber $k = 2\pi/r_c$ and its period is $\tau = 2\pi/\omega$, which gives

$$\tau \approx r_c \left(\frac{\rho\mu}{\{B_s^2\}}\right)^{1/2}. \tag{3.13}$$

Using $r_c = 3.5 \times 10^6$ m, $B_s = 0.5$ mT, $\rho = 10^4$ kg m^{-3} and $\mu = 4\pi \times 10^{-7}$, we obtain that $\tau \approx 25$ years. Exact eigenfunctions of the free modes depend on the spatial variations in B_s and also on the coupling with the mantle and inner core. Nevertheless, this simple order of magnitude analysis suggest that the normal modes of torsional oscillations should have periods of decades. Their velocity is

$$\frac{\partial\omega}{\partial k} = \left(\frac{\{B_s^2\}}{\rho\mu}\right)^{1/2} \tag{3.14}$$

which is the familiar Alfvén wave velocity.

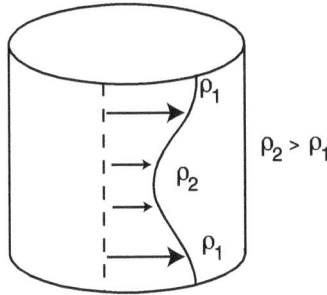

Fig. 4. Profile of u_ϕ of torsional oscillations allowing for changes in density in the core.

One last important note about torsional oscillations is that the "rigidity" in the azimuthal motions is only strictly correct when the density is considered a constant throughout the core. In truth, the "rigid" quantity is ρu_ϕ because

$$\frac{1}{\Sigma} \int \rho \frac{\partial u_\phi}{\partial t} \, d\Sigma = \frac{\partial}{\partial t} \{\rho u_\phi\}. \tag{3.15}$$

Since density increases with depth in the core, the real profile of u_ϕ involved in torsional oscillations should include slight variations along z as illustrated in Fig. 4. The Lorentz term should be modified to allow for $\partial u_\phi/\partial z$. The above development of torsional oscillations is correct only if ρ is assumed constant. In the Earth's core, $\Delta\rho/\rho \approx 20\%$, and so the approximation is relatively good. However, if a similar analysis is carried over for giant planets like Jupiter and Saturn, this approximation is no longer valid.

4. Taylor's constraint and numerical models of the geodynamo

Although Taylor's constraint has played a major role in the development of geo-dynamo theory, it is perhaps no longer the case today. This is because numerical solutions of the geodynamo have now been achieved and these solutions have provided a leap of progress in our understanding of the geodynamo. In a sense, Taylor's constraint establishes a constraint on only one component of the dynamics. Since the whole solution must satisfy many such constraints, one can argue that there is no reason to focus specifically on this one.

In fact, the numerical solutions that have been achieved so far are not in a Taylor state (though one study reports to have achieved a quasi-Taylor state [11]). The reason for this is two-fold. First, numerical resolution restricts computations to Ekman numbers of the order of 10^{-5}. As a result, even when one uses stress-free boundary conditions on the velocity to eliminate the Ekman pumping effect, the viscous torque between adjacent cylinders remains of comparable magnitude to the Lorentz torque. No large cancellation of the Lorentz torque on cylinders is required to occur and the solution is in a form of an Ekman state. The second reason is that, although all theoretical considerations of Taylor's constraint assumed that the Reynolds stresses term (the $\mathbf{u} \cdot \nabla \mathbf{u}$ term) is unimportant in the torque balance, numerical models of geodynamo have shown that they play a significant role, at least for the parameter regimes that have been reached up to date. Indeed, it is observed that the torque from the Reynolds stresses provides a very efficient way to balance the Lorentz torque [12]. Any imbalance between the viscous, Lorentz and Reynolds stresses torques can always be accommodated by rigid accelerations and the torque balance on cylindrical surfaces in the present-day numerical models involves all of the terms neglected from the magnetostrophic balance

$$\int \rho \frac{\partial u_\phi}{\partial t} d\Sigma + \int \rho \left(\mathbf{u} \cdot \nabla \mathbf{u}\right)_\phi d\Sigma$$
$$= \int (\mathbf{J} \times \mathbf{B})_\phi d\Sigma + \int \rho \nu (\nabla^2 \mathbf{u})_\phi d\Sigma. \tag{4.1}$$

Because the viscous torque and the torque from the Reynolds stresses are both large, the magnetic field does not have to organize itself in such a way that the Lorentz torque mostly cancels itself out. This is a situation very different than that of Taylor's constraint.

As the Ekman number is decreased toward more realistic values, the viscous torque is expected to decrease. However, it is not at all clear whether we should expect a similar behavior from the Reynolds stresses torque. If it decreases, then an $\mathcal{O}(1)$ magnetic field will require large cancellation in the Lorentz torque and we are back to a Taylor state. But perhaps the Reynolds stresses torque, even at small E, allows to balance a large part of an $\mathcal{O}(1)$ Lorentz torque. No theoretical study exists on the Reynolds stresses in the context of Taylor's constraint and it is difficult to forecast exactly what their role may be in the torque balance.

These considerations concerning Taylor's constraint, although they may appear of purely theoretical nature in the face of numerical solutions of the geodynamo, are actually of practical interest. The first reason is that Taylor's constraint is related to the axial angular momentum balance of the Earth's core. The evolution of the axial angular momentum L_z is determined by

$$\frac{\partial L_z}{\partial t} = \hat{z} \cdot \int_{Vol} \mathbf{r} \times \mathbf{F} \, dV = \int_{Vol} s F_\phi \, dV = \int_0^{r_c} s \left(\int F_\phi \, d\Sigma \right) ds, \tag{4.2}$$

where F_ϕ is the azimuthal component of all the different forcing terms in the momentum equation. If we separate the volume into a sum of thin cylinders aligned with the rotation axis, the angular momentum balance on each thin cylinder given by (4.2) is precisely that given by (4.1). This approach can also be extended for the fluid regions of other planets. For example, the torque balance that controls the atmosphere of giant planets is likely between the Reynolds stresses and viscous forces [13, 14]

$$\int (\mathbf{u} \cdot \nabla \mathbf{u})_\phi \, d\Sigma = \int \nu (\nabla^2 \mathbf{u})_\phi d\Sigma. \tag{4.3}$$

The second reason why the ideas behind Taylor's constraint are not purely of theoretical value is because observations may actually indicate that the geodynamo in a quasi-Taylor state with torsional oscillations, as we cover in the next section.

5. Observations of rigid flow and torsional oscillations

Braginsky's original idea that torsional oscillations can both explain the LOD changes while being consistent with the secular variation of the magnetic field has since received further support. Jault et al. [15] have reconstructed maps of the flow at the top of the core between 1969 and 1985 which best explain the secular variation of the magnetic field. Based on the axially symmetric and equatorially symmetric azimuthal component of this flow, and assuming that the motion extend rigidly inside the core, they calculated the changes in core angular momentum carried by these motions and showed that it correlates well with the changes in mantle angular momentum required to explain the LOD variations during that period. Jackson et al. [16] subsequently showed that this correlation extends back to 1900. Not only this confirmed that changes in LOD at decadal timescale were indeed due to core-mantle angular momentum exchanges, it also confirmed the presence of decadal timescale rigid azimuthal motions in the core. These can be interpreted as rigid flow adjustment to satisfy the torque balance constraints on individual cylinders, either from a viscous torque (see (2.11)) or from a rigid acceleration (see (2.12)).

Subsequent work has given more support to the rigid acceleration hypothesis. The variations in time of the rigid cylindrical surfaces suggest large exchanges of angular momentum inside the core and relatively little exchange with the mantle [17–19]. Furthermore, not only is the decade timescale of these waves consistent with the theory of torsional oscillations, it has been shown that the motion of cylinders can easily be explained by a superposition of free waves [17,20,21]. An example of the time-dependent rigid cylindrical flows in the Earth's core which possibly represent torsional oscillations is shown in Fig. 5 for the period between

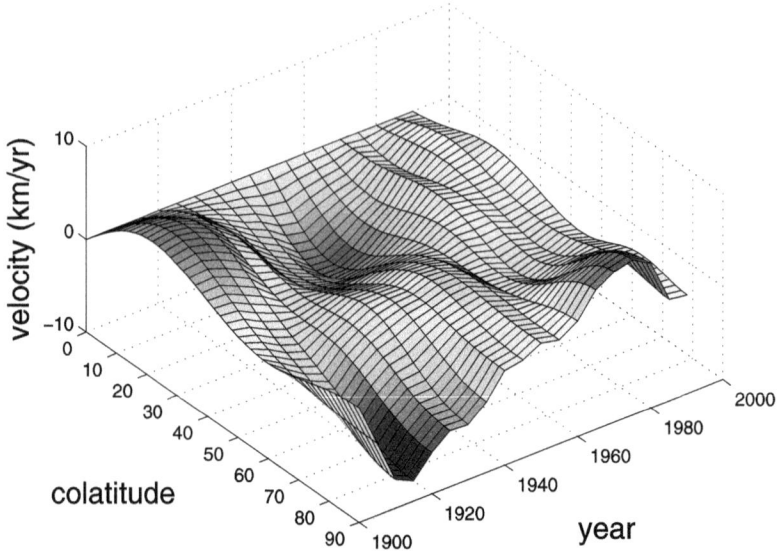

Fig. 5. An example of the time-dependent axisymmetric, equatorially symmetric part of the azimuthal velocity at the surface of the core between 1900 and 1990 inverted from magnetic field model *ufm1* [22]. Colatitudes (θ) are given in degrees. If this part of the velocity extends rigidly inside the core, then it represents time-dependent velocities of rigid cylindrical surfaces with radius $s = r_c \sin \theta$. The undulations in radius and time are suggestive of a propagating wave consistent with torsional oscillations.

1900 and 1990. Observations may thus corroborate the idea that the core in is a quasi-Taylor state with torsional oscillations.

   If this is correct, then torsional oscillations present an opportunity to illuminate physical quantities inside the core or at the CMB for which we otherwise have little or no information. This is because the free modes of torsional oscillations must obey

$$\frac{\partial^2 u_\phi}{\partial t^2} = \frac{1}{\rho \mu s^2 z_T} \frac{\partial}{\partial s} \left(s^3 z_T \{B_s^2\} \frac{\partial}{\partial s} \frac{u_\phi}{s} \right) - \frac{\partial f_\phi}{\partial t}, \tag{5.1}$$

where f_ϕ now represents the sum of the forces that the cylinder exerts on the boundary at the CMB. This means that we can use the observed free modes to recover information about the parameters that enter (5.1), in a similar way that seismology allows to retrieve information about the density and rheology inside the Earth. This idea has already been exploited to extract the possible amplitude and structure of B_s inside the core [17]. Alternatively, one can also use (5.1) to constrain the coupling mechanism between the core and the mantle that enter through the parameter f_ϕ [23–25].

While these represent exciting prospects, one has to be aware of the potential pitfalls of this exercise. This idea depends on the presence of free torsional oscillations in the core. However, it is not at all clear that the observed rigid cylindrical motions are necessarily free waves. For one thing, free modes need to be excited. The behavior described above, where torsional oscillations occur with respect to a steady equilibrium state, is correct only if the waves are excited by a mechanism which acts suddenly and then plays no further role in the dynamics. A perhaps more likely scenario is one where cylindrical surfaces are subject to a forcing which plays an active role in the dynamics at all times. If this latter view is correct, the resulting time-dependent rigid flows would consist of a combination of forced and free torsional oscillations. The nature of this forcing is at present unknown but perhaps it simply consists of the continual changes in the magnetic field, and hence of the Lorentz torque on any cylinder, produced by the convective dynamics in the core (recall that Braginsky's torsional oscillations were built assuming a steady background field). If Reynolds stresses are important, as suggested in Section 4, then they may also participate in the excitation process, and also very likely alter the behavior of the free modes of torsional oscillations. Another reason to suspect the presence of forced rigid cylindrical motions is that free modes may be attenuated very efficiently by coupling processes with the mantle. Indeed, recent work [26] suggests that if the radial field strength at the CMB and conductance of the lower mantle inferred by the Earth's forced nutation is correct, free torsional oscillations should be attenuated by electromagnetic coupling in less than a decade. This would suggest that the observed rigid motions cannot represent free modes and must instead be forced cylindrical motions.

6. Conclusions and future direction

The theory of torsional oscillations has been taken much further than the basic ideas presented in Section 3, with proper incorporation of various coupling effects at the CMB and inner core boundary [4, 27]. These include viscous, electromagnetic and topographic coupling as well as gravitational coupling between the mantle and the inner core. Recent investigations of free torsional oscillations have been performed both in the time [28] and frequency domain [24, 25]. Much less work exists on forced torsional oscillations, an issue related to the excitation mechanism which has never been thoroughly investigated. However, a recent theoretical model has been developed to investigate the excitation of torsional oscillations [29].

A promising future avenue to further our understanding of the torque balance on cylinders in the core and of torsional oscillations and their excitation is through numerical models of the geodynamo. Rigid cylindrical flows are one of our most robust observation of core dynamics so realistic numerical models of the geodynamo should include them. Oscillations of such flows have been observed in some numerical models [12, 30]. However, these oscillations are much more complex than simple free torsional oscillations for the reasons stated in Section 4. Nevertheless, the ability of the numerical models to produce geostrophic flows is a step in the right direction and these models can be used to study the excitation mechanism. With numerical models becoming increasingly closer to Earth-like conditions, there is hope that they will soon encompass realistic free and forced torsional oscillations.

References

[1] Fearn, D.R. 1994. Nonlinear planetary dynamos, in: *Lectures on Solar and Planetary Dynamos*, edited by M.R.E. Proctor and A.D. Gilbert, Cambridge University Press, pp. 219–244.

[2] Hollerbach, R. 1996. On the theory of the geodynamo, Phys. Earth Planet. Inter. **98**, 163–185.

[3] Bloxham, J. 1998. Dynamics of angular momentum in the Earth's core, Ann. Rev. Earth Planet. Sci. **26**, 501–517.

[4] Jault, D. 2003. Electromagnetic and topographic coupling, and lod variations, in: *Earth's core and lower mantle*, edited by C.A. Jones, A. Soward and K. Zhang, The fluid mechanics of astrophysics and geophysics, Taylor & Francis, London, pp. 56–76.

[5] Gubbins, D. and Roberts, P.H. 1987. Magnetohydrodynamics of the Earth's core, in: *Geomagnetism*, edited by J.A. Jacobs, vol. 2, Academic Press, London, pp. 1–184.

[6] Taylor, J.B. 1963. The magneto-hydrodynamics of a rotating fluid and the Earth's dynamo problem, Proc. R. Soc. Lond., A **274**, 274–283.

[7] Fearn, D.R. and Proctor, M.R.E. 1987. Dynamically consistent magnetic fields produced by differential rotation, J. Fluid Mech. **178**, 521–534.

[8] Jault, D. 1995. Model Z by computation and Taylor's condition, Geophys. Atrophys. Fluid Dyn. **79**, 99–124.

[9] Jault, D. and Cardin, P. 1999. On dynamic geodynamo models with imposed velocity as energy source, Phys. Earth Planet. Inter. **111**, 75–81.

[10] Braginsky, S.I. 1970. Torsional magnetohydrodynamic vibrations in the Earth's core and variations in day length, Geomag. Aeron. **10**, 1–10.

[11] Takahashi, F., Matsushima, M., and Honkura, Y. 2005. Simulations of a quasi-Taylor state geomagnetic field including polarity reversals on the Earth simulator, Science **309**, 459–461.

[12] Dumberry, M. and Bloxham, J. 2003. Torque balance, Taylor's constraint and torsional oscillations in a numerical model of the geodynamo, Phys. Earth Planet. Inter. **140**, 29–51.

[13] Aurnou, J. and Olson, P. 2001. Strong zonal winds from thermal convection in a rotating spherical shell, Geophys. Res. Lett. **28**, 2557–2559.

[14] Christensen, U.R. 2001. Zonal flow driven by deep convection in the major planets, Geophys. Res. Lett. **28**, 2553–2556.

[15] Jault, D., Gire, C. and Le Mouël, J.-L. 1988. Westward drift, core motions and exchanges of angular momentum between core and mantle, Nature **333**, 353–356.

[16] Jackson, A., Bloxham, J. and Gubbins, D. 1993. Time-dependent flow at the core surface and conservation of angular momentum in the coupled core-mantle system, in: *Dynamics of the Earth's Deep Interior and Earth Rotation*, edited by J.-L. Le Mouël, D.E. Smylie, and T. Herring, vol. 72, AGU Geophysical Monograph, Washington, DC, pp. 97–107.

[17] Zatman, S. and Bloxham, J. 1997. Torsional oscillations and the magnetic field within the Earth's core, Nature **388**, 760–763.

[18] Hide, R., Boggs, D.H. and Dickey, J.O. 2000. Angular momentum fluctuations within the Earth's liquid core and torsional oscillations of the core-mantle system, Geophys. J. Int. **143**, 777–786.

[19] Pais, A. and Hulot, G. 2000. Length of day decade variations, torsional oscillations and inner core superrotation: evidence from recovered core surface zonal flows, Phys. Earth Planet. Inter. **118**, 291–316.

[20] Bloxham, J., Zatman, S. and Dumberry, M. 2002. The origin of geomagnetic jerks, Nature **420**, 65–68.

[21] Amit, H. and Olson, P. 2006. Time-averaged and time-dependent parts of core flow, Phys. Earth Planet. Inter. **155**, 120–139.

[22] Bloxham, J. and Jackson, A. 1992. Time dependent mapping of the geomagnetic field at the core-mantle boundary, J. Geophys. Res. **97**, 19537–19564.

[23] Buffett, B.A. 1998. Free oscillations in the length of day: inferences on physical properties near the core-mantle boundary, in: *The Core-mantle Boundary Region*, edited by M. Gurnis, M.E. Wysession, E. Knittle and B.A. Buffett, vol. 28 of Geodynamics series, AGU Geophysical Monograph, Washington, DC, pp. 153–165.

[24] Mound, J.E. and Buffett, B.A. 2003. Interannual oscillations in the length of day: implications for the structure of mantle and core, J. Geophys. Res. **108(B7)**, 2334, doi:10.1029/2002JB002054.

[25] Mound, J.E. and Buffett, B.A. 2005. Mechanisms of core-mantle angular momentum exchange and the observed spectral properties of torsional oscillations, J. Geophys. Res. **110**, B08103, doi:10.1029/2004JB003555.

[26] Dumberry, M. and Mound, J.E. 2008. Constraints on core-mantle electromagnetic coupling from torsional oscillation normal modes, J. Geophys. Res. **113**, B03102, doi:10.1029/2007JB005135.

[27] Braginsky, S.I. 1984. Short-period geomagnetic secular variation, Geophys. Atrophys. Fluid Dyn. **30**, 1–78.

[28] Jault, D. and Légaut, G. 2005. Alfvén waves within the Earth's core, in: *Fluid Dynamics and Dynamos in Astrophysics and Geophysics*, edited by A.M. Soward, C.A. Jones, D.W. Hugues, and N.O. Weiss, The fluid mechanics of astrophysics and geophysics, Taylor & Francis, London, pp. 277–293.

[29] Buffett, B.A. and Mound, J.E. 2005. A green's function for the excitation of torsional oscillations in the Earth's core, J. Geophys. Res. **110**, B08104, doi:10.1029/2004JB003495.

[30] Busse, F.H. and Simitev, R. 2005. Convection in rotating spherical fluid shells and its dynamo states, in: *Fluid Dynamics and Dynamos in Astrophysics and Geophysics*, edited by A.M. Soward, C.A. Jones, D.W. Hugues, and N.O. Weiss, The fluid mechanics of astrophysics and geophysics, Taylor & Francis, London, pp. 359–392.

Course 8

WAVES IN THE PRESENCE OF MAGNETIC FIELDS, ROTATION AND CONVECTION

Christopher C. Finlay

Institute for Geophysics, ETH Zurich,
Switzerland

Ph. Cardin and L.F. Cugliandolo, eds.
Les Houches, Session LXXXVIII, 2007
Dynamos
© *2008 Published by Elsevier B.V.*

2007 8 22

Contents

Preamble

Geophysical and astrophysical bodies commonly possess self-sustained magnetic fields due to dynamo action in their fluid regions. These systems are often also rapidly-rotating and convecting. The fluid dynamics underlying the dynamo process is consequently a competition between the stabilizing influence of rotation and magnetic fields and the forcing away from equilibrium provided by convection and perhaps also precessional or tidal driving. The stability endowed by magnetic fields and rotation simultaneously provides a basis for wave motions; these are the focus here.

This chapter is designed to be a primer for students and researchers wishing for an introduction to the physics of the waves that can exist in rapidly-rotating, hydromagnetic systems. The magnetic tension waves of Alfvén, Inertial waves arising from the intrinsic stability of rotating fluids, and the Magnetic Coriolis (MC) waves arising when these phenomenon occur in concert are described. The influence of density stratification and convective instability leading to Magnetic Archimedes Coriolis (MAC) waves is reviewed and the importance of diffusive effects and spherical geometry are outlined. Connections to recent laboratory experiments, numerical dynamo simulations and geophysical and astrophysical observations are also discussed.

This written account is an extended version of a lecture given during the 'Dynamos' summer school held at Les Houches in August 2007. The organizing committee did an excellent job of charming me from my seat in the audience to the front of the lecture theatre. I thank them for that invitation and hope that this account is of use both to my fellow attendees and to future students. The basis for my lecture and this article was a review chapter from my Ph.D. thesis [43] and two articles that appeared recently in the Encyclopedia of Geomagnetism and Palaeomagnetism [44, 45].

1. Introduction

1.1. Some motivating thoughts concerning the study of waves

Laboratory experiments studying the influence of rotation and magnetic fields on electrically conducting fluids dramatically demonstrate the intrinsic stability that these influences can impart to the fluids. Some beautiful examples of such stability have been captured in National Committee for Fluid Mechanics films on Rotating Fluids [47] and Magnetohydrodynamics [92]. Excerpts from these films were shown during the summer school and they can also be found online; they vividly illustrate the fundamental phenomena involved and are strongly recommended for the physical insight they provide.

We can intuitively understand how stability might come about in magnetic and rotating fluids by considering how magnetic field lines and vorticity lines threading through fluid parcels will effectively impart elasticity to the fluid. If some disturbance, instability, or forcing then moves fluid parcels, the tension in the magnetic field lines and vortex lines will seek to return the parcels to their original position. Such reasoning leads us to expect that fluids that undergo rotation or that are electrically conducting and permeated by a magnetic field should support oscillations and waves.[1]

Waves occurring as a result of the stability imparted by fluid rotation are known as Inertial waves. They are observed in natural rotating fluid systems such as Earth's ocean. Waves in electrically conducting fluids occurring as a result of the stability imparted by magnetic fields are known as hydromagnetic or Alfvén waves. They are found in plasmas or fluids with high electrical conductivity, such as the solar corona and Earth's magnetosphere and core. It is therefore a matter of considerable geophysical and astrophysical importance to understand and be able to quantitatively model such waves and their generalizations that occur when both magnetic fields and rotation are present.[2] This is the primary motivation behind this present chapter.

One may wonder what can be gained from the study of waves. A glance at the progress made in seismology, helioseismology and oceanography provides a vivid illustration of what may be achieved. Waves properties, particularly their dispersion, and how these vary with location and time depend on the nature of the system supporting the waves. Study of waves thus allows underlying physical properties of often inaccessible geophysical and astrophysical systems to be investigated. Furthermore, waves are a mechanism by which energy and momentum can be transported; observations of waves can therefore tell us much about how the underlying system is operating and evolving.

---

[1] The waves discussed in this chapter fundamentally involve the momentum equation (often coupled to the induction and heat equations) and rely on restoring forces for their existence. Dynamo waves [76] which are oscillatory solutions to the induction equation are not considered here because they are a purely kinematic phenomenon involving the magnetic field alone.

[2] No account is taken here of compressibility effects and associated acoustic waves. Readers interested in systems where compressibility is important can if necessary generalize the treatment given here. They may also wish to consult the books by Lighthill [68] and Sturrock [97] for introductions to acoustic waves and their modification in the presence of magnetic fields respectively; the review by Roberts [87] gives further technical detail.

The subject of waves in the presence of rotation, magnetic fields and convection has a long and rich history. Before embarking on a detailed account of the physics of such waves, it is useful to sketch the history of investigations into these phenomena, to acknowledge the contributions of the scientists who have laboured on this subject, to collect useful references for further reading, and to introduce necessary terminology.

1.2. Historical sketch and literature survey

The first work on oscillations and waves arising from the intrinsic stability of a rotating fluid was carried out by Lord Kelvin in 1880 [58] in a cylindrical geometry. Poincaré [78], Bryan [12] and Cartan [18] later studied similar motions in rotating spheroids. Bjerknes and co-workers [8] independently re-discovered these rotation-reliant motions, naming them 'Elastoid-Inertia' oscillations. Much of the early work is collected in the monograph by Greenspan [49] where both Inertial waves in unbounded fluids and Inertial wave modes in contained geometries are described. Recently, considerable theoretical progress has been made towards finding explicit Inertial wave mode solutions in full sphere, spheroid and cylindrical geometries by Zhang and co-workers [108–110], and in spherical shells by Rieutord and co-workers [80, 81].

The basic mechanism underlying waves in electrically conducting fluids permeated by magnetic fields was elucidated by Alfvén [5]. He described a scenario whereby magnetic tension and inertial effects give rise to oscillations and travelling waves, which became known as Alfvén waves in his honour. In this account, any wave involving both magnetic field and fluid mechanical effects will be referred to a hydromagnetic wave; Alfvén waves are the simplest possible example of hydromagnetic waves.

Lehnert [66] deduced that rapid rotation of a hydromagnetic system would lead to the splitting of plane Alfvén waves into two circularly polarized, transverse, waves. He realized these would have very different timescales if the frequency of Inertial waves was much larger than that of pure Alfvén waves in the system. Here, such waves will be collectively be referred to as Magnetic Coriolis (MC) waves. Chandrasekhar [21] studied the effects of buoyancy on rotating hydromagnetic systems, though he focused primarily on axisymmetric motions. Braginsky [9, 10] described the influence of density stratification and convection driving non-axisymmetric waves naming these Magnetic Archimedes Coriolis (MAC) waves.

Hide [52] was the first to consider the influence of spherical geometry on two dimensional (being invariant parallel to the rotation axis) hydromagnetic waves in a rapidly-rotating fluid. He studied the effects of a linear variation of the Coriolis force with distance from the rotation axis, mimicking the effect of the variation in the Coriolis force with latitude in a spherical shell (known as the β effect in meteorology and oceanography). This is the essential ingredient needed for Rossby waves [77]. He found a similar mode splitting to that discovered by Lehnert; here Hide's waves arising from the magnetic modification of Rossby waves are referred to as MC Rossby waves (see Appendix A). Malkus [70] studied MC waves in a full sphere for the special case of a background magnetic field increasing in strength with distance from the rotation axis (see Appendix B). Stewartson [95] also developed an asymptotic theory of MC waves in a thin spherical shell geometry.

Eltayeb and Roberts [27, 29, 83] and more recently Roberts and Jones [55, 86] have studied MAC waves driven by thermal instability in a rotating plane layer permeated by a strong, uniform magnetic field. They demonstrated the importance of including magnetic and thermal diffusion in such systems, but nonetheless found that it is sometimes possible for thermal instability to drive diffusionless MC and MAC waves. Busse [14] and Soward [93] extended these studies to MAC waves in a quasi-geostrophic (QG) sloping annulus geometry. This involved a formally correct description of the arrangement suggested by Hide [52] for capturing the Rossby wave mechanism thought to be relevant in a thick spherical shell geometry (see Appendix C).

Acheson [1] pointed out that hydromagnetic waves could also be driven via magnetic field instability. Fearn [36–38] later studied this mechanism in more detail using numerical simulations. Roberts and Loper [85] further illustrated that magnetic field instability can excite hydromagnetic waves, even in the presence of stable stratification. An excellent discussion of the counter-intuitive means by which magnetic or thermal diffusion can facilitate instability is given by Acheson [2].

Eltayeb and Kumar [28] and Fearn [35] carried out the first numerical studies of thermally-driven hydromagnetic waves including the effects of both buoyancy and diffusion in a rapidly-rotating, full sphere geometry. Fearn and Proctor [40] and later Zhang [103, 106] went on to consider the effect of more general background magnetic field configurations. They found that if the imposed magnetic field drives a magnetic wind this can (via a Doppler shift) greatly change the observed hydromagnetic wave frequencies.

The overviews by Hide and Stewartson [53] and Braginsky [11] provide concise introductions to the theory of hydromagnetic waves in rapidly-rotating systems. The textbooks by Melchoir [73] and Davidson [22] and the lengthy review of Gubbins and Roberts [50] contain useful summaries of hydromagnetic wave theory in the context of more general surveys of the magnetohydrodynamics. A number of more focused technical reviews of the subject also exist including those by Roberts and Soward [82], Acheson and Hide [3], Eltayeb [30], Fearn Roberts and Soward [41], Proctor [79], Zhang and Schubert [107] and most recently by Soward and Dormy [94]. Finally, in the context of the role played by hydromagnetic waves in dynamo action (the central theme of the summer school!), chapter 10 of Moffatt's monograph [74] remains essential reading.

2. Inertial waves and intrinsic stability due to rotation

2.1. The Coriolis force, vortex lines and Inertial oscillations in rotating fluids

This section I begin by considering the purely hydrodynamic case of a homogeneous fluid undergoing rapid rotation, in the absence of magnetic fields and convection. Rotation imparts intrinsic stability to the fluid due to the action of the Coriolis force on fluid parcels. The Coriolis force is well known as being the origin of circulating eddies in rotating fluids, for example, in hurricanes in Earth's atmosphere. The Coriolis force occurs only in rotating reference frames where inertial motions follow curved trajectories. Acting at right angles to direction of fluid motion, the Coriolis force results in circular motions of fluid elements, so that they periodically return to their initial position after being perturbed. This is the origin of the intrinsic stability possessed by rotating fluids [7, 98].

An intuitive understanding of this stability may be gained by thinking in terms of vortex lines which point in the direction of the local vorticity [58, 67]. Vortex lines resulting from the presence of the Coriolis force are initially parallel to the rotation vector for a homogeneous fluid in solid body rotation. They act by imparting an effective elasticity to the fluid by resisting fluid motions that distort them. This resistance of vortex lines to distortion is what gives rise to Inertial oscillations and waves [67, 68]. An alternative introductory perspective on inertial waves is given by Tritton [98], while the monography by Greenspan [49] and the review article by Stewartson [96] provide more comprehensive surveys.

2.2. The Inertial wave equation

For an infinite plane layer of homogeneous fluid with density ρ undergoing rotation with angular speed Ω, considering small velocity perturbations $\boldsymbol{u}$ about a state of rest in the rotating frame the linearized momentum equation takes the form

$$
\underbrace{\frac{\partial \boldsymbol{u}}{\partial t}}_{\substack{\text{Inertial} \\ \text{acceleration}}} + \underbrace{2(\boldsymbol{\Omega} \times \boldsymbol{u})}_{\substack{\text{Coriolis} \\ \text{acceleration}}} = \underbrace{-\frac{1}{\rho}\nabla p}_{\substack{\text{Pressure} \\ \text{gradient}}}, \tag{2.1}
$$

where p is the mechanical pressure of the fluid, $\boldsymbol{\Omega} = \Omega \boldsymbol{z}$ and Cartesian coordinates have been used. An equation describing the waves supported by this system can be derived by taking the curl of (2.1) and which leads to an equation describing the evolution of the fluid vorticity $\boldsymbol{\xi} = \nabla \times \boldsymbol{u}$

$$\frac{\partial \boldsymbol{\xi}}{\partial t} = 2(\boldsymbol{\Omega} \cdot \nabla)\boldsymbol{u}. \tag{2.2}$$

Taking a further curl and also a time derivative and noting that $\nabla \times \boldsymbol{\xi} = -\nabla^2 \boldsymbol{u}$ for incompressible fluids yields

$$\frac{\partial^2(\nabla^2 \boldsymbol{u})}{\partial^2 t} = -2(\boldsymbol{\Omega} \cdot \nabla)\frac{\partial \boldsymbol{\xi}}{\partial t}. \tag{2.3}$$

Using (2.2) again leads to a wave equation for the perturbation of the velocity field,

$$\frac{\partial^2(\nabla^2 \boldsymbol{u})}{\partial^2 t} = -4(\boldsymbol{\Omega} \cdot \nabla)^2 \boldsymbol{u}. \tag{2.4}$$

This is the Inertial wave equation in a rotating incompressible fluid, so called because it relies only on inertia and the Coriolis force; the later is itself a consequence of inertia in a rotating reference frame.

2.3. Inertial wave dispersion relation and properties

The properties of Inertial waves can be determined using the plane wave ansatz

$$\boldsymbol{u} = \mathcal{Re}\{\widehat{\boldsymbol{u}} e^{i(\boldsymbol{k} \cdot \boldsymbol{r} - \omega t)}\}. \tag{2.5}$$

Substituting from (2.5) into (2.4) yields

$$\omega^2 |\boldsymbol{k}|^2 \widehat{\boldsymbol{u}} = -4(\boldsymbol{\Omega} \cdot \boldsymbol{k})^2 \widehat{\boldsymbol{u}}, \tag{2.6}$$

which simplifies to the classic inertial wave dispersion relation [49],

$$\omega = \pm \frac{2(\boldsymbol{\Omega} \cdot \boldsymbol{k})}{|\boldsymbol{k}|} = \pm 2\Omega \cos\theta, \tag{2.7}$$

where θ is the angle between $\boldsymbol{\Omega}$ and $\boldsymbol{k}$, so that the angular frequency of Inertial waves is restricted to lie between 0 and $\pm 2\Omega$ s^{-1}.

The phase velocity (the speed and direction in which individual peaks and troughs move) is ω/k in the direction of the wavevector $\widehat{\boldsymbol{k}}$; for Inertial waves this is

$$C_{\text{ph}} = \frac{\omega(k)}{|\boldsymbol{k}|}\widehat{\boldsymbol{k}} = \pm \frac{2(\boldsymbol{\Omega} \cdot \boldsymbol{k})}{|\boldsymbol{k}|^3}\boldsymbol{k}. \tag{2.8}$$

In contrast, the group velocity (the speed and direction at which energy is transferred, see the book of Lighthill [68]) is the derivative of ω with respect to $\boldsymbol{k}$,

$$C_g = \frac{\partial \omega(\boldsymbol{k})}{\partial \boldsymbol{k}} = \pm \frac{2(k^2 \boldsymbol{\Omega} - (\boldsymbol{\Omega} \cdot \boldsymbol{k})\boldsymbol{k}}{|\boldsymbol{k}|^3}. \tag{2.9}$$

An interesting consequence of this expression for the group velocity is that when the frequency of inertial waves is very small (i.e. $(\boldsymbol{\Omega} \cdot \boldsymbol{k}) \ll 1$) then C_g is approximately parallel to the rotation axis $\boldsymbol{\Omega}$, so that information concerning slow disturbances is communicated in this direction. This is the mechanism which underlies and mediates the seemingly magical Proudman–Taylor theorem whereby slow fluid motions tend to be invariant parallel to the rotation axis in a rapidly-rotating fluid. Finally, it is worth emphasizing that Inertial waves are dispersive, anisotropic and are characterized by circular particle motions.

3. Alfvén waves and magnetic tension

3.1. The Lorentz force, magnetic field lines and Magneto-Inertial oscillations

Next, let us consider how waves can arise due to the presence of a magnetic field permeating an electrically conducting fluid that is not rotating. A simple thought experiment [22] is useful for illustrating the physical mechanism at work. Imagine a uniform magnetic field is permeating a perfectly conducting fluid, and a uniform flow is initially normal to the magnetic field lines. According to the frozen flux theorem, fluid flow will distort the magnetic field lines so they become curved. The curvature of magnetic field lines produces a magnetic (Lorentz) force on the fluid in a direction opposing further curvature, as specified by Lenz's law. Following Newton's second law, this Lorentz force then changes the momentum of the fluid, pushing it (and consequently the magnetic field lines) back towards the original, undistorted state.

As the curvature of the magnetic field lines increases, so does the strength of the restoring force. Eventually the Lorentz force becomes strong enough to reverse the direction of the fluid flow. Magnetic field lines are pushed back to their original, undistorted configuration and the Lorentz force associated with their curvature weakens until the field lines become straight again. The sequence of flow causing field line distortion and field line distortion exerting a force on the fluid now repeats, but with the initial flow (a consequence of fluid inertia) now present in the opposite direction. In the absence of dissipation this Magneto-Inertial oscillation will continue indefinitely. Figure 1 shows one complete cycle of such an oscillation. If a magnetic field line is disturbed in this manner at one location, then a disturbance will travel along the field line away from this point

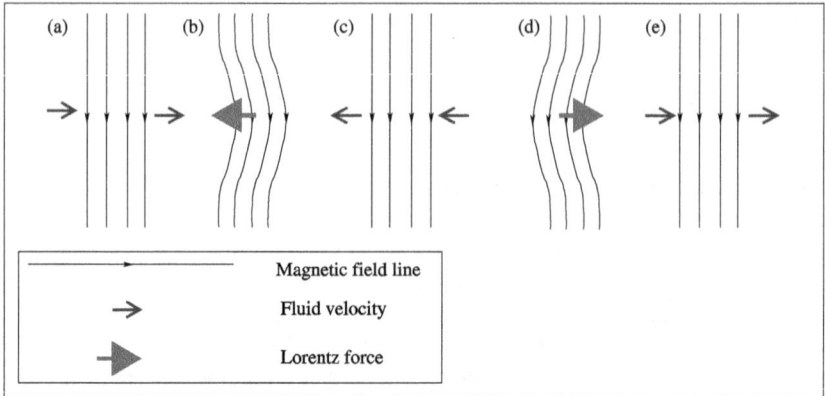

Fig. 1. Schematic illustration of a complete cycle of the basic oscillation mechanism underlying Alfvén waves.

due to the coupling provided between adjacent fluid parcels by the magnetic field line which resists strong curvature. Such travelling disturbances arising from the balance between magnetic (Lorentz) forces and inertial effects are known as Alfvén waves after Hannes Alfvén who discovered them [5].

Additional physical intuition concerning Alfvén waves can be obtained by an analogy between the response of a magnetic field line distorted by fluid flow and the response of an elastic string when plucked. Both rely on tension as a restoring force, elastic tension in the case of the string and magnetic tension in the case of the magnetic field line. Both result in transverse waves propagating in directions perpendicular to their displacement. When visualizing Alfvén waves it can be helpful to think of a fluid as being endowed with a pseudo-elastic nature by the presence of a magnetic field, and consequently supporting transverse waves. Lucid accounts of the basic principles of Alfvén waves can be found in the books by Alfvén and Fälthammar [6] and Davidson [22].

3.2. The Alfvén wave equation

To deduce the properties of Alfvén waves in a more quantitative manner, one follows the classical approach of deriving a wave equation. Consider a uniform, steady, magnetic field $\boldsymbol{B_0}$ imposed in infinite, homogeneous, incompressible, non-rotating electrically conducting fluid of density ρ and magnetic permeability μ. For simplicity, viscosity is neglected and the fluid is assumed to be perfectly conducting.

Now imagine that the fluid is perturbed by an infinitesimally small flow $\boldsymbol{u}$ inducing a perturbation magnetic field $\boldsymbol{b}$. Ignoring terms that are quadratic in small quantities, the linearized equations describing conservation of momentum and the evolution of the magnetic fields (by frozen flux advection) are

$$\underbrace{\frac{\partial u}{\partial t}}_{\text{Inertial acceleration}} = \underbrace{-\frac{1}{\rho}\nabla p}_{\substack{\text{Combined mechanical and} \\ \text{magnetic pressure gradient}}} + \underbrace{\frac{1}{\rho\mu}(B_0 \cdot \nabla) b}_{\substack{\text{Lorentz acceleration} \\ \text{due to magnetic tension}}}. \tag{3.1}$$

$$\underbrace{\frac{\partial b}{\partial t}}_{\substack{\text{Change in the} \\ \text{magnetic field}}} = \underbrace{(B_0 \cdot \nabla) u}_{\substack{\text{Stretching of magnetic} \\ \text{field by fluid motion}}}. \tag{3.2}$$

Taking the curl $(\nabla\times)$ of (3.1) gives an equation describing how the fluid vorticity $\xi = \nabla \times u$ evolves

$$\frac{\partial \xi}{\partial t} = \frac{1}{\rho\mu}(B_0 \cdot \nabla)(\nabla \times b). \tag{3.3}$$

$\nabla\times$ (3.2) gives

$$\nabla \times \frac{\partial b}{\partial t} = (B_0 \cdot \nabla)\xi. \tag{3.4}$$

To find the wave equation, take a further time derivative of (3.3) so that

$$\frac{\partial^2 \xi}{\partial t^2} = \frac{1}{\rho\mu}(B_0 \cdot \nabla)\left(\nabla \times \frac{\partial b}{\partial t}\right), \tag{3.5}$$

and then eliminate b using an expression for $\frac{1}{\rho\mu}(B_0 \cdot \nabla)(\nabla \times \partial b/\partial t)$ in terms of ξ obtained by operating with $\frac{1}{\rho\mu}(B_0 \cdot \nabla)$ on (3.4)

$$\frac{1}{\rho\mu}(B_0 \cdot \nabla)\left(\nabla \times \frac{\partial b}{\partial t}\right) = \frac{1}{\rho\mu}(B_0 \cdot \nabla)^2 \xi. \tag{3.6}$$

When this is substituted into (3.5) the Alfvén wave equation is obtained

$$\frac{\partial^2 \xi}{\partial t^2} = \frac{1}{\rho\mu}(B_0 \cdot \nabla)^2 \xi. \tag{3.7}$$

The term on the right hand side arises from the restoring force caused by the stretching of magnetic field lines.

3.3. Alfvén wave dispersion relation and properties

Substituting a simple plane wave solution of the form $\boldsymbol{\xi} = \mathcal{R}e\{\widehat{\boldsymbol{\xi}}\, e^{i(\mathbf{k\cdot r}-\omega t)}\}$ into the Alfvén wave equation (3.7), valid solutions are possible provided that

$$\omega^2 = \frac{B_0^2(\mathbf{k} \cdot \widehat{\boldsymbol{B_0}})^2}{\rho\mu}, \tag{3.8}$$

where $\widehat{\boldsymbol{B_0}} = \boldsymbol{B_0}/|B_0|$.

(3.8) is the dispersion relation for Alfvén waves. It is a quadratic equation in ω, so the well known quadratic formula can be used to find explicit solutions for ω, which are

$$\omega = \pm v_A(\mathbf{k} \cdot \widehat{\boldsymbol{B_0}}), \tag{3.9}$$

where v_A known is the Alfvén velocity

$$v_A = \frac{B_0}{(\rho\mu)^{1/2}}. \tag{3.10}$$

This derivation illustrates that the Alfvén velocity is the speed at which an Alfvén wave propagates along magnetic field lines. Alfvén waves are non-dispersive because their phase velocity is independent of $|k|$ and equal to their group velocity are equal. Alfvén waves are however anisotropic, with their properties dependent on the angle between the background magnetic field and the wave propagation direction. Finally, it should be remarked that particle motions associated with simple Alfvén waves are linearly polarized.

4. Magnetic Coriolis (MC) waves

4.1. Force balances in rapidly-rotating, hydromagnetic fluids

Considering a system that is both rapidly-rotating and strongly influenced by a magnetic field, both magnetic field tension and vortex tension must be considered leading to a combination of Alfvén waves and Inertial waves. Lehnert [66] demonstrated that in the simplest case when the rotation axis and the background magnetic field are aligned, two scenarios are possible:

1. **Lorentz and Coriolis forces act together: Fast MC waves**
   Circularly polarized motion of the velocity perturbation, caused by the rotation of the fluid, occurs in an anti-clockwise direction. Therefore the Coriolis force acts in the same direction as the restoring Lorentz force caused by the

deviation of the fluid element (and magnetic field) from its undisturbed position. The restoring force due to the sum of the Lorentz and Coriolis forces is therefore strong and causes rapid motion and significant inertial acceleration. The resulting hydromagnetic waves are typically faster than pure Alfvén waves and are therefore known as fast MC waves.

2. **Lorentz and Coriolis forces in opposition: Slow MC waves**
   On the other hand, if the circularly polarized motion of the velocity perturbation occurs in an clockwise direction, the Coriolis force acts in the opposite direction to the restoring Lorentz force. The restoring force is thus weakened and the resulting particle motions are slow and do not involve significant inertial acceleration. These hydromagnetic waves are typically slower than pure Alfvén waves and are known as slow MC waves. Note that the kinetic energy associated with slow MC waves is much less than that associated with fast MC waves, though the slow waves still involve significant magnetic energy.

In general, the rotation axis and background magnetic field will not be parallel and the situation will be more complex. Nevertheless, the physical picture of circular particle motions involving Lorentz and Coriolis forces combining to produce fast and slow MC modes remains intact.

4.2. The MC wave equation

To derive the MC wave equation for a rapidly-rotating fluid permeated by a strong magnetic field in the absence of viscous and magnetic diffusion, the starting point is the linearized momentum equation including Coriolis, Lorentz and inertial acceleration

$$\frac{\partial \boldsymbol{u}}{\partial t} + 2(\boldsymbol{\Omega} \times \boldsymbol{u}) = -\frac{1}{\rho}\nabla p + \frac{1}{\rho\mu}(\boldsymbol{B_0} \cdot \nabla)\boldsymbol{b}, \tag{4.1}$$

and the frozen flux induction equation

$$\frac{\partial \boldsymbol{b}}{\partial t} = (\boldsymbol{B_0} \cdot \nabla)\boldsymbol{u}. \tag{4.2}$$

Taking the curl of each of these, using the solenoidal properties of the magnetic and velocity fields, and recognizing the vorticity $\boldsymbol{\xi} = \nabla \times \boldsymbol{u}$ leads to

$$\frac{\partial \boldsymbol{\xi}}{\partial t} - 2(\boldsymbol{\Omega} \cdot \nabla)\boldsymbol{u} = \frac{1}{\rho\mu}(\boldsymbol{B_0} \cdot \nabla)(\nabla \times \boldsymbol{b}), \tag{4.3}$$

and

$$\frac{\partial(\nabla \times \boldsymbol{b})}{\partial t} = (\boldsymbol{B_0} \cdot \nabla)\boldsymbol{\xi}. \tag{4.4}$$

A further time derivative of (4.3) and substituting from (4.4) enables these to be combined into a single vorticity equation

$$\frac{\partial^2 \boldsymbol{\xi}}{\partial t^2} - 2(\boldsymbol{\Omega} \cdot \nabla)\frac{\partial \boldsymbol{u}}{\partial t} = \frac{1}{\rho \mu}(\boldsymbol{B_0} \cdot \nabla)^2 \boldsymbol{\xi}. \tag{4.5}$$

Taking the curl of this and using the property $\nabla \times \boldsymbol{\xi} = \nabla \times (\nabla \times \boldsymbol{u}) = -\nabla^2 \boldsymbol{u}$ gives

$$-2(\boldsymbol{\Omega} \cdot \nabla)\frac{\partial \boldsymbol{\xi}}{\partial t} = \left(\frac{\partial^2}{\partial t^2} - \frac{1}{\rho \mu}(\boldsymbol{B_0} \cdot \nabla)^2 \right) \nabla^2 \boldsymbol{u}. \tag{4.6}$$

Finally taking $\partial/\partial t$ (4.5) gives an expression for $\partial \boldsymbol{\xi}/\partial t$ that can be substituted into (4.6) to eliminate $\boldsymbol{\xi}$ leaving a single equation for the perturbation velocity $\boldsymbol{u}$. This is known as the Magnetic Coriolis (MC) or Alfvén-Inertial wave equation [3, 22, 66],

$$\left(\frac{\partial^2}{\partial t^2} - \frac{1}{\rho \mu}(\boldsymbol{B_0} \cdot \nabla)^2 \right)^2 \nabla^2 \boldsymbol{u} = -4(\boldsymbol{\Omega} \cdot \nabla)^2 \frac{\partial^2 \boldsymbol{u}}{\partial t^2}. \tag{4.7}$$

4.3. MC wave dispersion relation and properties

The MC wave equation is considerably more complex than the Alfvén and Inertial wave equations: it is 4th order in the time derivative, leading us to expect the existence of 4 different modes. These modes can be identified as before by substituting a plane wave anstatz $\boldsymbol{u} = \mathcal{R}e\{\hat{\boldsymbol{u}}e^{i(\boldsymbol{k}\cdot\boldsymbol{r}-\omega t)}\}$ into the wave equation (4.7) which yields

$$\left(\frac{(\boldsymbol{B_0} \cdot \boldsymbol{k})^2}{\rho \mu} - \omega^2 \right)^2 k^2 - 4(\boldsymbol{\Omega} \cdot \boldsymbol{k})^2 \omega^2 = 0. \tag{4.8}$$

Taking the final term to the right hand side, dividing through by k^2 and taking the square root it can be seen that the possible solutions are

$$\frac{(\boldsymbol{B_0} \cdot \boldsymbol{k})^2}{\rho \mu} - \omega^2 = \pm \frac{2(\boldsymbol{\Omega} \cdot \boldsymbol{k})}{|k|}\omega. \tag{4.9}$$

These are two quadratic equations in ω: each can be solved using the usual quadratic formula to obtain the four solutions

$$\omega_{MC} = \pm \frac{(\boldsymbol{\Omega} \cdot \boldsymbol{k})}{k} \pm \left(\frac{(\boldsymbol{\Omega} \cdot \boldsymbol{k})^2}{k^2} + \frac{(\boldsymbol{B_0} \cdot \boldsymbol{k})^2}{\rho \mu} \right)^{1/2}. \tag{4.10}$$

When the two signs are the same polarity, then 2 fast MC waves (with Lorentz and Coriolis forces reinforcing each other) travelling in opposite directions are obtained. When the two signs are of different polarity, 2 slow MC waves (with Lorentz and Coriolis forces opposing each other) travelling in opposite directions are obtained.

Note that in the limit where rotation becomes unimportant ($\mathbf{\Omega} \to 0$) the Alfvén wave dispersion relation $\omega = \pm(\mathbf{B_0} \cdot \mathbf{k})/\sqrt{\rho\mu}$ is recovered, while in the opposite limit when magnetic fields are unimportant ($\mathbf{B_0} \to 0$) the Inertial wave dispersion relation $\omega = \pm 2(\mathbf{\Omega} \cdot \mathbf{k})/k$ is obtained.

If the frequency of Inertial waves is much larger than the frequency of Alfvén waves so that

$$|2(\mathbf{\Omega} \cdot \mathbf{k})/k| \gg |(\mathbf{B_0} \cdot \mathbf{k})/\sqrt{\rho\mu}| \tag{4.11}$$

(i.e. if rotation is sufficiently rapid) then it is possible to carry out a Taylor series expansion of (4.10) in the small quantity $k^2(\mathbf{B_0}\cdot\mathbf{k})^2/4(\mathbf{\Omega}\cdot\mathbf{k})^2\rho\mu$. One then finds a very clear splitting of the fast and slow wave frequencies such that the leading order expressions for the dispersion relations are

$$\omega_{MC}^f = \pm\frac{(2\mathbf{\Omega} \cdot \mathbf{k})}{k}\left(1 + \frac{k^2(\mathbf{B_0} \cdot \mathbf{k})^2}{4(\mathbf{\Omega} \cdot \mathbf{k})^2\rho\mu}\right), \tag{4.12}$$

and

$$\omega_{MC}^s = \pm\frac{k(\mathbf{B_0} \cdot \mathbf{k})^2}{2(\mathbf{\Omega} \cdot \mathbf{k})\rho\mu}. \tag{4.13}$$

Remembering that $k^2(\mathbf{B_0} \cdot \mathbf{k})^2/4(\mathbf{\Omega} \cdot \mathbf{k})^2\rho\mu$ is a small quantity, it is observed in this limit that the fast MC wave (ω_{MC}^f) is essentially an Inertial wave slightly modified by the presence of a magnetic field such that the wave frequency is higher than that of a pure inertial wave. Thus frequencies greater than 2Ω that are impossible for pure inertial waves are possible for fast MC waves. This fact can be a useful diagnostic property when trying to determine the type of waves present in an experiment or simulation.

The slow MC wave (ω_{MC}^s) that emerges in this limit emerges signifies a new fundamental timescale for rapidly-rotating, hydromagnetic systems

$$\tau_{MC} = \frac{2\Omega\rho\mu L^2}{B_0^2}, \tag{4.14}$$

where L is the lengthscale associated with a slow MC wave disturbance. Note that this is the square of the Alfvén wave timescale divided by the inertial wave timescale. Thus, the faster the rotation, the longer τ_{MC} becomes, while the stronger the magnetic field is, the shorter τ_{MC} becomes. Physically, this is be-cause the Lorentz force is the fundamental restoring mechanism, with the Corio-lis force acting to oppose it. It is worth reiterating that inertia plays no role in the slow MC waves in this limit: they are a consequence of a slowly evolving push and pull between the Lorentz force and the Coriolis force. Slow MC waves are therefore sometimes also called magnetostrophic waves [74].

Slow MC waves are both anisotropic and dispersive with their phase and group velocity taking the form

$$C_{\mathrm{ph}} = \frac{(\boldsymbol{B_0} \cdot \boldsymbol{k})^2}{2(\boldsymbol{\Omega} \cdot \boldsymbol{k})k\rho\mu} \boldsymbol{k},$$ (4.15)

$$C_{\mathrm{g}} = \frac{k(\boldsymbol{B_0} \cdot \boldsymbol{k})^2}{2(\boldsymbol{\Omega} \cdot \boldsymbol{k})\rho\mu} \left(\frac{\boldsymbol{k}}{k} + \frac{2\boldsymbol{B_0}}{(\boldsymbol{k} \cdot \boldsymbol{B_0})} - \frac{\boldsymbol{\Omega}}{(\boldsymbol{k} \cdot \boldsymbol{\Omega})} \right).$$ (4.16)

The slow MC wave has attracted the attention of geophysicists because $\tau_{MC} \sim$ 300 years for plausible geophysical parameters of $2\pi/k \sim 2.75 \times 10^6$ m and $B_0 \sim 5 \times 10^{-3}$ T). This is coincident with timescale of observed wave-like geomagnetic secular variation signals [42].

Hide (see Appendix A) and Malkus (see Appendix B) have extended models of MC waves to take account of spherical geometry. Hide used a β-plane ap-proach similar to that commonly employed in meteorology and oceanography. Malkus used a special cylindrically symmetric, force free, imposed magnetic field in full sphere geometry to simply the spherical MC wave problem to a com-plex variable generalization of the well known Inertial wave problem in a sphere.

5. Magnetic Archimedes Coriolis (MAC) waves

5.1. Influence of density stratification and convective instability

In this section, the generalization of MC waves to the case including density strat-ification that can drive convection is described. Density variations are introduced by imposing a background temperature gradient, this can produce both stable stratification or convective instability. The Boussinesq approximation [21] to the heat and momentum equations is also employed for simplicity. The wave mo-tions resulting in this scenario, when magnetic fields, density gradient and rapid rotation are present, are known as Magnetic Archimedes Coriolis (MAC) waves.

The dimensional linearized equations describing Boussinesq rotating magne-toconvection in an infinite plane layer geometry are

$$2\boldsymbol{\Omega} \times \boldsymbol{u} = -\frac{1}{\rho_0}\nabla p + \frac{1}{\mu\rho_0}(\boldsymbol{B_0} \cdot \nabla)\boldsymbol{b} + \gamma\alpha\Theta\hat{\boldsymbol{z}},$$ (5.1)

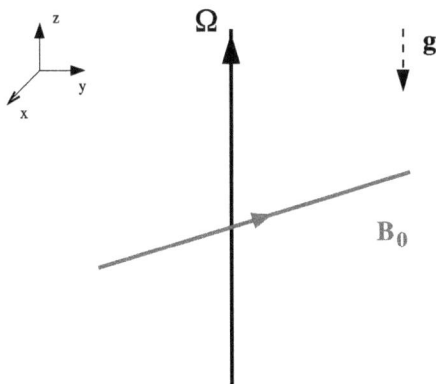

Fig. 2. Infinite plane layer setup for MAC waves. Gravity is imposed parallel to the rotation axis (as is the imposed temperature gradient, not shown). The imposed magnetic field is uniform but at an arbitrary angle to the rotation axis.

$$\frac{\partial \boldsymbol{b}}{\partial t} = (\mathbf{B_0} \cdot \nabla)\, \boldsymbol{u}, \tag{5.2}$$

$$\frac{\partial \Theta}{\partial t} = -\beta'(\widehat{\boldsymbol{z}} \cdot \boldsymbol{u}). \tag{5.3}$$

Cartesian co-ordinates $(\widehat{\boldsymbol{x}}, \widehat{\boldsymbol{y}}, \widehat{\boldsymbol{z}})$ have been employed with the axis of rotation chosen to lie along $\widehat{\boldsymbol{z}}$ (see Fig. 2). The imposed magnetic field is chosen to be uniform (i.e. $\nabla \mathbf{B_0} = 0$) but in an arbitrary direction, so $\mathbf{B_0} = (B_{0x}\widehat{\boldsymbol{x}} + B_{0y}\widehat{\boldsymbol{y}} + B_{0z}\widehat{\boldsymbol{z}})$. The uniform background temperature gradient is $\nabla T_0 = \beta'\widehat{\boldsymbol{z}}$. In a gravity field $\boldsymbol{g} = -\gamma\widehat{\boldsymbol{z}}$, the buoyancy force is therefore $\rho_0 \gamma \alpha \Theta$ in the $\widehat{\boldsymbol{z}}$ direction, where $T = T_0 + \Theta$ and α is the thermal expansivity. Note that $\beta' \gg 0$ implies less dense fluid overlies more dense fluid (stable stratification) while $\beta' \ll 0$ implies colder and more dense fluid overlies hotter and less dense fluid, a situation that is unstable to convection.

Note that in (5.1) to (5.3) the inertial term $(\partial \boldsymbol{u}/\partial t)$ as well as viscous, magnetic and thermal diffusion have been ignored, in order to focus attention on the essential physics involving the Lorentz (Magnetic), buoyancy (Archimedes) and Coriolis forces. Ignoring the inertial term in the previous section would have resulted in only the slow MC mode being obtained, a similar filtering out of the fast mode occurs here. ρ_0 rather than ρ is now used to represent the fluid density to emphasize that the Boussinesq approximation has been made.

Taking $\frac{\partial}{\partial t} \nabla \times$ (5.1) to eliminate pressure gives

$$2(\boldsymbol{\Omega} \cdot \nabla)\frac{\partial \boldsymbol{u}}{\partial t} = \frac{(\mathbf{B_0} \cdot \nabla)}{\mu \rho_0}\frac{\partial}{\partial t}(\nabla \times \boldsymbol{b}) + \gamma \alpha \frac{\partial}{\partial t}(\nabla \times \Theta \widehat{\boldsymbol{z}}), \tag{5.4}$$

while taking $\nabla \times$ (5.2) yields

$$\frac{\partial}{\partial t}(\nabla \times \boldsymbol{b}) = (\mathbf{B_0} \cdot \nabla)(\nabla \times \boldsymbol{u}). \tag{5.5}$$

Substituting from (5.5) into (5.4) for $\frac{\partial}{\partial t}(\nabla \times \boldsymbol{b})$ gives a vorticity equation quantifying the MAC balance with terms arising from Coriolis force on the left hand side and terms arising from the magnetic and buoyancy forces on the right hand side

$$2(\boldsymbol{\Omega} \cdot \nabla)\frac{\partial \boldsymbol{u}}{\partial t} = \frac{(\mathbf{B_0} \cdot \nabla)^2}{\mu \rho_0}(\nabla \times \boldsymbol{u}) + \gamma \alpha \frac{\partial}{\partial t}(\nabla \times \Theta \widehat{\boldsymbol{z}}). \tag{5.6}$$

Operating on this with $\frac{(\mathbf{B_0} \cdot \nabla)^2}{\mu \rho_0}\nabla \times$ gives

$$2(\boldsymbol{\Omega} \cdot \nabla)\frac{\partial}{\partial t}\frac{(\mathbf{B_0} \cdot \nabla)^2}{\mu \rho_0}(\nabla \times \boldsymbol{u}) = \left[\frac{(\mathbf{B_0} \cdot \nabla)^2}{\mu \rho_0}\right]^2(\nabla \times (\nabla \times \boldsymbol{u}))$$

$$+ \gamma \alpha \frac{(\mathbf{B_0} \cdot \nabla)^2}{\mu \rho_0}\frac{\partial}{\partial t}(\nabla \times (\nabla \times \Theta \widehat{\boldsymbol{z}})). \tag{5.7}$$

(5.7) can be simplified by noting that $\frac{(\mathbf{B_0} \cdot \nabla)^2}{\mu \rho_0}(\nabla \times \boldsymbol{u})$ can be eliminated by using (5.6) again, and remembering that for an incompressible fluid $\nabla \times (\nabla \times \boldsymbol{u}) = -\nabla^2 \boldsymbol{u}$. Then taking the dot product of (5.7) with $\widehat{\boldsymbol{z}}$ one also recognizes that $\widehat{\boldsymbol{z}} \cdot (\nabla \times \Theta \widehat{\boldsymbol{z}}) = 0$ and $\widehat{\boldsymbol{z}} \cdot (\nabla \times (\nabla \times \Theta \widehat{\boldsymbol{z}})) = -(\frac{\partial^2}{\partial x^2} + \frac{\partial^2}{\partial y^2})\Theta = -\nabla_H^2 \Theta$. Carrying out these operations it is found that

$$4(\boldsymbol{\Omega} \cdot \nabla)^2\frac{\partial^2}{\partial t^2}u_z = -\left[\frac{(\mathbf{B_0} \cdot \nabla)^2}{\mu \rho_0}\right]^2\nabla^2 u_z - \gamma \alpha \frac{(\mathbf{B_0} \cdot \nabla)^2}{\mu \rho_0}\nabla_H^2\frac{\partial \Theta}{\partial t}. \tag{5.8}$$

Finally, making use of the linearized heat equation (5.3) to eliminate $\frac{\partial \Theta}{\partial t}$, a sixth order equation in u_z is obtained. This is the diffusionless MAC wave equation [9]

$$\left(4(\boldsymbol{\Omega} \cdot \nabla)^2\frac{\partial^2}{\partial t^2} + \left[\frac{(\mathbf{B_0} \cdot \nabla)^2}{\mu \rho_0}\right]^2\nabla^2 - \gamma \alpha \beta'\frac{(\mathbf{B_0} \cdot \nabla)^2}{\mu \rho_0}\nabla_H^2\right)u_z = 0. \tag{5.9}$$

Properties of diffusionless MAC waves can now be deduced by substitution of plane wave solutions of the form $u_z = \mathcal{R}e\{\widehat{u}_z e^{i(\mathbf{k}\cdot\mathbf{r}-\omega t)}\}$, where $\mathbf{k}$ is the wavevector and ω is the angular frequency

$$4(\mathbf{\Omega}\cdot\mathbf{k})^2\omega^2 - \left[\frac{(\mathbf{B_0}\cdot\mathbf{k})^2}{\mu\rho_0}\right]^2 k^2 - \frac{(\mathbf{B_0}\cdot\mathbf{k})^2}{\mu\rho_0}\gamma\alpha\beta'(k_x^2 + k_y^2) = 0. \tag{5.10}$$

This expression can be written more concisely by observing that terms in it correspond to characteristic angular frequencies for Alfvén waves (see Section 3), internal gravity waves in a thermally stratified fluid, and Inertial waves (see Section 2), respectively defined as

$$\omega_M^2 = \frac{(\mathbf{B_0}\cdot\mathbf{k})^2}{\mu\rho_0}, \qquad \omega_A^2 = \frac{\gamma\alpha\beta'(k_x^2 + k_y^2)}{k^2}, \qquad \omega_C^2 = \frac{4(\mathbf{\Omega}\cdot\mathbf{k})^2}{k^2}. \tag{5.11}$$

Using these simplifies (5.10) to

$$\omega_C^2\omega^2 - \omega_M^4 - \omega_M^2\omega_A^2 = 0. \tag{5.12}$$

Solving for ω gives the MAC wave dispersion relation which must be satisfied by the angular frequency and wavevectors of plane MAC waves [9,93]

$$\omega_{\text{MAC}} = \pm\frac{\omega_M^2}{\omega_C}\left(1 + \frac{\omega_A^2}{\omega_M^2}\right)^{1/2} = \pm\omega_{MC}^s\left(1 + \frac{\omega_A^2}{\omega_M^2}\right)^{1/2}. \tag{5.13}$$

Due to the importance of this result, for completeness it is also worth stating the full expression which is

$$\omega_{\text{MAC}} = \pm\frac{k(\mathbf{B_0}\cdot\mathbf{k})^2}{2\rho_0\mu(\mathbf{\Omega}\cdot\mathbf{k})}\left(1 + \frac{\gamma\alpha\beta'\rho_0\mu(k_x^2 + k_y^2)}{k^2(\mathbf{B_0}\cdot\mathbf{k})^2}\right)^{1/2}. \tag{5.14}$$

An alternative derivation of this relation is given by Soward and Dormy [94] who follow Braginsky [10] in using an elegant formulation in terms of the displacement vector of the wave perturbations. This approach is useful for clearly showing how MAC waves are associated with elliptical particle motions. The more pedestrian derivation given above is preferred here because it involves quantities that are more straight-forward to interpret physically, and because links to simpler forms of waves are more transparent.

Note that (5.14) is singular if $\mathbf{B_0}\cdot\mathbf{k} = 0$ or if $\mathbf{\Omega}\cdot\mathbf{k} = 0$, so diffusionless MAC waves cannot propagate normal to magnetic field lines or to the rotation axis. The frequency of MAC waves depends strongly on their wavelength (i.e., they are highly dispersive) and on their direction compared to the direction of the rotation axis and the magnetic field direction (i.e., they are anisotropic). When buoyancy forces are absent ($\alpha = 0$), the dispersion relation simplifies to that for slow MC waves (remember the inertial terms were filtered out in this derivation).

The important extra ingredient introduced in the presence of thermal stratification is that if the background temperature gradient $\beta' \ll 0$ then ω_A^2 will be negative so ω_{MAC} will have an imaginary part, thus there will a growing or unstable mode. This is of course a thermally-driven (convective) instability: MAC waves in this case are the form in which rotating magnetoconvection is manifested. On the other hand, if the thermal stratification is stable with $\beta' \gg 0$ then this imaginary part is absent and the MAC waves are stable free waves. In the latter case the stable stratification acts increase the frequency of the MAC waves compared to MC waves. A complete and rigorous analysis of the growth rates of MAC waves requires one to assume that ω is complex and to look at the real and imaginary parts of the resulting dispersion relation.

5.2. Influence of diffusion

Thus far, the influence of any form of dissipation (viscous, magnetic or thermal diffusion) has been neglected in order to simplify both the mathematical analysis and the physical picture. In this section these effects are re-introduced. Naively, the presence of dissipation might be expected to merely damp disturbances and irreversibly transform energy to an unusable form. Although such processes certainly occur, the presence of diffusion also has more unexpected consequences.

Perhaps most importantly, diffusion adds extra degrees of freedom to the system, permitting exchange of heat, momentum and magnetic fields with the surroundings. This can aid the destabilization of waves that were stable in the absence of diffusion (see [84, 85] and [2]). Mathematically, the diffusion coefficients appear as part of the complex dispersion relation determining the growth rate, and not just in a simple dissipative manner. The result is that MAC/MC wave instability can occur for smaller unstable density gradients (or even for stable density gradients) compared to when no diffusion is present. Since diffusion is crucial for determining the fastest growing modes, it must be included if one wishes to determine which MAC wave modes will dominate the solution for a specified temperature gradient.

The diffusive instability mechanism works most effectively when the oscillation frequency matches the rate of the dominant diffusion process. The timescale of the most unstable waves is therefore often similar to that of the diffusion process which is facilitating the instability [2]. Diffusion thus introduces new preferred timescales for unstable MAC waves: those of thermal or magnetic diffusion, in addition to the diffusionless MC/MAC wave timescale.

To include diffusion in the mathematical description of MAC waves, it is necessary to replace the operator $\frac{\partial}{\partial t}$ by $\left(\frac{\partial}{\partial t} - \nu\nabla^2\right)$ in the momentum equation, $\left(\frac{\partial}{\partial t} - \eta\nabla^2\right)$ in the induction equation, and $\left(\frac{\partial}{\partial t} - \kappa\nabla^2\right)$ in the heat equation. Retaining the acceleration (inertial) term in the momentum equation and including the Laplacian diffusion terms before the substitution of plane wave solutions results in a rather more complicated, explicitly complex, dispersion relation for diffusive MAC waves in a plane layer [74, 93, 94],

$$\left(\omega_C^2(\omega + i\eta k^2)^2 - \left[(\omega + i\nu k^2)(\omega + i\eta k^2) - \omega_M^2\right]^2\right)(\omega + i\kappa k^2)$$
$$+ \omega_A^2(\omega + i\eta k^2)\left[(\omega + i\nu k^2)(\omega + i\eta k^2) - \omega_M^2\right] = 0. \tag{5.15}$$

More detail on diffusive MC waves with and without the inclusion of inertia, and their troublesome impact on geodynamo simulations can be found in [99] and [54].

6. MAC waves in spherical geometry

6.1. Quasi-geostrophic (QG) models of MAC waves

The simple models presented in previous sections have ignored the presence of any boundaries, in an attempt to focus on the essential physics. However, many geophysical and astrophysical systems involve spherical confining geometry. In this section, an analytic 'quasi-geostrophic' (QG) MAC wave model capturing the important latitudinal variations in planetary vorticity (the origin of Rossby waves) is outlined, before MAC wave solutions in full spherical geometry derived by numerical computations are presented in the next section.

Quasi-geostrophic (QG) models of MAC waves (also known in the literature as QG models of rotating magnetoconvection) were first developed by Busse [14] and Soward [93] building Hide's pioneering study of MC Rossby waves [52] (also see Appendix A). The setup of the QG model is shown in Fig. 3. The mathematical details of the model are set out in Appendix C.

The crucial new physics in the QG model compared to plane layer models arises from its sloping upper and lower boundaries. This geometry mimics the changes experienced by a columnar disturbance[3] as it moves latitudinally in a spherical shell geometry [13, 15, 52]. The sloping boundaries cause stretching or compression of a columnar disturbance as it moves in the direction perpendicular to the rotation axis. But compressed columns acquire anticyclonic vorticity, while stretched columns acquire cyclonic vorticity due to conservation of potential vorticity [77]. The acquired or lost vorticity causes displaced columns to

---

[3]The natural structure for slow motions in a rapidly-rotating fluid by virtue of the Proudman–Taylor theorem [49].

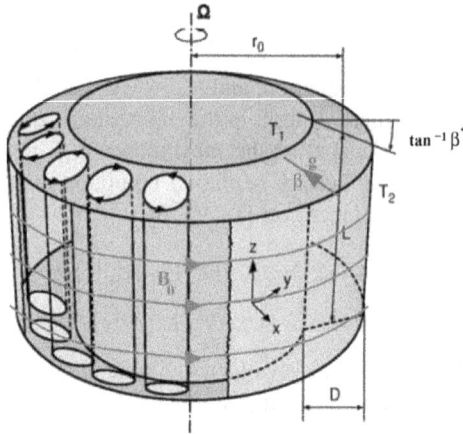

Fig. 3. Setup of Quasi-Geostrophic sloping annulus model for studying 2D MAC waves that are structurally invariant parallel to the rotation axis. It includes radial gravity, a radial temperature gradient and an azimuthal imposed magnetic field. Analysis is carried out in a local co-ordinate system with $\hat{z}$ in the direction of the rotation axis, $\hat{y}$ in the prograde azimuthal direction and $\hat{x}$ in the cylindrically radial direction. This figure is modified from Busse (2002) [15].

drift back towards their initial position, while also drifting azimuthally. This is the basis of the Rossby wave propagation mechanism (e.g. see Busse [15], his Fig. 2) and is the reason Hide's model (see Appendix A) also supports Rossby waves. The QG model therefore includes the Rossby wave mechanism important for slow motions in a spherical geometry via its sloping boundaries: this important physical effect is ignored in plane layer models.

Formally, in the QG approximation, the momentum, induction and heat equations are integrated in the direction parallel to the rotation axis, along which only small variations are expected due to the rapid rotation of the system. This results in equations for the evolution of z-averaged axial vorticity perturbation, electric current density perturbation and temperature perturbation (see Appendix C for details) in a 2D plane. The effect of sloping boundaries and the associated deviations from geostrophy are taken into account by respecting conservation of mass at the boundaries. Manipulating the resulting equations to produce a single wave equation, and substituting a simple wave anstatz $\sim e^{i(kx+ky-\omega t)}$, assuming the wavenumber in the $\hat{x}$ and $\hat{y}$ direction to be the same, leads to the following complex dispersion relation,

$$E(-i\omega + k^2) + \frac{i\beta^*}{k} = \frac{ERa}{-iPr\,\omega + k^2} - \frac{\Lambda k^2}{-iPr_m\omega + k^2}, \qquad (6.1)$$

where non-dimensionalization by the depth of the annulus L and the viscous diffusion timescale has been carried out and the resulting control parameters are

$$\Lambda = \frac{B_0^2}{2\Omega\mu\rho_0\eta} = \frac{\text{Magnetic diffusion timescale}}{\text{slow MC timescale}}, \tag{6.2}$$

$$Pr = \frac{\nu}{\kappa} = \frac{\text{Thermal diffusion timescale}}{\text{Viscous diffusion timescale}}, \tag{6.3}$$

$$Pr_m = \frac{\nu}{\eta} = \frac{\text{Magnetic diffusion timescale}}{\text{Viscous diffusion timescale}}, \tag{6.4}$$

$$E = \frac{\nu}{2\Omega L^2} = \frac{\text{Rotation timescale}}{\text{Viscous diffusion timescale}}, \tag{6.5}$$

$$Ra = \frac{\gamma \alpha |\nabla T_0| L^4}{\kappa\nu} = \frac{\text{Buoyant rise timescale in viscous fluid}}{\text{Thermal diffusion timescale}}, \tag{6.6}$$

and β^* is the tangent of the boundary slope. For a full derivation, interested readers should consult Appendix C.

The first term in dispersion relation (6.1) comes from the inertial and viscous effects, the second term results from the Coriolis force (actually it is the β-effect due to the sloping boundary altering the vorticity of columnar disturbances described above). The third term represents thermal (convective) forcing due to the action of gravity on density gradients, and the fourth term describes the influence of the uniform azimuthal magnetic field.

By looking at (6.1) in different limits, different types of waves can be isolated; these give insight into the forms of dynamics contained within the QG model.

1. **Rossby waves.**
   Ignoring thermal driving and the influence of magnetic fields leaving inertia, viscosity and the β-effect arising from column stretching in balance,

$$E(-i\omega + k^2) = -\frac{i\beta^*}{k} \quad => \quad \omega = \frac{\beta^*}{Ek} - i\frac{k^2}{E}. \tag{6.7}$$

This is the viscous timescale non-dimensionalization version of the dispersion relation expected for free Rossby waves, damped by viscous diffusion effects. These waves travel in the prograde (eastward) direction in the sloping annulus geometry mimicking the scenario outside the tangent cylinder in a thick spherical shell.

2. **Hide's slow MC-Rossby waves.**

Ignoring viscosity, inertia, thermal driving and magnetic diffusion, the β-effect is balanced by the magnetic field which is evolving only through frozen-flux effects,

$$\frac{i\beta^*}{k} = \frac{\Lambda k^2}{-iPr_m\omega} \quad => \quad \omega = -\frac{\Lambda k^3}{\beta^* Pr_m}. \tag{6.8}$$

This is the dispersion relation expected for Hide's slow MC-Rossby waves, travelling in the retrograde (westward) direction with frequency inversely proportional to the β-effect and proportional to the square of the magnetic field strength (i.e. to Λ).

3. **Magnetically-modified thermal Rossby waves.**

Including the magnetic field but neglecting its time changes, (assuming $\omega Pr_m \ll 1$) while retaining buoyancy force (thermal driving), the dispersion relation reduces to

$$E(-i\omega + k^2) + \frac{i\beta^*}{k} = \frac{E\,Ra}{-iPr\,\omega + k^2} - \Lambda. \tag{6.9}$$

Eliminating Ra from the imaginary part of this equation and using the expression for Ra obtained from the real part leads to the relation

$$\omega = \frac{\frac{\beta^* k}{E\,Pr}}{(1 + Pr^{-1})k^2 + \frac{\Lambda}{E}}. \tag{6.10}$$

When the magnetic field is absent ($\Lambda = 0$), this is the dispersion relation for Busse's thermal Rossby waves [13]. For a weak magnetic field, when $\Lambda \ll 1$, the character of the waves remains essential that of a thermal Rossby wave, but slightly slowed by the influence of the magnetic field.

4. **Thermal magneto-Rossby waves.**

For larger Λ the dispersion relation (6.10) is dominated by Λ / E and it reduces to the form

$$\omega = \frac{\beta^* k}{\Lambda Pr}. \tag{6.11}$$

This type of wave relies crucially on the magnetic field for its existence: it is not just a small correction to thermal Rossby waves, but a new type of motion where the magnetic forces take part in the leading order force balance and viscous forces are less important compared to the case for thermal Rossby waves [14, 34, 93]. It is commonly referred to as a thermal magneto-Rossby wave because it involves a balance between magnetic field effects, the Rossby wave restoring mechanism due to the sloping boundary (β-effect) and is thermally driven. As the magnetic field strength increases these waves have low frequency (in contrast to Hide's slow MC waves, indicating a different role of the magnetic field in the restoring mechanism), while the waves with short wavelengths (larger k) have higher frequency.

5. Thermal MC-Rossby waves.

If viscous and inertial terms are again ignored, but if $\omega Pr_m \gg 1$, for example when magnetic diffusion is negligible, then (6.9) reduces to

$$\omega = \frac{\Lambda k^3 (Pr - Pr_m)}{\beta^* Pr_m^2}. \tag{6.12}$$

This type of wave is essentially Hide's slow MC Rossby wave (hence involves the frozen flux approximation) but thermally driven, so it operates on a much slower timescale than free MC Rossby waves. It is known as the thermal MC Rossby wave. Unfortunately, these waves an unlikely to be relevant for liquid metals because their magnetic diffusion time is much shorter than the thermal diffusion timescale on which these waves operate, so the frozen-flux approximation necessary to obtain them will not be valid. Note however that MC Rossby waves could still occur in liquid metals if they are forced by mechanisms (other than thermal instability) with characteristic timescales that are short compared to that of magnetic diffusion.

To summarize, the QG model includes the essential ingredients for modelling slow motions of a rapidly-rotating fluid in a spherical geometry: motions are then essentially 2D (invariant parallel to the rotation axis) and they undergo changes in vorticity as they move towards and away from the rotation axis (the β-effect). It supports a remarkably rich variety of dynamics; three distinct forms of thermally-driven 2D MAC waves are possible: magnetically-modified thermal Rossby waves, thermal magneto-Rossby waves, and thermal MC Rossby waves. Free Rossby and MC Rossby waves are also possible in the absence of density stratification. When the magnetic field is strong and Pr_m is small, as is the case for liquid metals in planetary cores, thermal magneto-Rossby waves are the most likely form of QG wave to be excited by convection.

The limitations of the QG model should however also be remembered. In particular, it will break down if motions are no longer predominantly 2D (invariant parallel to the rotation axis), for example, (i) where very strong magnetic fields are present; (ii) where the boundary slope approaches infinity in the equatorial regions of a thick spherical shell (i.e. $\beta^* \gg 1$); (iii) when fluid motions are so rapid that inertia becomes dominant and fast, fully 3D, inertial waves rather than slower 2D Rossby waves occur; (iv) when the background magnetic field is no longer uniform, then additional terms are necessary to represent the Lorentz force and magnetically-driven 3D instabilities also become possible. The full spherical model sketched in the next section performs better under all of these circumstances, but must be solved numerically and lacks the understanding power provided by concise analytic solutions provided by the QG framework.

6.2. *MAC waves in full sphere geometry*

In order to obtain detailed predictions that can be directly related to observations of spherical systems such as Earth's core, it is perhaps best to move to full spherical geometry, due to the limitations of the QG model noted at the end of the previous section.

The problem now becomes a question of efficiently computing the required solutions. This chapter, being focused on physical understanding, is not a suitable place to expound the full numerical details. Only a sketch, giving a flavour of how the spatial structure, frequency and growth rates of wave modes may be obtained is given. Those readers desiring further details should consult chapter 7 of Finlay (2005) [43].

The example calculation presented here involves an imposed magnetic field of the form proposed by Malkus [70] (see Appendix B) with $B_0 = B_0 r \sin\theta \widehat{\phi}$. This is a force-free field for which the Lorentz force takes a particularly simple form, which does not drive magnetic instabilities, and is suitable for studying thermally-driven waves. The momentum, induction and heat equations are solved in a full sphere geometry subject to electrically insulating, stress-free and isothermal boundary conditions using a version of the spherical magnetoconvection code of Worland and Jones [57,101] modified by Finlay [43]. The governing equations are non-dimensionalized using the lengthscale $L = r_0$ (radius of the sphere) and the viscous diffusion timescale. The control parameters are therefore again E, Λ, Ra, Pr and Pr_m as defined in (6.2) to (6.6). The equations are linearized around the background magnetic field and a zero velocity field. Working in spherical polar co-ordinates the $\widehat{z}$ component of the curl and double curl of the momentum and induction equations are then taken.

Considering trial wave solutions of a particular azimuthal wavenumber and frequency, the poloidal and toroidal scalars representing the velocity and the magnetic fields and the temperature field are expanded in terms of spherical harmonics horizontally and Chebyshev polynomials radially. The collocation method is then used to formulate the equations on a radial grid. The resulting linear equations can be written in matrix form as a generalized complex eigenvalue problem

$$\lambda \mathcal{B}x = \mathcal{A}x, \tag{6.13}$$

where λ is the complex eigenvalue representing the frequency and growth rate of a particular mode, $\mathcal{B}$ is a matrix of the factors pre-multiplying the $\frac{\partial}{\partial t}$ terms in the system of equations and $\mathcal{A}$ is a matrix containing the pre-factors of all the other terms in the governing equations. x is a vector containing the unknown coefficients. The eigenvalue problem for complex λ (for a choice of m and the control parameters) can be solved, for example using ARPACK routines [43]. One then iterates varying Ra to locate Ra_c for the onset of thermally-driven instability (i.e. the marginally critical mode for which the growth rate is zero), then iterates over m to find that m_c with lowest Ra_c: this then constitutes the most unstable mode for a given set of control parameters. Once a mode of interest is found, its eigenvector defines the spatial structure of the wave while its eigenvalue determines the wave frequency and growth rate.

Here, in contrast to previous sections where the focus was on dispersion relations, the aim is to discuss the spatial structure thermally-driven MAC waves in a sphere. Figures 4, 5 and 6 show examples of the structure of examples of such waves, a magnetically-modified thermal Rossby wave and a thermal magneto-Rossby wave respectively. I have chosen these waves because they were discussed previously in the context of the QG model (hence they are predominantly 2D), but other fully 3D thermal MAC wave modes are also possible at larger Λ in this spherical system.

All these waves involve spatially periodic disturbances of positive and negative anomalies in the velocity, magnetic and temperature fields, though their detailed structure is different. It is worth remarking that the thermal magneto-Rossby modes that are most likely to be driven by convection in planetary core involve large amplitude perturbations of the radial magnetic field close to the outer boundary at low latitudes.

7. Limitations of linear models and towards nonlinear models

All the models discussed so far are linear in the perturbation velocity, magnetic and temperature fields. As such they are fundamentally limited since they ignore: (i) saturation mechanisms, (ii) wave-wave interactions, (iii) wave-mean flow interactions. Nonlinear effects are often responsible for development of more complex spatial and temporal structure of disturbances as the forcing of the system is increased. Furthermore, the models presented above ignore the complications introduced by phase mixing [51]. This occurs when gradients in background fields lead to frequency (and wavelength) detuning between neighbouring perturbation fields; it is known to have important implications for both Alfvén waves [51] and (magnetically-modified) thermal Rossby waves in a sphere [56, 57]. Although simple linear models are useful for determining the most unstable modes in a

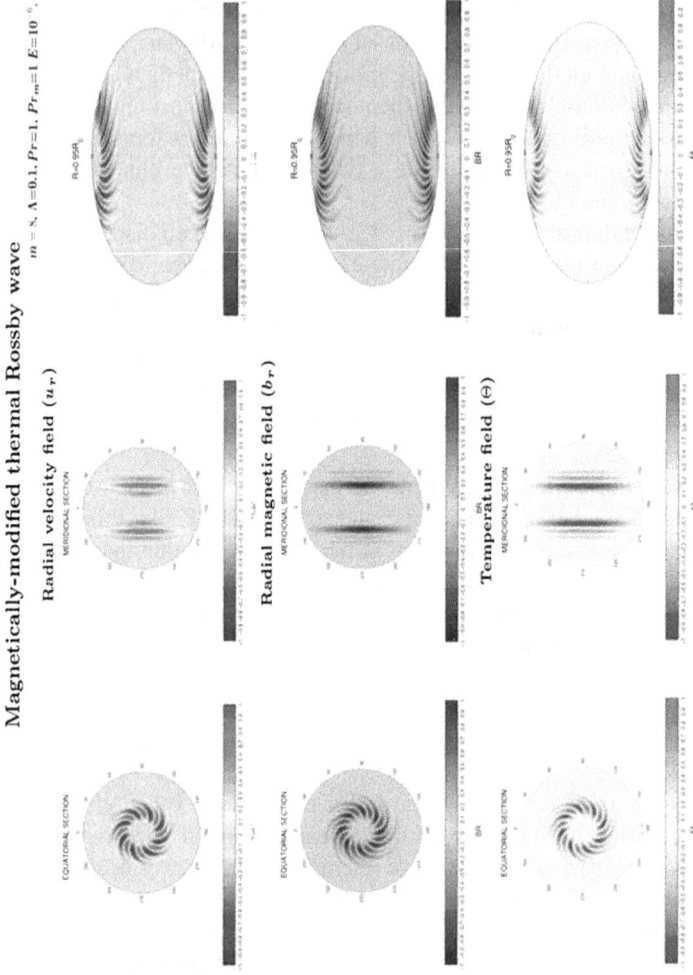

Fig. 4. Example of the structure of a $m = 8$, marginally-critical, magnetically-modified thermal Rossby wave for $E = 10^{-6}$, $Pr = 1$, $Pr_m = 1$, $\Lambda = 0.1$. Top row shows the perturbation radial velocity field, middle row shows the perturbation radial magnetic field and bottom row shows the perturbation temperature field. Left hand column shows a section through the equatorial plane, centre column shows a north-south meridional section, right hand column shows the fields just below the outer boundary. Taken from Finlay (2005) [43].

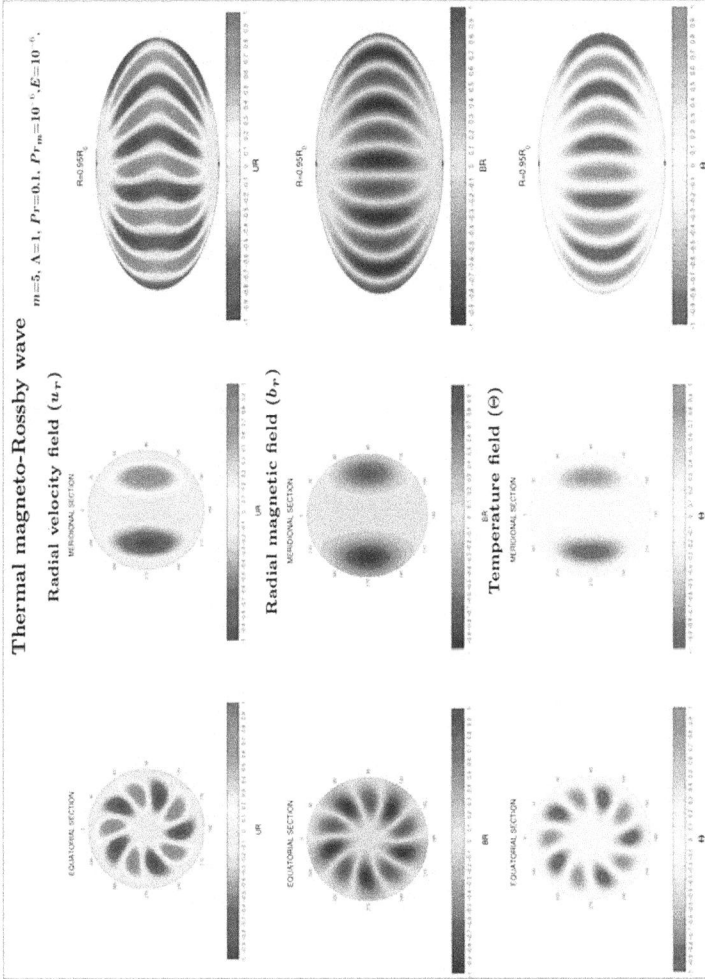

Fig. 5. Example of the structure of a $m = 5$, marginally-critical, thermal magneto- Rossby wave for $E = 10^{-6}$, $Pr = 0.1$, $Pr_m = 10^{-6}$, $\Lambda = 1$. Top row shows the perturbation radial velocity field, middle row shows the perturbation radial magnetic field and bottom row shows the perturbation temperature field. Left hand column shows a section through the equatorial plane, centre column shows a north-south meridional section, right hand column shows the fields just below the outer boundary. Taken from Finlay (2005) [43].

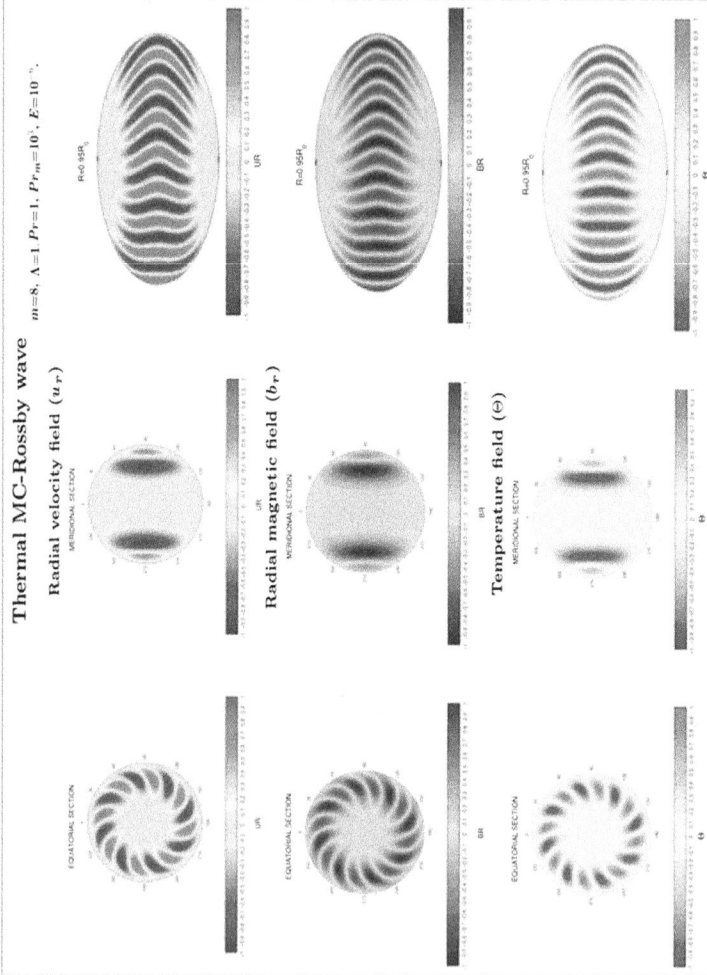

Fig. 6. Example of the structure of a $m = 8$, marginally critical, thermal MC Rossby wave $E = 10^{-6}$, $Pr = 1$, $Pr_m = 10^3$, $\Lambda = 1$. Top row shows the perturbation radial velocity field, middle row shows the perturbation radial magnetic field and bottom row shows the perturbation temperature field. Left hand column shows a section through the equatorial plane, centre column shows a north–south meridional section, right hand column shows the fields just below the outer boundary. Taken from Finlay (2005) [43].

particular parameter regime, the neglect of nonlinear processes and gradients in background fields makes detailed comparisons of model results with geophysical and astrophysical observations a perilous activity.

Some limited progress has been made on developing more sophisticated non-linear models. Roberts and Stewartson [83] performed a weakly nonlinear analysis of MAC waves in a plane layer geometry. Eltayeb and co-workers [26, 30] derived equations for diffusionless MAC waves in a plane layer with a slowly varying background mean state. Their equations describe the evolution of the wave amplitude via the conservation of wave action (wave energy divided by wave frequency per unit volume). Ewen and Soward [31–33] performed a weakly nonlinear analysis of a QG cylindrical annulus model of MAC waves. They derived and analysed equations describing the amplitude modulation of thermal magneto-Rossby waves, finding that a geostrophic flow is driven nonlinearly by the Lorentz forces resulting from magnetic field perturbations, and is linearly damped by the Ekman boundary layer. They also found that the nonlinear interaction resulting from the geostrophic flow can redistribute energy amongst wave modes. Cardin and Olson [19] used the QG approximation to numerically simulate fully non-linear magnetoconvection and found that imposing a strong magnetic field reduced the observed wavenumber and reduced the amplitude of zonal flows driven by nonlinear Reynolds stresses.

In full spherical geometry Walker and Barenghi [100] considering only the leading order nonlinearity of the geostrophic flow found that the resulting flow strongly influenced the frequency and propagation direction of waves. Zhang [104] studied nonlinear rotating magnetoconvection in a spherical shell. He found a nonlinear bifurcation occurred as the Rayleigh number was increased, that involved simultaneous breaking of temporal and azimuthal symmetry leading to vacillating wave motions.

Theoretical study of nonlinear waves in rapidly-rotating, electrically conducting fluids in the presence of strong magnetic fields is still in its infancy. In particular efforts have focused on convection-driven systems to the neglect of other possible driving mechanisms such as elliptical instabilities, shear instabilities, boundary layer instabilities and magnetic instabilities. Future investigations into these possibilities will likely yield important insights enabling better comparisons with observations.

8. Waves in experiments

Having reviewed the properties of waves in the presence of magnetic fields, rotation and convection using a range of theoretical models, in this section a link is made to physical reality in the form of laboratory experiments.

Alfvén waves were first studied experimentally in liquid mercury by Lundquist [69] and in liquid sodium by Lehnert [65]. These studies confirmed the existence of Alfvén waves, but also highlighted the difficulty in studying the simplest plane waves experimentally since at low frequencies their wavelength becomes larger than the apparatus, while at high frequencies the inertia of the liquid destroys ideal behaviour. Most recent experimental work on Alfvén waves has focused on plasmas rather than liquid metals.

Inertial waves have been more intensively studied experimentally. It is likely that Kelvin himself first observed them in a public demonstration [59]. Interest in the late 20th century was stimulated by the beautiful study of Schultz [46] and by the photographs shown in the monograph of Greenspan [49]. Much recent work has focused on how Inertial waves might be generated by elliptical instability, for example by tidal straining or precessional forcing (see the studies by Aldridge and co-workers [4, 91] and LeGal and co-workers [25, 62, 63]) and on the nonlinear interactions and evolution of Inertial waves, for example leading to the phenomenon of resonant growth and collapse cycles [62, 71, 72]. In general, when viscous corrections and appropriate boundary conditions are employed, the predictions of linear theory for inertial wave frequencies, wavenumbers and growth rates have proved spectacularly successful.

In contrast to the numerous studies of Alfvén and Inertial waves, there have been only a very small number of studies of wave motions in rapidly-rotating electrically conducting fluids in the presence of magnetic fields. Lacaze et al. [64] have studied the induction produced by Inertial waves in a Galinstan cylinder in the presence of a weak axial magnetic field. Kelley et al. [60] have identified the induction signature of Inertial waves in a liquid sodium spherical Couette experiment with an applied axial magnetic field. In both cases, the Lorentz force was not strongly influencing the dynamics, and the magnetic field was simply acting as a tracer for the Inertial waves. In contrast the DTS experiment at Grenoble [20, 75], also a spherical Couette experiment using liquid sodium, is designed to operate in a regime where the Lorentz force constitutes an important part of the leading order force balance. Preliminary results show non-axisymmetric motions in this experiment to be dominated by wave motions [90]. It seems likely that these must are some form of MC wave, but the analysis is complicated by the coincidence of timescales (magnetic diffusion, Alfvén and inertial) in the experiment, the non-uniform imposed magnetic field and the difficulty in imaging the interior flow of the liquid metal. Nonetheless, this challenging experiment is very exciting as it opens up a new avenue for experimental tests of MC wave theory that promises many further surprises.

At this time, few experiments on MAC waves have yet been reported. Experiments on rotating magnetoconvection in a weak applied field have recently been reported by Gillet et al. [48]. The rotating magnetoconvection experiment at UCLA of Aurnou, King and co-workers promises to approach the interesting regime where the Lorentz forces play a leading role in the dynamics and heat transport.

Many possibilities exist for future experiments probing waves in the presence of magnetic fields, rotation and convection. For example, it would be particularly interesting to build a simple experiment (e.g. with cylindrical geometry and uniform applied magnetic field parallel to the rotation axis) designed to obtain a clear splitting of fast and slow MC modes. This would allow the intruiging slow MC mode to be isolated so its properties might be studied in detail. At the moment it is a significant engineering challenge to produce splitting of MC modes due to the large rotation rates required, but such technical issues are not insurmountable.

9. Waves in numerical dynamo simulations

One of the most powerful tools for understanding the dynamics of geophysical and astrophysical systems is full numerical simulation of the governing (MHD) equations. Although currently limited in the regions of parameter space in which they can accurately operate, such simulations are capable of capturing important aspects of solar and (especially) terrestrial magnetic field generation and evolution.

Since they solve the full rotating MHD equations (i.e. the induction equation and Navier-Stokes equation including Coriolis, Lorentz and full non-linear inertial terms), and typically drive motions via thermal convection, these simulations can in principle capture Alfvén, inertial, MC and MAC waves. In fact, such waves are often viewed as a nuisance by modellers since they place tight requirements on time-steps necessary to capture the dynamics [54,99]. However, in present simulations the influence of diffusion (especially viscous and thermal diffusion) is stronger than in true geophysical and astrophysical systems due to computational constraints. As a consequence many potentially interesting waves are often over damped and are not able to be studied.

Nonetheless, a few preliminary studies suggest that simulations are beginning to reach the regimes where they can provide interesting information concerning waves. Glatzmaier has reported 2D simulations of magnetoconvection where Alfvén waves are clearly evident, while Dumberry [24] and Wicht have independently studied the special case of geostrophic Alfvén waves (Torsional Oscillations) in fully 3D geodynamo simulations. In my Ph.D. thesis [43], I showed that in a rather simple 3D, convection-driven, dynamo with Rayleigh number only 3 times critical, thermally-driven magneto-Rossby waves modes of convection could be identified. In simulations with more realistic parameters (lower Ekman

number and more super-critical Rayleigh number) more localised wave-like features were found in both the velocity field and the magnetic field close to the
outer boundary.

Investigations of waves in dynamo simulations with diffusivities as small as
possible, including detailed study of the evolution of the force balance involved
in wave-like features, certainly seems to be a fruitful direction for future research.
The detailed diagnostics provided by simulations should allow many hypotheses
concerning wave motions and their nonlinear evolution to be tested. One of the
biggest and most important questions that simulations can answer is what role
MC/MAC waves play when the magnetic field structure is not uniform, but is
dominated by compact regions of strong magnetic field where dynamo action
occurs. There Lorentz forces will be particularly dominant and phase mixing of
waves is likely to occur. Such questions cannot be answered by simple linear
theories with uniform applied fields such as those presented in previous sections.

10. Concluding remarks on waves in geophysical and astrophysical systems

In these lecture notes, I have reviewed the theory of waves in the presence of
magnetic fields, rotation and convection and discussed how experiments and numerical simulations are becoming important tools for furthering our understanding of these phenomena.

Are the waves discussed above of geophysical and astrophysical interest?
Since these systems often possess magnetic fields and undergo rapid rotation
they are certainly likely to support waves. Furthermore, they are usually characterized by large hydrodynamic and hydromagnetic Reynolds numbers resulting
from weak viscous and magnetic diffusivities, so it seems plausible that waves
will not be damped too rapidly in these systems. The key question is therefore
whether mechanisms exist for exciting such hydromagnetic waves.

Personally, I find it is hard to imagine how waves could not be important in
many geophysical and astrophysical fluid systems, because they are the natural
response of a hydromagnetic system to perturbation. For example, turbulent convection could continuously excite a very broadband spectrum of waves, similar to
the scenario for the driving helioseismic waves on the sun. Moreover, waves are
the natural manifestation of a variety of instability mechanisms including convective instability [9], magnetic instability [1], shear instability [39], boundary layer
instability [23] topographical forcing [52] and elliptical (tidal or precessional)
instability [61]. Which (if any) of these mechanisms are relevant depends on the
details of the particular system being studied.

In astrophysical systems involving plasmas or giant gaseous planets, compressibility effects will not be negligible and the theory outlined above will have

to be modified accordingly. Nonetheless, there is already considerable astrophysical interest in waves similar to those discussed above, for example regarding the role hydromagnetic waves might play in the dynamics of the tachocline [89]. Considering planetary cores (including that of the Earth), incompressibility is likely to be a reasonable first approximation, and observations of geomagnetic field evolution [42] provide intruiging evidence in favour of the existence of wave-like features. Thermally-driven magneto-Rossby waves advected by background zonal flows or topographically-driven MC waves appear the most likely candidates for explaining the geophysical observations [43]. It is presently unclear whether elliptical instability could occur in Earth's core [61] or whether it could excite slow MC/MAC waves.

Though the basic theory of waves in the presence of magnetic fields, rotation and convection exists, much remains to be done in order to apply it to geophysical and astrophysical systems. The input from experimental studies in coming years together with development of nonlinear theories of wave-wave and wave-mean flow interactions informed by study of waves in numerical simulations, are likely to be crucial in the future development of the subject.

Acknowledgements

I would like to express my gratitude to Philippe Cardin and the scientific committee for organizing the 'Dynamos' summer school and to thank the principal lecturers for their efforts in delivering informative and thought-provoking lectures. I am also happy to acknowledge stimulating conversations regarding waves with my fellow students, particularly with Doug Kelley, Wietze Herreman and Mathieu Dumberry. Thanks also to Nadège Gagniere for reading and commenting on these notes.

Appendix A. Hide's β-plane model of MC Rossby waves

Hide [52] developed a simple analytical model for MC waves in a spherical shell geometry. He assumed that the waves would have a two dimensional form, being to a first approximation invariant parallel to the rotation axis. He was therefore able to focus on local motions of fluid columns in the eastward $\hat{x}$ and northward $\hat{y}$ directions in spherical geometry and focused on the evolution of the vertical $\hat{z}$ component of vorticity. He followed Rossby [88] and assumed that the Coriolis force could be approximated as varying linearly in the north-south direction at a fixed latitude. This results in the a simplification of the Coriolis force to the form $-f u_y$ in the eastward direction and $f u_x$ in the northward direction, where $f_c = 2\Omega \sin\theta_{lat} + \frac{2\Omega \cos\theta_{lat}}{c} y = f_0 + \beta y$ and c is the outer radius of the spherical shell. Though $\beta = \frac{2\Omega \cos \theta_{lat}}{c}$ in thin spherical shell, Hide argued that it should take the opposite sign $\left(\beta = -\frac{2\Omega \cos \theta_{lat}}{c}\right)$ in thick spherical shells.

Hide studied a scenario where the imposed magnetic field $\boldsymbol{B_0}$ was uniform but inclined at an arbitrary angle to the eastward direction; here I present his model for a special case when $\boldsymbol{B_0}$ is purely eastwards. The background velocity field is also assumed to be zero for simplicity. Magnetic and viscous diffusion and all thermal and density stratification effects are also neglected. The linearized equations governing the evolution of the velocity field and magnetic field disturbances $(\boldsymbol{u}, \boldsymbol{b})$ on the β-plane, after the subtraction of the leading order geostrophic balance between pressure and the constant part of the Coriolis force $(f_0 = 2\Omega \sin \theta_{lat})$, are then

$$\frac{\partial u_x}{\partial t} - \beta y u_y = -\frac{1}{\rho_0}\frac{\partial p}{\partial x}, \tag{A.1}$$

$$\frac{\partial u_y}{\partial t} + \beta y u_x = -\frac{1}{\rho_0}\frac{\partial p}{\partial y} + \frac{B_0}{\rho_0\mu_0}\left(\frac{\partial b_y}{\partial x} - \frac{\partial b_x}{\partial y}\right), \tag{A.2}$$

$$\frac{\partial b_x}{\partial t} = B_0 \frac{\partial u_x}{\partial x}, \tag{A.3}$$

$$\frac{\partial b_y}{\partial t} = B_0 \frac{\partial u_y}{\partial x}, \tag{A.4}$$

where B_0 is the strength of the background magnetic field. Note for this special case of $\boldsymbol{B_0} = B_0\hat{\boldsymbol{x}}$ there is no Lorentz force term in (A.1).

Taking the curl of the momentum equations ($\frac{\partial}{\partial y}$ of (A.1) minus $\frac{\partial}{\partial x}$ of (A.2)) yields the $\hat{\boldsymbol{z}}$ component of the vorticity equation

$$\frac{\partial \zeta}{\partial t} + \beta u_y = \frac{B_0}{\rho_0\mu_0}\frac{\partial}{\partial x}\left(\frac{\partial b_y}{\partial x} - \frac{\partial b_x}{\partial y}\right), \tag{A.5}$$

where $\zeta = \left(\frac{\partial u_x}{\partial y} - \frac{\partial u_y}{\partial x}\right)$ is the z component of vorticity. Taking the time derivative of (A.5), substituting from $\frac{\partial}{\partial y}$ (A.3) and $\frac{\partial}{\partial x}$ (A.4) and operating with $\nabla_H^2 = \left(\frac{\partial}{\partial x^2} + \frac{\partial}{\partial y^2}\right)$ to eliminate u_y, an equation for the evolution of ζ on the β plane is obtained

$$\left(\frac{\partial^2}{\partial t^2} - \frac{B_0^2}{\rho_0\mu_0}\frac{\partial^2}{\partial x^2}\right)\nabla_H^2\zeta + \beta\frac{\partial}{\partial t}\left(\frac{\partial \zeta}{\partial x}\right) = 0. \tag{A.6}$$

Plane wave solutions of the form $\zeta = \hat{\zeta}e^{i(kx+ky-\omega t)}$, where k is the wavenumber of the disturbance (for simplicity assumed to be the same in the northward and eastward directions), and ω is the angular frequency of the waves, can now be considered. Substituting the plane wave solutions into (A.6) yields the dispersion relation

$$\omega^2 + \frac{\beta\omega}{k} - \frac{B_0^2 k^2}{\rho_0 \mu_0} = 0. \tag{A.7}$$

This quadratic equation can be solved to give an expression for ω in terms of k,

$$\omega = -\frac{\beta}{2k} \pm \frac{\beta}{2k} \left(1 + \frac{4B_0^2 k^4}{\rho_0 \mu_0 \beta^2} \right)^{1/2}. \tag{A.8}$$

For long wavelength disturbances in a rapidly-rotating fluid $\left(\frac{4B_0 k^4}{\rho_0 \mu_0 \beta^2} \right)$ will be small and a Taylor expansion in this parameter shows that two rather different solutions for ω are possible to leading order

$$\omega_r = -\frac{\beta}{k} \quad \text{and} \quad \omega_m = \frac{B_0^2 k^3}{\mu_0 \rho_0 \beta}. \tag{A.9}$$

The mode ω_r is recognizable as a Rossby wave on a β-plane [49, 88]. Rossby waves are the special low-frequency, columnar, uni-directional inertial waves that arise because of the latitudinal variation of the Coriolis force in spherical shell. ω_m on the other hand corresponds to a wave very similar in form to the slow MC wave found in a rotating plane layer, but inversely proportional to $\beta = \frac{2\Omega \cos \theta_{lat}}{c}$ rather than 2Ω. It is often referred to as a (slow) MC Rossby wave or sometimes as Hide's wave. Note that if the wavenumbers in the $\hat{x}$ and $\hat{y}$ directions are different then slightly more complicated dispersion relations are obtained, but the essential physics remains the same.

Appendix B. Malkus' model of MC waves in a full sphere

Malkus [70] analytically studied MC waves in a full sphere geometry using a special toroidal and purely azimuthal imposed magnetic field of the form $B_0 = B_0 r \sin\theta \, \hat{\phi}$ where $0 \leq r \leq 1$. Here, $r = r'/c$ with r' being the dimensional spherical polar radius, c is the dimensional radius of the outer spherical boundary, θ is the co-latitude, and B_0 is the maximum magnitude of the imposed field at the outer boundary in the equatorial plane. This field (often called the Malkus field) is invariant on cylindrical surfaces aligned with the rotation axis and increases in strength linearly with distance from the rotation axis. It can be conveniently written in cylindrical polar co-ordinates as $B_0 = B_0 s \, \hat{\phi}$ where $0 \leq s \leq 1$ where s is the normalized cylindrical radius. It is also force free (i.e. $\nabla \times (\nabla \times B_0) = 0$), so $U_0 = 0$ is a consistent choice for the co-existing background velocity field.

Defining the rotation axis to be along $\hat{z}$ and working in cylindrical polar co-ordinates $(\hat{s}, \hat{z}, \hat{\phi})$, the linearized governing equations for the velocity and magnetic field perturbations ignoring density stratification and all diffusive processes are

$$\rho_0 \frac{\partial \boldsymbol{u}}{\partial t} + 2\rho_0 \Omega (\hat{z} \times \boldsymbol{u}) = -\nabla p + \frac{1}{\mu_0} [(\boldsymbol{B_0} \cdot \nabla)\boldsymbol{b} + (\boldsymbol{b} \cdot \nabla)\boldsymbol{B_0}], \tag{B.1}$$

$$\frac{\partial \boldsymbol{b}}{\partial t} = \nabla \times (\boldsymbol{u} \times \boldsymbol{B_0}). \tag{B.2}$$

The major difference of the Malkus model compared to studies of uniform imposed magnetic fields is in the inclusion of the final term in the momentum equation $(\boldsymbol{b} \cdot \nabla)\boldsymbol{B_0}$ resulting from the non-zero spatial gradient in $\boldsymbol{B_0}$.

For the Malkus field, the Lorentz force and the advection term in the induction equation take very simple forms. Recognizing that the advection term can be rewritten as $\nabla \times (\boldsymbol{u} \times \boldsymbol{B_0}) = (\boldsymbol{B_0} \cdot \nabla)\boldsymbol{u} - (\boldsymbol{u} \cdot \nabla)\boldsymbol{B_0}$ and making use of the following relations that hold for the Malkus field

$$(\boldsymbol{b} \cdot \nabla)\boldsymbol{B_0} = B_0(\hat{z} \times \boldsymbol{b}), \tag{B.3}$$

$$(\boldsymbol{B_0} \cdot \nabla)\boldsymbol{b} = B_0 \left(\frac{\partial \boldsymbol{b}}{\partial \phi} + \hat{z} \times \boldsymbol{b} \right), \tag{B.4}$$

$$(\boldsymbol{B_0} \cdot \nabla)\boldsymbol{u} = B_0 \left(\frac{\partial \boldsymbol{u}}{\partial \phi} + \hat{z} \times \boldsymbol{u} \right), \tag{B.5}$$

$$(\boldsymbol{u} \cdot \nabla)\boldsymbol{B_0} = B_0(\hat{z} \times \boldsymbol{u}) \tag{B.6}$$

the governing equations become

$$\rho_0 \frac{\partial \boldsymbol{u}}{\partial t} + 2\rho_0 \Omega (\hat{z} \times \boldsymbol{u}) = -\nabla p + \frac{B_0}{\mu_0} \left[\frac{\partial \boldsymbol{b}}{\partial \phi} + 2\hat{z} \times \boldsymbol{b} \right], \tag{B.7}$$

$$\frac{\partial \boldsymbol{b}}{\partial t} = B_0 \left(\frac{\partial \boldsymbol{u}}{\partial \phi} \right). \tag{B.8}$$

Substituting in azimuthally travelling wave solutions of the form

$$(\boldsymbol{u}, \boldsymbol{b}) = (\hat{\boldsymbol{u}}(s, z)e^{i(m\phi - \omega t)}, \hat{\boldsymbol{b}}(s, z)e^{i(m\phi - \omega t)}), \tag{B.9}$$

where m is the azimuthal wavenumber of the wave, and ω is its angular frequency leads to the relations

$$-i\omega\rho_0 \boldsymbol{u} + 2\omega\rho_0 (\hat{z} \times \boldsymbol{u}) = -\nabla p + \frac{B_0}{\mu} \left(im\boldsymbol{b} + 2(\hat{z} \times \boldsymbol{b}) \right), \tag{B.10}$$

$$-i\omega\boldsymbol{b} = im B_0 \boldsymbol{u}. \tag{B.11}$$

Substituting from (B.11) into (B.10) for $\boldsymbol{b}$ and collecting like terms yields a single equation

$$\mathcal{L}_1(\hat{z} \times \boldsymbol{u}) + \mathcal{L}_2\boldsymbol{u} + \nabla p = 0, \tag{B.12}$$

where

$$\mathcal{L}_1 = 2\left(\Omega\rho_0 + \frac{B_0^2 m}{\mu_0 \omega}\right), \qquad \mathcal{L}_2 = \left(-i\omega\rho_0 + i\frac{B_0^2 m^2}{\mu_0 \omega}\right). \tag{B.13}$$

This has the same form as the momentum equation governing the evolution of Inertial waves in a sphere [73, 102, 108], but with $\mathcal{L}_2$ rather than $-i\omega\rho_0$ and $\mathcal{L}_1$ rather than $2\rho_0\Omega$. The consequence of the presence of the Malkus background magnetic field is therefore only that the strength of the latitude-dependent Coriolis force and the timescale of the inertial response of the fluid have been changed.

(B.12) can be re-written in terms of pressure only (for details of this manipulation consult [70]) as

$$\left(\mathcal{L}_2^2\nabla^2 + \mathcal{L}_1^2\frac{\partial}{\partial z}\right)p = 0. \tag{B.14}$$

The appropriate rigid spherical boundary condition ($\boldsymbol{u} \cdot \hat{\boldsymbol{n}} = 0|_{r=c}$ where $\hat{\boldsymbol{n}}$ is a normal to the spherical surface), in a form applicable to pressure can be obtained using the link between $\boldsymbol{u}$ and p in (B.12) and turns out to be

$$\left(\mathcal{L}_2^2 s\frac{\partial}{\partial s} + (\mathcal{L}_1^2 + \mathcal{L}_2^2)z\frac{\partial}{\partial z} + i\mathcal{L}_1\mathcal{L}_2\right)p = 0. \tag{B.15}$$

Next, by choosing

$$\lambda_{in} = \frac{2\mathcal{L}_2}{i\mathcal{L}_1} \quad \text{so that} \quad -\lambda_{in}^2 = \frac{4\mathcal{L}_2^2}{\mathcal{L}_1^2} \tag{B.16}$$

the governing equations in p are transformed into the standard form of the Poincaré equation for Inertial waves in a sphere (see, for example, [102])

$$\left(\nabla^2 - \frac{4}{\lambda_{in}^2}\frac{\partial}{\partial z}\right)p = 0, \tag{B.17}$$

and the associated boundary condition

$$\left(s\frac{\partial}{\partial s} + \frac{2m}{\lambda_{in}} - \frac{4}{\lambda_{in}^2}z\frac{\partial}{\partial z}\right)p = 0 \quad \text{on } s^2 + z^2 = 1. \tag{B.18}$$

The solution to the Inertial wave problem in the sphere involves finding eigenvalues λ_{in} and associated eigenvectors which satisfy (B.17) and (B.18). The solutions are very complicated in general and have only recently been written down explicitly by Zhang and co-workers [108], though Malkus [70] and Zhang [102] had earlier studied some simple cases.

Regarding λ_{in} as known, the angular frequencies ω of the solutions to the hydromagnetic wave problem in a sphere in the presence of the Malkus magnetic field are now defined. By substituting expressions for $\mathcal{L}_1$, $\mathcal{L}_2$ into (B.16) the relation between λ_{in} and ω is found to be

$$\lambda_{in} = \frac{2\left(-\omega\rho_0 + \frac{B_0^2 m^2}{\mu_0 \omega}\right)}{2\left(\Omega\rho_0 + \frac{B_0^2 m}{\mu_0 \omega}\right)}, \tag{B.19}$$

which can be rearranged into a quadratic equation in the hydromagnetic wave angular frequency ω

$$\omega^2 + \Omega\lambda_{in}\omega + \frac{B_0^2 m}{\rho_0 \mu_0}(\lambda_{in} - m) = 0, \tag{B.20}$$

that has solutions

$$\omega = \frac{\Omega\lambda_{in}}{2}\left[-1 \pm \left(1 - \frac{4B_0^2 m(\lambda_{in} - m)}{\Omega^2 \lambda_{in}^2 \rho_0 \mu_0}\right)^{1/2}\right]. \tag{B.21}$$

In the case of large wavelength disturbances (small m) and rapid rotation, when $\frac{4B_0^2 m}{\Omega^2 \lambda_{in}^2 \rho_0 \mu_0}$ is small, a Taylor series expansion shows that the two possible solutions are to leading order [70]

$$\omega_i \sim -\Omega\lambda_{in} \quad \text{and} \quad \omega_m \sim \frac{B_0^2}{\Omega\rho_0\mu_0}\frac{m(m - \lambda_{in})}{\lambda_{in}}. \tag{B.22}$$

ω_i is essentially an Inertial wave, that can travel both eastward and westward depending on the sign of λ_{in}, while ω_m is a slow MC wave where the inertial term is unimportant and the Lorentz and Coriolis forces balance each other to leading order.

The relation between Malkus' slow MC wave and Hide's slow MC-Rossby waves becomes clear if one considers a λ_{in} corresponding to a quasi-geostrophic Inertial wave (QGIW) [17, 105, 108]. Such QGIWs are unidirectional (always travel eastward because λ_{in} is < 0 in a full sphere), and columnar in structure parallel to the rotation axis: they are in fact none other than Rossby waves in a sphere. For such QGIWs with $\lambda_{in} < 0$, the associated slow MC waves (i.e. the MC-Rossby wave) will travel westward as predicted by Hide's analysis.

The ingenious analysis of Malkus, based around a clever choice of imposed field, thus permits the hydromagnetic wave problem in the full sphere to reduce to the classic problem of Inertial waves in a full sphere. It should however be remembered that the Malkus field is rather atypical in its cylindrical symmetry. In general the effect of the Lorentz forces from magnetic fields will not be able to be absorbed completely into modified Coriolis and inertial terms.

Appendix C. Busse and Soward's QG model of MAC waves

Busse [14] and Soward [93] developed a formally precise quasi-geostrophic (QG) model of 2D (structurally invariant parallel to the rotation axis), thermally-driven MAC waves. In a cylindrical annulus with sloping upper and lower boundaries, a uniform azimuthal magnetic field is imposed together with a cylindrically ra-dial background temperature gradient and gravity field that combine to give a cylindrically radial buoyancy force (see Fig. 3). The gap D between the inner and outer cylinders is assumed to be small compared to the height of the annu-lus L, making it possible to work in a local Cartesian co-ordinate system. In this scenario it is possible to write $\widehat{\boldsymbol{\Omega}} = \widehat{\boldsymbol{z}}$, $\widehat{\boldsymbol{g}} = -\widehat{\boldsymbol{x}}$ and $\widehat{\nabla T_0} = -\widehat{\boldsymbol{x}}$ so that $(\boldsymbol{u} \cdot \nabla)T_0 = -\beta' u_x$ while $\boldsymbol{B_0} = B_0 \widehat{\boldsymbol{y}}$ so $(\boldsymbol{b} \cdot \nabla)\widehat{\boldsymbol{B_0}} = 0$, $(\widehat{\boldsymbol{B_0}} \cdot \nabla)\boldsymbol{b} = \frac{\partial \boldsymbol{b}}{\partial y}$ and $\nabla \times (\boldsymbol{u} \times \widehat{\boldsymbol{B_0}}) = \frac{\partial \boldsymbol{u}}{\partial y}$. With these simplifications the linearized momentum, induc-tion and heat transport equations governing the evolution the velocity, magnetic and temperature field can be written as,

$$E\left(\frac{\partial}{\partial t} - \nabla^2\right)\boldsymbol{u} + (\widehat{\boldsymbol{z}} \times \boldsymbol{u}) = -\nabla p + ERa\Theta\widehat{\boldsymbol{x}} + \Lambda\frac{\partial \boldsymbol{b}}{\partial y}, \tag{C.1}$$

$$\left(\nabla^2 - Pr_m\frac{\partial}{\partial t}\right)\boldsymbol{b} = \frac{\partial \boldsymbol{u}}{\partial y}, \tag{C.2}$$

$$\left(\nabla^2 - Pr\frac{\partial}{\partial t}\right)\Theta = u_x, \tag{C.3}$$

where maximum height of the annulus L has been used as the unit of length and the viscous diffusion time has been used as the unit of time. The non-dimensional control parameters and their physical meanings are defined in (6.2) to (6.6) in the main text.

Taking the $\widehat{z}$ component of the curl of (C.1) and (C.2) while defining the axial component of perturbation vorticity as $\zeta = \widehat{z} \cdot (\nabla \times \boldsymbol{u})$ and the axial component of the perturbation electric current density as $j = \widehat{z} \cdot (\nabla \times \boldsymbol{b})$ gives

$$E \left(\frac{\partial}{\partial t} - \nabla^2 \right) \zeta + \frac{\partial u_z}{\partial z} = -E \, Ra \frac{\partial \Theta}{\partial y} + \Lambda \frac{\partial j}{\partial y}, \tag{C.4}$$

$$\left(\nabla^2 - Pr_m \frac{\partial}{\partial t} \right) j = \frac{\partial \zeta}{\partial y}. \tag{C.5}$$

The assumption of quasi-geostrophy is next implemented by first integrating equations (C.4), (C.5) and (C.3) with respect to z and dividing by the depth of the fluid. ζ, j, and Θ can then be interpreted as the vertically averaged perturbations in axial vorticity, axial electrical current and temperature respectively. This is reasonable because geostrophy (z independence) holds to leading order when the slope of the top boundary is small (see, for example, [16] for a formal development of the QG model and [15] for a discussion of its utility). However, because of the presence of the sloping top and bottom boundaries, the term $\frac{\partial u_z}{\partial z}$ arising from the Coriolis force cannot be z independent if mass is conserved. Conservation of mass for the incompressible fluid ($\nabla \cdot \boldsymbol{u} = 0$) applied at the boundary requires that[4]

$$\int_0^1 \frac{\partial u_z}{\partial z} dz = [u_z]_0^1 = -\beta^* u_x, \tag{C.6}$$

where the non-dimensional parameter β^* is the tangent of the angle between the boundary slope and the equatorial plane, and is related to the dimensional β-plane parameter appearing in Hide's model (see Appendix A) by $\beta = 2\Omega\beta^*/L$. To leading order, the velocity fields in the QG approximation are 2D so they can be represented by a stream function χ where $u_x = \frac{\partial \chi}{\partial y}$ and $u_y = -\frac{\partial \chi}{\partial x}$ so the $\widehat{z}$ component of vorticity is $\zeta = -\nabla \chi$. The governing equations for the QG system in terms of χ, j and Θ are then,

$$-E \left(\frac{\partial}{\partial t} - \nabla^2 \right) \nabla^2 \chi + \beta^* \frac{\partial \chi}{\partial y} = -E \, Ra \frac{\partial \Theta}{\partial y} + \Lambda \frac{\partial j}{\partial y}, \tag{C.7}$$

$$\left(\nabla^2 - Pr \frac{\partial}{\partial t} \right) \Theta = -\frac{\partial \chi}{\partial y}, \tag{C.8}$$

$$\left(\nabla^2 - Pr_m \frac{\partial}{\partial t} \right) j = -\frac{\partial}{\partial y} \nabla^2 \chi. \tag{C.9}$$

---

[4]This expression ignores the influence of viscous boundary layers and associated Ekman pumping. It only rigorously applies when free slip boundary conditions are implemented.

Plane wave solutions for χ, j and Θ proportional to $e^{i(kx+ky-\omega t)}$ (invariant in the $\hat{z}$ direction and taking k to be an estimate of the wavenumber in both the x and y directions for simplicity) can then be substituted into (C.7) to (C.9). This yield relations between j and χ, and between Θ and χ

$$\Theta = \frac{-ik\chi}{-k^2 + i\,Pr\omega} \quad \text{and} \quad j = \frac{ik^3\chi}{-k^2 + i\,Pr_m\,\omega}, \tag{C.10}$$

which when substituted into (C.7) give a dispersion relation for the complex frequency ω

$$E(-i\omega + k^2) + \frac{i\beta^*}{k} = \frac{E\,Ra}{-i\,Pr\,\omega + k^2} - \frac{\Lambda k^2}{-i\,Pr_m\omega + k^2}. \tag{C.11}$$

This is the QG dispersion relation discussed in the main text that describes various types of 2D, thermally-driven, Rossby and MC/MAC waves.

References

[1]   D.J. Acheson, J. Fluid. Mech. **52**, 529 (1972).

[2]   D.J. Acheson, J. Fluid. Mech. **96**, 723 (1980).

[3]   D.J. Acheson and R. Hide, Rep. Prog. Physics **36**, 159 (1973).

[4]   K. Aldridge, B. Seyed-Mahmoud, G. Henderson and W. Van Wijngaarden, Phys. Earth Plan. Int. **103**, 365 (1997).

[5]   H. Alfvén, Nature **150**, 405 (1942).

[6]   H. Alfvén and C.-G. Fälthammer, *Cosmical Electrodynamics*, Oxford University Press, 1963.

[7]   G.K. Batchelor, *An Introduction to Fluid Mechanics*, Cambridge University Press, 1967.

[8]   V. Bjerknes, J. Bjerknes, H. Solberg and T. Bergeron, *Physikalische Hydrodynamik*, Berlin, J. Springer, 1933.

[9]   S.I. Braginsky, Geomag. Aeron. (English Translation) **4**, 698 (1964).

[10] S.I. Braginsky, Geomag. Aeron. (English Translation) **7**, 851 (1967).

[11] S.I. Braginsky, in Encycl. Solid. Earth. Geophys. (1989).

[12] G.H. Bryan, Phil. Trans. Roy. Soc. Lond. A. **180**, 187 (1889).

[13] F.H. Busse, J. Fluid Mech. **44**, 441 (1970).

[14] F.H. Busse, Phys. Earth. Planet. Int. **12**, 350 (1976).

[15] F.H. Busse, Phys. Fluids **14**, 1301 (2002).

[16] F.H. Busse and A.C. Or, J. Fluid Mech. **166**, 173 (1986).

[17] F.H. Busse, K. Zhang and X. Liao, Astrophys. J. **631**, L171 (2005).

[18] M.E. Cartan, Bull. Sci. Math. **46**, 317 (1922).

[19] P. Cardin and P. Olson, Earth. Planet. Sci. Lett. **132**, 167 (1995).

[20] P. Cardin, D. Brito, D. Jault, H.C. Nataf and J.P. Masson, Magnetohydrodynamics **38**, 177 (2002).

[21] S. Chandrasekhar, *Hydrodynamic and Hydromagnetic Stability*, Oxford University Press, 1961.

[22] P.A. Davidson, *An Introduction to Magnetohydrodynamics*, Cambridge University Press, 1978.

[23] B. Desjardins, E. Dormy and E. Grenier, Phys. Earth Plan. Int. **124**, 283 (2001).

[24] M. Dumberry and J. Bloxham, Phys. Earth Plan. Int. **140**, 29 (2003).

[25] C. Eloy, P. Le Gal and S. Le Dizés, J. Fluid. Mech. **476**, 357 (2003).

[26] M. El Sawi and I.A. Eltayeb Quart. App. Math. **34**, 187 (1981).

[27] I.A. Eltayeb and P.H. Roberts, Astrophys. J. **162**, 699 (1970).

[28] I.A. Eltayeb, Proc. Roy. Soc. Lond. A. **353**, 145 (1977).

[29] I.A. Eltayeb, Proc. Roy. Soc. Lond. A. **326**, 229 (1972).

[30] I.A. Eltayeb, Phys. Earth. Planet. Int. **24**, 259 (1981).

[31] S.A. Ewen and A.M. Soward, Geophys. Astrophys. Fluid Dyn. **77**, 209 (1994).

[32] S.A. Ewen and A.M. Soward Geophys. Astrophys. Fluid Dyn. **77**, 231 (1994).

[33] S.A. Ewen and A.M. Soward, Geophys. Astrophys. Fluid Dyn. **77**, 263 (1994).

[34] D.R. Fearn, Proc. Roy. Soc. Lond. A. **369**, 227 (1979).

[35] D.R. Fearn, Geophys. Astrophys. Fluid. Dyn. **14**, 102 (1979).

[36] D.R. Fearn, Geophys. Astrophys. Fluid. Dyn. **27**, 137 (1983).

[37] D.R. Fearn, Geophys. Astrophys. Fluid. Dyn. **30**, 227 (1984).

[38] D.R. Fearn, Geophys. Astrophys. Fluid. Dyn. **44**, 55 (1988).

[39] D.R. Fearn, Geophys. Astrophys. Fluid. Dyn. **49**, 173 (1989).

[40] D.R. Fearn and M.R.E. Proctor, J. Fluid. Mech. **128**, 1 (1983).

[41] D.R. Fearn, P.H. Roberts and A.M. Soward, Convection, Stability and the Dynamo in Pitman Research Notes in Mathematics Series, **168**, 60 (1983).

[42] C.C. Finlay and A. Jackson, Science **300**, 2084 (2003).

[43] C.C. Finlay, Ph.D. Thesis, University of Leeds, U.K., 2005.

[44] C.C. Finlay, Alfvén waves, in: *Encyclopedia of Geomagnetism and Paleomagnetism*, edited by D. Gubbins and E. Herraro-Bervera, 2007, p. 3.

[45] C.C. Finlay, Magnetohydrodynamic waves, in: *Encyclopedia of Geomagnetism and Paleomagnetism*, edited by D. Gubbins and E. Herraro-Bervera, 2007, p. 632.

[46] D. Fultz, J. Meteorology **16**, 199 (1959).

[47] D. Fultz, *Rotating Flows: National Committee for Fluid Mechanics Films*, http://web.mit.edu/fluids/www/Shapiro/ncfmf.html, 1969.

[48] N. Gillet, D. Brito, D. Jault and H.-C. Nataf J. Fluid Mech. **580**, 123 (2007).

[49] H.P. Greenspan, *The Theory of Rotating Fluids*, Cambridge University Press, 1968.

[50] D. Gubbins and P.H. Roberts, Geomagnetism, ed. J.A. Jacobs **2**, 1 (1987).

[51] J. Hayvaerts and E.R. Priest, Astron. Astrophys. **117**, 220 (1983).

[52] R. Hide, Philos. Trans. Roy. Soc. Lond. A. **259**, 155 (1966).

[53] R. Hide and K. Stewartson, Rev. Geophys. **10**, 579 (1973).

[54] R. Hollerbach, The core-mantle boundary region, AGU Geodynamical Monograph, **31**, 181 (2003).

[55] C.A. Jones and P.H. Roberts, Geophys. Astrophys. Fluid. Dyn. **93**, 173 (2000).

[56] C.A. Jones, A.M. Soward and A.I. Mussa, J. Fluid Mech. **405**, 157 (2000).

[57] C.A. Jones, A.I. Mussa and S.J. Worland, Proc. Roy. Soc. A. **459**, 773 (2003).

[58] Lord Kelvin, Philosoph. Mag. **10**, 155 (1880).

[59] Lord Kelvin, *Proc. Roy. Inst. 4th March 1881* (see also Collected Papers of Lord Kelvin, Vol. 4. pp. 169) 1881.

[60] D.H. Kelley, S.A. Triana, D.S. Zimmerman, A. Tilgner and D.P. Lathrop, Geophys. Astrophys. Fluid Dyn. **101**, 469 (2007).

[61] R.R. Kerswell, J. Fluid. Mech. **274**, 219 (1994).

[62] L. Lacaze, P. Le Gal, and S. Le Dizés, J. Fluid. Mech. **475**, 1 (2004).

[63] L. Lacaze, P. Le Gal, and S. Le Dizés, Phys. Earth Plan. Int. **151**, 194 (2005).

[64] L. Lacaze, W. Herreman, M. Le Bars, S. Le Dizés and P. Le Gal, Geophys. Astrophys. Fluid Dyn. **100**, 299 (2006).

[65] B. Lehnert, Phys. Rev. **94**, 815 (1953).

[66] B. Lehnert, Astrophys. J. **119**, 647 (1954).

[67] M.J. Lighthill, J. Fluid. Mech. **26**, 411 (1966).

[68] M.J. Lighthill, *Waves in Fluids*, Cambridge University Press, 1978.

[69] S. Lundquist, Phys. Rev. **107**, 1805 (1949).

[70] W.V.R. Malkus, J. Fluid Mech. **28**, 793 (1967).

[71] W.V.R. Malkus, Geophys. Astrophys. Fluid Dyn. **48**, 123 (1989).

[72] A.D. McEwan, J. Fluid Mech. **40**, 603 (1970).

[73] P. Melchoir, *Physics of the Earth's Core*, Pergamon Press, 1986.

[74] H.K. Moffatt, *Magnetic Field Generation in Electrically Conducting Fluids*, Cambridge University Press, 1978.

[75] H.-C. Nataf, T. Alboussiére, D. Brito, P. Cardin, N. Gagniére, D. Jault, J.P. Masson and D. Schmitt, Geophys. Astrophys. Fluid Dyn. **100**, 281 (2006).

[76] E.N. Parker, Astrophys. J. **122**, 293 (1955).

[77] J. Pedlosky, *Geophysical Fluid Dynamics*, Springer-Verlag, New York, 1987.

[78] H. Poincaré, Acta Math. **7**, 259 (1885).

[79] M.R.E. Proctor, *Lectures in Solar and Planetary Dynamos*, edited by M.R.E. Proctor and A.D. Gilbert, 1994, p. 97.

[80] M. Rieutord and L. Valdettaro, J. Fluid Mach. **341**, 77 (1997).

[81] M. Rieutord, B. Georgeot and L. Valdettaro, J. Fluid Mech. **435**, 103 (2001).

[82] P.H. Roberts and A.M. Soward, Ann. Rev. Fluid. Mech. **4**, 117 (1972).

[83] P.H. Roberts and K. Stewartson, Proc. Roy. Soc. Lond. A. **277**, 287 (1974).

[84] P.H. Roberts, Reprinted in Magnetohydrodynamics and Earth's core: Selected works of Paul Roberts, ed. A.M. Soward, Published by Taylor and Francis (2003), **90**, 261 (1977).

[85] P.H. Roberts and D.E. Loper, J. Fluid. Mech. **90**, 641 (1979).

[86] P.H. Roberts and C.A. Jones, Geophys. Astrophys. Fluid. Dyn. **92**, 289 (2000).

[87] B. Roberts, *In Solar System Magnetohydrodynamics*, eds. E.R. Priest and A.W. Hood, Published by Cambridge University Press, 1991, p. 37.

[88] C.-G., Rossby, J. Mar. Res. **2**, 33 (2000).

[89] D.A. Schecter, J.F. Boyd and P.A. Gilman, Astrophys. J. **551**, L185 (2001).

[90] D. Schmitt, T. Alboussiére D. Brito, P. Cardin, N. Gagniére, D. Jault, H.-C. Nataf, J. Fluid Mech. **604**, 175–197 (2008).

[91] B. Seyed-Mahmoud, K. Aldridge and G. Henderson, Phys. Earth Plan. Int. **142**, 257 (2004).

[92] J.A. Shercliff, *Magnetohydrodynamics: National Committee for Fluid Mechanics Films*, http://web.mit.edu/fluids/www/Shapiro/ncfmf.html, 1965.

[93] A.M. Soward, Phys. Earth. Planet. Int. **20**, 134 (1979).

[94] A.M. Soward and E. Dormy, Dynamics of rapidly rotating fluids: Waves and boundary layers, in: *Mathematical Aspects of Natural Dynamos* (Published by Taylor and Francis), 2007, p. 151.

[95] K. Stewartson, Philos. Trans. Roy. Soc. Lond. A. **299**, 173 (1967).

[96] K. Stewartson, in: *Rotating Fluids in Geophysics*, edited by A.M. Soward and P.H. Roberts, (1978), p. 67.

[97] P.A. Sturrock, *Plasma Physics*, Cambridge University Press, 1994.

[98] D.J. Tritton, *Physical Fluid Dynamics*, Oxford University Press, 1987.

[99] M.R. Walker, C.F. Barenghi and C.A. Jones, Geophys. Astrophys. Fluid. Dyn. **8**, 261 (1998).

[100] M.R. Walker and C.F. Barenghi, Phys. Earth. Planet. Int. **111**, 35 (1999).

[101] S.J. Worland, Ph.D. Thesis, University of Exeter, U.K., 2004.

[102] K. Zhang, J. Fluid Mech. **248**, 203 (1993).

[103] K. Zhang, Proc. Roy. Soc. Lond. A. **448**, 245 (1995).

[104] K. Zhang, Phys. Earth. Planet. Int. **111**, 93 (1999).

[105] K. Zhang and X. Liao, J. Fluid Mech. **518**, 319 (2004).

[106] K. Zhang and D. Gubbins, Math. and Comp. Model. **36**, 389 (2002).

[107] K. Zhang and G. Schubert, Ann. Rev. Fluid. Mech. **32**, 409 (2000).

[108] K. Zhang, P. Earnshaw, X. Liao and F.H. Busse, J. Fluid Mech. **437**, 103 (2001).

[109] K. Zhang, X. Liao and P. Earnshaw, J. Fluid Mech. **504**, 1 (2004).

[110] K. Zhang, X. Liao and F.H. Busse, J. Fluid Mech. **575**, 449 (2007).

Course 9

DYNAMOS OF THE ICE GIANTS

S. Stanley

Department of Physics, University of Toronto,
Toronto, ON, Canada

Ph. Cardin and L.F. Cugliandolo, eds.
Les Houches, Session LXXXVIII, 2007
Dynamos

Contents

453

1. Introduction

The planets Uranus and Neptune have much in common. They are the two furthest planets in our Solar System (now that Pluto has been demoted from planet status). They have similar radii, masses and rotation periods. Although their magnetic fields are not similar in exact form, both planets have fields with significant power in non-dipolar non-axisymmetric components, in contrast to the magnetic fields of Earth, Jupiter and Saturn which are dominated by their axial-dipolar components. In this paper we will give an overview of the ice giants' magnetic fields, discuss proposed explanations for their unusual fields, highlight insights that have been provided by numerical dynamo models and mention future prospects for understanding these planets' fields.

Uranus and Neptune are named the 'ice giants' because their main constituents (H_2O, NH_3 and CH_4) are solid 'ices' at temperatures characteristic of the region of the solar system where they are located. However, in the planets' interiors the temperatures are much higher, and hence the constituents are certainly fluid. The planets also contain thin hydrogen/helium atmospheres and likely also have rocky cores made up of silicates and metals. Interior profiles of density and pressure versus radius are generated that match the planets' observed mass, radius, moment of inertia and zonal gravitational coefficients (for review papers on interior models see [1,2], more recent models include [3,4]). The profiles are highly non-unique and can match the observational data with different proportions of gas (hydrogen and helium), ice (water, ammonia, methane) and rock (silicates and metals). Layered models can also match the profiles as can models where the different constituents are mixed to some degree.

Shock experiments on a 'synthetic' ice giant (a mixture of water, ammonia and methane) carried out at characteristic temperatures and pressures of the ice giant interiors have determined that these planets can attain a large ionic conductivity [5]. Electrical conductivities of order 10^3 S/m are reached at pressures higher than 40 GPa which corresponds to regions within $0.7 R_p$ for Uranus and $0.8 R_p$ for Neptune (where R_p is the planetary radius). Therefore, a large region of their interiors are significantly conducting and hence a potential region for the dynamo to operate.

Uranus and Neptune have been visited by one spacecraft, Voyager II, which performed a flyby of Uranus in 1986 and of Neptune in 1989. During these flybys, magnetic field measurements were taken and magnetic models generated [6,7]. Amazingly, both Voyager spacecraft are still operating today and continue to explore the furthest reaches of our solar system.

Planetary magnetic field observations are used to constrain coefficients of the magnetic potential spherical harmonic expansion:

$$\vec{B}(r, \theta, \phi) = -\nabla V(r, \theta, \phi)$$

$$V = a \sum_{l=1}^{\infty} \sum_{m=0}^{l} \left(\frac{a}{r}\right)^{l+1} \left[g_l^m \cos(m\phi) + h_l^m \sin(m\phi)\right] P_l^m(\cos\theta) \tag{1.1}$$

where $\vec{B}$ is the magnetic field, V is the magnetic potential, r is radius, θ is co-latitude, ϕ is longitude, a is the radius of the planet's surface, g_l^m and h_l^m are the Gauss coefficients of the expansion and $P_l^m(\cos\theta)$ are associated Legendre polynomials. Table 1 lists the Gauss coefficients for Uranus and Neptune (coefficients

Table 1

Gauss coefficients for Uranus and Neptune in nT. Only the dipole, quadrupole and octupole coefficients are listed. Data from [8]

Gauss coefficient	Uranus	Neptune
g_1^0	11855	10336
g_1^1	11507	3359
h_1^1	−15812	−9772
g_2^0	−5877	8566
g_2^1	−13085	−406
g_2^2	−605	4644
h_2^1	5851	11139
h_2^2	4185	−743
g_3^0	4183	−5749
g_3^1	−1336	11632
g_3^2	−6776	−1889
g_3^3	−4021	−2920
h_3^1	−5817	−3905
h_3^2	−357	903
h_3^3	−2265	−245

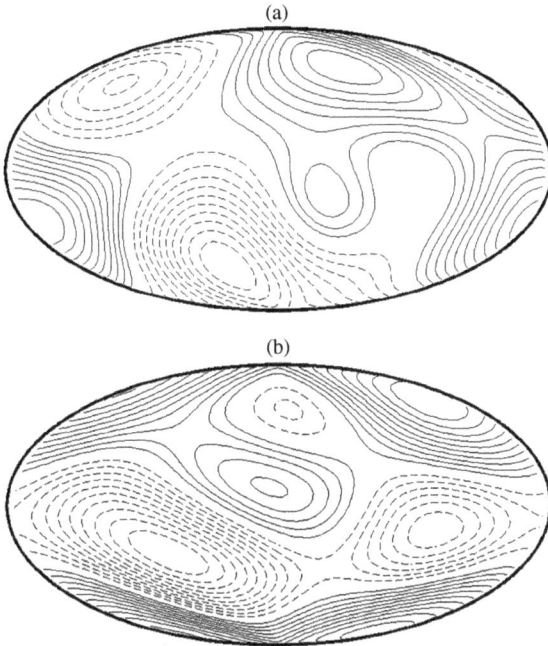

Fig. 1. Contours of the radial component of the magnetic field for Uranus (a) and Neptune (b) plotted at the planetary 'surface' (1 bar pressure level). Positive (negative) radial field is shown in solid (dashed) contours.

for Earth, Jupiter and Saturn can be found in other lectures in this publication). The dominance of the axial ($m = 0$) dipole ($l = 1$) in Earth, Jupiter and Saturn is evident by how much larger the g_1^0 coefficient is in these planets relative to the other coefficients. In contrast, Uranus' and Neptune's magnetic fields are not dominated by the axial dipole as several other coefficients are just as large as the g_1^0 coefficient. The non-axial-dipolar nature of their fields is also evident in maps of the radial components of the fields shown in Fig. 1.

In our quest to understand planetary dynamos, it is important to determine why Uranus and Neptune have different magnetic field morphologies than the other planets. There are two ways to interpret the fact that their surface magnetic field observations are not directly similar to each other: (1) each planet's surface field is fairly stable, so the magnetic field of Uranus is really different from Neptune, in which case we need two dynamo mechanisms (one to explain each planet), or (2) both planets' fields are non-dipolar and non-axisymmetric, but the morphology is unstable, so the surface field snapshots are time variable. In this case, we may only need one dynamo mechanism to explain both magnetic fields. Since Uranus and Neptune are so similar in their interior structural properties (i.e. similar masses, radii and rotation rates), the second option is more appealing if we believe the dynamo mechanism depends on the structural properties of a planet's interior. This seems likely since key dynamo diagnostic parameters, such as the magnetic Reynolds number ($Re_m = \sigma \mu_0 U L$ where σ is electrical conductivity, μ_0 is the magnetic permeability of free space, U is a characteristic velocity and L is a characteristic lengthscale) and the Elsasser number ($\Lambda = B^2/2\Omega\rho\sigma$ where B is the characteristic magnetic field strength, Ω is the planetary rotation rate and ρ is density) depend on these properties.

In order to understand why the fields are different, a first step may be to look at the dynamo source regions and what properties may be different about these regions. In the Earth, the dynamo is generated in the fluid iron outer core, in Jupiter and Saturn it is generated in the metallic hydrogen regions, and in Uranus and Neptune it is most likely that they occur in the fluid ionic ice regions. All of these regions are relatively thick shells and therefore the geometries of the dynamo regions appear similar. The compositions, however, are different and we should explore how these affect the dynamo. One key physical parameter for the dynamo is the electrical conductivity of the region. The iron in Earth's core and the metallic hydrogen in Jupiter and Saturn have conductivities of $O(10^5 \text{ S/m})$. In contrast, Uranus and Neptune's ionic conductivity is $O(10^3 \text{ S/m})$. The question arises then as to whether the difference in conductivity values can explain Uranus' and Neptune's fields. To answer this question we look at where the electrical conductivity comes into play in the equations governing the dynamo. One place is the magnetic induction equation (equation (1.2), where $\vec{B}$ is magnetic field, t is time and $\vec{u}$ is velocity), since the magnetic diffusivity ($\lambda = 1/\sigma\mu_0$) is inversely proportional to electrical conductivity. This means that the ohmic dissipation is larger in Uranus and Neptune than in the other planets which are better conductors.

$$\frac{\partial \vec{B}}{\partial t} = \nabla \times (\vec{u} \times \vec{B}) + \lambda \nabla^2 \vec{B} \qquad (1.2)$$

However, the magnetic Reynolds number, which gives a measure of the ratio of the generation of magnetic field to its dissipation (its basically a ratio of the first term to the second term on the right-hand side of equation (1.2)) also depends on velocities and length scales. Therefore, Uranus and Neptune may still be able to generate the same dynamo activity as the other planets if they have similar magnetic Reynolds numbers. In this case, the specific value of the electrical conductivity is not in itself a determining factor.

Another place that electrical conductivity enters into the governing equations is in the comparison of the different diffusive processes. Planetary dynamos are subject to three diffusive mechanisms: the diffusion of momentum through viscous forces, the diffusion of heat through thermal conduction and the diffusion of magnetic fields through ohmic dissipation. Comparisons between these diffusions is usually quantified by the ratios of the diffusivities (a series of Prandtl numbers). Only 2 of the following 3 numbers are independent:

1. $Pr = \nu/\kappa$ where ν is the viscosity and κ is the thermal diffusivity,

2. $q_\kappa = \kappa/\eta$ where η is the magnetic diffusivity,

3. $Pm = \nu/\eta$.

If the thermal diffusivity and viscosity are the same in all the planets (these are not well known values even for Earth, but assuming they are similar) then their would be a difference in Uranus' and Neptune's Prandtl numbers compared to the other planets due to the different magnetic diffusivity. However, if the convection generating the magnetic fields is turbulent in nature, then one could make an "eddy diffusivity" argument and say the dominant diffusivity (which is the magnetic diffusivity) replaces the molecular diffusivities in the problem, in which case all the Prandtl numbers are equal to 1. In this case, the fact that Uranus and Neptune have smaller electrical conductivities may not be as important.

Another property of Uranus and Neptune which is different from the other planets is their heat flow. Observations by Voyager II give their internal heat fluxes as 0.042 ± 0.047 W/m^2 for Uranus and 0.433 ± 0.046 W/m^2 for Neptune [9]. These values are significantly lower than the heat fluxes out of Jupiter and Saturn. We will discuss the importance of this different heat flow in relation to explaining their different dynamos in Section 2.

2. Proposed explanations

Different explanations for Uranus' and Neptune's non-dipolar non-axisymmetric fields have been proposed. After the Uranus flyby, Schulz and Paulikas [10] suggested Uranus is undergoing a field reversal since magnetic fields lose their axial dipole dominance during a reversal. However, after the Neptune flyby, this became an unlikely explanation if Uranus' and Neptune's reversal frequencies are similar to those of Earth. Essentially, the chance of catching two planets in reversals at the same time is extremely small.

Also after the Uranus observations, Podolak et al. [1] proposed Uranus' field is a result of its unusually large obliquity (the spin axis is tilted by 98 degrees with respect to the plane of the solar system, so it is essentially rotating on its side). However, since Neptune does not have a large obliquity, one would need a separate mechanism to explain Neptune's field.

Connerney et al. [6,7] suggest that the non-axisymmetry of the fields is due to fewer and larger convection cells in Uranus and Neptune, however their argument does not explain the non-dipole nature of the fields.

Holme and Bloxham [8] give two potential explanations. The first is that the fields are the result of a lack of magnetostrophic balance in the source regions (i.e. weak field dynamos are produced), unlike the other giant planets and Earth which are believed to have strong field dynamos. However, dynamo models that produce weak fields can readily produce axial dipole dominated fields. There would still need to be an explanation for why the weak fields are not axially dipolar. The second possibility they discuss is that the lack of a large solid conducting inner core in Uranus and Neptune results in a lack of stabilization of an axially-dipolar dominated field. The stabilizing effect of a solid conducting inner core was first demonstrated through kinematic modelling by Hollerbach and Jones [11,12]. However, Jupiter and Saturn do not contain large solid inner cores and yet have axial-dipole dominated fields, so it does not appear to be a sufficient condition.

Ruzmaikin and Starchenko [13] suggest that the dynamo operates in a thin shell of metallized carbon at the base of the ice layer deep inside the planet. They argue that the thinness of the shell would result in more complex, small-scale fields. However, this dynamo region is so far from the surface that the power in the non-dipolar terms would decay significantly with distance from the source region, and so one would expect much weaker power in the non-dipolar components than is observed.

Hubbard et al. [2] employ the mechanism suggested by [13] (i.e. a thin shell dynamo), however they propose a different location for the shell. They argue that the observed heat flows from Uranus and Neptune are not consistent with whole-planet convection and therefore conclude that only the outer regions of the ice layers are convecting and hence the dynamo is only generated in this thin outer shell. They are therefore able to explain both the anomalous heat flows and the anomalous magnetic fields.

3. Insights from numerical dynamo models

The majority of numerical planetary dynamos in the literature produce axially-dipolar dominated fields. A valid question is if this is a result of the fact that the planetary dynamo these models usually try to simulate (Earth's) is an axially dipolar dominated field. The bias may occur because studies intend to compare models to observations (and hence it only makes sense to include axial-dipolar dominated model results) or because of the chosen initial conditions. Geodynamo models are likely to be initiated from a strong axial dipolar morphology since they are trying to simulate the geodynamo and starting from a field similar to the expected field may reduce the transient decay time.

Kinematic dynamo studies which are able to cover a larger area of parameter space, find non-axial, non-dipolar dominated fields for certain classes of imposed velocity fields [14, 15]. These velocity field morphologies are not too different from velocity fields that produce axial dipolar dominated fields (e.g. [16]), suggesting non-axially-dipolar dynamos may not require significantly different generation mechanisms than axial-dipolar dynamos.

There are several numerical studies of planetary dynamos that produce non-axially-dipolar dominated fields. For example, Ishihara and Kida [17, 18] and Aubert and Wicht [19] find equatorial dipolar dynamos. Although these very simple morphologies are not directly comparable to the complex fields of Uranus and Neptune, these models allow important investigation of the mechanisms generating specific components. Grote and Busse [20–22] examine the effect of varying the magnetic Prandtl number while keeping the other non-dimensional parameters fixed at values similar to other studies. They find axial-quadrupolar and hemispherical (equal contributions from dipole and quadrupole) dominated dynamos for magnetic Prandtl numbers Pm of order 10^{-1}–1 and axial-dipolar solutions for larger magnetic Prandtl numbers. This may be relevant to Uranus and Neptune since their magnetic Prandtl numbers are smaller than the other planets. Gomez-Perez and Heimpel [23] find more complex non-dipolar, non-axisymmetric magnetic fields in models which cover the same range in magnetic Prandtl number as [20–22] but have larger Rayleigh numbers. Although this is encouraging for a potential explanation involving the higher magnetic diffusivity of Uranus and Neptune, one must keep in mind that all the planets' magnetic Prandtl numbers are significantly smaller (of $O(10^{-6})$) than any of the simulation values and so a direct extrapolation of these results (i.e. magnetic Prandtl numbers <1 give non-axial-dipolar results) would imply all planets should have non-axisymmetric, non-dipolar fields. Several studies find that non-dipolar dynamos occur when the buoyancy force becomes large enough [24–26]. However, it is then difficult to resolve why Jupiter, which is believed to have a much larger driving buoyancy force, produces an axial-dipolar field.

Roberts and Glatzmaier [27] examine the effect of changing shell thickness on generated magnetic fields in dynamo models. The motivation for the paper is to examine the Earth's dynamo throughout its history as the inner core grows, however they mention the resulting conclusions for Uranus and Neptune as well. They find that the dipole tilt increases with shell thin-ness, however they do not produce the level of non-dipolarity or non-axisymmetry seen in Uranus and Neptune.

Models that take into account the geometry of the convection regions discussed by [2] in order to explain the heat flow constraints (i.e. a dynamo operating in a thin shell surrounding a stably-stratified interior) do produce Uranus and Neptune like fields [28, 29]. These models can reproduce the relative power in the non-dipole and non-axisymmetric components compared to the axial dipole power. It appears that the geometry of the dynamo region does have a significant effect on the morphology of the generated fields. In these models, the thin shell promotes smaller scale structure, but another important component is the fact that the region interior to the convecting shell is fluid rather than solid. Having a solid conductor interior to the convecting shell acts to stabilize the field, as demonstrated by [11, 12]. By replacing the conducting solid with a conducting stable fluid, the magnetic fields can respond to electromagnetic stress by moving the fluid (rather than being frozen in to a solid conductor). This results in less communication between the hemispheres and hence a more non-dipolar, non-axisymmetric field.

4. Conclusions and future prospects

The cause of Uranus' and Neptune's unusual fields is intriguing. So far, only one of the dynamo models adheres to the geometry required to explain the low heat flow observation. However, if alternative explanations for the low heat flows can be found that allow whole planet convection, then there are other potential explanations for the unusual fields.

No missions are planned to Uranus and Neptune in the near future and therefore the prospect of gathering better data to answer the question is grim. Perhaps the best test of the various theories will be new observations of other planetary dynamos. For example, spacecraft observations demonstrate that Mercury and Ganymede also have active dynamos [30, 31] (although it is also possible to explain Mercury's observed field as a crustal remanent field [32]). The data are fairly sparse, and although for both planets, they can be modeled as fields dominated by an axial dipole, better data may determine that there are significant higher degree and order structure in these fields. For the planets that we have resolved dipole, quadrupole and octupole components, the current score is axially-dipolar dominated fields:3, non-axial, non-dipolar dominated fields:2 (note that the score is much better than it was before the Uranus and Neptune observations when it was 3:0). The initial 3:0 score may have resulted in the mindset that planetary dynamos are dominated by axial-dipole components. With a score of 3:2, the dominance of axial-dipolar fields may be called into question. Also, if Mercury or Ganymede increase the non-axial, non-dipole score further, the mindset may need to be changed altogether.

Future observations of Mercury and Ganymede have implications for Uranus and Neptune. If Mercury and Ganymede both have axial-dipole dominated fields, then appealing to the unusual geometry of the convection region in Uranus and Neptune may be the favourable explanation for their unusual fields. If instead, Mercury or Ganymede have non-axial, non-dipolar fields, then perhaps planetary dynamos don't have a preference for axial-dipole solutions and appealing to the convective region geometry is not necessary. In this case, the other possibilities mentioned in Sections 2 and 3 may prove correct. We look forward to a preliminary answer soon as the MESSENGER mission will provide new data on Mercury's magnetic field in the near future.

References

[1] M. Podolak, W.B. Hubbard and D. Stevenson, in: *Uranus*, eds. J. Bergstralh, E. Minor, and M.S. Matthews, University of Arizona Press, Tucson, 1991, p. 29.

[2] W.B Hubbard, M. Podolak and D. Stevenson, in: *Neptune and Triton*, ed. D. Cruickshank, University of Arizona Press, Tucson, 1995, p. 109.

[3] M. Podolak, A. Weizman and M. Marley, Planet. Space Sci. **43**, 1517 (1995).

[4] M. Podolak, J. Podolak and M. Marley, Planet. Space Sci. **48**, 143 (2000).

[5] W.J. Nellis, D. Hamilton, N. Holmes, H. Radousky, F. Ree, A. Mitchell and M. Nicol, Science **240**, 779 (1988).

[6] J.E.P. Connerney, M. Acuna and N. Ness, J. Geophys. Res. **92**, 15329 (1987).

[7] J.E.P. Connerney, M. Acuna and N. Ness, J. Geophys. Res. **96**, 19023 (1991).

[8] R. Holme and J. Bloxham, J. Geophys. Res. **101**, 2177 (1996).

[9] J.C. Pearl and B.J. Conrath, J. Geophys. Res. **96**, 18921 (1991).

[10] M. Schulz and G.A. Paulikas, Adv. Space Res. **10**, 155 (1990).

[11] R. Hollerbach and C.A. Jones, Nature **365**, 541 (1993).

[12] R. Hollerbach and C.A. Jones, Phys. Earth Planet. Int. **87**, 171 (1995).

[13] A.A. Ruzmaikin and S.V. Starchenko, Icarus **93**, 82 (1991).

[14] R. Holme, Phys. Earth Planet. Int. **102**, 105 (1997).

[15] D. Gubbins, C.N. Barber, S. Gibbons and J.J. Love, Proc. Roy. Soc. Lond. A **456**, 1669 (2000).

[16] D. Gubbins, C.N. Barber, S. Gibbons and J.J. Love, Proc. Roy. Soc. Lond. A **456**, 1333 (2000).

[17] N. Ishihara and S. Kida, J. Phys. Soc. Jpn. **69**, 1582 (2000).

[18] N. Ishihara and S. Kida, Fluid Dynam. Res. **31**, 253 (2002).

[19] J. Aubert and J. Wicht, Earth Planet. Sci. Lett. **221**, 409 (2004).

[20] E. Grote, F. Busse and A. Tilgner, Phys. Rev. E **60**, R5025 (1999).

[21] E. Grote, and F. Busse, Phys. Rev. E **62**, 4457 (2000).

[22] E. Grote, F. Busse and A. Tilgner, Phys. Earth Planet. Int. **117**, 259 (2000).

[23] N. Gomez-Perez and M. Heimpel, Icarus, 2007, in press.

[24] C. Kutzner and U. Christensen, Phys. Earth Planet. Int. **131**, 29 (2002).

[25] U.R. Christensen and J. Aubert, Geophys. J. Int. **166**, 97 (2006).

[26] P. Olson and U. Christensen, Earth Planet. Sci. Lett. **250**, 561 (2006).

[27] P.H. Roberts and G.A. Glatzmaier, Geophys. Astrophys. Fluid Dynam. **94**, 47 (2001).

[28] S. Stanley and J. Bloxham, Nature **428**, 151 (2004).

[29] S. Stanley and J. Bloxham, Icarus **184**, 556 (2006).

[30] J.E.P. Connerney and N.F. Ness, in: *Mercury*, eds. F. Vilas, C. Chapman and M. Matthews, University of Arizona Press, Tucson, 1988, p. 494.

[31] M. Kivelson, K. Khurana, C. Russell, R. Walker, J. Warnecke, F. Coroniti, C. Polansky, D. Southwood and G. Schubert, Nature **384**, 537 (1996).

[32] O. Aharonson, M.T. Zuber and S.C. Solomon, Earth Planet. Sci. Lett. **218**, 261 (2004).